DSC/HISS Modeling Applications for Problems in Mechanics, Geomechanics, and Structural Mechanics

Understanding the mechanical behavior of solids and contacts (interfaces and joints) is vital for the analysis, design, and maintenance of engineering systems. Materials may simultaneously experience the effects of many factors such as elastic, plastic, and creep strains; different loading (stress) paths; volume change under shear stress; and microcracking leading to fracture and failure, strain softening, or degradation. Typically, the available models account for only one factor at a time; however, the disturbed state concept (DSC) with the hierarchical single-surface (HISS) plasticity is a unified modeling approach that can allow for numerous factors simultaneously, and in an integrated manner. *DSC/HISS Modeling Applications for Problems in Mechanics, Geomechanics, and Structural Mechanics* provides readers with comprehensive information including the basic concepts and applications for the DSC/HISS modeling regarding a wide range of engineering materials and contacts. Uniformity in format and content of each chapter will make it easier for the reader to appreciate the potential of using the DSC/HISS modeling across various applications.

Features:

- Presents a new and simplified way to learn characterizations and behaviors of materials and contacts under various conditions
- Offers modeling applicable to several different materials including geologic (clays, sands, rocks), modified geologic materials (structured soils, overconsolidated soils, expansive soils, loess, frozen soils, chemically treated soils), hydrate-bearing sediments, and more.

DSC/HISS Modeling Applications for Problems in Mechanics, Geomechanics, and Structural Mechanics

Edited by
Chandrakant S. Desai, Yang Xiao, Musharraf Zaman, and
John Phillip Carter

CRC Press
Taylor & Francis Group
Boca Raton London New York

CRC Press is an imprint of the
Taylor & Francis Group, an **informa** business

Cover Image: Courtesy of Meghan Vallejo and Rafi Tarefder.

First edition published 2024
by CRC Press
6000 Broken Sound Parkway NW, Suite 300, Boca Raton, FL 33487-2742

and by CRC Press
4 Park Square, Milton Park, Abingdon, Oxon, OX14 4RN

CRC Press is an imprint of Taylor & Francis Group, LLC

ISBN: 978-1-032-42284-8 (hbk)
ISBN: 978-1-032-42285-5 (pbk)
ISBN: 978-1-003-36208-1 (ebk)

DOI: 10.1201/9781003362081

Typeset in Times
by codeMantra

Contents

SECTION 1 Introduction

SECTION 2 Geologic Materials

SECTION 3   Structured and Bio- or Chem-Treated Soils

SECTION 4   Fracturing and Crushing Grains

SECTION 5   Geotechnical Structures and Engineering

SECTION 6   Concrete, Pavement Materials, and Structural Engineering

SECTION 7   Interfaces and Joints

Editors

Chandrakant S. Desai is a Regents' Professor (Emeritus), Department of Civil and Architectural Engineering and Mechanics, University of Arizona, Tucson, Arizona. He has made original and significant contributions in basic and applied research in material-constitutive modeling, laboratory testing, and computational methods for a wide range of problems in engineering mechanics, civil engineering (geomechanics, structural dynamics, earthquake engineering), flow through porous media, and mechanical engineering (electronic packaging). One of the main topics of Dr. Desai's research involves development of the new and innovative disturbed state concept (DSC) for constitutive modeling of materials and interfaces/joints, e.g., soils, rocks, glacial tills, concrete, asphalt, metals, alloys, silicon with impurities, polymers, and lunar ceramic composites. DSC connects mechanics, physics (thermodynamics), and philosophy. In conjunction with nonlinear computer (finite element) methods, it provides a new and alternative procedure for analysis, design, and reliability for challenging and complex problems of modern technology. The DSC is a major and significant contribution that has been adopted for research, teaching, and application in several engineering disciplines. He has authored/edited about 25 books, 20 book chapters, and has authored/coauthored over 350 technical papers in refereed journals and conferences. He is the founding Editor-in-Chief of the *International Journal of Geomechanics*, ASCE. He has received several awards and recognitions, e.g. The Distinguished Member Award by the American Society of Civil Engineers (ASCE); The Nathan M. Newmark Medal by Engineering Mechanics and Structural Engineering Institutes, ASCE; and The Karl Terzaghi Award by Geo Institute (ASCE).

Yang Xiao is Full Professor of the School of Civil Engineering, Chongqing University. He received his Ph.D. degree from the College of Civil and Transportation Engineering, Hohai University, Nanjing, China. He is a recognized researcher in the area of rockfill dam, railway geotechnics, biotreatment geotechnics, recycled waste, and granular flow mechanics. His research focuses on the testing and theoretical modeling of clay, silt, sand, rockfill, ballast, and biocemented soils. He has an H-Index of 41 in Scopus. He has 5722 citations and 100+ publications. He has got many awards, such as 2018 and 2019 Most Cited Chinese Researcher from Elsevier, 2020 World's Top 2% Scientists (Singleyr), John Carter Award and Excellent Contributions (Regional) in IACMAG, and Excellent Doctoral Dissertation of Chinese Society for Rock and Mechanics and Engineering. He is also Associate Editor of *International Journal of Geomechanics*, ASCE, and Editor Board Member of *Acta Geotechnica*, *Canadian Geotechnical Journal*, and *Soils and Foundations*.

Musharraf Zaman holds the Aaron Alexander Professorship in Civil Engineering and Alumni Chair Professorship in Petroleum Engineering at the University of Oklahoma (OU) in Norman. He received his Ph.D. in Civil Engineering from the University of Arizona, Tucson. He has been serving as the Director of the Southern Plains Transportation Center (SPTC) – a consortium of eight universities in U.S. DOT Region 6 – for more than 7 years. He served as the Associate Dean for Research and Graduate Programs in OU Gallogly College of Engineering for more than 8 years. His research interests are in geomechanics, constitutive modeling, soil–structure interaction, laboratory and field testing, pavement materials and systems, and intelligent compaction. At OU, he has received more than $30 Million in external funding, developed two new asphalt laboratories, and supervised more than 80 theses and dissertations to completion. He has published more than 200 journals, 275 peer-reviewed conference proceedings papers, and 12 books and book chapters. Several of his papers have won prestigious awards from international societies and organizations. In 2011, he won the prestigious Outstanding Contribution Award, given by IACMAG, for life-long contributions in geomechanics.

John Carter is an Emeritus Professor at the University of Newcastle. He is a former Pro Vice-Chancellor and Dean of Engineering at the University of Newcastle, NSW, and a former Consultant Director of Advanced Geomechanics Pty Ltd (now Fugro AG Pty Ltd). He is a graduate of the Australian Institute of Company Directors and a former Director of the Newcastle Port Corporation. He is currently the President, International Association for Computer Methods and Advances in Geomechanics. John is a geotechnical engineer with more than 40 years of experience in teaching, research, and consulting in civil and geotechnical engineering. He is the author of more than 400 technical articles in peer-reviewed journals and international conferences. He has experience as an expert witness in legal cases in Queensland, New South Wales, and Victoria, and he has also been an advisor and consultant to industry and government. In December 2019, he was appointed by the Governor of Queensland as a Commissioner of the Paradise Dam Inquiry, which concluded with a report presented to the Queensland government at the end of April 2020. He is a Fellow of the Australian Academy of Science, the Australian Academy of Technology and Engineering, the Royal Society of NSW, the EU Academy of Sciences, Engineers Australia, and the Australian Institute of Building. In January 2006, he was appointed as a Member of the Order of Australia (AM) for his contributions to civil engineering through research into soil and rock mechanics and as an adviser to industry.

Contributors

Amir Akbari Garakani
Power Industry Structures Research
    Department
Niroo Research Institute (NRI)
Tehran, Iran

Rasika Athukorala
Geotechnical Engineer
Golder Associates
Melbourne, Victoria, Australia

John P. Carter
University of Newcastle
Newcastle, New South Wales, Australia

Meng Chen
School of Civil Engineering
Chongqing University
Chongqing, China

Hao Cui
School of Civil Engineering
Chongqing University
Chongqing, China

An Deng
School of Civil, Environment and Mining
    Engineering
The University of Adelaide
Adelaide, Australia

Chandrakant S. Desai
Department of Civil and Architectural
    Engineering and Mechanics
University of Arizona
Tucson, AZ

Mohd Firoj
Indian Institute of Technology Roorkee
Roorkee, India

Tariq Hamid
Dulles Geotechnical and Materials Testing
    Services, Inc.
Chantilly, Virginia

Arash Hassanikhah
Transportation/Geotechnical Engineer
Hushmand Associates, Inc.
Irvine, California

Kianoosh Hatami
School of Civil Engineering and Environmental
    Science
University of Oklahoma
Norman, Oklahoma

Massoud Hosseinali
Department of Civil and Environmental
    Engineering
University of Utah
Salt Lake City, Utah

Ming Huang
College of Civil Engineering
Fuzhou University
Fuzhou, China

Buddhima Indraratna
Transport Research Centre
School of Civil and Environmental Engineering
University of Technology Sydney (UTS)
Sydney, Australia

Yufeng Jia
The State Key Laboratory of Coastal and
    Offshore Engineering
Dalian University of Technology
Dalian, China

Song Jiang
School of Civil Engineering
Fujian University of Technology
Fuzhou, China

Xiang Jiang
School of Materials Science and Engineering
Chongqing Jiaotong University
Chongqing, China
and
College of Aerospace Engineering
Chongqing University
Chongqing, China

Charbel Khoury
KCI Technologies, Inc.
Sparks, Maryland

Changwon Kwak
Department of Civil and Environmental
    Engineering
Inha Technical College
Michuhol-gu, South Korea

Fang Liang
School of Civil Engineering
Chongqing University
Chongqing, China

Enlong Liu
Department of Geotechnical and Underground
    Engineering, College of Water Resource and
    Hydropower
Sichuan University
Chengdu, China

Martin D. Liu
University of Wollongong
Wollongong, Australia

Bal Krishna Maheshwari
Indian Institute of Technology Roorkee
Roorkee, India

Gerald Miller
School of Civil Engineering and Environmental
    Science
University of Oklahoma
Norman, Oklahoma

Ahad Ouria
Civil Engineering Department
University of Mohaghegh Ardabili
Ardabil, Iran

Innjoon Park
Department of Infrastructure Systems
Hanseo University
Seosan, South Korea

Shivani Rani
City of Lawton Engineering Division
Lawton, Oklahoma

Zengchun Sun
School of Civil Engineering
Chongqing University
Chongqing, China

Mengjie Tang
The State Key Laboratory of Coastal and
    Offshore Engineering
Dalian University of Technology
Dalian, China

Vahab Toufigh
Civil Engineering Department
Sharif University of Technology
Tehran, Iran

Sai K. Vanapalli
Department of Civil Engineering
University of Ottawa
Ottawa, Canada

Jayan S. Vinod
School of Civil, Mining, Environmental and
    Architectural Engineering, Faculty of
    Engineering and Information Sciences
University of Wollongong (UOW)
Wollongong, Australia

Chenggui Wang
School of Civil Engineering
Chongqing University
Chongqing, China

Lei Wang
School of Civil Engineering
Chongqing University
Chongqing, China

Huanran Wu
School of Civil Engineering
Chongqing University
Chongqing, China

Yang Xiao
School of Civil Engineering
Chongqing University
Chongqing, China

Chaoshui Xu
School of Civil, Environment and Mining
   Engineering
The University of Adelaide
Adelaide, Australia

Dexiang Xu
Urban Planning Design Institute of Ganzhou
Ganzhou, China

Xiuhan Yang
Department of Civil Engineering
University of Ottawa
Ottawa, Canada

Musharraf Zaman
University of Oklahoma
Norman, Oklahoma, USA

Jinwu Zhan
School of Civil Engineering
Fujian University of Technology
Fuzhou, China

Jian-feng Zhu
School of Civil Engineering and Architecture
Zhejiang University of Science and Technology
Hangzhou, China

Section I

Introduction

1 Disturbed State Concept (DSC) with Hierarchical Single Surface (HISS) Plasticity
Theory and Application

Chandrakant S. Desai
Dept. of Civil and Architectural Engineering and
Mechanics, University of Arizona

Zengchun Sun
School of Civil Engineering, Chongqing University

1.1 PREFACE

This book contains contributions by prominent researchers and practitioners who have been involved in developing the DSC/HISS approach. The contributions are related to a wide range of materials and interfaces for problems in geotechnical, structural, earthquake, mining and mechanical engineering, and composites in electronic packaging. The major advantage of the DSC/HISS modeling is that it has greater versatility than other available models. It has opened a new and simplified way to learn and teach characterization of the behavior of materials and contacts. This chapter presents details of the DSC/HISS modeling and a few applications. Many other applications are presented in the subsequent chapters.

Comprehensive details of the history, fundamentals and theory, testing, specimen-level validations, computer (finite element) implementation, practical boundary value problem validations, and comparison with other approaches such as the critical state concept, self-organized criticality (SOC), and acoustic emissions have been presented in the book by Desai [1]. Several relevant publications for the DSC/HISS modeling are available for the reader and have been included in the References section with brief titles regarding the contents of the publication, as listed in **Supplementary 1**. Only brief details are presented below, which can also be used by the authors of other chapters.

1.2 INTRODUCTION

Understanding and characterizing the mechanical behavior of solids and contacts (interfaces and joints) play vital roles for analysis, design, and maintenance of engineering systems. A deforming material may experience, simultaneously, the effect of many factors such as elastic, plastic, and creep strains; different loading (stress) paths; volume change under shear stress; microcracking leading to fracture and failure; strain softening or degradation; and healing or strengthening including environmental (temperature, moisture, chemical, etc.) effects. Most of the available models account for usually one factor at a time. However, for realistic prediction of the behavior, there is a need for developing unified constitutive models that account for these factors simultaneously. The disturbed state concept with the hierarchical single surface DSC/HISS plasticity is a unified modeling approach that can allow the above factors simultaneously and in an integrated manner [1]. Its hierarchical nature permits adoption of specific versions according to the foregoing factors.

DOI: 10.1201/9781003362081-2

The DSC/HISS approach has been developed by Chandrakant S. Desai and his coworkers over four decades or so. The DSC is based on the basic physical consideration that the observed response of a material can be expressed in terms of the responses of its constituents, connected by the coupling or disturbance function. In simple words, the observed material state is represented by disturbance or deviation with respect to the behavior of the material for appropriately defined reference states. This approach is consistent with the idea that the current state of a material system, animate or inanimate, can be the disturbed state with respect to its initial and final state(s).

Because of its generality and unified nature, the DSC/HISS modeling has been developed and used for a wide range of materials in engineering, e.g., geologic materials (clays, sands, rocks), modified geologic materials (structured soils, over-consolidated soils, expansive soils, loess, frozen soils, chemically treated soils), hydrate-bearing sediments, grain and particle crushing, application for problems like seepage and consolidation, earth pressures, reinforced earth, soil–structure interaction, piles, pavements, concrete and masonry structures, composites in electronic packaging, and interfaces and joints for various conditions, e.g., dry, saturated and unsaturated [1,2,21].

The HISS model is a hierarchical and unified framework in which the common constitutive models [1], e.g., Mohr–Coulomb, Drucker–Prager, Von Mises, critical state (Cam clay and Cap), Matsuoka-Nakai, Lade-Duncan, Lade-Kim, Vermeer, Cam Clay, etc., are engaged as a specific state. As indicated above, the DSC is based on the fundamental idea that the behavior of an engineering material can be defined by an appropriate connection that characterizes the interaction between material components. In the DSC model, it is assumed that the applied mechanical and environmental forces cause disturbances or changes in the material's microstructure with respect to its reference states. Usually, the two reference states consist of a continuum part called the relative intact (RI) state and a disturbed part called the fully adjusted (FA) state, as shown in Figure 1.1(a). The RI state refers to the material components that exclude the influences of factors such as microcracking, damage, softening and stiffening, which are defined and formulated based on the theories of continuum mechanics. The FA state refers to the asymptotic state, which is defined and formulated based on the approximations of the ultimate asymptotic response of the material. The material in the RI state is transformed continuously into the FA state through a process of natural self-adjustment of its microstructure, as shown in Figure 1.1(b). At any stage during deformation, the material is treated as the coupled mixture of the RI and FA states, which are distributed (randomly) over the material elements. The observed behavior of the material is defined in terms of the behaviors at the reference states connected through the disturbance function, D.

1.2.1 Relative Intact (RI) State

The response of the material part in the RI state excludes the influence of factors that cause the disturbance. It represents an "intact" state relative to the state of material affected by the disturbance. Schematics of RI and FA behaviors in terms of various measured quantities: stress vs. strain, volume or void ratio response, nondestructive behavior (velocity), and effective stress (or pore water pressure), are shown in Figure 1.2. For some conditions, the behavior of RI state can be assumed to be linear elastic defined by the initial slope. However, such an assumption may not be valid if the material behavior is nonlinear and affected by factors such as coupled volume change behavior, e.g., volume change under shear loading. Hence, very often, conventional or continuous yield or hierarchical single surface (HISS) plasticity is adopted as the RI response.

1.2.2 Fully Adjusted (FA) State

The material part in the FA state can be considered to have reached an asymptotic state, in which its structure reaches a condition that its properties are different from those in the RI state, and it may not be able to respond to the imposed shear and hydrostatic stresses as does the RI material. As a simple approach, it can be assumed that the material in the FA state has no strength, just like

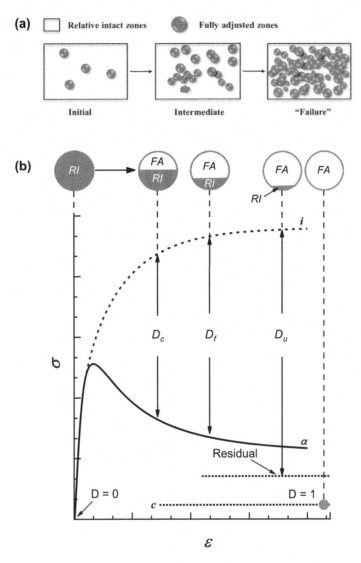

FIGURE 1.1 Schematics of DSC: (a) relative intact (RI) and fully adjusted (FA) states during deformation and (b) schematic of stress–strain behavior in the disturbed state concept (DSC).

in the classical damage model [3–5]. This assumption ignores the interaction between the RI and FA states, may lead to local models, and may cause computational difficulties. The second assumption is to consider that the material in the FA state can bear hydrostatic stress like a constrained liquid and the bulk modulus (K) can be used to define the response. In the critical state [1,6], the FA material can be considered to be similar to liquid–solid. The material approaches a state at which no change in volume or density or void ratio occurs under increasing shear stress after continuous yield. The equations for the strength of the material in the critical state (FA) are expressed as:

$$\sqrt{J_{2D}^c} = \bar{m} J_1^c \tag{1.1}$$

$$e^c = e_0^c - \lambda \ln\left(J_1^c / 3p_a\right) \tag{1.2}$$

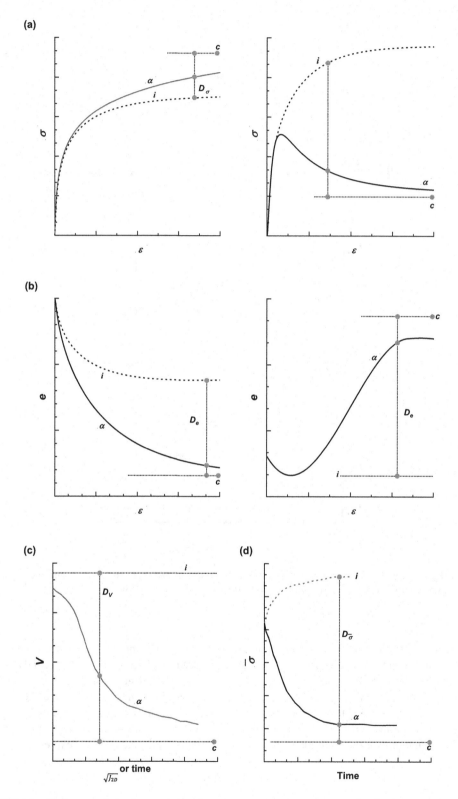

FIGURE 1.2   Various test behaviors to define disturbance: (a) stress–strain, (b) void ratio, (c) nondestructive velocity, and (d) effective stress.

where superscript c denotes the critical state; J_{2D} is the second invariant of the deviatoric stress tensor; $\bar{m}$ is the slope of the critical state line; J_1 is the first invariant of the stress tensor; e is the void ratio; e_0 is the initial void ratio; λ is the slope of the consolidation line; and p_a is the atmospheric pressure constant.

1.2.3 Disturbance

The idea of disturbance can be derived from the residual flow procedure proposed by Desai [7] and Desai and Li [8] for free surface seepage in two- and three-dimensional problems. As mentioned before, the disturbance that connects the interacting responses of the RI and FA parts in the same material (or of the components as reference materials) denotes the deviation of the observed response from the responses of the reference states. The disturbance can be determined in two ways, i.e., (i) from observed stresses, volumetric response, pore fluid pressure, entropy, and/or nondestructive (ultrasonic) velocities during loading, unloading, and reloading and (ii) by mathematical description in terms of internal variables.

From measurements, for example, the disturbance function D can be expressed as [9,10]:

$$D_\sigma = \frac{\sigma^{RI} - \sigma^a}{\sigma^{RI} - \sigma^{FA}} \qquad \left(\text{stress-strain behavior}\right) \tag{1.3}$$

$$D_v = \frac{V^{RI} - V^a}{V^{RI} - V^{FA}} \qquad \left(\text{nondestructive behavior}\right) \tag{1.4}$$

where σ^a is the observed or average stress and V^a is the observed or average nondestructive velocity.

For mathematical description, the disturbance function D can be expressed using the Weibull distribution function in terms of internal variable such as accumulated (deviatoric) plastic strain $\left(\xi_D\right)$ or plastic work [1,11]:

$$D = D_u\left[1 - \exp\left(-A\xi_D^Z\right)\right] \tag{1.5}$$

where A, Z, and D_u are the parameters. The value of D_u is obtained from the ultimate FA state. The expression in Eq. (1.5) is like those used in various areas, such as in biology to simulate birth to death, or growth and decay, and in engineering to define damage in classical damage mechanics. In addition, the disturbance function D for over-consolidated clays can be defined as [12]:

$$D = 1 - \left(1 / O_\delta\right)^\chi \tag{1.6}$$

where O_δ is the ratio of pre-consolidated stress to the current-consolidated stress and χ is the material constant that controls the effect of O_δ on the disturbance function. The parameter O_δ is changed during the loading process, and $O_{\delta 0}$, the initial value of O_δ, is equal to the overconsolidation ratio (OCR) of over-consolidated clays.

In the residual flow procedure [7,8], a residual or correction fluid "load" is evaluated based on the difference between permeabilities k_s or the degree of saturation S_r, in the saturated and unsaturated conditions. The permeability and saturation decrease with an increase in suction, and approach asymptotically those in the dry state. The disturbance D can be written as [1]:

$$D_k = \frac{k_s - k_{us}}{k_s - k_f} \tag{1.7}$$

and

$$D_s = \frac{1 - S_r}{1 - S_{rf}} \qquad (1.8)$$

Initially, in the saturated or RI state, $k_{us} = k_s$ and $S_r = 1$; finally, $k_{us} = k_f$ and $S_r = S_{rf}$. Hence, D varies from 0 to 1. Thus, if the values of permeabilities or saturation are measured during deformation of a specimen, the values of D can be determined.

However, the concept of disturbance is much different from damage. The former defines deviation of observed response from the RI (or FA) state, in the material treated as a mixture of interacting components, while the latter represents the effect of physical damage or cracks. Once the RI and FA states and disturbance are defined, the incremental DSC equations based on equilibrium of a material element can be derived as:

$$d\sigma_{ij}^a = (1 - D)d\sigma_{ij}^{RI} + Dd\sigma_{ij}^{FA} + dD\left(\sigma_{ij}^{FA} - \sigma_{ij}^{RI}\right) \qquad (1.9)$$

$$d\sigma_{ij}^a = (1 - D)C_{ijkl}^{RI}d\varepsilon_{kl}^{RI} + DC_{ijkl}^{FA}d\varepsilon_{kl}^{FA} + dD\left(\sigma_{ij}^{FA} - \sigma_{ij}^{RI}\right) \qquad (1.10)$$

where σ_{ij} and ε_{kl} denote the stress and strain tensors, respectively; C_{ijkl} is the constitutive tensor; and dD is the increment or rate of the disturbance function.

Eq (1.10) is general from which conventional continuum (elasticity, plasticity, creep, etc.) models can be derived as special cases by setting $D = 0$, as:

$$d\sigma_{ij}^a = C_{ijkl}^{RI}d\varepsilon_{kl}^{RI} \qquad (1.11)$$

If $D \neq 0$, Eqs. (1.9) and (1.10) account for microstructural modifications in the material leading to fractures and instabilities like failure and liquefaction. It is more general and effective to use the HISS plasticity model to define the RI behavior, which is described below.

1.2.4 Hierarchical Single Surface (HISS) Plasticity

The need for a unified and general plasticity model that can account for the factors mentioned before was the driving force for the development of the HISS plasticity model [13–20]. The yield surface, F, in HISS associative plasticity is expressed as:

$$F = \bar{J}_{2D} - \left(-\alpha \bar{J}_1^n + \gamma \bar{J}_1^2\right)\left(1 - \beta S_r\right)^{-0.5} = 0 \qquad (1.12)$$

where $\bar{J}_{2D} = J_{2D} / p_a^2$ is the nondimensional second invariant of the deviatoric stress tensor; $\bar{J}_1 = (J_1 + 3R) / p_a$ is the nondimensional first invariant of the total stress tensor; R is the term related to the cohesive (or tensile) stress, $\bar{c}$; $S_r = \sqrt{27}J_{3D} / 2J_{2D}^{1.5}$; n is the parameter related to the transition from compressive to dilative volume change; γ and β are the parameters associated with the ultimate surface; and α is the hardening or growth function, which can be given by:

$$\alpha = \frac{a_1}{\xi^{\eta_1}} \qquad (1.13)$$

where α_1 and η_1 are the hardening parameters and ξ is the accumulated or trajectory of plastic strains, given by:

$$\xi = \xi_v + \xi_D \qquad (1.14)$$

The accumulated volumetric plastic strain is given by:

$$\xi_v = \frac{1}{\sqrt{3}}\left|\varepsilon_{ii}^p\right| \tag{1.15}$$

The accumulated deviatoric plastic strain is given by:

$$\xi_D = \int \left(d\varepsilon_{ij}^p d\varepsilon_{ij}^p\right)^{0.5} \tag{1.16}$$

where ε_{ii}^p is the plastic volumetric strain and $d\varepsilon_{ii}^p$ is the increment of plastic shear strain. The yield surfaces of the HISS model in the stress spaces with different parameters β, α, γ, and n are shown in Figures 1.3–1.6.

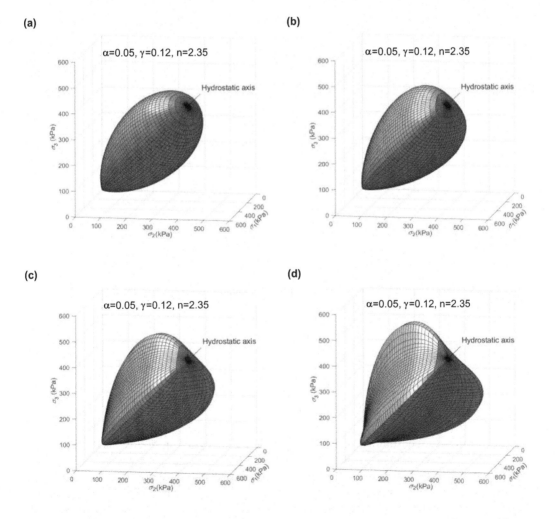

FIGURE 1.3 Hierarchical single surface (HISS) yield surfaces in three-dimensional stress space: (a) $\beta = 0.0$, (b) $\beta = 0.45$, (c) $\beta = 0.75$, and (d) $\beta = 0.90$.

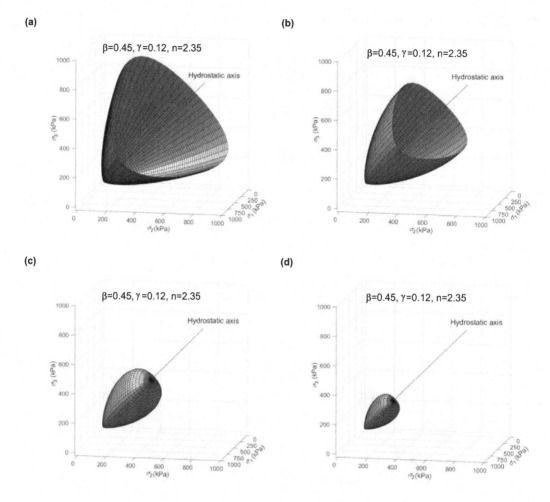

FIGURE 1.4 Hierarchical single surface (HISS) yield surfaces in three-dimensional stress space: (a) $\alpha = 0.03$, (b) $\alpha = 0.04$, (c) $\alpha = 0.05$, and (d) $\alpha = 0.06$.

According to the plasticity theory, the following incremental relationship between the stress vectors (σ_{ij}) and strain vectors (ε_{kl}) is inferable:

$$d\sigma_{ij} = C_{ijkl}^{ep} d\varepsilon_{kl} \tag{1.17}$$

$$C_{ijkl}^{ep} = C_{ijkl}^{e} - \frac{C_{ijrs}^{e} \dfrac{\partial F}{\partial \sigma_{rs}} \left(\partial F / \partial \sigma_{pq}\right)^{T} C_{pqkl}^{e}}{\left(\partial F / \partial \sigma_{mn}\right)^{T} C_{mnpq}^{e} \left(\partial F / \partial \sigma_{pq}\right) - \left(\partial F / \partial \xi\right)} \tag{1.18}$$

where C_{ijkl}^{e} is the elastic constitutive matrix and C_{ijkl}^{ep} is the elastoplastic constitutive matrix. The HISS plasticity model allows for continuous yielding, volume change including the dilation before peak, stress-path dependent strength, and effect of both volumetric and deviatoric strains on the yield behavior, and it does not contain any discontinuities in the yield surface. The HISS surface, Eq. (1.12), represents a unified yield surface, and most of the previous conventional and continuous yield surfaces can be derived as special cases. Moreover, the HISS model can be used for nonassociative and anisotropic hardening responses, etc.

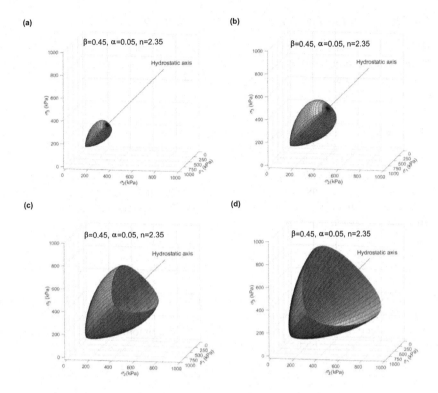

FIGURE 1.5 Hierarchical single surface (HISS) yield surfaces in three-dimensional stress space: (a) $\gamma = 0.10$, (b) $\gamma = 0.12$, (c) $\gamma = 0.14$, and (d) $\gamma = 0.16$.

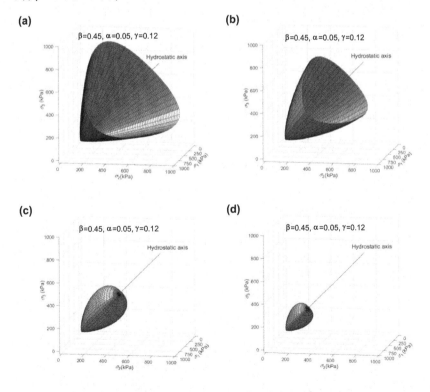

FIGURE 1.6 Hierarchical single surface (HISS) yield surfaces in three-dimensional stress space: (a) $n = 2.15$, (b) $n = 2.25$, (c) $n = 2.35$, and (d) $n = 2.45$.

1.3 APPLICATIONS OF DISTURBED STATE CONCEPT

The DSC/HISS theory has been used for a wide range of problems in geomechanics, geotechnical engineering, and other disciplines of engineering. A comprehensive list of publications related to geomechanics and geotechnical engineering by the author and coworkers is presented in **Supplementary 1**.

1.3.1 DSC for Saturated and Unsaturated Materials

Although the research about the modeling of the behavior of partially saturated soils just became popular recently, its importance for various engineering problems has been recognized for a long time. Many researchers have considered the basic mechanisms and proposed analytical and numerical models, including critical state theory, HISS plasticity, DSC, and computer implementation. The DSC for saturated soil mechanics can be expressed as [1,14,22]:

$$\sigma^a = (1-D)p^f + D\sigma^s \tag{1.19}$$

where σ^a is the total stress, σ^s is the stress at contacts, p^f is the fluid stress, and D is the disturbance.

A symbolic representation of an element of partially saturated material is shown in Figure 1.7. The responses are decomposed here into those of the solid skeleton, fluid (water), and gas (air). If the material is fully saturated, the force equilibrium gives:

$$\sigma^a = \bar{\sigma}(u_w) + u_w(\bar{\sigma}) \tag{1.20}$$

where u_w and $\bar{\sigma}$ are the pore water pressure and effective stress in the solid skeleton, respectively.

For partially saturated materials, the equilibrium can be expressed as:

$$\sigma^a = \bar{\sigma}(s) + u_w(\bar{\sigma},s) + u_g(\bar{\sigma},s) \tag{1.21}$$

where u_g is the pore gas or air pressure and $s = u_g - u_w$ is the matrix suction; if the gas happens to be air, $s = u_a - u_w$, where u_a is the pore air pressure.

The microstructure (solid skeleton or matrix) experiences relative particle motions as affected by irreversible strains, air and pore water pressures, and suction. Hence, the disturbance can be considered to occur in the solid skeleton and can be incorporated through the effective stress. For the saturated case, the observed effective stress, $\bar{\sigma}^a$ (in the solid skeleton), can be expressed as [1]:

$$\bar{\sigma}_{ij}^a = (1-D)\bar{\sigma}_{ij}^i + D\bar{\sigma}_{ij}^c \tag{1.22}$$

where $\bar{\sigma}_{ij}^a$, $\bar{\sigma}_{ij}^i$, and $\bar{\sigma}_{ij}^c$ are the vectors of stresses in the observed, RI, and FA states in the solid skeleton, respectively, and D is the disturbance in the solid skeleton.

The incremental form of Eq. (1.22) can be written as:

$$d\bar{\sigma}_{ij}^a = (1-D)d\bar{\sigma}_{ij}^i + Dd\bar{\sigma}_{ij}^c + dD\left(\bar{\sigma}_{ij}^c - \bar{\sigma}_{ij}^i\right) \tag{1.23}$$

or

$$d\bar{\sigma}_{ij}^a = (1-D)\bar{C}^i d\bar{\varepsilon}_{ij}^i + D\bar{C}^c d\bar{\varepsilon}_{ij}^c + dD\left(\bar{\sigma}_{ij}^c - \bar{\sigma}_{ij}^i\right) \tag{1.24}$$

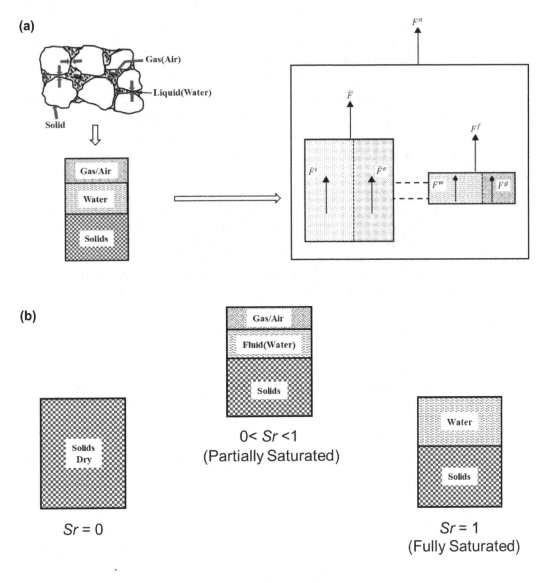

FIGURE 1.7 Dry and saturated materials: (a) equilibrium of forces, and (b) symbolic representation.

where $\bar{C}^i$ and $\bar{C}^c$ are the constitutive matrices for the responses of the RI and FA parts of the solid skeleton, respectively; $\bar{\varepsilon}_{ij}^i$ and $\bar{\varepsilon}_{ij}^c$ are the strains in the RI and FA parts, respectively; and dD is the increment or rate of D.

In the case of partially saturated materials, Eq. (1.24) can be written as a function of s (or S_r) as [1,22]:

$$d\bar{\sigma}_{ij}^a = [1 - D(s)]\bar{C}^i(s)d\bar{\varepsilon}_{ij}^i + D(s)\bar{C}^c(s)d\bar{\varepsilon}_{ij}^c + dD(s)\left(\bar{\sigma}_{ij}^c - \bar{\sigma}_{ij}^i\right) \qquad (1.25)$$

Disturbance can be evaluated using laboratory test data such as stress–strain, volumetric (void ratio), effective (pore water pressure), or nondestructive properties. Then, it can be expressed as:

$$D(s, \bar{p}_0) = D_{u1}(s, \bar{p}_0)\left\{1 - \exp\left[-A_1(s, \bar{p}_0)\xi_D^{Z_1(s, \bar{p}_0)}\right]\right\} \qquad (1.26)$$

or

$$D(s,\overline{p}_0) = D_{u2}(s,\overline{p}_0)\left\{1 - \exp\left[-A_2(s,\overline{p}_0)w^{Z_1(s,\overline{p}_0)}\right]\right\} \tag{1.27}$$

where s and $\overline{p}_0$ are the suction and initial effective mean pressure, respectively, for a given test, and ξ_D and w are the (deviatoric) plastic strain trajectory and plastic work, respectively.

1.3.2 DSC FOR STRUCTURED SOILS

Materials such as soils can develop a structure due to natural deposition processes during the geological history or artificial methods such as an application of mechanical loading. Such a structure can be defined with respect to that of the reconstituted or remolded state, which can be considered to represent the basic state before the factors causing structural changes were imposed.

In the DSC, the behavior of the reconstituted soil can be treated as the FA state, and then the effect of the structure developed can be introduced as the change in disturbance with respect to the disturbance at the reconstituted state. Then, the modified disturbance, $\overline{D}$, counting in the structure, which can result in stiffening with respect to the FA state, can be expressed as follows [1,23,24]:

$$\overline{D} = D_{ub}\left[1 - \exp\left(-A_b\xi^{Z_b}\right)\right] + f(s)\exp\left(-A_s\xi^{Z_s}\right) \tag{1.28}$$

where the function $f(s)$ denotes the change (increase or decrease) in disturbance due to the structure, ξ is the (deviatoric) plastic strain trajectory, and D_u, A, and Z are the material parameters.

The incremental strain equations for the DSC are given by:

$$d\varepsilon_{ij}^a = (1 - D_\varepsilon)d\varepsilon_{ij}^i + D_\varepsilon d\varepsilon_{ij}^c + dD_\varepsilon\left(\varepsilon_{ij}^c - \varepsilon_{ij}^i\right) \tag{1.29}$$

where D_ε is the disturbance function.

The disturbance function for the structured soil is expressed as [25]:

$$\overline{D}_\varepsilon = D_\varepsilon + \left(\frac{\partial D_\varepsilon}{\partial \varepsilon_{ij}^c}\right)^T \varepsilon_{ij}^c \tag{1.30}$$

Hence, the observed strain is given by:

$$d\varepsilon_{ij}^a = \overline{D}_\varepsilon d\varepsilon_{ij}^c \tag{1.31}$$

The disturbance function can be decomposed in two parts:

$$d\varepsilon_v^a = \overline{D}_{\varepsilon v}d\varepsilon_v^c \tag{1.32}$$

and

$$d\varepsilon_d^a = \overline{D}_{\varepsilon d}d\varepsilon_d^c \tag{1.33}$$

where $\bar{D}_{\varepsilon v}$ and $\bar{D}_{\varepsilon d}$ are disturbance functions related to volumetric and deviatoric behaviors, respectively, and

$$\varepsilon_v = \varepsilon_1 + \varepsilon_2 + \varepsilon_3 \tag{1.34}$$

$$\varepsilon_d = \frac{\sqrt{2}}{3}\sqrt{\left(\varepsilon_1 - \varepsilon_2\right)^2 + \left(\varepsilon_2 - \varepsilon_3\right)^2 + \left(\varepsilon_1 - \varepsilon_3\right)^2} \tag{1.35}$$

1.3.3 DSC FOR INTERFACES AND JOINTS

Behavior at contacts or interfaces between two (different) materials plays a significant role in the overall response of an engineering system [1,26–28]. The contact problem occurs in a number of disciplines in engineering and physics. One of the main advantages of the DSC is that its mathematical framework can be specialized for contacts (interfaces and joints). The adoption of the same framework for both solids and contacts provide the necessary consistency.

Figure 1.8 shows the schematic of two- and three-dimensional contacts, disturbed states, and deformation modes. As the interface is represented by an equivalent thickness, t, it can be treated as a deforming material element that is composed of the RI and FA parts. The RI behavior in the interface can be simulated by various models such as nonlinear elastic and plastic (conventional or continuous yield) models. If the HISS plasticity model is adopted for the RI part, the yield function specialized from Eq. (1.12) for a two-dimensional interface is given by [28]:

$$F = \tau^2 + \alpha\sigma_n^n - \gamma\sigma_n^q = 0 \tag{1.36}$$

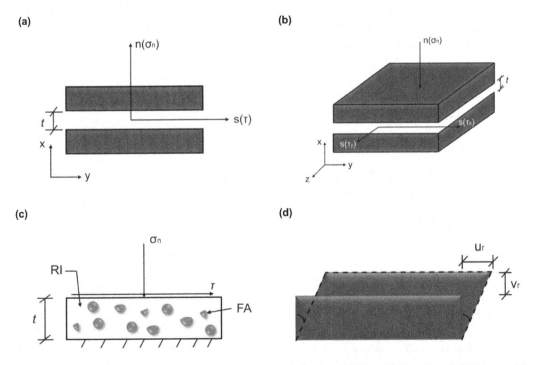

FIGURE 1.8 Thin-layer element and material parts in relative intact (RI) and fully adjusted (FA) states: (a) two-dimensional thin-layer zone element, (b) three-dimensional thin-layer element, (c) interface zone with RI and FA parts, and (d) deformation models in two-dimensional zone.

where τ is the shear stress, σ_n is the normal stress, which can be modified as $\sigma_n + R$, R is the intercept along the σ_n axis, γ is the slope of ultimate response, n is the phase change parameter, which designates the transition from compressive to dilative response, q governs the curve of the ultimate envelope (if the ultimate envelope is linear, $q = 2$), and a is the growth or yield function given by:

$$\alpha = h_1 / \xi^{h_2} \tag{1.37}$$

where h_1 and h_2 are hardening parameters and ξ is the trajectory of plastic relative shear and normal displacements (u_r and v_r), respectively, given by :

$$\xi = \int \sqrt{\left(du_r^p \cdot du_r^p + dv_r^p \cdot dv_r^p\right)} = \xi_D + \xi_v \tag{1.38}$$

where ξ_D and ξ_v denote accumulated plastic shear and normal displacements, respectively, and p denotes plastic.

The idea of the critical state for the behavior of joints and interfaces can be built on their observed responses. Figure 1.9 shows schematics of τ vs. u_r and normal displacement during shear (v_r) vs. u_r for typical interfaces or joints. Just as in the case of solids, the contact can reach the critical state irrespective of the initial roughness and normal stress (σ_n), when the relative normal displacement v_r remains invariant.

The expression for critical shear stress and normal stress is:

$$\tau^c = c_0 + c_1 \sigma_n^{(c)c_2} \tag{1.39}$$

where c_0 is related to the adhesive strength and denotes the critical value of τ^c when $\sigma_n = 0$; σ_n^c is the normal stress at the critical state; and c_1 and c_2 are parameters related to the critical state.

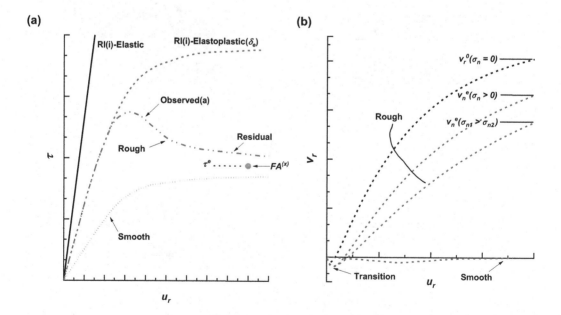

FIGURE 1.9 Test data for contact (interface or joint): (a) τ vs. u_r and (b) v_r vs. u_r.

The relation between the normal stress σ_n and the relative normal displacement at the critical state v_r^c is expressed as follows:

$$v_r^c = v_r^0 \exp(-\bar{\lambda}\sigma n) \tag{1.40}$$

where v_r^0 is the critical or ultimate dilation and $\bar{\lambda}$ is a material parameter.

Combining Eq. (1.5), the disturbance function D can be defined based on the shear stress vs. relative shear displacement and/or relative normal vs. relative shear displacement curves:

$$D = \frac{\tau^i - \tau^a}{\tau^i - \tau^c} \tag{1.41}$$

1.4 DETERMINATION OF MATERIAL PARAMETERS

The general DSC/HISS model with microcracking, fracture, and softening involves the following parameters. Brief statements for their determination from (standard) laboratory tests are given below; comprehensive procedures are given elsewhere (1):

1. Elastic parameters: e.g., Young's modulus, E, and Poisson's ratio, v (or shear modulus, G, and bulk modulus, K). These can be determined from the unloading curves or slopes at initial stress, in terms of (deviator) stress vs. (axial) strain, and volumetric strain vs. strain, from triaxial tests, respectively. These parameters can be treated as variable functions of stress such as mean pressure, p, and shear stress, $\sqrt{J_{2D}}$. The resilient modulus, M_R, can be considered to be a special case of such nonlinear model.
2. Plasticity parameters: they depend on the model used; e.g., for von Mises model, one parameter, cohesive strength, c (or yield stress σ_y), and for the Mohr–Coulomb model, cohesive strength, c, and angle friction, ϕ. For the HISS model, five parameters are required:
   i.   Ultimate yield γ and β (Eq. 1.12) are determined from the plot of the ultimate envelope based on the asymptotic stress in the ultimate region of the stress–strain curve, Figures 1.3 and 1.5.
   ii. Phase change (transition from compaction to dilation), n, is determined from the stress condition at which transition from compaction to dilation occurs, usually before the peak stress on the (triaxial) stress–strain curve, Figure 1.6.
   iii. Hardening or yielding parameters, a_1 and η_1 (Eq. 1.13), are determined by computing the accumulated plastic strain, ξ (Eq. 1.14), from increments on the stress–strain curve. Then, value of α for the specific increment is obtained from Eq. (1.12) by substituting the stress on the stress–strain curve and ploting $\ln\alpha$ vs. $\ln\xi$, yielding the values of a_1 and η_1.
   iv. Parameter R related to the cohesive strength, $\bar{c}$, can be obtained by using the following expression: $R = \bar{c} / (3\sqrt{\gamma})$.
3. Critical state parameters: The value of parameter $\bar{m}$, Eq. (1.1), is obtained as the slope of the critical state line. Parameters λ and κ are obtained as loading and unloading slopes in the plot of the void ratio at the critical state (Eq. 1.2) vs. mean pressure at the critical state, $J_1^c / 3p_a$, i.e., e^c vs. $\ln(J_1^c / 3p_a)$.
4. Disturbance parameters: The parameter D_u can be obtained from the residual response (Figure 1.1), with often using a value near unity. Parameters A and Z are obtained by first determining various values of D from the test data by using Eq. (1.3), and then plotting the logarithmic form (Eq. 1.5).

Procedures similar to soils and rocks are followed for determination of parameters for interfaces and joints. Further details of the procedures are presented by Desai [1].

The foregoing parameters for the general DSC/HISS model are needed only if elasto-visco-plastic and disturbance (microcracking, fracture, softening) characterization is desired. However, it is important to note that only the parameters for a specific option are needed.

REFERENCES

1. Desai, C. S. *Mechanics of Materials and Interfaces: The Disturbed State Concept*. Boca Raton, FL: CRC Press; 2001.
2. Desai, C. S. "Disturbed state concept as unified constitutive modeling approach." *J. Rock Mech. Geotech. Eng.*, 8, 3, 2016, 277–293.
3. Kachanov, L. M. *Introduction to Continuum Damage Mechanics*. Dordrecht, The Netherlands: Martinus Nijhoft Publishers; 1986.
4. Desai, C. S., and Woo, L. "Damage model and implementation in nonlinear dynamic problems." *Comput. Mech.*, 11, 2, 1993, 189–206.
5. Bažant, Z. P., and Lin, F.-B. "Non-local yield limit degradation." *Int. J. Numer. Meth. Eng.*, 26, 8, 1988, 1805–1823.
6. Roscoe, K. H., Schofield, A., and Wroth, C. P. "On yielding of soils." *Géotechnique*, 8, 1, 1958, 2–53.
7. Desai, C. S. "Finite element residual schemes for unconfined flow." *Int. J. Numer. Meth. Eng.*, 10, 6, 1976, 1415–1418.
8. Desai, C. S., and Li, G. C. "A residual flow procedure and application for free surface flow in porous media." *Adv. Water Resour.*, 6, 1, 1983, 27–35.
9. Desai, C. S., and Toth, J. "Disturbed state constitutive modeling based on stress-strain and nondestructive behavior." *Int. J. Solids Struct.*, 33, 11, 1996, 1619–1650.
10. Desai, C. S. "Constitutive modeling of materials and contacts using the disturbed state concept: Part 1 – Background and analysis." *Comput. Struct.*, 146, 2015, 214–233.
11. Desai, C. S., Basaran, C., and Zhang, W. "Numerical algorithms and mesh dependence in the disturbed state concept." *Int. J. Numer. Meth. Eng.*, 40, 16, 1997, 3059–3083.
12. Xiao, Y., and Desai, C. S. "Constitutive modeling for over-consolidated clays based on disturbed state concept. I: Theory." *Int. J. Geomech.*, 19, 9, 2019, 04019101.
13. Desai, C. S. "A general basis for yield, failure and potential functions in plasticity." *Int. J. Numer. Anal. Methods Geomech.*, 4, 4, 1980, 361–375.
14. Desai, C. S., Somasundaram, S., and Frantziskonis, G. "A hierarchical approach for constitutive modelling of geologic materials." *Int. J. Numer. Anal. Methods Geomech.*, 10, 3, 1986, 225–257.
15. Shao, C., and Desai, C. S. "Implementation of DSC model and application for analysis of field pile tests under cyclic loading." *Int. J. Numer. Anal. Methods Geomech.*, 24, 6, 2000, 601–624.
16. Athukorala, R., Indraratna, B., and Vinod, J. S. "Disturbed state concept-based constitutive model for lignosulfonate-treated silty sand." *Int. J. Geomech.*, 15, 6, 2015, 11.
17. Toufigh, V., Shirkhorshidi, S. M., and Hosseinali, M. "Experimental investigation and constitutive modeling of polymer concrete and sand interface." *Int. J. Geomech.*, 17, 1, 2017, 11.
18. Abyaneh, M. J., and Toufigh, V. "Softening behavior and volumetric deformation of rocks." *Int. J. Geomech.*, 18, 8, 2018, 11.
19. Baghini, E. G., Toufigh, M. M., and Toufigh, V. "Analysis of pile foundations using natural element method with disturbed state concept." *Comput. Geotech.*, 96, 2018, 178–188.
20. Baghini, E. G., Toufigh, M. M., and Toufigh, V. "Application of DSC model for natural-element analysis of pile foundations under cyclic loading." *Int. J. Geomech.*, 19, 7, 2019, 17.
21. Desai, C. S., and Zhang, W. "Computational aspects of disturbed state constitutive models." *Comput. Methods Appl. Mech. Eng.*, 151, 3, 1998, 361–376.
22. Geiser, F., Laloui, L., Vulliet, L., and Desai, C. (1997). "Disturbed state concept for partially saturated soils." In: *Proceedings of the 6th International Symposium on Numerical Models in Geomechanics*, Montreal, Canada.
23. Liu, M., Carter, J., and Desai, C. "Modeling compression behavior of structured geomaterials." *Int. J. Geomech.*, 3, 2003, 191–204.
24. Ouria, A. "Disturbed state concept-based constitutive model for structured soils." *Int. J. Geomech.*, 17, 7, 2017, 10.

25. Liu, M. D., Carter, J. P., Desai, C. S., and Xu, K. J. "Analysis of the compression of structured soils using the disturbed state concept." *Int. J. Numer. Anal. Methods Geomech.*, 24, 8, 2000, 723–735.
26. Fakharian, K., and Evgin, E. "Elasto-plastic modelling of stress-path-dependent behaviour of interfaces." *Int. J. Numer. Anal. Methods Geomech.*, 24, 2, 2000, 183–199.
27. Desai, C. S., Zaman, M. M., Lightner, J. G., and Siriwardane, H. J. "Thin-layer element for interfaces and joints." *Int. J. Numer. Anal. Methods Geomech.*, 8, 1, 1984, 19–43.
28. Desai, C. S., and Fishman, K. L. "Plasticity-based constitutive model with associated testing for joints." *Int. J. Rock Mech. Min. Sci. Geomech. Abstr.*, 28, 1, 1991, 15–26.

Section 2

Geologic Materials

2 Liquefaction in Geologic Materials

Chandrakant S. Desai
Dept. of Civil and Architectural Engineering and
Mechanics, University of Arizona

2.1 INTRODUCTION AND BACKGROUND

Materials and interfaces/joints are composed of microstructure or matrix (Figure 2.1) and are connected through friction, adhesion, or bonding at contacts. The microstructure consists of solid particles and pore spaces, which are empty for dry materials but often infiltrated with fluid (water) and/or gas (air).

During deformations, the matrix can undergo reorganization of particles involving deformations without sliding, with sliding, rotation, and opening or closing at contacts [1,2]. The grains may experience oscillating motions at the microlevel (Figure 2.1) while average or near continuous response is measured at the macro level.

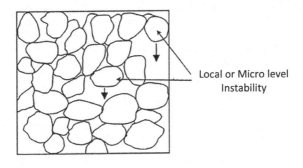

Local or Micro level
Instability

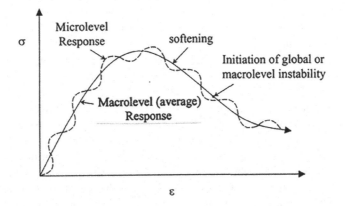

FIGURE 2.1 Particle matrix and instabilities.

DOI: 10.1201/9781003362081-4

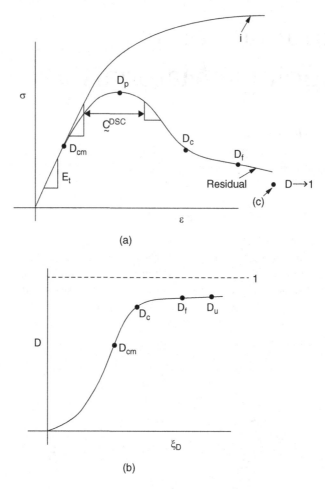

FIGURE 2.2 Threshold or critical disturbances: (a) on stress–strain response and (b) D vs. accumulated plastic strains, ξ_D.

The matrix (grains) can experience instantaneous unstable conditions leading to different threshold states that may or may not involve failure or collapse or liquefaction. As loading and deformation progress, the particles experiencing (local) instability may assume threshold states at which various behavioral features are exhibited, e.g., transition from volume compression to dilation (D_{cm}), peak stress (D_p), initiation of failure or liquefaction (D_c), and final failure or liquefaction (D_f) (Figure 2.2). Disturbance, D, in the disturbed state concept (DSC) is described later.

Materials like metals and certain alloys may exhibit nonlinear response, but very little or no volume change, whereas granular materials like sands can exhibit different behavior, for initially loose and dense states (Figure 2.3). The responses include various critical or threshold states such as volume transition, peak, and failure or liquefaction in the ultimate stages.

Most materials contain discontinuities like microcracks, faults, and dislocations in the initial state and they change, e.g., grow, during deformation. They contribute to particle readjustment and can play a significant role in inducing the unstable states. As discontinuities grow, the material can transform gradually from the continuum state to the discontinuous state; in the limit, the entire material can approach a fully discontinuous state. In the intermediate state containing both continuous and discontinuous states, the material has a distributed or diffused condition identified by irreversible (plastic) strains or work or disturbance. The issue of common shear band formations,

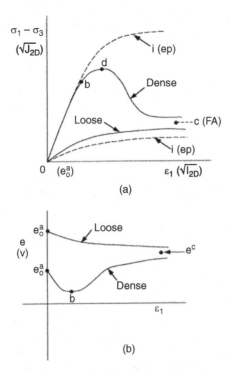

FIGURE 2.3 Responses for initially loose and dense cohesionless materials.

which can cause instability, is a special case of the more general distributed condition in the DSC. The volume of material elements at the distributed condition reaches certain values corresponding to the threshold of unstable states stated above. For instance, critical disturbance, D_c, denotes initiation of failure or liquefaction (Figure 2.2).

The instabilities within the material cause various threshold states including failure and liquefaction. In this chapter, liquefaction instability is the main topic of interest. For a fundamental approach for modeling liquefaction such internal mechanisms are significant and need to be considered. The DSC and energy approaches are capable of considering the *internal* mechanisms [3]. In this chapter, the energy approach is described briefly, while the DSC is described in detail because it provides simpler and more direct modeling of liquefaction. The conventional *external* approaches are first described briefly, as they are often used in practice.

The instabilities within the material cause various threshold states including failure and liquefaction. In this chapter, liquefaction instability is the main topic of interest. For a fundamental approach for modeling liquefaction, such internal mechanisms are significant and need to be considered. The DSC and energy approaches are capable of considering the *internal* mechanisms. In this chapter, the energy approach is described briefly while the DSC is described in detail because it provides simpler and more direct modeling for liquefaction. The conventional *external* approaches are first described briefly, as they are often used in practice.

2.2 REVIEW OF OTHER APPROACHES FOR LIQUEFACTION

2.2.1 CONVENTIONAL APPROACHES

The conventional approaches are based on factors such as effective stress and the results of standard penetration tests (SPT), cyclic stress ratio (CSR), and cyclic resistance ratio (CRR). They and their modifications are described and reviewed in various publications [4–8]. Only a brief description of the CSR is presented below, which is expressed as:

$$\text{CSR} = \tau_{\text{avg}} / \sigma_v' = 0.65 \left(a_{\max} / g \right) \left(\sigma_v / \sigma_v' \right) r_d \qquad (2.1)$$

where $\tau_{\text{avg}} = 0.65\, \tau_{\max}$, $\tau_{\max}$ = maximum shear stress, σ_v = total vertical stress at the bottom of soil (column), σ_v' = vertical effective stress $(\sigma_v - u)$, u = pore water pressure, g = acceleration due to gravity, $a_{\max}$ = maximum horizontal ground acceleration due to earthquake, and r_d = reduction factor. To connect liquefaction with SPT results, a relation between the shear stress ratio and SPT blow count has been established.

Although the above conventional methods often yield useful results for liquefaction, they are simply phenomenological and do not relate to the internal mechanisms that are responsible for liquefaction.

2.2.2 ENERGY APPROACHES

There are many publications on the energy approaches for identification of liquefaction that relate to the internal mechanisms. A few are referenced here [3,9–13]. Energy approaches involve consideration of factors such as critical dissipated energy, the relation between dissipated energy during cyclic loading and excess pore water pressures leading to liquefaction, strain energy required for initiation of liquefaction, and liquefaction potential (LP) evaluated using the energy capacity for liquefaction.

Desai [1,3] has shown the relation between the energy approach and the DSC. Based on the torsional shear tests, Figueroa et al. [12] investigated dissipated energy and liquefaction. Detailed analyses of these test data and the DSC have been presented by Desai [1]; a typical figure comparing the DSC and accumulated energy for the test data is shown in Figure 2.4.

Figure 2.4a shows the variation of D vs. N (1 cycle = 10 s) [13,14], and Figure 2.4b shows accumulated energy (w) vs. N computed by using measured data from Ref. [12]. The critical disturbance at the initiation of liquefaction is found to be $D_c = 0.87$. The corresponding critical energy density, w_c, is about 1500 J/m^3. The critical disturbance, D_c, has been used for identification of initial liquefaction for several problems including those presented later.

2.3 LIQUEFACTION USING DSC

Details of the DSC/HISS modeling including the background, basic equations, and some applications are presented in Chapter 1. Hence, only brief details are presented below.

As described before, the DSC/HISS model relates to the internal mechanisms during deformation that can cause instabilities like failure and liquefaction. Various threshold conditions during deformation can be identified by corresponding measured disturbances (Figure 2.2).

Figure 2.5 shows some of the conditions such as D_c for critical disturbance when initial liquefaction and D_f when final liquefaction can occur. At complete disintegration of the material, the disturbance can be identified as ultimate, D_u.

As indicated previously, initiation and final liquefaction can occur at critical disturbances, D_c and D_f, respectively (Figure 2.5). The laboratory test data in terms of stress–strain, volumetric, effective stress, and nondestructive velocities can be used to find the disturbance (Figure 2.2). The critical disturbance, D_c, can be found at the intersection of tangents to the graph of D versus time (or cycles) or D vs. accumulated plastic strains, N, or time (Figure 2.5). Then, the LP can be evaluated based on the critical values of D_c. In Section 2.4.1, the initiation of liquefaction at depth = 12 m at the Port

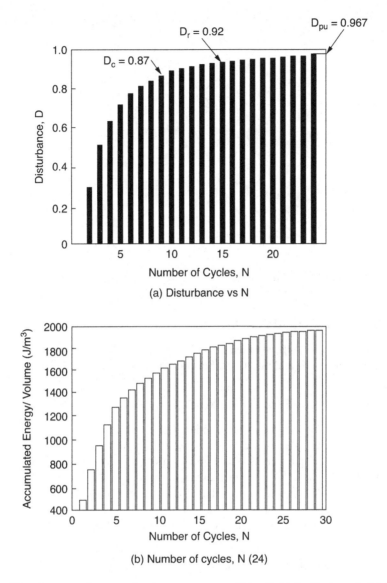

FIGURE 2.4 Disturbance and accumulated energy versus N for Reid Bedford sand: (a) D vs. N and (b) Energy vs. N (Figueroa et al., 1994). N = number of cycles.

Island site is indicated at the critical disturbance $D_c = 0.905$ at about 14.6 s (Figure 2.8). Then, at about 18 s the liquefied zone grows toward depth = 16 m, and at 21 s, toward depth = 8 m. At about 30 s, the liquefied zone covers depths below 4 m and up to 20 m.

The energy approach provides a fundamental way to relate to the internal mechanisms of liquefaction. However, as discussed in Desai [1,3], it is not easy to compute energy from field and laboratory test data and in computer procedures. On the other hand, it is simpler to compute the disturbance including in computer (finite element) procedures, as shown in the examples presented below. In the computer (finite element) procedures, D, is found at every increment or time just like displacements, stresses, strain, and pore water pressures. Hence, it is convenient to plot contours of D in the mesh and identify elements where liquefaction has been initiated. Thus, the progress of liquefaction in the mesh can be computed, viewed, and plotted.

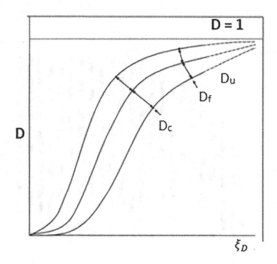

FIGURE 2.5 Disturbance vs. ξ_D (or time or no. of cycles) under required factors.

2.4 VALIDATIONS AND APPLICATIONS

Desai and coworkers [1,3,13–19] and others, e.g., Park et al. [20], have used the DSC for liquefaction analysis in geologic materials and interfaces. Several typical examples are presented below.

2.4.1 LIQUEFACTION IN THE FIELD DURING KOBE EARTHQUAKE

The DSC has been used for analysis of liquefaction in the field at Port Island, Kobe, Japan, during the Hyogoken-Nanbu earthquake. Figure 2.6 shows the soil profile and the locations of strong motion detection instruments at different depths at the Port Island site, which were adopted from various publications cited in [9,10,21]. Figure 2.7 shows measured shear wave velocities at the Kobe site. These shear wave velocities were used to compute the disturbance (see Chapter 1). Figure 2.8 shows the disturbance vs. time at different depths. The time at liquefaction = 14.6 s was indicated for D_c = 0.901 (Figure 2.8); this time to liquefaction compares excellently with the measured value from the field data.

2.4.2 LIQUEFACTION IN THE SHAKE TABLE TEST

Liquefaction analysis using the DSC/HISS modeling has been conducted for the test results from a laboratory shake table test presented by Akiyoshi et al. [22]. The shake table was instrumented with various measurement devices, and Fuji River sand was used. The DSC/HISS model parameters were not available for the Fuji River sand. Hence, because the index properties of the Fuji River and Ottawa sands were found to be very similar, the model parameters for the Ottawa sand were used for the analysis. The DSC parameters for the Ottawa sand were found from comprehensive cyclic trial tests [23].

The finite element mesh with 160 elements and 90 nodes involving 120 elements for soil and 40 elements for the steel box is shown in Figure 2.9. The DSC-DYN2D code was used [24] for the incremental dynamic analysis. The applied loading consists of horizontal displacements, X, applied at the bottom nodes, given by

$$X = u_x \sin(2\pi \, ft) \tag{2.2}$$

where u_x = amplitude of displacement (0.0013 m), f = frequency (5 Hz), and t = time.

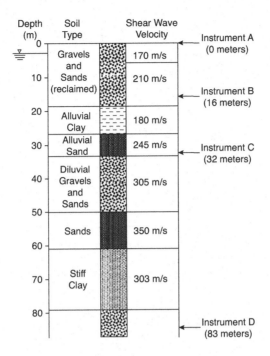

FIGURE 2.6 Details of soils and instruments at Port Island, Kobe site.

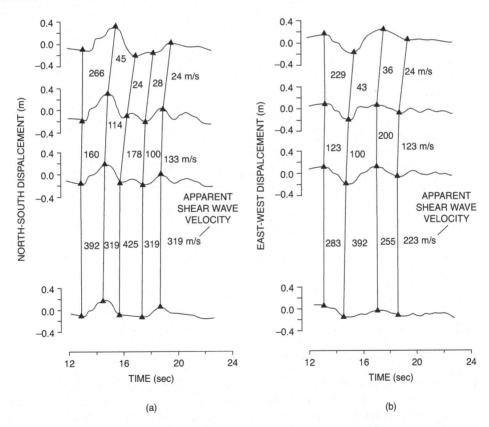

FIGURE 2.7 Measured shear wave velocities with time (Davis and Berrill 1982): (a) North-South and (b) East-West.

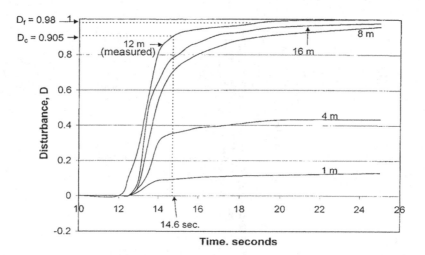

FIGURE 2.8 Disturbance versus time at different depths and times at liquefaction.

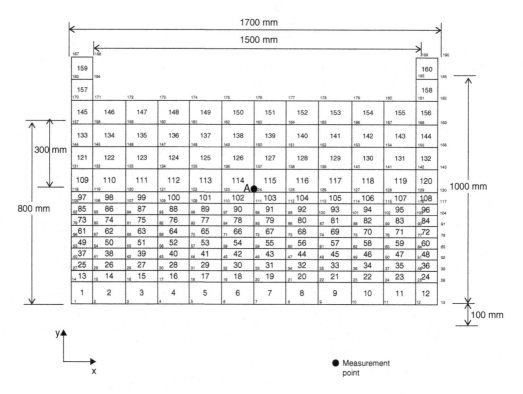

FIGURE 2.9 Finite element mesh for a shake table test.

Figure 2.10 shows comparisons between measured and computed excess pore water pressure with time at the point (depth = 300 mm) shown as A and a dot in Figure 2.9. The test data show that the initial liquefaction occurred at 2.0 s when the pore water pressure became equal to the initial effective stress. The finite element predictions are in very good agreement with the measured time for initial liquefaction.

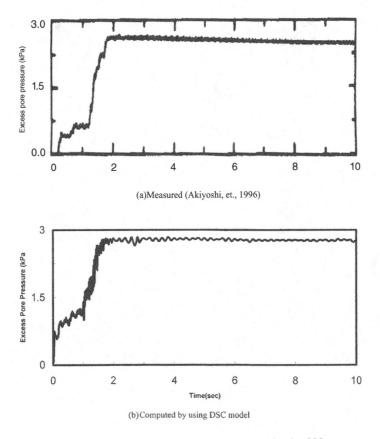

(a)Measured (Akiyoshi, et., 1996)

(b)Computed by using DSC model

FIGURE 2.10   Measured and computed excess pore water pressure at depth = 300 mm.

Figure 2.11a–d shows growth of disturbance in different elements at various times = 0.5, 1.0, 2.0 and 10 s. The computed plot of disturbance at depth = 300 mm is shown in Figure 2.11e. The critical disturbance $D_c = 0.84$ at initial liquefaction is shown in laboratory tests data [14,16]. At time = 0.5 s, the computed D in the sand is well below the critical disturbance in all elements. At time = 2.0 s, the disturbance has reached values higher than 0.80 at and below the depth = 300 mm, which indicates the onset of liquefaction at about 2.0 s. At time = 10 s, disturbance above the critical value (0.84) has been reached in most of the sand volume indicating final liquefaction. These results compare well with the observation in the laboratory [22].

2.4.3 LIQUEFACTION IN ROUGH INTERFACES

Testing and DSC/HISS modeling for sand–concrete interfaces with one single type of roughness were reported by Desai et al. [15]. For the study presented here, comprehensive monotonic and cyclic simple shear experiments were conducted on Ottawa sand–steel interfaces with three types of roughness under undrained conditions using the Cyclic Multi-degree of Freedom shear device (CYMDOF-P) with pore water pressure measurements [19]. The effect of various parameters such as normal stress, surface roughness of steel, type of loading, and the amplitude and frequency of the applied displacement in two-way cyclic loading were investigated. The effects of roughness on the interface behavior are a unique contribution of this study because such results are not readily available.

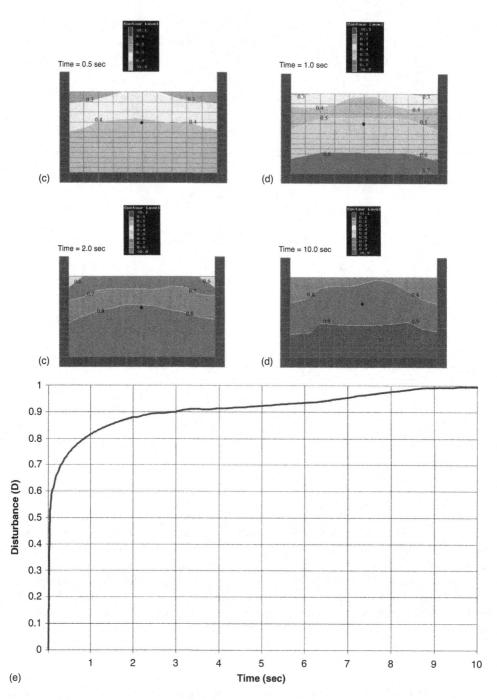

FIGURE 2.11 Growth of disturbance in sand at depth = 300 mm: (a) Time=0.5 s; (b) Time = 1.0 s; (c) Time = 2.0 s; (d) Time = 10.0 s; and (e) Disturbance versus time at depth = 300 mm.

Three different type of roughness, smooth (RI), slightly textured (RII), and moderately textured (RIII) were considered. Several definitions of roughness were reviewed, and one called, Rs, was used. It denotes the ratio of actual surface area and projected area. It was measured by using a ST400 30 Optical Profilometer.

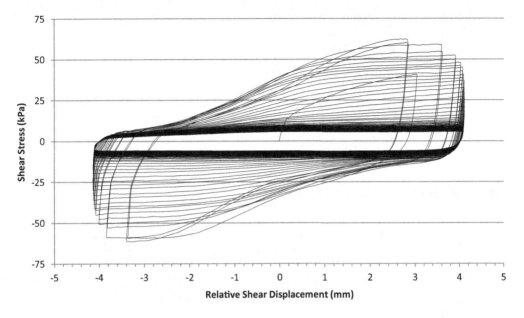

FIGURE 2.12 Two-way cyclic shear test (σ_n = 100 kPa, Roughness: RII, Amplitude = 5.0 mm, Frequency=0.1 Hz).

Figure 2.12 shows a typical test result for cyclic tests. Table 2.1 shows the time to liquefaction, t_{liq}, and number of cycles to liquefaction, N_{liq}, from laboratory tests in comparison with time to liquefaction, t_c, and cycles to liquefaction, N_c, by using the DSC, for the three type of roughness. Table 2.1 shows excellent correlation between the DSC predictions and tests results.

It is interesting to note that the times to liquefaction and cycles to liquefaction are influenced significantly by the interface roughness.

TABLE 2.1
Comparisons of Liquefaction from Test Data and DSC Predictions for Two-Way Cyclic Shear Tests

Test	Roughness	Based on Lab. Test Results		Based on Critical Disturbance Concept		
		t_{liq} (s)	N_{liq} (cycles)	D_c	t_c (s)	N_c (cycles)
SS25	RI	2600	104	0.92	2614	105
SS23	RI	1000	100	0.93	975	98
SS26	RI	360	90	0.92	357	90
SS28	RI	400	40	0.90	448	45
SS24	RI	440	44	0.91	450	45
SS30	RI	680	68	0.93	695	70
SS31	RII	200	20	0.94	166	17
SS32	RII	320	32	0.88	346	35
SS33	RII	440	44	0.91	445	45
SS34	RIII	200	20	0.93	215	22
SS35	RIII	320	32	0.88	348	35
SS36	RIII	400	40	0.91	410	41

(Dc) average. = 0.91

TABLE 2.2

Liquefaction Results for Joomujin Sand

	Initial Confining Pressure σ_o' (kPa)			Relative Density, D_r		
	70	**100**	**150**	**50**	**60**	**70**
D_c	0.935	0.949	0.955	0.947	0.949	0.957
N_c (DSC)	18	35	40	33	35	40
N_{liq} (Lab Tests)	19	35	40	33	35	40

2.4.4 LIQUEFACTION IN SATURATED KOREAN SAND

To study liquefaction using the DSC, Park et al. [20] conducted triaxial tests on specimens of saturated Joomujin sand (Korea). The tests were conducted with three initial confining pressures ($\sigma_o' = 70$, 100 and 150 kPa) and three relative densities ($D_r = 50\%$, 60% and 70%). The number of cycles to liquefaction was identified from the test data (N_{liq}) and the DSC modeling (N_c).

Table 2.2 shows these values for different σ_o' and D_r. It is seen that the DSC predictions for cycles to liquefaction and those from the test data are in excellent correlation.

2.4.5 LIQUEFACTION: PILES UNDER DYNAMIC LATERAL LOADING

Predicting the dynamic (earthquake) behavior of a pile foundation in liquefiable sand is a complex and challenging problem in geomechanics. Pradhan and Desai [17] presented a study of liquefaction around piles in sand by analyzing centrifuge tests for axially loaded piles under dynamic loading. Behavior under lateral loading is not fully available. Nevertheless, analyses of laterally loaded piles including modeling and field verifications are presented here.

Numerical dynamic analysis of a laterally loaded pile in the field [25] has been conducted by Essa and Desai [18] using finite element analysis with the DSC/HISS constitutive model. Two independent finite element analyses, *with* and *without an interface,* were carried out to demonstrate the effect of the interface on the pile–soil behavior. Material properties of the sand and the sand–steel interface were derived for the DSC/HISS model based on the laboratory tests results reported in Refs. [16,18,23], respectively. The interface parameters for the DSC/ISS model were found from the comprehensive tests using the CYMDO shear device [18,19].

Harmonic sinusoidal wave displacement was applied in the lateral direction 0.61 m below the pile top (Figure 2.13). The effect of boundary conditions has been essentially eliminated by using a suitable mesh dimension ($w = 70d$ and $h = 15d$), where w is the distance from the side boundary

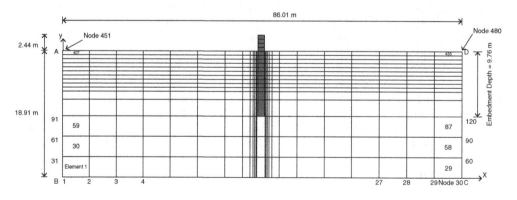

FIGURE 2.13   Finite element mesh (not to scale).

to the pile face, h is the depth of mesh below the pile tip, and d is the pile diameter [18]. The finite element mesh used in this problem is shown in Figure 2.13.

Porewater pressure variations with time were evaluated at three locations (elements 339, 398, and 399) corresponding to the locations of piezometers installed in the field. The results from the numerical model for both cases with an interface and without an interface are compared with the field measurements. The comparisons between the DSC predictions and field measurements are shown at a typical location in 398 (Figure 2.14). The predicted variations for both with and without an interface are comparable with the field measurements. However, the numerical results for the case of no interface show irregular variation and different amplitudes from the measured variation.

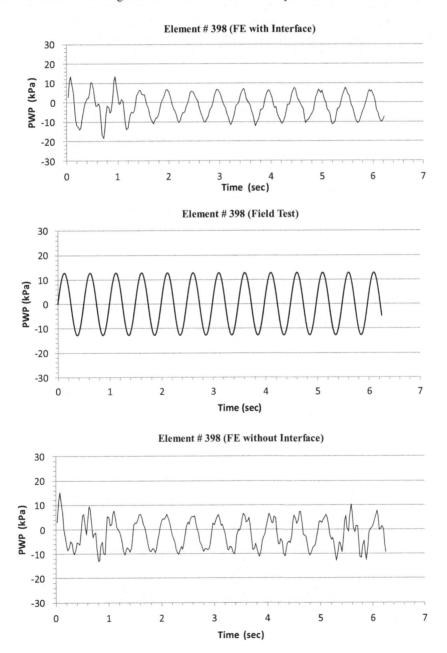

FIGURE 2.14   Excess pore pressures in element 398.

2.4.6 LIQUEFACTION POTENTIAL AND DISTURBANCE

In conventional methods, if the pore pressure in the soil approaches the initial effective stress, the soil loses its strength completely and is liquefied [3].

Desai et al. [14] and Park and Desai [16] proposed the method based on the DSC to identify lique-faction. A new parameter called the critical disturbance parameter (D_c) is defined at which instabil-ity leading to further or final collapse occurs in the microstructure. For Ottawa sand, they identified the initiation of liquefaction at $D_c=0.84$. For Ottawa sand–steel interface, the value of $D_c=0.91$ was identified by Essa and Desai [19] at the initiation of liquefaction.

Two opposite peaks of the eleventh applied displacement cycle are chosen to identify the soil liquefaction for the laterally loaded pile (step 204: displacement=6.58 mm, step 214: displace-ment=−6.58 mm). The distribution of excess pore water pressure generated at the times of the two peaks is shown in Figure 2.15. When the pile head was subjected to positive displacement

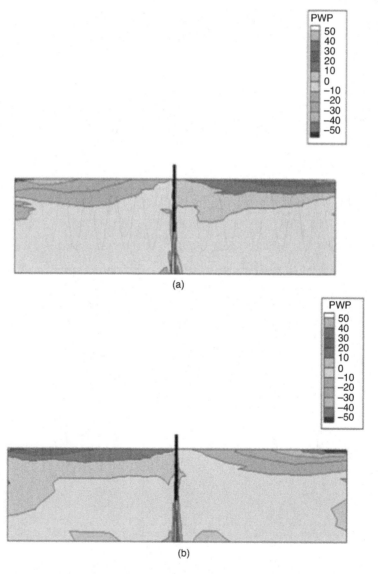

(a)

(b)

FIGURE 2.15   Distribution excess pore water pressures (a) at step 204 and (b) at step 214.

(step 204), a positive excess pore pressure was generated in the soil mass located in the front (right) side of the pile while the soil mass in the back (left) side of the pile experiences negative pore pressure (Figure 2.15a). However, at step 214 where the displacement is negative, the signs of excess pore pressure are reversed (Figure 2.15b). The higher generation in the excess pore pressure occurs mostly at shallow depths around the pile head where liquefaction occurs.

Figure 2.16 shows the distribution of LP predicted by conventional methods for the two peaks. The value of LP becomes equal to or greater than 1.0 at shallow depths, which indicates soil liquefaction due to the generation of positive excess pore pressure. However, the value of LP reduces with depth due to the increased overburden pressure, and hence, the soil does not show the tendency to liquefy at depth.

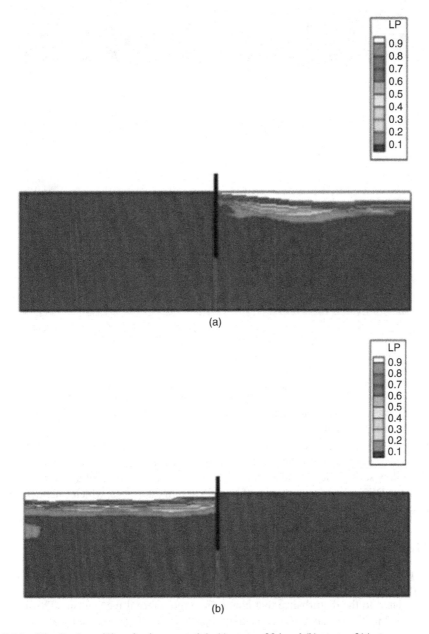

FIGURE 2.16 Distribution of liquefaction potentials (a) at step 204 and (b) at step 214.

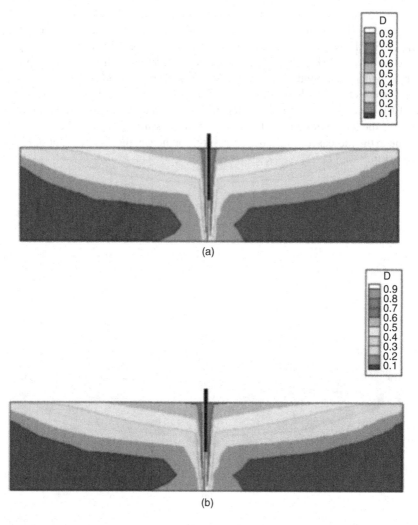

FIGURE 2.17 Disturbance contours (a) at step 204 and (b) at step 214.

Figure 2.17 presents the disturbance (D) contours at the two peaks. The value of D varied between 0.7 and 1.0 immediately around the pile and gradually reduced in the far regions. By considering the critical disturbance values for Ottawa sand ($D_c=0.84$) and for Ottawa sand–steel interface ($D_c=0.91$), it can be concluded from Figure 2.17 that liquefaction occurs around the pile and in the shallow soil and it coincides with the field observation reported by ERTEC [25]. It may be noted that the DSC allows computation of liquefaction in interfaces, which can be different from that in the soil.

2.4.7 USEFULNESS OF DSC/HISS MODELING FOR LIQUEFACTION

Liquefaction denotes a state of microstructural instability in a deforming material. It is a basic mechanism that is internal to the material. The DSC defines the liquefaction based on the critical parameters related to the disturbance that connects to the internal mechanism. This is in contrast with the classical methods that depend on external parameters. Thus, the DSC can be called a fundamental approach.

The procedure for evaluating liquefaction based on the DSC is simple and can be derived from normal tests conducted with triaxial, multiaxial, and shear conditions. In computer methods, liquefaction can be identified based on computed disturbances during incremental nonlinear analysis at various steps and/or times.

2.5   CONCLUSIONS

Microstructural instabilities, identified by threshold states, occur during deformations in most materials. The grains or particles in granular soils undergo such instabilities, some of which lead to initiation and final liquefaction. Conventional methods for liquefaction analysis do not account for the internal mechanisms that lead to liquefaction. The DSC and energy methods, on the other hand, allow for the internal mechanisms and lead to fundamental models for liquefaction. It has been found that the DSC can lead to simpler modeling compared to the energy method because it is easier to compute the disturbance in computer (finite element) procedures.

In this chapter, the DSC was described in detail including its theory, parameters, and validations. Several validations with respect to field and laboratory simulation of practical geomechanical problems were presented. It can be concluded that the DSC is perhaps the only fundamental and easiest procedure for liquefaction analysis. It can be used in practice by evaluating the critical disturbance form standard tests and by using appropriate computer procedures.

REFERENCES

1. Desai, C.S. Disturbed state concept and energy approaches for liquefaction analysis. Report, CEEM Department, Univ. of AZ, Tucson AZ, USA, 1999.
2. Desai, C.S. *Mechanics of Materials and Interfaces: The Disturbed State Concept.* CRC Press, Boca Raton, FL, 2001.
3. Desai, C.S. "Evaluation of liquefaction using disturbed state and energy approaches." *J. Geotech. Geoenviron. Eng. ASCE*, 126, 7, 2000, 619–631.
4. Seed, H.B., and Lee, K.L." Liquefaction of saturated sands during cyclic loading." *J. Soil Mech. Found. Div., ASCE*, SM6, 1966, 105–134.
5. Idriss, I.M., Boulanger, R.W. "SPT-based liquefaction triggering procedures." Report No UCD/CGM-10102, *Center for Geotechnical Modeling, Univ. of California, Davis, CA*, 2010.
6. Ishihara, K. "Liquefaction and flow during earthquakes." *Geotechnique*, 43, 3, 1993, 351–415.
7. National Research Council (NRC). "Liquefaction of soils during earthquakes." Report No. CETS-EE001, *National Academic Press, Washington, DC*, 1995.
8. Youd, T.L., and Idriss, I.M. "Evaluation of Liquefaction resistance of soils." Report, NCEET Workshop, *National Centre of Earthquake Eng., State Univ. of New York, Buffalo, NY*, 1997.
9. Davis, R.O., and Berrill, J.B." Energy dissipation and seismic liquefaction in sands." *Earthq. Eng. Soil Dyn.*, 10, 1982, 59–68.
10. Davis, R.O., and Berrill, J.B. "Pore pressures and dissipated energy in earthquakes-field verification." *J. Geotech. Geoenviron. Eng., ASCE* 127, 3, 2001, 269–274.
11. Law, K.T., Cao, Y.L., and He, G.W. "An energy approach for assessing seismic liquefaction potential." *Can. Geotech. J.*, 27, 3, 1990, 320–329.
12. Figueroa, J.L., Saada, A., Liang, L. and Dahisaria, N.M. "Evaluation of soil liquefaction by energy principles." *J. Geotech. Eng., ASCE*, 120, 9,1994, 1554–1569.
13. Desai, C.S., Shao, C.M., Rigby, D. Discussion on evaluation of liquefaction by energy procedures. *J. Geotech. Eng., ASCE*, 122, 1996, 3.
14. Desai, C.S., Park, I.J., and Shao, C.M. "Fundamental yet simplified model for liquefaction instability." *Int. J. Num. Anal. Methods Geomech.*, 22, 9, 1998, 721–748.
15. Desai, C.S., Pradhan, S.K., and Cohen, D. "Cyclic testing and constitutive modeling of saturated sand-concrete interfaces using the disturbed state concept." *Int. J. Geomech., ASCE*, 5, 4, 2005, 286–294.
16. Park, I.J., Desai, C.S. "Cyclic behavior and liquefaction of sand using disturbed state concept." *J. Geotech. Geoenviron. Eng., ASCE*, 126, 9, 2000, 834–846.
17. Pradhan, S.K., Desai, C. S. "DSC model for soil and interface including liquefaction." *J. Geotech. Geoenviron. Eng., ASCE*, 132, 2, 2006, 214–222.

18. Essa, M.J.K., and Desai, C.S. "Dynamic soil-pile interaction using DSC constitutive model." *Indian Geotech. J.*, 47, 2, 2017, 137–149.

19. Essa-Alyounis, M.E., and Desai, C.S. "Testing and modeling of saturated interfaces with effect of surface roughness: Part I: test behavior, Part II: Modeling and validations-II." *Int. J. Geomech., ASCE*, 19, 2019, 8. https://doi.org/10.1061/(ASCE)GM.1943-5622.0001459.

20. Park, I.J., Kim, S.I., and Choi, J.S. "Disturbed state modeling of saturated sand under dynamic loads." *Proceedings of 12 World Conferences on Earthquake Engineering*, Auckland, New Zealand, 2000.

21. Iwasaki, Y., Tai, M. "Strong motion records at Kobe Port Island." *Soils and Foundn, Tokyo*, 36, 1996, 29–40.

22. Akiyoshi, T., Fang, H.L., Fuchida, K., and Matsumoto, H. "A nonlinear seismic response analysis method for saturated soil-structure systems with absorbing boundary." *Int. J. Num. Anal. Methods Geomech.*, 20, 5, 1996, 307–329.

23. Gyi, M.M., and Desai, C.S. "Multiaxial cyclic testing of saturated Ottawa sand." Dept. of Civil Eng. & Eng. Mech, Report., University of Arizona, Tucson, Ariz, 1996.

24. Desai, C.S. "DSC-DYN2D-computer code for static and coupled consolidation and dynamic analysis." Manuals Parts I, II and III, C. Desai, Tucson, AZ, 2000.

25. ERTEC Western, Inc. "Full-scale pile vibration tests." Report No. PB82–192378. *Earth Technology Corp. California*, 1981.

3 Disturbed State Concept–Based Compression Model for Structured Clays

Hao Cui, Fang Liang, Zengchun Sun,
Huanran Wu, Chenggui Wang, and Yang Xiao
School of Civil Engineering, Chongqing University

3.1 INTRODUCTION

Structured clays exist widely in nature. The structure of clays refers to the comprehensive characteristics of mineral composition, grain arrangement, and cementation formed in the stress history and deposition process [1]. Due to the existence of structure, the mechanical properties of natural structured clays are usually quite different from those of the corresponding reconstituted clays, which has a great influence on the engineering properties. Compared with reconstituted soils with the same composition and void ratio, structured clays have higher pre-consolidation pressure, yield stress, shear strength, and stiffness [2]. At the same time, the structured clay formed a larger void ratio during the deposition process, leading to its greater compressibility, which brings great challenges to the stability of the foundation.

To explore the compression behaviors of structured clays, many tests have been carried out, including one-dimensional compression tests [3–6] and isotropic compression tests [2,7]. In addition, some theoretical models for structured clays have been developed. Liu and Carter [8] used the additional voids ratio as the structure parameter and introduced it into the modified Cam Clay model. The proposed model can simulate the phenomenon that the compression line of structured clay approaches the compression line of reconstituted clay with increasing stress after the initial yield. Based on the theoretical framework of Liu and Carter [9], the Modified Structured Cam Clay (MSCC) model was proposed by introducing the cementation degradation and its effect on the yield surface of structured soils [10]. Then, Suebsuk et al. [11] extended the MSCC model to the Bounding Surface (MSCC-B) model to simulate the mechanical behaviors of overconsolidated structured soils.

In addition, the constitutive model of structured soils based on the disturbed state concept (DSC) proposed by Desai and Toth [12] has been widely studied. The DSC is a constitutive simulation method that is developed for the mechanical disturbance of materials. It combines the theory of continuum mechanics and damage mechanics and can comprehensively analyze the mechanical behaviors of materials [13–15]. Liu et al. [16,17] analyzed the compression behavior of structured geomaterials using the DSC. Ouria [18] regarded the state in which the structured soils began to yield as a relatively intact (RI) state, and the state in which the structure was completely lost as a fully adjusted (FA) state.

Based on the DSC, a new disturbance factor is introduced according to the change of void ratio on the compression curve of structured clays, and the expression of disturbed function is obtained by fitting test data. Finally, the proposed compression model is used to simulate one-dimensional compression tests and isotropic compression tests of structured clays, and good results are obtained.

DOI: 10.1201/9781003362081-5

3.2 COMPRESSION MODEL FOR STRUCTURED CLAYS BASED ON DSC

3.2.1 Determination of Disturbance Factor

In the DSC framework, the observed mechanical behavior of materials at any state under loading is considered as a mixture of two reference states: the RI state and the FA state. The disturbance function, D, represents a gradual transition between two reference states. Using the DSC, the behavior of the material at any state between the RI and FA states can be described as [12]:

$$F_i = (1 - D)\mathrm{RI} + D\mathrm{FA} \tag{3.1}$$

Typical compression curves of the structured clay [19] and the corresponding reconstituted clay are shown in Figure 3.1, which can be idealized in Figure 3.2.

According to the DSC theory, the RI state of the structured clay can be described by the elastic model. If no structure degradation occurs, the compression line of the structured clay will continue to develop along the dashed line after exceeding the yield stress p'_y, as shown in Figure 3.2. The FA state of the structured clay can be considered as the complete loss of structure, and the compression line of the structured clay coincides with the reconstituted clay.

The compression line of the structured clay in the RI state in the $e - \ln p'$ plane can be expressed as:

$$e_i = e_0 - \kappa \ln p' \tag{3.2}$$

where e_i is the void ratio in RI state; p' is the mean effective stress; e_0 is the void ratio at $p' = 1$ kPa; and κ is the slope of the initial elastic part of the RI soil in the $e - \ln p'$ plane.

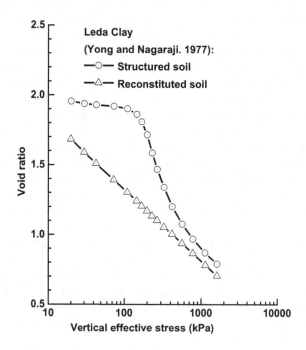

FIGURE 3.1 Compression tests' data of structured clay.

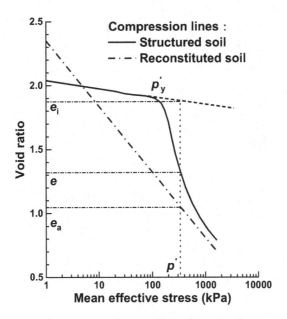

FIGURE 3.2   Compression line of structured clay and disturbance diagram.

The compression line of the structured clay in the FA state in the $e - \ln p'$ plane can be expressed as:

$$e_a = e_N - \lambda \ln p' \tag{3.3}$$

where e_a is the void ratio in the FA state; e_N is the void ratio of the reconstituted soil at $p' = 1$ kPa; and λ is the compression index of the soil in the FA state.

According to the relative position relationship of the compression curve in Figure 3.2, it is assumed that the disturbance factor of the structured soil is:

$$D = \frac{e_i - e}{e_i - e_a} \tag{3.4}$$

When $e = e_i$, the structured soil is in a RI state, $D = 0$; when the structure is sufficiently disturbed, the compression curve of the structured clay is infinitely close to the compression line of reconstituted clay and e is infinitely close to e_a, D gradually approaching 1.

3.2.2   DISTURBANCE FUNCTION

The specific expression of the disturbance function is obtained by fitting test data, which can truly reflect the change process of structured clay from a RI state to a FA state after yield. According to the compression curve of structured soil in Figure 3.2, the disturbance factor D is related to the mean effective stress p', and the disturbance occurs after the soil yield. Therefore, the relationship between D and p', p_y' can be established by fitting the test data. Figures 3.3 and 3.4 show the test data and fitting curves of isotropic compression tests on stiff Vallericca clay [20] and one-dimensional compression tests on soft Osaka clay [21], respectively.

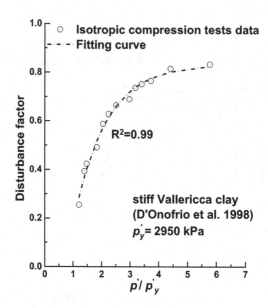

FIGURE 3.3 Relationships between D and p' / p'_y.

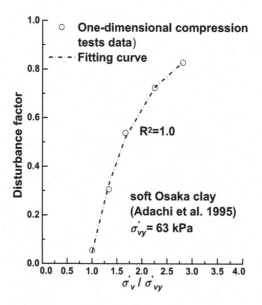

FIGURE 3.4 Relationships between D and σ'_v / σ'_{vy}.

The curves in Figures 3.3 and 3.4 can be fitted with the following function, respectively:

$$D_{p'} = \alpha - \beta \xi^{(p'/p'_y)} \tag{3.5}$$

$$D_{\sigma'_v} = \alpha - \beta \xi^{(\sigma'_v/\sigma'_{vy})} \tag{3.6}$$

where p'_y is mean effective yield stress; σ'_v is vertical effective stress; σ'_{vy} is vertical effective yield stress; and α, β, and ξ are material parameters.

3.2.3 STRESS–DEFORMATION RELATIONSHIP

The deformation of structured soils is elastic when $p' < p'_y$ or $\sigma'_v < \sigma'_{vy}$ and can be expressed as:

$$de = -\kappa \frac{dp'}{p'} \tag{3.7}$$

$$de = -\kappa \frac{d\sigma'_v}{\sigma'_v} \tag{3.8}$$

After yielding, according to the DSC, the void ratio at the current state can be expressed as:

$$e = (1 - D)e_i + De_a \tag{3.9}$$

By taking the derivative of the above equation, the expressions of structured clays' deformation for isotropic and one-dimensional compression tests can be obtained as follows, respectively:

$$de = dD_{p'}(e_a - e_i) - \left[(1 - D_{p'})\kappa + D_{p'}\lambda\right]\frac{dp'}{p'} \tag{3.10}$$

$$de = dD_{\sigma'_v}(e_a - e_i) - \left[(1 - D_{\sigma'_v})\kappa + D_{\sigma'_v}\lambda\right]\frac{d\sigma'_v}{\sigma'_v} \tag{3.11}$$

where $dD_{p'}$ and $dD_{\sigma'_v}$ can be, respectively, expressed as follows:

$$dD_{p'} = -\beta \xi^{(p'/p'_y)} \ln \xi dp' / p'_y \tag{3.12}$$

$$dD_{\sigma'_v} = -\beta \xi^{(\sigma'_v/\sigma'_{vy})} \ln \xi d\sigma'_v / \sigma'_{vy} \tag{3.13}$$

3.3 MODEL VALIDATION

The DSC–based compression model for structured clays contains eight parameters, which are e_0, e_N, λ, k, $p'_y (\sigma'_{vy})$, α, β, and ξ. To verify the performance of the proposed model to describe the compression behaviors of structured clays, compression tests of stiff Vallericca clay [20], soft Pisa clay [17], soft Osaka clay [21], and Leda clay [19] are selected for simulation. The model parameters determined from test data are shown in Table 3.1.

TABLE 3.1
Values of Model Parameters

Parameters	e_0	e_N	λ	κ	$p'_y (\sigma'_{vy})$	α	β	ξ
Stiff Vallericca clay	0.91	1.58	0.122	0.026	2,950 kPa	0.83	1.25	0.43
Soft Pisa clay –test a	1.625	1.635	0.126	0.065	125 kPa	1.07	1.304	0.78
Soft Pisa clay –test b	1.805	1.635	0.126	0.068	185 kPa	0.84	1.544	0.54
Soft Osaka clay	1.941	2.351	0.243	0.063	63 kPa	0.951	2.757	0.329
Leda clay	2.040	2.347	0.223	0.029	169 kPa	0.913	2.627	0.307

The simulation results of the model are shown in Figures 3.5–3.8. It can be seen that the compression model of structured clay based on the DSC can very well simulate the isotropic compression test and one-dimensional compression tests of structured clays.

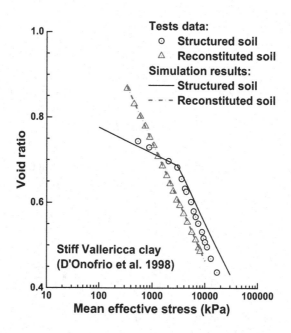

FIGURE 3.5 Isotropic compression tests and simulation results for stiff Vallericca clay.

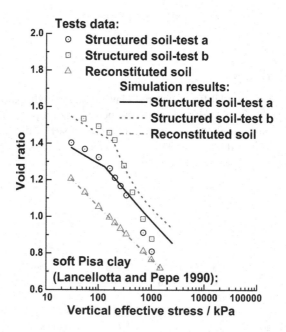

FIGURE 3.6 One-dimensional compression tests and simulation results for soft Pisa clay.

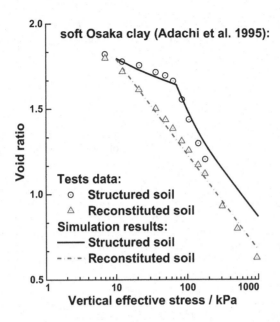

FIGURE 3.7   One-dimensional compression tests and simulation results for soft Osaka clay.

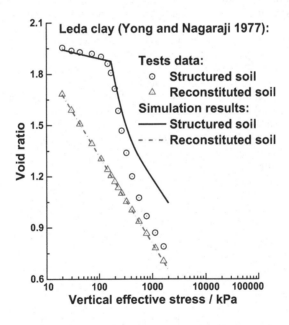

FIGURE 3.8   Oedometer tests and simulation results for Leda clay.

3.4   CONCLUSIONS

1. Based on the DSC, a model is proposed to describe the compressive behaviors of structured clays.
2. A new disturbance factor is introduced according to the change of void ratio on the compression curves of structured clays, and the expression of disturbed function is obtained by fitting test data.

3. The comparison between the calculated results of the proposed compression model and the test results shows that the model can well simulate the compression behaviors of structured clays.

REFERENCES

1. Liu, M. D., and Carter, J. P. "Virgin compression of structured soils." *Géotechnique*, 49, 1, 1999, 43–57.
2. Cuccovillo, T., and Coop, M. R. "On the mechanics of structured sands." *Géotechnique*, 49, 6, 1999, 741–760.
3. Ekinci, A., Ince, C., and Ferreira, P. M. V. "An experimental study on compression and shrinkage behavior of cement-treated marine deposited clays." *Int. J. Geosynth. Ground Eng.*, 5, 3, 2019, 17.
4. Horpibulsuk, S., Suddeepong, A., Chinkulkijniwat, A., and Liu, M. D. "Strength and compressibility of lightweight cemented clays." *Appl. Clay Sci.*, 69, 2012, 11–21.
5. Locat, J., and Lefebvre, G. "The compressibility and sensitivity of an artificially sedimented clay soil: The Grande-Baleine Marine Clay, Québec, Canada." *Mar. Geores. Geotechnol.*, 6, 1, 1985, 1–28.
6. Mesri, G., Rokhsar, A., and Bohor, B. F. "Composition and compressibility of typical samples of Mexico City clay." *Géotechnique*, 25, 3, 1975, 527–554.
7. Ng, C. W. W., Akinniyi, D., and Zhou, C. "Influence of structure on the compression and shear behaviour of a saturated lateritic clay." *Acta Geotech.*, 15, 3433–3441, 2020.
8. Liu, M. D., and Carter, J. P. "A structured Cam Clay model." *Can. Geotech. J.*, 39, 6, 2002, 1313–1332.
9. Liu, M. D., and Carter, J. P. "Modelling the destructuring of soils during virgin compression." *Géotechnique*, 50, 4, 2000, 479–483.
10. Suebsuk, J., Horpibulsuk, S., and Liu, M. D. "Modified Structured Cam Clay: A generalised critical state model for destructured, naturally structured and artificially structured clays." *Comput. Geotech.*, 37, 7–8, 2010, 956–968.
11. Suebsuk, J., Horpibulsuk, S., and Liu, M. D. "A critical state model for overconsolidated structured clays." *Comput. Geotech.*, 38, 5, 2011, 648–658.
12. Desai, C. S., and Toth, J. "Disturbed state constitutive modeling based on stress-strain and nondestructive behavior." *Int. J. Solids Struct.*, 33, 11, 1996, 1619–1650.
13. Desai, C. S. "Constitutive modeling of materials and contacts using the disturbed state concept: Part 1-Background and analysis." *Comput. Struct.*, 146, 2015, 214–233.
14. Desai, C. S. "Constitutive modeling of materials and contacts using the disturbed state concept: Part 2-Validations at specimen and boundary value problem levels." *Comput. Struct.*, 146, 2015, 234–251.
15. Desai, C. S. "Disturbed state concept as unified constitutive modeling approach." *J. Rock Mech. Geotech. Eng.*, 8, 3, 2016, 277–293.
16. Liu, M. D., Carter, J. P., Desai, C. S., and Xu, K. J. "Analysis of the compression of structured soils using the disturbed state concept." *Int. J. Numer. Anal. Meth. Geomech.*, 24, 8, 2000, 723–735.
17. Liu, M. D., Carter, J. P., and Desai, C. S. "Modeling Compression Behavior of Structured Geomaterials." *Int. J. Geomech.*, 3, 2, 2003, 191–204.
18. Ouria, A. "Disturbed state concept-based constitutive model for structured soils." *Int. J. Geomech.*, 17, 7, 2017, 10.
19. Yong, R. N., and Nagaraj, T. S. "Investigation of fabric and compressibility of sensitive clay." *Proceedings of the International Symposium on Soft Clay*, Asian Institute of Technology, 327–334.
20. D'Onofrio, A., Santucci De Magistris, F., and Olivares, L. (1998). "Influence of soil structure on the behaviour of two natural stiff clays in the pre-failure range." *International Symposium on Hard Soils - Soft Rocks (2; Naples 1998-10-12)*, A. A. Balkema, Rotterdam, 497–505.
21. Adachi, T., Fusao, O., Takehiro, H., Tadashi, H., Junichi, N., Mamoru, M., and Pradhan, T. B. S. "Stress-strain behavior and yielding characteristics of Eastern Osaka clay." *Soils Found.*, 35, 3, 1995, 1–13.

4 Sands and Structural Clays

Jian-feng Zhu
School of Civil Engineering and Architecture, Zhejiang
University of Science and Technology

4.1 INTRODUCTION

With the rapid development of urbanization in China, civil engineering construction such as tunneling and pile foundation and excavation has become more and more frequent. Meanwhile, problems of geo-environment caused by construction have become more and more serious. Construction inevitably leads to disturbance of soils, which will change the property of soils. Unfortunately, the physical and mechanical parameters obtained from the laboratory tests only reflect the properties of soils at a certain state but not the true properties of soils during the whole process of construction. Therefore, the changes of the properties caused by disturbance are possible through utilizing the disturbed state concept (DSC) [1], which can provide micromechanical understanding of the problem. The DSC is a unified approach for constitutive modeling of the materials and interfaces, and is presented in a number of publications together with its applications in a wide range of materials such as clays [2–4], sands [5–7], stabilized soil [8], and tunnel engineering [9,10].

The initial idea of DSC was first introduced by Desai [1] to characterize the softening response of an overconsolidated soil. In the DSC, the observed behavior of a material element can be expressed in terms of its behavior in two reference states: relative intact (RI) and fully adjusted (FA). At any stage during deformation under mechanical and/or environmental loadings, the material element is considered to be composed of a mixture of randomly disturbed material parts in the RI and FA states, Figure 4.1.

It is generally assumed that applied forces will cause a disturbance in the material's microstructure. As a result, an initially RI material transforms continuously through a process of natural self-adjustment and a part of it reaches the FA state. During the transformation, particles in the microstructure may experience local instability or transitions of their states. At certain critical

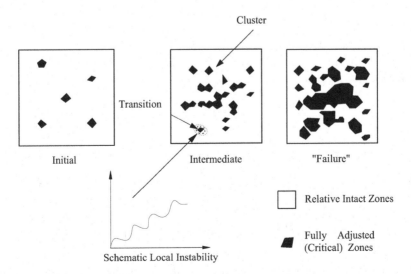

FIGURE 4.1 Schematic of relative intact and fully adjusted clusters in disturbed state concept (DSC) [11].

DOI: 10.1201/9781003362081-6

locations during deformation, the transitions may indicate threshold transitions involving significant or abrupt changes at states such as contractive to dilative volume change, peak stress, and initiation of residual stress conditions [11]. In view of this, the main work of this chapter is to introduce a couple of constitutive models to predict the mechanical properties of sand, clay, or overconsolidated soil.

4.2 GENERAL DISTURBANCE FUNCTION

In the DSC, a deforming material element was treated as a mixture of the initial continuum or RI part and the FA part. The observed behavior of the material was expressed in terms of the coupled behavior of the materials in the RI and FA states by utilizing the disturbance function, D, which acted as an interpolation and coupling mechanism between the RI and FA states.

Schmertmann [12] developed a disturbance function for sampling as follows:

$$D = \frac{\Delta e}{\Delta e_0} \tag{4.1}$$

where Δe presents the ratio of the void-ratio change between mechanical compression curve and practical compression curve, and Δe_0 presents the ratio of the void-ratio change between mechanical compression curve and fully disturbed compression curve.

Ladd and Lambe [13] believed that undrained modulus of saturated soil was most sensitive to disturbance; then, the following sampling disturbance was developed.

$$D = \frac{[E_u] - E_{50}}{[E_u] - [E_{50}]} \tag{4.2}$$

where E_{50} and $[E_{50}]$ present the undrained modulus of practical and remolded specimens, respectively, and the superscript "50" presents the undrained modulus of soil when the strain approaches the 50% of failure strain. $[E_u]$ presents the undrained modulus of intact soil.

Zhang [14] developed the following disturbance function of construction based on the failure surface of soil in p-q-e dimension.

$$D = \frac{\sqrt{\left(\Delta p^2 + \Delta q^2 + \Delta e^2\right)}}{\sqrt{\left(p_f^2 + q_f^2 + e_f^2\right)}} \tag{4.3}$$

where p_f, q_f, and e_f present the average stress, deviatoric stress, and void ratio, respectively.

Xu [15] believed that the construction disturbance could be divided into stress and strain disturbance. Then, the following disturbance function was developed.

$$D = 1 - \frac{M_d}{M_0} \tag{4.4}$$

where M_0 and M_d present the mechanical parameters before and after construction disturbance, respectively.

Desai [16] developed the following disturbance function as shown in Figure 4.2:

$$D = D(\varepsilon) = D_u\left[1 - \exp\left(-A\varepsilon^z\right)\right] \tag{4.5}$$

where D_u is the ultimate disturbance value, A and Z are material parameters, and ε is material stain, which includes the volume strain and shear strain.

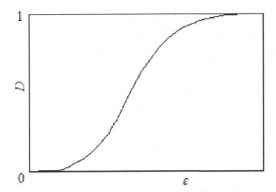

FIGURE 4.2 Relationship between disturbed degree and strain [16].

The disturbance could represent the effect of microcracking and fracture leading to degradation in which the value of D ranged from 0 to 1, and enhanced particle bonding leading to stiffening in which the value of D ranged from −1 to 0. Unfortunately, most disturbance functions only considered the degradation effects with the correspondent value of D ranging from 0 to 1.

Zhu [17] believed the disturbance to result in the decrease and increase of relative density of sand as the "positive disturbance" and "negative disturbance", respectively. And, the general disturbance function of sand can be developed as following:

1. "positive disturbance" $\left(D_r \leq D_{r0}\right)$

$$D = \frac{2}{\pi} \arctan\left(\frac{D_{r0} - D_r}{D_r - D_{r\min}}\right) \tag{4.6a}$$

2. negative disturbance" $\left(D_r > D_{r0}\right)$

$$D = \frac{2}{\pi} \arctan\left(\frac{D_{r0} - D_r}{D_{r\max} - D_r}\right) \tag{4.6b}$$

where D_r presents the current relative density of sand $\left(0 \leq D_r \leq 1\right)$, D_{r0} presents the initial relative density of sand, $D_{r\min}$ presents the loosest relative density of sand, which is always considered to be 0, and $D_{r\max}$ presents the densest relative density of sand, which is always considered to be 1. According to Eq. (4.6), the relationship between D and D_r is shown in Figure 4.3.

4.3 MODIFIED CAM-CLAY MODEL OF STRUCTURAL CLAY BASED ON DSC

Soil structure is one of the most important factors influencing the mechanical properties of intact clay. When constructing a building, construction disturbance to clay will alter soil structure and thus the stress–strain relationship. Substantial engineering practice show that disturbance to soils will result in extra settlement of the soil, and further increase the settlement of foundation soil. Unfortunately, traditional constitutive models for structural clay seldom consider the effect of structure disturbance, which might be the reason that the calculated settlement is far from the actual values [18]. Thus, on the basis of DSC, Wang et al. [3] developed the modified Cam-Clay model to investigate the soil behavior of structural clay at the disturbed state.

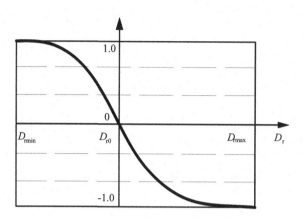

FIGURE 4.3 Relationship between disturbed degree and relative density [17].

1. Reference State

According to the modified Cam-Clay model [19,20], the constitutive relationship of soil elastic-strain in incremental form can be:

$$\begin{bmatrix} d\varepsilon_v^i \\ d\varepsilon_q^i \end{bmatrix} = \begin{bmatrix} 1/K & 0 \\ 0 & 1/3G \end{bmatrix} \begin{bmatrix} dp' \\ dq \end{bmatrix} \tag{4.7}$$

where superscript i represents the RI state, K presents bulk modulus, G presents shear modulus, which can be:

$$K = \frac{v^i p'}{\kappa^i} \tag{4.8}$$

$$G = \frac{3(1-2\mu)K}{2(1+\mu)} \tag{4.9}$$

$$v^i = 1 + e^i \tag{4.10}$$

$$e^i = e_0 - \kappa^i \ln p' \tag{4.11}$$

where κ^i presents the gradient of e^i -lnp' of soil at the RI state, which can be estimated by the slope of e^i -lnp' of intact soil at the initial elastic state, v^i presents the specific volume of soil at the RI state, e^i presents the void ratio of soil at the RI state, e_0 presents the void ratio of soil, μ presents Poisson's ratio, p' presents the effective stress, and q presents the generalized shear stress.

2. Fully Adjusted State

For structural clay at the FA state, its mechanical behavior can be investigated by the modified Cam-Clay model as it could be considered as remolded soil.

$$\frac{p'}{p_0'} = \frac{M^2}{M^2 + \eta^2} \tag{4.12}$$

where p_0' presents the harden parameter, η presents the stress ratio, and $\eta = q/p'$. Taking the differential form of Eq. (4.12), then

$$\left(\frac{M^2 - \eta^2}{M^2 + \eta^2}\right)\frac{dp'}{p'} + \left(\frac{2\eta}{M^2 + \eta^2}\right)\frac{dq}{p'} - \frac{dp_0'}{p_0'} = 0 \tag{4.13}$$

In Cam-Clay model, the soil is considered to be work harden material and follows the associated flow rule, in which the plastic potential function is the same as the yield surface function, and thus

$$\frac{d\varepsilon_q^p}{d\varepsilon_v^p} = \frac{\partial F/\partial q}{\partial F/\partial p'} = \frac{2q}{M^2\left(2p' - p_0'\right)} = \frac{2\eta}{M^2 - \eta^2} \tag{4.14}$$

1. Volumetric strain

$$d\varepsilon_v^c = d\varepsilon_v^e + d\varepsilon_v^p = \kappa\frac{dp'}{v^c p'} + \left(\lambda - \kappa\right)\frac{dp_0'}{v^c p_0'} \tag{4.15}$$

2. Deviator strain

$$d\varepsilon_q^c = d\varepsilon_q^e + d\varepsilon_q^p = \frac{2(1+v)}{9(1-2v)}\kappa\frac{dq'}{v^c q'} + \left(\frac{2\eta}{M^2 - \eta^2}\right)\left(\lambda - \kappa\right)\frac{dp_0'}{v^c p_0'} \tag{4.16}$$

$$v^c = 1 + e^c \tag{4.17}$$

$$e^c = e_0 - \left(\lambda - \kappa\right)\ln p_0' - \kappa\ln p' \tag{4.18}$$

where superscript c presents the fully adjusted state, λ and κ present the compression index and rebound index of fully adjusted soil, respectively, and e_0 presents the initial void ratio, in which e_0 at the RI state is considered to be the same as that with the FA state. Substitute Eq. (4.13) into Eqs. (4.15) and (4.16), then

$$\begin{bmatrix} d\varepsilon_v^c \\ d\varepsilon_d^c \end{bmatrix} = \frac{2(\lambda - \kappa)\eta}{v^c\left(M^2 + \eta^2\right)p'}\begin{bmatrix} \dfrac{\lambda M^2 + (2\kappa - \lambda)\eta^2}{2(\lambda - \kappa)\eta} & 1 \\ 1 & \dfrac{2(1+v)\kappa}{9(1-2v)}\left(\dfrac{M^2 + \eta^2}{2(\lambda - \kappa)\eta}\right) + \left(\dfrac{2\eta}{M^2 - \eta^2}\right) \end{bmatrix}\begin{bmatrix} dp' \\ dq \end{bmatrix} \tag{4.19}$$

For the clay at a specific disturbed state, the total deformation of soil can be determined by the summation of the deformation of soil at RI state and FA state as the following:

$$\varepsilon_{ij} = (1 - D)\varepsilon_{ij}^i + D\varepsilon_{ij}^c \tag{4.20}$$

Taking the differential form of Eq. (4.20), the incremental equation of strain of soil based on the DSC model can be:

$$d\varepsilon_{ij} = (1-D)d\varepsilon_{ij}^i + Dd\varepsilon_{ij}^c + dD\left(\varepsilon_{ij}^c - \varepsilon_{ij}^i\right) \tag{4.21}$$

where ε_{ij} presents the strain tensor and D presents the disturbance function.

3. Disturbance Function

Unlike traditional solid material, the disturbance of soil should cover the volumetric strain effect apart from the influence of deviator strain. It is supposed that the disturbance on soil–volumetric strain and soil–deviator strain mainly depends on the volumetric deviator strain, respectively. Then, according to the disturbance function Eq. (4.5) developed by Desai [12], the following disturbance function of volumetric and deviator strain can be built as:

1. Volumetric strain

$$D_v = 1 - \exp\left(-A_v\varepsilon_v^{Z_v}\right) \tag{4.22}$$

2. Deviator strain

$$D_q = 1 - \exp\left(-A_q\varepsilon_q^{Z_q}\right) \tag{4.23}$$

where A_v, Z_v, A_q, and Z_q are material parameters. Then, Eq. (4.22) can be replaced with Eqs. (4.24) and (4.25).

1. Volumetric strain

$$d\varepsilon_v = (1-D_v)d\varepsilon_v^i + D_v d\varepsilon_v^c + dD_v\left(\varepsilon_v^c - \varepsilon_v^i\right) \tag{4.24}$$

2. Deviator strain

$$d\varepsilon_v = (1-D_q)d\varepsilon_v^i + D_q d\varepsilon_v^c + dD_q\left(\varepsilon_v^c - \varepsilon_v^i\right) \tag{4.25}$$

Substitute Eqs. (4.7) and (4.19) into Eqs. (4.24) and (4.25), respectively. The elasto-plastic constitutive relationship of structural clay based on DSC can be:

$$\begin{bmatrix} d\varepsilon_v \\ d\varepsilon_q \end{bmatrix} = \begin{bmatrix} \dfrac{(1-D_v)}{K} + D_v K_f K_k & D_v K_f \\[2ex] D_q K_f & \dfrac{(1-D_q)}{3G} + D_q K_f K_k \end{bmatrix} \begin{bmatrix} dp' \\ dq \end{bmatrix}$$

$$+ \begin{bmatrix} \varepsilon_v^c - \varepsilon_v^i & 0 \\ 0 & \varepsilon_q^c - \varepsilon_q^i \end{bmatrix} \begin{bmatrix} dD_v \\ dD_q \end{bmatrix} \tag{4.26}$$

As a result, it can be simplified as

$$
\begin{bmatrix} dp' \\ dq \end{bmatrix} = H \begin{bmatrix} K_{vv} & K_{vq} \\ K_{qv} & K_{qq} \end{bmatrix} \begin{bmatrix} d\varepsilon_v \\ d\varepsilon_q \end{bmatrix}
\tag{4.27}
$$

where

$$
H = \frac{1}{\left(\dfrac{1-D_v}{K} + D_v K_f K_k \right) \left(\dfrac{1-D_q}{3G} + D_q K_f K_g \right) - D_v K_q K_f^2}
\tag{4.28a}
$$

$$
K_{vv} = \left(\frac{1-D_q}{3G} + D_q K_f K_g \right) \left\{ 1 - \left(\varepsilon_v^c - \varepsilon_v^i \right) \cdot \left[A_v Z_v \varepsilon_v^{Z_v-1} \exp\left(-A_v \varepsilon_v^{Z_v} \right) \right] \right\} - D_v K_f
\tag{4.28b}
$$

$$
K_{vq} = \left(\frac{1-D_q}{3G} + D_q K_f K_g \right) - D_v K_f \left\{ 1 - \left(\varepsilon_q^c - \varepsilon_q^i \right) \cdot \left[A_q Z_q \varepsilon_q^{Z_q-1} \exp\left(-A_q \varepsilon_q^{Z_q} \right) \right] \right\}
\tag{4.28c}
$$

$$
K_{qv} = \left(\frac{1-D_v}{K} + D_v K_f K_k \right) - D_q K_f \left\{ 1 - \left(\varepsilon_v^c - \varepsilon_v^i \right) \cdot \left[A_v Z_v \varepsilon_v^{Z_v-1} \exp\left(-A_v \varepsilon_v^{Z_v} \right) \right] \right\}
\tag{4.28d}
$$

$$
K_{qq} = \left(\frac{1-D_v}{K} + D_v K_f K_k \right) \left\{ 1 - \left(\varepsilon_q^c - \varepsilon_q^i \right) \cdot \left[A_q Z_q \varepsilon_q^{Z_q-1} \exp\left(-A_q \varepsilon_q^{Z_q} \right) \right] \right\} - D_q K_f
\tag{4.28e}
$$

4.4 MODIFIED DUNCAN–CHANG MODEL OF SAND BASED ON DSC

Kondner et al. [21,22] suggested that the nonlinear behavior of soils such as clay and sand can be effectively estimated by a hyperbola function, which was expressed as:

$$
\sigma_1 - \sigma_3 = \frac{\varepsilon_1}{a + b\varepsilon_1}
\tag{4.29}
$$

where a and b are model parameters for determining the initial tangent modulus and critical stress state when the stress–strain curve approaches infinite strain $(\sigma_1 - \sigma_3)_{\text{ult}}$.

In traditional triaxial test, Eq. (4.29) can be:

$$
\frac{\varepsilon_1}{\sigma_1 - \sigma_3} = a + b\varepsilon_1
\tag{4.30}
$$

In traditional triaxial test, $d\sigma_2 = d\sigma_3 = 0$, and the initial tangent modulus, E_0, can be:

$$
E_0 = 1/a
\tag{4.31}
$$

Let $\varepsilon_1 \to \infty$ in Eq. (4.29), then

$$
(\sigma_1 - \sigma_3)_{\text{ult}} = 1/b
\tag{4.32}
$$

As seen, a and b behave the reciprocal relationship with E_0 and $(\sigma_1-\sigma_3)_{\text{ult}}$, respectively.

Duncan and Chang [23] suggested that a and b could be determined as:

$$b = \frac{1}{(\sigma_1-\sigma_3)_{\text{ult}}} = \frac{\left[\varepsilon_1/(\sigma_1-\sigma_3)\right]_{95\%} - \left[\varepsilon_1/(\sigma_1-\sigma_3)\right]_{70\%}}{(\varepsilon_1)_{95\%} - (\varepsilon_1)_{70\%}} \tag{4.33}$$

$$a = \frac{1}{E_0} = \frac{\left[\varepsilon_1/(\sigma_1-\sigma_3)\right]_{95\%} + \left[\varepsilon_1/(\sigma_1-\sigma_3)\right]_{70\%}}{2} - \frac{1/(\sigma_1-\sigma_3)_{\text{ult}}\left[(\varepsilon_1)_{95\%} + (\varepsilon_1)_{70\%}\right]}{2} \tag{4.34}$$

where the subscripts 95% and 70% represent the ratio between the value of stress difference $(\sigma_1-\sigma_3)$ and their peak value of strength $(\sigma_1-\sigma_3)_f$, respectively.

The relationship between $(\sigma_1-\sigma_3)_f$ and $(\sigma_1-\sigma_3)_{\text{ult}}$ is established by the failure ratio R_f as

$$R_f = \frac{(\sigma_1-\sigma_3)}{(\sigma_1-\sigma_3)_{\text{ult}}} \tag{4.35}$$

Based on the laboratory tests, Janbu [24] suggested that the relationship between the initial tangent modulus and confining pressure could be expressed as:

$$E_0 = Kp_a\left(\frac{\sigma_3}{p_a}\right)^n \tag{4.36}$$

where p_a is the atmospheric pressure, K is a model constant, n is a dimensionless parameter related to the rate of variation of E_0 and σ_3.

Another form of Eq. (4.36) can be stated as:

$$\lg(E_0/p_a) = \lg K + n\lg(\sigma_3/p_a) \tag{4.37}$$

With regard to the Mohr–Coulomb failure criterion, the peak failure strength can be derived as [17,23]:

$$(\sigma_1 - \sigma_3)_f = \text{M}p_a\left(\frac{\sigma_3}{p_a}\right)^l \tag{4.38}$$

where M and l are the dimensionless parameters. Then, Eq. (4.38) can be further simplified as:

$$\lg\left[(\sigma_1-\sigma_3)_f/p_a\right] = \lg M + l\lg\left[\sigma_3/p_a\right] \tag{4.39}$$

The tests results [17] show that K and M have the following relationship with D_r:

$$\ln K = c + dD_r \tag{4.40}$$

$$M = \varsigma + gD_r \tag{4.41}$$

where, c, d, ς, and g are model parameters.

As mentioned earlier, the physical parameter such as D_r is liable to change due to the construction disturbance, which will further alter the mechanical parameter of sand such as K and M. Finally, the stiffness and strength of sand before construction will behave significantly different from those of sand after construction. Then, Zhu [17] developed the following relationship between K, M and disturbed degree, D.

1. For "positive disturbance" $\left(D_r \leq D_{r0}\right)$

    Supposing that $D_{rmin}=0$, Eq. (4.6a) can be:

$$D_{r0} - D_r = D_r \tan\left(\pi D/2\right) \tag{4.42}$$

Substituting Eq. (4.6a) into Eq. (4.40), then

$$K_D = \frac{K_0}{\exp\left(dD_r \tan\left(\pi D/2\right)\right)} \tag{4.43}$$

Substituting Eq. (4.6a) into Eq. (4.41), then

$$M_D = M_0 - gD_r \tan\left(\pi D/2\right) \tag{4.44}$$

where K_0 and M_0 present the parameters K and M at the initial state, K_D and M_D present the parameter K and M under the disturbed state.

2. For "negative disturbance" $\left(D_r > D_{r0}\right)$

    Taking $D_{rmax}=1$, Eq. (4.6b) can be:

$$D_{r0} - D_r = \left(1 - D_r\right)\tan\left(\pi D/2\right) \tag{4.45}$$

Substituting Eq. (4.6b) into Eq. (4.40), then

$$K_D = \frac{K_0}{\exp\left(d\left(1 - D_r\right)\tan\left(\pi D/2\right)\right)} \tag{4.46}$$

Substituting Eq. (4.6b) into Eq. (4.41), then

$$M_D = M_0 - g\left(1 - D_r\right)\tan\left(\pi D/2\right) \tag{4.47}$$

As a result, K_D and M_D at a specific value of D_r can be:

1. When $D_r = D_{r0}$, the sand is at the initial state without any disturbance

$$K_D = K_0, \quad M_D = M_0 \tag{4.48a}$$

2. When $D_r = D_{rmin} = 0$, the sand will approach the loosest state, then $D = 1$, and

$$K_D = K_0 \exp\left(-dD_{r0}\right), M_D = M_0 - gD_{r0} \tag{4.48b}$$

3. When $D_r = D_{r\max} = 1$, the sand will approach the densest state, then $D = -1$, and

$$K_D = K_0 \exp\big(d(1 - D_{r0})\big), M_D = M_0 - g(1 - D_{r0}) \tag{4.48c}$$

Consequently, the modified Duncan–Chang model based on DSC can be:

1. For "positive disturbance"

$$\sigma_1 - \sigma_3 = \cfrac{\varepsilon_1}{\cfrac{\exp\big(dD_r \tan(\pi D/2)\big)}{K_0 p_a \left(\cfrac{\sigma_3}{p_a}\right)^{n_0}} + \cfrac{R_{f0}\varepsilon_1}{\big(M_0 - gD_r \tan(\pi D/2)\big)\sigma_3}} \tag{4.49a}$$

2. For "negative disturbance"

$$\sigma_1 - \sigma_3 = \cfrac{\varepsilon_1}{\cfrac{\exp\big(d(1 - D_r) \tan(\pi D/2)\big)}{K_0 p_a \left(\cfrac{\sigma_3}{p_a}\right)^{n_0}} + \cfrac{R_{f0}\varepsilon_1}{\big(M_0 - g(1 - D_r) \tan(\pi D/2)\big)\sigma_3}} \tag{4.49b}$$

4.5 OTHER CONSTITUTIVE MODEL OF SAND AND STRUCTURAL CLAY BASED ON DSC

1. Sand

A constitutive model is established by Samieh and Wong [25] to simulate both the experimental and homogeneous deformational responses of Athabasca oil sand. The model is based on describing the evolution of internal microstructural changes with shear loading through a scalar disturbance function. The deformational response of the material is expressed in terms of the responses of its reference states, namely, the virgin and fully disturbed states, through the scalar disturbance function. The virgin state of the material is modeled by a generalized single surface plasticity model, whereas the fully disturbed state is assumed to be the critical state. The parameters required to define the model were identified and evaluated. Comparisons between the predicted results and experimental data were made for model performance evaluation. In 2011, Zhu et al. [17] established a general disturbance function(D), in which the disturbance parameter could be deduced from the relative density of sands. Then, an elasto-plastic model considering disturbance, which could reflect the contribution of disturbance to the strength-deformation property of sands, was developed to predict the stress–strain–volume change relations of sands at a randomly disturbed state. The test results show that the proposed model can well describe the strength–strain–volume change relations of sands at a given disturbed state.

2. Structural clay

Katti and Desai [2] developed a new constitutive model for predicting the undrained stress-deformation and pore water pressure response of saturated cohesive soils subjected to cyclic loading developed using the DSC. The model takes into account inelastic response during loading (virgin) and unloading–reloading (nonvirgin) behavior, which is verified with respect to the observed behavior of undisturbed clay samples. Overall, the proposed model provides highly satisfactory predictions of the observed behavior during virgin and nonvirgin loading. Based on DSC and hypoplasticity theory, Zhou et al. [26] developed the elasto-plastic and damage model by means of the method from simple loading to complex loading. The model could be used to describe the strength and deformation characteristics of structured soils under simple and complex loading. The corresponding elasto-plastic and damage matrix are derived so that the model can be implemented into finite element code. According to the results of undrained triaxial shear tests and electrical resistivity tests, Yu et al. [27] took electrical resistivity indexes as the disturbed variable to reflect the changes in soil structures and build a disturbed function. In addition, the rotational hardening parameter is taken to describe the disturbed state of natural soft clays; accordingly, a DSC constitutive model taking into account stress-induced anisotropy for soft clays is presented. As a result, the traditional methods measuring disturbed variable and establishing disturbed function may be optimized, and the applicability and accuracy of the model may be improved. Results from undrained triaxial shear tests show that the model can describe the stress–strain characteristics fairly well and could be applied to practice.

4.6 CONCLUSION

Accurate solutions to geotechnical engineering problems using conventional or advanced methods are dependent significantly on the responses of soil. Hence, constitutive modeling of geological material such as sands and structural clay plays a vital role in reliable solutions to geomechanical problems. However, most of the existing constitutive models merely account for specific characteristics of soil behavior and will be invalid to predict the response of soil during construction disturbance. In view of this, the DSC theory is firstly presented in the present chapter. Then, a couple of disturbance functions with different disturbance parameters are elaborated to evaluate the disturbed degree of sand or clay. Based on DSC, the modified Cam–Clay model and Duncan–Chang model are developed to investigate the mechanical behavior of structural clay and sand at a specific disturbed state, respectively. In addition, a couple of prospective constitutive models, which behave well to address the response of sand and clay at the disturbed state, are briefly reviewed to further demonstrate the wide application of DSC theory.

REFERENCES

1. Desai, C. S., "A consistent finite element technique for work softening behavior." *Proc. Int. Conf. Comp. Meth. Nonlinear Mech.* University of Texas, Austin, TX. 1974.
2. Katti, D. R., and Desai, C. S. "Modeling and testing of cohesive soil using disturbed state concept." *J. Eng. Mech., ASCE*, 121, 5, 1995, 648–658.
3. Wang, G. X., Xiao, S. F., Huang H. W., and Wu, C. Y. "Study of constitutive model of structural clay based on the disturbed state concept." *Acta Mech. Solida Sin.* 25, 2, 2004, 191–197 (in Chinese).
4. Pradhan, S. K., and Desai, C. S. "DSC model for soil and interface including liquefaction and prediction of centrifuge test." *J. Geotech. Geoenviron.*, 132, 2, 2006, 214–222.
5. Park, I. J., and Desai, C. S. "Cyclic behavior and liquefaction of sand using disturbed state concept." *J. Geotech. Geoenviron.*, 126, 9, 2000, 834–846.
6. Zhu, J. F., Xu, R. Q., Li, X. R., and Chen, Y. K. "Calculation of earth pressure based on disturbed state concept theory." *J. Central South Uni. Tech.*, 18, 4, 2011, 1240–1247.
7. Zhu J. F., Xu, R. Q., Wang X. C., Zhang, J., Ma S. G., and Liu, X. "An elastoplastic model for sand considering disturbance." *Chin. J. Rock Mech. Eng.*, 30, 1, 2011, 193–201 (in Chinese).
8. Rao, C. Y., Zhu, J. F., Pan, B. J., Liu, H. X., and Zhou, Z. J. "One-dimensional compression model solidified silt based on theory of disturbed state concept." *Chin. J. Geotech. Eng.* 41, 2019, 173–176 (in Chinese).
9. Zhu J. F., Xu, R. Q., and Liu G. B. "Analytical prediction for tunnelling-induced ground movements in sands considering disturbance." *Tunn. Undergr. Sp. Tech.* 41, 2014, 165–175.

10. Zhu J. F., Zhao, H. Y., Xu, R. Q., Luo, Z. Y., and Jeng, D. S. "Constitutive modeling of physical proper-
    ties of coastal sand during tunneling construction disturbance." *J. Mar. Sci. Eng.* 9, 2021, 167. https://
    doi.org/10.3390/jmse9020167.

11. Desai, C. S., and Toth, J. "Disturbed state constitutive modeling based on stress-strain and nondestruc-
    tive behavior." *Int. J. Solids Struct.*, 33, 11, 1996, 1619–1650.

12. Schmertmann, J. H. "The undisturbed consolidation behavior of clay." *Trans. Am. Soc. Civ. Eng.*, 20,
    1955, 1201–1233.

13. Ladd, C. C., and Lambe, T. W. "The strength of undisturbed clay determined from undrained tests."
    ASTM, Laboratory Shear Testing of Soils, STP 361, 1963, 342–371.

14. Zhang, M. X. "Study on engineering properties of soil masses disturbed due to construction." PhD
    Dissertation. Shanghai, Tongji University, 1999 (in Chinese).

15. Xu, Y. F. Sun, D. A., Sun, J., Fu, D. M., and Dong, P. "Soil disturbance of Shanghai silty clay during EPB
    tunneling." *Tunn. Under. Sp. Tech.* 18, 2003, 537–545.

16. Desai, C. S. *Mechanics of Materials and Interfaces: The Disturbed State Concept*," CRC Press, Boca
    Raton, FL, 2001.

17. Zhu, J. F. "Study on properties of soil considering disturbance." *PhD Dissertation*. Hangzhou, Zhejiang
    University, 2011 (in Chinese).

18. Yu, X. J., and Shi, J. Y. "Research on the disturbed state concept for soft clay roadbed." *Proc. GeoHunan
    Int. Conf. Soil Rock Instrumentation, Behavior Modeling*, Changsha, Hunan, China. 2009. pp. 92–98.

19. Wood, D. M. *"Soil Behavior and Critical State Soil Mechanics,"* Cambridge University Press,
    Cambridge, 1990.

20. Liu, M. D., and Carter, J. P. "A structured Cam Clay Model." Center for Geotechnical Research, Sydney
    University, 2002.

21. Kondner, R. L. and Zelasko, J. S. "A hyperbolic stress-strain formulation for sands." *Proc. 2nd Pan-Am.
    Conf. Soil Mech. Found. Eng.*, Brazil, 1963, 1, 289–324.

22. Kondner, R. L., and Horner, J. M. "Triaxial compression of a cohesive soil with effective octahedral
    stress control." *Can. Geotech. J.* 2, 1, 1965, 40–52.

23. Duncan, J. M., and Chang, C. Y. "Nonlinear analysis of stress and strain in soils." *J. Soil Mech. Found.
    Div.* 96, 5, 1970, 1629–1653.

24. Janbu, N. "Soil compressibility as determined by oedometer and triaxial tests." *Proc., Eur. Conf. Soil
    Mech. & Found. Eng.*, Wiesbaden, Germany, 1963, 1, 19–25.

25. Samieh, A. M., and Wong, R. C. K. "Modelling the responses of Athabasca oil sand in triaxial compres-
    sion tests at low pressure." *Can. Geotech. J.*, 38, 1995, 395–406.

26. Zhou, C., Shen, Z. J., Chen, S. S., and Chen, T. L. "A hypoplasticity disturbed state model for structured
    soils." *Chi. J. Geotech. Eng.*, 26, 4, 2004, 435–439 (in Chinese).

27. Yu, X. J., Shi, J. Y., and Xu, Y. B. "Modelling disturbed state and anisotropy of natural soft clays." *Rock
    Soil Mech.*, 30, 11, 2009, 3307–3312 (in Chinese).

5 A New Creep Constitutive Model for Soft Rocks and Its Application in Prediction of Time-Dependent Deformation Disaster in Tunnels

Ming Huang
College of Civil Engineering, Fuzhou University

Jinwu Zhan
College of Civil Engineering, Fujian University of Technology

Chaoshui Xu
School of Civil, Environment and Mining Engineering,
University of Adelaide

Song Jiang
College of Civil Engineering, Fujian University of Technology

5.1 INTRODUCTION

Tunnels' stability in soft (weak) rocks can be affected significantly by the creep properties of the surrounding rocks. Examples of time-dependent deformation failure of tunnels include the Simplon Tunnel between Switzerland and Italy (Steiner 1996), the Bolu Mountain Tunnel in Turkey (Dalgıç 2002), the Jiazhuqing Tunnel (Zhang 2003), the Yacambú–Quibor Tunnel in Venezuela (Hoek and Guevara 2009), and the Zhegu Mountain Tunnel (Meng et al. 2013) in China. Because of the time-dependent behavior of soft rocks, these engineering projects were prone to dangers both during and after they were built. The behavior of weak rocks over time has been studied in many ways and various creep theories have been proposed (Gasc-Barbier et al. 2004; Ma and Daemen 2006; Fabre and Pellet 2006; Sterpi and Gioda 2007). Creep constitutive models for weak rocks can be divided into four main groups (Zhou et al. 2011): empirical models, component-based models, creep damage models, fracture models (Zhao 1998; Mazzotti and Savoia 2003; Shao et al. 2003, 2006; Hamiel et al. 2004), and those models based on endochronic creep theory (Khoei et al. 2003). Compared with the theoretical model, the empirical model is based on experiment and requires fewer parameters, so it is widely used to predict the creep of weak rocks (Yang et al. 1999). For example, a nonlinear creep model for soft rocks was developed after the viscoelastic-plastic creep tests of ore rocks from the Jinchuan No. 3 mine. Zhang et al. (2012) studied the time-dependent behavior of strongly weathered sandstone in Xiangjiaba by means of triaxial creep tests. The experiments demonstrated that the rock underwent notable time-dependent deformation. The results of this study were later used to analyze the long-term stability of dam foundations. However, due to the limited experimental time, the empirical method may have errors in long-term creep deformation prediction compared to actual engineering projects.

DOI: 10.1201/9781003362081-7

The component-based model is composed of standard elements such as Newtonian dampers, Hooke springs, and friction elements, and has the flexibility to adapt to different types of creep behavior (Shao et al. 2003; Metzler and Nonnenmacher 2003; Zhou et al. 2013). Wang (2004) developed a model of constitutive creep damage in salt rocks and verified it experimentally. The results demonstrated that the constitutive model can not only accurately describe the creep damage behavior of salt rock under high stress conditions but also describe the basic creep characteristics and stable creep characteristics of salt rock under low stress conditions. Zhou et al. (2011) proposed a creep constitutive model based on the time-based fractional derivative by replacing a Newtonian dashpot in the classical Nishihara model with the fractional derivative Abel dashpot. Yang and Cheng (2011) proposed a nonstationary nonlinear viscoelastic shear creep model based on the viscoelastic shear creep test data of shale samples under four different shear stresses. It can predict the long-term shear deformation behavior of shale more accurately.

There are some problems in the use of the relatively mature rheological modeling method to model the creep behavior of rocks. The research of rock mass rheological model ignores the macroscopic characteristics determined by the test data, and the empirical model often cannot reflect the inherent creep mechanism and characteristics of rock, so it is difficult to determine the damage and fracture properties of rock materials in the modeling process. In the creep model based on internal time theory, it is difficult to obtain the model parameters under complex stress conditions.

At present, the research on the creep mechanisms of soft rocks is mainly based on the time-dependent damage during the creep. It is generally believed that under the long-term action of external forces, microfractures will further expand and extend, and these cracked or damaged rocks are considered to have lost their bearing capacity. However, the stress–strain curve of soft rock shows that the damaged soft rock still has some residual strength, which indicates that the damaged rock still has some bearing capacity. For rocks, the mechanical properties of the damaged and undamaged parts are different and therefore require different constitutive relationships. To deal with this problem, this paper adopts the theory of interference dynamic concept.

Past studies based on the disturbed state concept (DSC) proposed by Desai and Toth (1996) have made progress in proposing new types of constitutive models. Existing constitutive models based on DSC were comprehensively reviewed by Varadarajan et al. (2003, 2006). DSC, based on the idea that a deforming material can be viewed as a mixture of the relatively intact (RI) and the fully adjusted (FA) parts (states), is considered by researchers to be a general method applicable to both soil and rock materials. Under external loading, the material undergoes internal changes in the microstructure because of the self-adjustment within the material, which leads to a continuous transformation of the components of the RI state to the FA state (Figure 5.1).

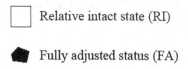

☐ Relative intact state (RI)

◆ Fully adjusted status (FA)

Original state   Intermediate state   Final state

FIGURE 5.1 Translation from the relatively intact (RI) to the fully adjusted (FA) state.

In this study, the model of creep behavior was established by introducing DSC, and a method is proposed to describe the RI state of rocks by the Burgers model and the FA state by the generalized Bingham model. A new creep constitutive model to describe the time-dependent deformation behavior of weak rocks is proposed and called the "Disturbance-Burgers (DBurgers) model." The tertiary creep behavior can be stimulated by the DBurgers model because it can reveal the plastic characteristics that the Burgers model does not have, and the DBurgers model can accurately capture the time evolution process of the transition from the RI state to the FA state. The application of the proposed DBurgers model has been exemplified in the Shilin tunneling project and the model parameters of this case are determined by curve fitting of the time-dependent deformation experimental data of argillaceous shale in the construction site.

5.2 ONE-DIMENSIONAL CONSTITUTIVE CREEP MODEL

5.2.1 DISTURBANCE FUNCTION AND ITS EVOLUTION

The basic principle of DSC (Desai 2001) was that the material could be regarded as a mixture of components in either the RI or the FA state. Under the condition of the rock creep model, the problem can be simplified as (i) describing weak rock with viscoelastic model and (ii) describing the FA state with an ideal viscoplastic model.

Figure 5.2 shows the notional representation model of rocks connected in series under two mixed states. In this model, ε is the total strain, and ε^{RI} and ε^{FA} are the strain contributions from the RI and FA parts, respectively. This conceptual representation can be expressed in the following relation:

$$\varepsilon = (1-D)\cdot\varepsilon^{RI} + D\cdot\varepsilon^{FA} \tag{5.1}$$

where D is the disturbance ratio based on the DSC (Desai 2001) defined as the proportion of total strain contributed by the FA part of the rock in the context of this study. Thus, D ranges from zero, where the rock consists of only RI components, to one, where only FA components were present. What should be noted is that the disturbance function D was used to describe the transformation of the material from the RI to the FA state according to the DSC (Desai 2001).

When the rock was subjected to external loads, the evolution of D can be defined in terms of plastic strain (Bažant 1994) because irreversible damage within the rock was related to strain. In the context of creep, in the case of creep similar to plastic strain, the creep strain can be used to estimate D, which in this study was modeled using a usually used exponential function in terms of time tg:

$$D = D\left(\xi^C\right) = D\left(f(t)\right) = 1 - e^{-At^B} \tag{5.2}$$

in which A and B are constants and $\xi^c = f(t)$ is the cumulative creep strain. This relationship essentially shows that there was no creep damage ($D=0$) at the initial stage (components of the rock are all in the RI state at $t=0$) and $D \rightarrow 1$ as $t \rightarrow \infty$ (components of the rock are all in the FA state eventually). In traditional damage mechanics, the strength of components of the rock was commonly

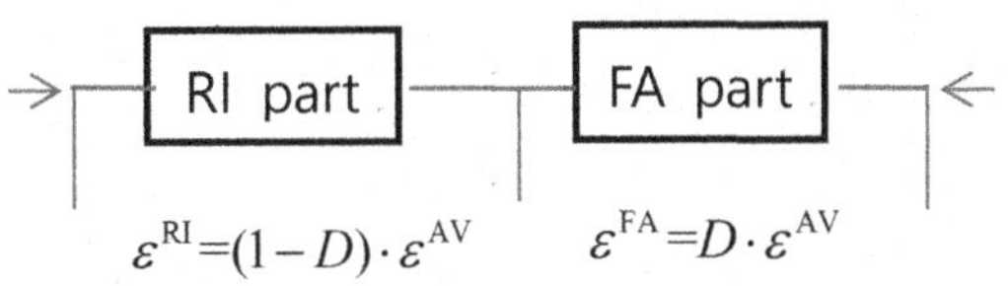

$$\varepsilon^{RI} = (1-D)\cdot\varepsilon^{AV} \qquad \varepsilon^{FA} = D\cdot\varepsilon^{AV}$$

FIGURE 5.2 Relationship between ε^{AV}, ε^{FA}, and ε^{RI}.

assumed to follow the Weibull distribution conceptually (Huang et al. 2018). The expression for the rock damage follows the same equation shown previously based on this assumption, but the parameters are different.

5.2.2 NEW CREEP MODEL BASED ON DSC

The creep properties of the rock are generally a nonlinear, time-dependent stress–strain relationship. As discussed previously, D is a variable parameter that varies with time. The irreversible change of rock composition from the RI to the FA state is described based on the DSC. In addition, RI and FA components have their own creep behavior in creep modeling.

1. According to the DSC (Desai 2001), experimental tests based on the complete material can obtain the constitutive relationship of the RI state. The Burgers viscoelastic model (Figure 5.3, left component) describes the characteristic of elasticity, stable creep, and complete relaxation, and it was adopted to describe the creep behaviors of the RI state in this study. The one-dimensional form of this model can be can be expressed as (Sun 1999):

$$E_1 \ddot{\varepsilon}^{RI} + \frac{E_1 E_2}{\eta_2} \dot{\varepsilon}^{RI} = \ddot{\sigma}^{RI} + \left(\frac{E_1}{\eta_1} + \frac{E_1}{\eta_2} + \frac{E_2}{\eta_2} \right) \dot{\sigma}^{RI} + \frac{E_1 E_2}{\eta_1 \eta_2} \sigma^{RI} \tag{5.3}$$

where η_1 and η_2 are Newtonian dashpot elements, the viscosity coefficients, E_1 and E_2, are the elastic constants of the two Hookean spring elements, σ^{RI} and ε^{RI} are the stress and strain in the RI state, and the one- and two-dot accents above the variable represent the first derivative (rate of change) and the second derivative against time.

2. In this paper, the generalized Bingham model (Figure 5.3, right component, Sun 1999) was used to describe the creep behavior of FA states:

$$\begin{cases} \varepsilon^{FA} = \dfrac{\sigma^{FA}}{E_3} & \sigma^{FA} < \sigma_s \\[2ex] \dot{\varepsilon}^{FA} = \dfrac{\dot{\sigma}^{FA}}{E_3} + \dfrac{\sigma^{FA} - \sigma_s}{\eta_3} & \sigma^{FA} \geq \sigma_s \end{cases} \tag{5.4}$$

where E_3 is the elastic modulus of the spring element, η_3 is the viscosity coefficient of the Newtonian damper, and σ_s is the yield stress of the generalized Bingham model.

The rock was represented by the two state models connected in series (Figures 5.2 and 5.3) for creep analysis, $\sigma^{RI} = \sigma^{FA}$. Combining Eqs. (5.1) to (5.4) and removing the superscripts of RI and FA, the creep model based on DSC can be expressed as:

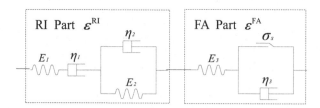

FIGURE 5.3 Proposed creep model (DBurgers model).

$$
\begin{cases}
\varepsilon(t) = e^{-At^B} \cdot \left[\dfrac{\sigma}{E_1} + \dfrac{\sigma}{\eta_1}t + \dfrac{\sigma}{E_2}\left(1 - e^{-\frac{E_2}{\eta_2}t}\right) \right] + \left(1 - e^{-At^B}\right) \cdot \dfrac{\sigma}{E_3} & \sigma^{FA} < \sigma_s \\[4mm]
\varepsilon(t) = e^{-At^B} \cdot \left[\dfrac{\sigma}{E_1} + \dfrac{\sigma}{\eta_1}t + \dfrac{\sigma}{E_2}\left(1 - e^{-\frac{E_2}{\eta_2}t}\right) \right] + \left(1 - e^{-At^B}\right)\left(\dfrac{\sigma - \sigma_s}{\eta_3}t + \dfrac{\sigma}{E_3} \right) & \sigma^{FA} \geq \sigma_s
\end{cases}
\tag{5.5}
$$

where A, B, E_1, E_2, E_3, η_1, η_2, η_3, and σ_s are as defined previously.

5.2.3 DETERMINATION OF PARAMETERS OF THE PROPOSED CREEP MODEL

Creep testing. In this study, a GDS triaxial instrument was used to perform creep tests on soft rocks (argillaceous shales from from East China Jiaotong University's Shilin Tunnel), so as to obtain creep model parameters (Figure 5.4). Cylindrical specimens with a radius of 25 mm and a height of 100 mm were used in the experiments. Uniaxial and multistage creep experiments were carried out. According to the uniaxial compressive strength (UCS) of the rocks, which are 3.24, 7.55, 11.32, and 14.56 MPa, the experimental creep loads at different stages are estimated. If the rate of creep displacement increase is less than 0.001 mm/h, the next stage of the load is applied. The data were recorded every 10 minutes and Chen's method (Sun 2007) was used to process experimental data. Figures 5.5 and 5.6 showed typical loading and unloading results of the argillaceous shale at natural moisture content.

(a)   (b)

FIGURE 5.4   GDS apparatus and creep testing of a rock sample.

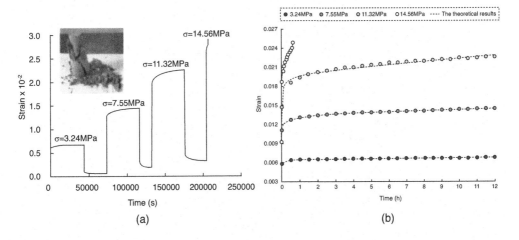

(a)   (b)

FIGURE 5.5   Strain–time curves of a rock sample during loading and unloading: (a) test curve before treating and (b) after treating with Chen's method (Sun 2007).

As shown in Figure 5.6, the load strain reached 0.0062 when the stress was 3.24 MPa, while the creep strain did not change significantly, and after 12 h, its value was only 0.0068, and it showed no signs of acceleration. At a stress of 7.55 MPa, the creep deformation value increases slightly and it reached 0.0145 after 12 h, which is 0.0035 higher compared to the instantaneous strain caused by the applied load condition. When the stress was 11.32 MPa, the creep strain reached 0.0050 after 12 h. Based on this, it is known that the creep deformation was more pronounced. When the loading stress is 14.56 MPa, the creep strain increases rapidly from 0.0162 to 0.0253, and then failure occurs (Figure 5.5).

The advantages of the proposed model in describing creep behavior were demonstrated by further comparing the developed model with the Burgers model based on laboratory test results. As shown in Figure 5.7, both models were the same in terms of instantaneous deformation, but there are significant differences in creep strain observed with time. This was caused by the notable effect of time-dependent plastic properties of the FA component in the supposed model presented in the Burgers model. In addition, the proposed model expressed a nonlinear behavior, which better demonstrated the initial stage, stable stage, and acceleration stage of creep deformation of commonly observed soft rocks in practice.

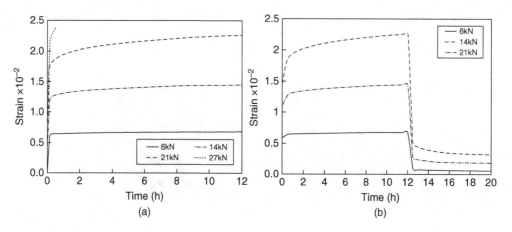

FIGURE 5.6   Strain–time curves under different axial loads (a). Time-dependent strain curves during loading and unloading (b).

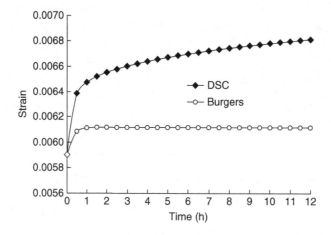

FIGURE 5.7   Analysis of the disturbed state concept (DSC) model and Burgers model with the same stress level and parameters.

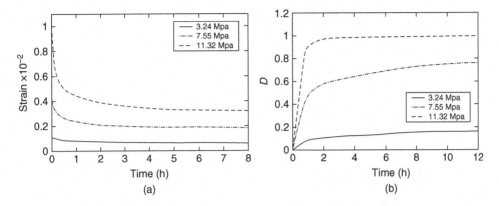

FIGURE 5.8   Curves of strain and disturbance factor D with time.

TABLE 5.1
Disturbance Factor Parameters under Different Axial Stresses

Sample	Stress(MPa)	A	B
Argillaceous shale	3.24	0.0898	0.2835
	7.55	0.6629	0.3185
	11.32	2.3607	0.3296
	14.56	10.2964	0.1054

TABLE 5.2
Parameters of the Disturbed State Constitutive Model

		Model Parameter						Fitting Effect
Sample	Stress (MPa)	E_1 MPa	E_2 MPa	E_3 MPa	η_1 MPa·h	η_2 MPa·h	η_3 MPa·h	R^2
Argillaceous	3.24	549	14,973	315	1.48E17	3,986	2.87E17	0.992
shale	7.55	685	14,574	516	34,519	8,971	4.99E19	0.999
	11.32	805	2.55E16	564	3.57E21	3.92E18	34,264	0.962
	14.56	1,582	4.58E13	734	3.65E18	1.56E23	1,391	0.891

5.2.4   DETERMINATION OF THE DISTURBANCE FACTOR D

Curves of strain and disturbance factor D with time (Figure 5.8). The variables A and B associated with the interference elements were found using the data fitting method presented by Huang et al. (2017) (Table 5.1). For Eq. (5.2), the interference factor D was calculated for a time interval of 1.0h after the start of the load.

5.2.5   PARAMETERS OF THE CREEP MODEL

The parameters of the new creep model using the nonlinear least-square multiparameter optimization method (Huang et al. 2012) were obtained from the creep test results of the argillaceous shale, which are presented in Table 5.2. Figure 5.9 demonstrates the adapted curves and clearly shows that Eq. (5.5) is sufficient to give a description of the creep properties of the soft rock that will be simulated in our case study (see the subsequent chapters).

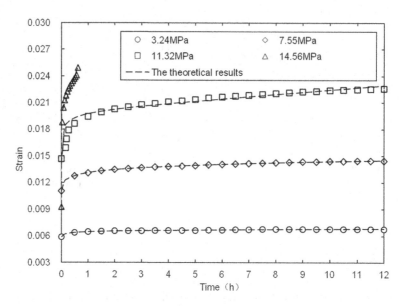

FIGURE 5.9 Theoretical fitting results with sample creep laboratory test data.

5.2.6 3D COMPOSITIONAL MODEL OF PERMANENT BULK MODULUS K

The surrounding rock is often under very complex three-dimensional stress conditions, so we need a suitable three-dimensional model to simulate it numerically. There are two ways to construct the model: one is to assume that the Poisson's ratio (μ) is constant, and the other is to assume that the bulk modulus (K) is constant. In this paper, the second assumption is chosen, and the aggregate strain is the sum of the volumetric strain and the deviatoric strain. Stress–strain relationship is expressed as:

$$\varepsilon_{ij} = e_{ij} + \varepsilon_m \delta_{ij} \tag{5.6}$$

and in Eq. (5.7),

$$\varepsilon_{ij} = \frac{S_{ij}}{2G}, \varepsilon_m = \frac{\sigma_m}{3K} \tag{5.7}$$

the σ_m is the averaged stress $(1/3)(\sigma_1 + \sigma_2 + \sigma_3)$, S_{ij} is the deviatoric stress tensor, G is the shear modulus, and K is the bulk modulus. On this basis, we consider that creep will only have an effect on the deviatoric strain. The bulk modulus K is constant at this point. Equation (5.7) is at this point:

$$\varepsilon_{ij}(t) = J'(t)S_{ij}, \quad \varepsilon_m = \frac{\sigma_m}{3K} \tag{5.8}$$

where $J'(t)$ is the creep modulus and $J'(t)$ is the modulus of creep.

$$e_{ij} = (1-D)\left(e_{ij}^K + e_{ij}^M\right) + D\left(e_{ij}^B + e_{ij}^H\right) \tag{5.9}$$

where the overall deviation strain e_{ij} is the composite strain of the Maxwell component $\left(e_{ij}^M\right)$, Kelvin component $\left(e_{ij}^K\right)$, Bingham component $\left(e_{ij}^B\right)$, and Hooker component $\left(e_{ij}^H\right)$.

5.3 FINITE DIFFERENCE PROGRAM AND NUMERICAL VALIDATION OF THE CREEP MODEL

5.3.1 FINITE DIFFERENCE VERSION OF THE PROPOSED MODEL

Equation (5.9) can also be written at this point as:

$$\Delta e_{ij} = (1-D)\left(\Delta e_{ij}^M + \Delta e_{ij}^K\right) + D\left(\Delta e_{ij}^B + \Delta e_{ij}^H\right) \tag{5.10}$$

For the Maxwell component:

$$\Delta e_{ij}^M = \frac{\Delta S_{ij}}{2G_1} + \frac{\overline{S}_{ij}}{2\eta_1}\Delta t \tag{5.11}$$

The relationship between incremental strains is:

$$\overline{S}_{ij}\Delta t = 2\eta_2 \Delta \varepsilon_{ij}^K + 2G_2 \overline{\varepsilon}_{ij}^K \Delta t \tag{5.12}$$

$\overline{S}_{ij}$ and $\overline{e}_{ij}^K$ are the average deviation stress and strain in one time step.

With respect to the Bingham component, the increasing deviation strain can be described as (Sun 1999):

$$\dot{e}_{ij}^B = \frac{\langle F \rangle}{\eta_3}\frac{\partial Q}{\partial \sigma_{ij}} - \frac{1}{3}\dot{e}_{vol}^B \delta_{ij} \tag{5.13}$$

where F is the Mohr–Coulomb yield function $F = \sigma_m \sin\varphi + \sigma\cos\theta_\sigma - \frac{J_2}{\sqrt{3}}\sin\theta_\sigma \sin\varphi - c\cos\varphi$, $\langle F \rangle$ is a switch operator given as $\langle F \rangle = \begin{cases} 0 & (F < 0) \\ F & (F \geq 0) \end{cases}$, Q is the plasticity potential function, $\dot{e}_{vol}^B$ represents the volumetric strain rate in the Bingham model, and δ_{ij} represents the Kronecker delta.

By substituting Eqs. (5.11) to (5.13) into Eq. (5.10), we get:

$$S_{ij}^N = \frac{1}{(1-D)a}\left[\Delta\varepsilon_{ij} - D\Delta\varepsilon_{ij}^B + (1-D)bS_{ij}^t - (1-D)\left(\frac{W}{V}-1\right)\varepsilon_{ij}^{K,t}\right] \tag{5.14}$$

where

$$a = \frac{1}{2G_1} + \frac{D}{2(1-D)G_3} + \frac{\Delta t}{4}\left(\frac{1}{\eta_1} + \frac{1}{V\eta_2}\right)$$

$$b = \frac{1}{2G_1} + \frac{D}{2(1-D)G_3} - \frac{\Delta t}{4}\left(\frac{1}{\eta_1} + \frac{1}{V\eta_2}\right)$$

$$V = 1 + \frac{G_2\Delta t}{2\eta_2}, \quad W = 1 - \frac{G_2\Delta t}{2\eta_2}$$

Here t and $t+1$ indicate the iteration time step. According to the theory of plastic deformation, the ball stress does not cause plastic deformation in the proposed model, so the average stress $\sigma_m^{t+1} = \sigma_m^t + K\left(\Delta\varepsilon_{vol} - \Delta\varepsilon_{vol}^{vp}\right)$, where K is the bulk modulus.

5.3.2 VERIFICATION OF THE PROPOSED CREEP MODEL

5.3.2.1 Viscoelastic Properties

In order to verify the validity of the new creep model, uniaxial compression creep tests on Ø 50 mm×H 100 mm cylindrical specimens were numerically simulated.

To check the viscoelasticity with the model, we set the values of A and B to zero to make sure that $D=0$ [Eq. (5.2)], when only the Burger model (RI state part) applies. The parameters used in the model at this point are: $\sigma=7.5$ MPa, $G_1=228.3$ MPa, $G_2=4858.0$ MPa, $\eta_1=11.5$ GPa·h, and $\eta_2=3.0$ GPa·h, and the deformation at the point [0, 0, 0] is shown in Figure 5.10. This shows that the viscoelastic energy of the model is the same as the Burgers model for an original displacement of 1.286 mm and a 12-h creep displacement of 1.60 mm.

5.3.2.2 Impact of Disturbance Factors

The same numerical simulations were performed in order to investigate the variation of D and its impact on the creep behavior. Figure 5.11 represents the variation of D for $B=0.1$ and $A=0.01$,

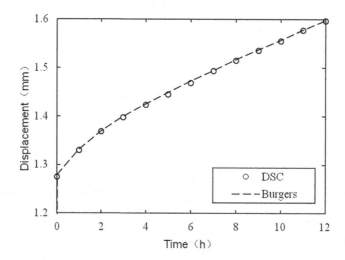

FIGURE 5.10 Time-dependent y-displacement of the validation model.

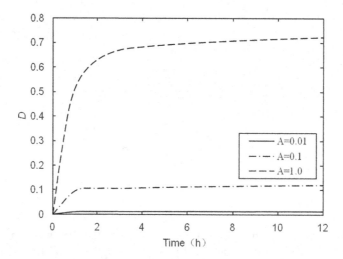

FIGURE 5.11 Variation of D with time at various amounts of A.

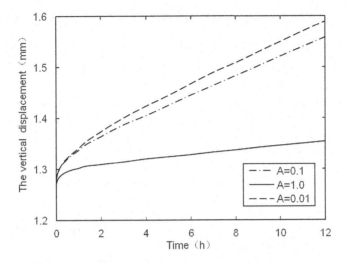

FIGURE 5.12 Variation of vertical displacement with time at various amounts of A.

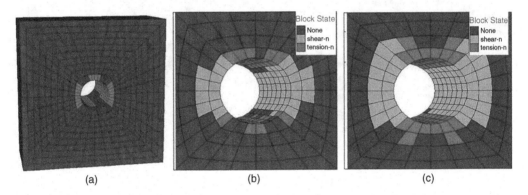

FIGURE 5.13 Deformation characteristics after 12 h: (a) calculation model result, (b) with cvisc model, and (c) result with the DBurgers model.

0.1, and 1.0. Figure 5.12 shows the corresponding creep properties with an initial displacement of 1.27 mm for the three cases. As time increases, different viscoelastic displacements result due to the increase of D at different rates after 12 h; the displacements are 1.59, 1.56 and 1.36 mm when A is 0.01, 0.1 and 1.0, respectively.

Creep displacements account for the bulk of the displacement when A is large, and conversely when A is small, D's low total displacement is mainly caused by the viscoelastic component, as calculated by the Burgers model, close to 1.6 mm. Therefore, it was possible to conclude that the creep model can produce the desired viscoplastic behavior.

5.3.2.3 Modeling Viscoplastic Deformations

A simple case is used to further validate the model proposed in this paper. The parameters are $c = 3.0$ MPa, $\varphi = 25°$, and $R_t = 0.5$ MPa for the proposed model, $G_3 = 172.0$ MPa, $\eta_3 = 1.6 \times 10^{16}$ GPa·h, $A = 0.66$, and $B = 0.32$. The size of the validation model is 1.0 m $\times 1.0$ m $\times 1.0$ m, and the diameter of the middle hole is 0.2 m. For the boundary conditions, a fixed constraint is applied to the bottom of the model, and the upper surface is a free surface. The deformation characteristics after 12 h are shown in Figure 5.13, but the two models were basically similar.

5.4 EXAMPLE STUDY: DEFORMATION OF THE SHILIN TUNNEL OVER TIME

5.4.1 GENERAL OVERVIEW OF LARGE-SCALE DEFORMATION IN THE SHILIN TUNNEL

The Shilin tunnel serving the Yongan-Ninghua Expressway in China was used as an example to simulate deformation of soft rock tunnels over time. The tunnel is 12.52 m wide and buried 460 m. Significant deformation and damage over time occurred during construction (Figure 5.14), causing great difficulties for subsequent tunnel excavation. We selected the section from YK14+293 to YK14+313 for our study and analysis (Figure 5.15). Through field tests and analysis of the tunnel damage characteristics, we found that the surrounding rock of this tunnel is a typical muddy shale soft rock, which produces more pronounced deformation over time. The purpose of this case study is to establish an intrinsic structural model with time, which can be used to analyze the Shilin tunnel.

1. The lining was severely deformed along YK14+260 to YK14+290 (Figure 5.16). Excavation from YK14+312 to YK14+320 produced a large settlement (Figure 5.17). Monitoring and measurement data confirmed that after excavation, the settlement rate of the vault reached a maximum of 24 mm/day and the horizontal convergence was 22 mm/day. The arch

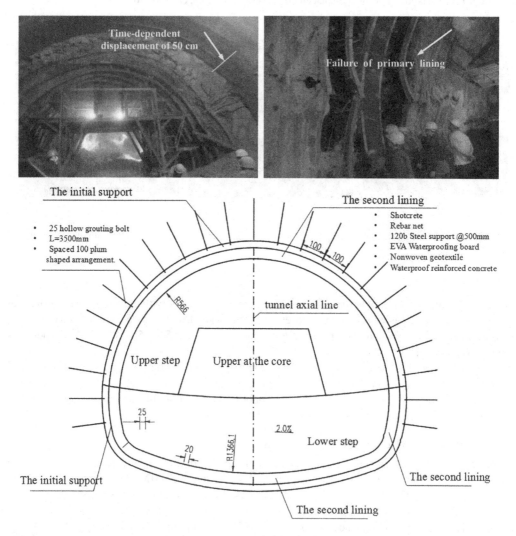

FIGURE 5.14 The problem of large-scale deformation in tunnel construction in the Shilin tunnel project.

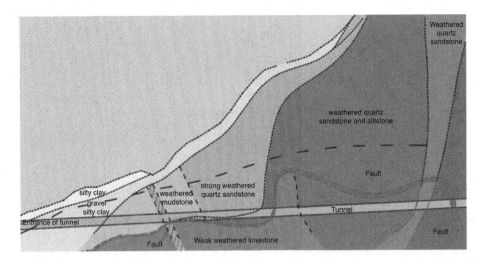

FIGURE 5.15 Geological cross section.

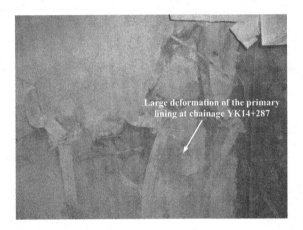

FIGURE 5.16 Lining damage.

settlement and horizontal convergence continued to increase until the tunnel became unstable; monitoring of the initial pier YK14+313 showed a large deformation of the initial support (Figure 5.17).

2. Characteristics of deformation after construction of secondary lining. After the completion of the elevation arch construction, the bottom slab of section YK14+260–YK14+320 bulged. As a result, cracks of larger size appeared in the tunnel floor (Figure 5.18). The monitoring of the arch settlement in sections YK14+301 and YK14+292 after the completion of secondary lining construction showed that the tunnel settlement had increased quickly to 32 mm within 10 days. After 80 days, the deformation rate gradually decreased. The total arch settlement of these two sections did not exceed 111 and 168 mm, respectively.

5.4.2 ESTABLISHMENT OF THE MODEL

5.4.2.1 Geological Model

In this study, a typical section of the Shilin tunnel with large deformations was selected for study. During construction, an alteration zone was identified at the base of the tunnel. In the numerical

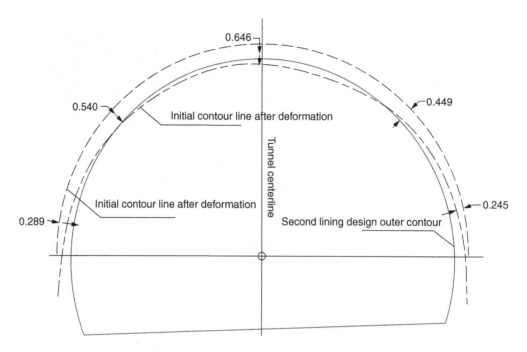

FIGURE 5.17   Deformation and invasion limits.

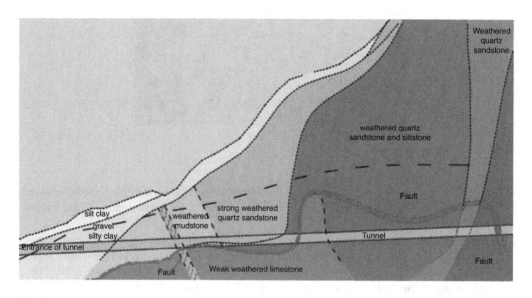

FIGURE 5.18   Cracks of larger size appeared in the tunnel floor.

model, the alteration zone was reduced to a 3 m thick horizontal fracture zone at a depth of 5 m from the tunnel floor. The upper boundary of the model is the ground surface 90 m from the tunnel floor, the lower boundary is 50 m below the tunnel floor to reduce boundary effects, the left and right boundaries are 50m from the tunnel centerline, and the model boundary is constrained by displacement boundary conditions. The top surface of the model is defined as the free boundary. In total, the 3D numerical tunnel model has 47,340 elements and 201,958 nodes (Figure 5.19).

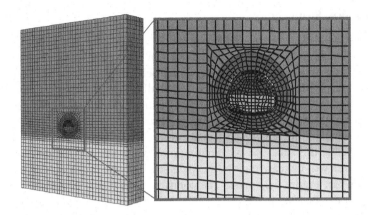

FIGURE 5.19 3D digital models.

TABLE 5.3
Mechanical Properties of Rocks

Type	E	μ	γ	c	Φ
	MPa	—	kN/m³	MPa	°
Phyllite sandstone	—	0.3	20	0.4	30
Fault fracture zone	7	0.30	18	0.05	28
Quartz sandstone	5,000	0.30	25	0.8	35

TABLE 5.4
Rheological Parameters of Phyllite Sandstone

Type	E_1	E_2	E_3	η_1	η_2	η_3	A	B
	MPa	MPa	MPa	MPa·d	MPa·d	MPa·d		
Phyllite sandstone	62.5	234.8	60.0	5,782.6	243.3	50,000	0.001	0.45

TABLE 5.5
Mechanical Properties' Parameters of the Tunnel Supports

Material Designation	E	μ	γ	d	K_{bond}	S_{bond}
	GPa	—	kN/m³	mm	MPa	N/m
Reinforced concrete support of steel arch	24	0.25	24	280	—	—
Anchor bolt	210	—	77	25	5.66×10^3	2.62×10^5
Secondary lining	29.5	0.25	25	450	—	—

5.4.2.2 Model Parameters

Based on the geological setting (Figure 5.17), the upper rocks consist mainly of strongly weathered mudstones and gabbroic sandstones. The lower rocks of the fault zone are mainly weakly weathered quartz sandstones. The mechanical properties of these rocks are given in Table 5.3. The creep intrinsic model proposed in this paper was used to analyze the deformation of the upper rock layer, and the Mora–Coulomb intrinsic model was used to analyze the deformation of the lower rock layer, which was simulated numerically. The parameters used in the creep model are given in Table 5.4, and the mechanical properties parameters of the tunnel supports are given in Table 5.5.

5.4.3 Deformation Simulations

Figure 5.20 shows the numerical model displacement curves of the measured points around the tunnel before the secondary lining construction. Within 10 days of excavation, the relative horizontal displacement on both sides of the surrounding rock increased rapidly, but after a period of large deformation lasting 10 days, the deformation gradually transitioned to a relatively stable phase. During the next 30 days, the rate of increase of deformation was very low and almost unchanged. The displacement finally stabilized at about 0.24 m. The arch enters the accelerated deformation stage, and the deformation rate reaches 0.05 m/day after 40 days of excavation. Unless the secondary lining is installed at this time, the surrounding rock will enter the unstable deformation stage, the sinking of the arch will increase rapidly, creep damage may eventually occur, and the surrounding rock may collapse. Figure 5.21 shows the comparison of vault settlement with and without secondary lining. It can be seen that with secondary lining, the surrounding rock mass maintains a stable creep, which is different from the accelerated deformation without secondary lining. The deformation rate of the surrounding rock mass can be controlled by applying secondary lining at

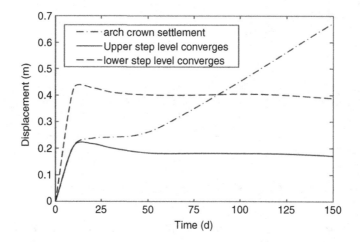

FIGURE 5.20   Numerical model displacement curves before construction of the secondary lining.

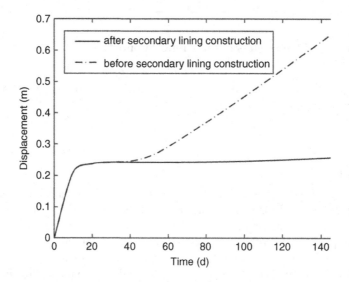

FIGURE 5.21   Comparison of vault settlement with and without secondary lining.

the appropriate time. Therefore, based on the numerical simulation results, the optimal time to arrange the secondary lining is between 20 and 40 days after excavation, which is consistent with the construction schedule of the Shilin tunnel. In this case, the secondary lining can suppress the final settlement time of the vault to within 0.25 m.

5.5 CONCLUSION

1. Based on the DSC, the formula and terms of the perturbation function D due to creep deformation can be determined; the D function allows to characterize the soft rock with respect to creep damage, where the required values of parameters A and B can be identified by simple creep experiments.
2. In this study, the Burgers and extensive Bingham models are used to characterize the creep behavior of the RI and FA states, respectively. By introducing a perturbation function D, a new intrinsic creep model for soft rocks is proposed, which combines these two models. The model has been validated by experiments on shale samples and published studies. Compared with the Burgers model, it can be used for simulating the creep behavior of soft rocks in the early, stable, and accelerated phases of deformation.
3. A finite difference version of the model was implemented in FLAC3D and its ability to simulate viscoelastic, viscoplastic, and creep damage was evaluated and proved to be correct.
4. The simulation results show that without secondary lining, the deformation rate is large within 10 days after excavation and declined slowly after 20 days to reach a stable creep state. The final stable arch settlement can be controlled within 0.25 m with the application of secondary lining after 20 days of excavation. The simulation results showed that the optimal construction time for the secondary lining is between 20 and 40 days after excavation. In general, the results of the numerical simulation match well with the measured data and the construction schedule of the tunnel lining.

REFERENCES

Bažant, Z. P. (1994). "Nonlocal damage theory based on micromechanics of crack interactions", *Journal of Engineering Mechanics*, 120(3), 593–617.

Desai, C. S., and Toth, J. (1996). "Disturbed state constitutive modeling based on stress-strain and nondestructive behavior", *International Journal of Solids and Structures*, 33(11), 1619–1650.

Desai, C. S. (2001). *Mechanics of Materials and Interfaces — the Disturbed State Concept*, Boca Raton, FL: CRC Press.

Dalgıç, S. (2002). "Tunneling in squeezing rock, the Bolu tunnel, Anatolian Motorway, Turkey", *Engineering Geology*, 67(1), 73–96.

Fabre, G., and Pellet, F. (2006). "Creep and time-dependent damage in argillaceous rocks", *International Journal of Rock Mechanics and Mining Sciences*, 43(6), 950–960.

Gasc-Barbier, M., Chanchole, S., and Bérest, P. (2004). "Creep behavior of Bure clayey rock", *Applied Clay Science*, 26(1), 449–458.

Hamiel, Y., Liu, Y., Lyakhovsky, V., Ben-Zion, Y., and Lockner, D. (2004). "A viscoelastic damage model with applications to stable and unstable fracturing", *Geophysical Journal International*, 159(3), 1155–1165.

Hoek, E., and Guevara, R. (2009). "Overcoming squeezing in the Yacambú-Quibor tunnel, Venezuela", *Rock Mechanics and Rock Engineering*, 42(2), 389–418.

Huang, M., Jiang, S., Xu, D. X., et al. (2018). "Load transfer mechanism and theoretical model of step tapered hollow pile with huge diameter", *Chinese Journal of Rock Mechanics and Engineering*, 37(10), 2370–2383.

Huang, M, Zhang, B. Q., Chen, F. Q., et al. (2017). "A new incremental load transfer model of pile-soil interaction based on disturbed state concep", *Rock and Soil Mechanics*, 38(S1), 173–178.

Huang, M., Liu, X. R. and Deng, T. (2012). "Numerical calculation of time effect deformations of tunnel surrounding rock in terms of water degradation", *Rock and Soil Mechanics*, 33(6), 1876–1882.

Khoei, A. R., Bakhshiani, A., and Mofid, M. (2003). "An implicit algorithm for hypoelasto-plastic and hypoelasto-viscoplastic endochronic theory in finite strain isotropic–kinematic-hardening model", *International Journal of Solids and Structures*, 40(13), 3393–3423.

Mazzotti, C., and Savoia, M. (2003). "Nonlinear creep damage model for concrete under uniaxial compression", *Journal of Engineering Mechanics*, 129(9), 1065–1075.

Metzler, R., and Nonnenmacher, T. F. (2003). "Fractional relaxation processes and fractional rheological models for the description of a class of viscoelastic materials", *International Journal of Plasticity*, 19(7), 941–959.

Ma, L., and Daemen, J. J. K. (2006). "An experiment study on creep of welded tuff", *International Journal of Rock Mechanics and Mining Sciences*, 43, 282–291.

Meng, L., Li, T., Jiang, Y., Wang, R., and Li, Y. (2013). "Characteristics and mechanisms of large deformation in the Zhegu mountain tunnel on the Sichuan–Tibet highway", *Tunnelling and Underground Space Technology*, 37, 157–164.

Sterpi, D., and Gioda, G. (2007). "Visco-plastic behaviour around advancing tunnels in squeezing rock", *Rock Mechanics and Rock Engineering*, 23(3), 292–299.

Shao, J. F., Zhu, Q. Z., and Su, K. (2003). "Modeling of creep in rock materials in terms of material degradation", *Computers and Geotechnics*, 30(7), 549–555.

Shao, J. F., Chau, K. T., and Feng, X. T. (2006). "Modeling of anisotropic damage and creep deformation in brittle rocks", *International Journal of Rock Mechanics and Mining Sciences*, 43(4), 582–592.

Steiner, W. (1996). "Tunnelling in squeezing rocks: Case histories", *Rock Mechanics and Rock Engineering*, 29(4), 211–246.

Sun, J. (1999). *Rheological Behavior of Geomaterials and Its Engineering Applications*, Beijing: China Architecture and Building Press.

Sun, J. (2007). "Rock rheologyical mechanics and its advance in engineering applications", *Chinese Journal of Rock Mechanics and Engineering*, 26(6), 1081–1106.

Varadarajan, A., Sharma, K. G., Venkatachalam, K., and Gupta, A. K. (2003). "Testing and modeling two rock-fill materials", *Journal of Geotechnical and Geoenvironmental Engineering*, 129(3), 206–218.

Varadarajan, A., Sharma, K. G., Abbas, S. M., and Dhawan, A. K. (2006). "Constitutive model for rockfill materials and determination of material constants", *International Journal of Geomechanics*, 6(4), 226–237.

Wang, G. (2004). "A new constitutive creep-damage model for salt rock and its characteristics", *International Journal of Rock Mechanics and Mining Sciences*, 41, 61–67.

Yang, C., Daemen, J. J. K, and Yin, J. H. (1999). "Experimental investigation of creep behavior of salt rock", *International Journal of Rock Mechanics and Mining Sciences*, 36(2), 233–242.

Yang, S. Q., and Cheng, L. (2011). "Non-stationary and nonlinear visco-elastic shear creep model for shale", *International Journal of Rock Mechanics and Mining Sciences*, 48(6), 1011–1020.

Zhang, Z. D. (2003). "Discussion and study on large deformation of tunnel in squeezing ground", *Modern Tunnelling Technology*, 2, 6–12.

Zhang, Z. L., Xu, W. Y., Wang, W., and Wang, R. B. (2012). "Triaxial creep tests of rock from the compressive zone of dam foundation in Xiangjiaba Hydropower Station", *International Journal of Rock Mechanics and Mining Sciences*, 50, 133–139.

Zhao, Y. (1998). "Crack pattern evolution and a fractal damage constitutive model for rock", *International Journal of Rock Mechanics and Mining Sciences*, 35(3), 349–366.

Zhou, H. W., Wang, C. P., Han, B. B., and Duan, Z. Q. (2011). "A creep constitutive model for salt rock based on fractional derivatives", *International Journal of Rock Mechanics and Mining Sciences*, 48(1), 116–121.

Zhou, H. W., Wang, C. P., Mishnaevsky, L., Duan, Z. Q., and Ding, J. Y. (2013). "A fractional derivative approach to full creep regions in salt rock", *Mechanics of Time-Dependent Materials*, 17(3), 413–425.

6 Hydro-Mechanical Behavior of Unsaturated Collapsible Loessial Soils

Amir Akbari Garakani
Power Industry Structures Research Dept.,
Niroo Research Institute (NRI)

6.1 INTRODUCTION

Loess is a well-known Aeolian deposit, which is characterized by some distinguished engineering properties including high initial void ratio, relatively low initial density and water content, high percentage of fine-grained particles, and low cohesion.[1–8] Loess sediments typically contain a mineralogical composition of quartz (30%–65%), feldspars (5%–25%), micas (1%–5%), calcite (7%–30%), and some clay minerals such as montmorillonite, illite, and kaolinite.[3,9] Under certain circumstances during formation, some of these minerals such as calcite may act as crystalline cementing materials, forming bridges between original sediment grains and binding them together. As a result, there usually exists a significant degree of cementation in these Aeolian deposits. In addition to this brittle, crystalline-type cementation, due to the low initial moisture content of loess sediments, the contribution of suction to the soil behavior can become quite significant, especially when a considerable amount of fine fractions such as silt or clay exists in the soil. Because of this complicated nature, loess sediments appear to have a strong and stable structure in their natural unsaturated state.[4,7] However, when they are subjected to wetting from heavy and continuous rainfall, excessive irrigation, broken water or sewer lines, or ground water rise, these generally dry deposits may get wet enough to experience a large loss of shear strength. In addition to the effect of decreasing suction, the water can dissolve or soften the bonds between the particles, allowing a great increase in compressibility with comparatively small changes in stress.[5,10] This mechanism is referred to as wetting-induced collapse, and its occurrence may require a high enough initial level of external stress to destroy the metastable soil structure within the soil deposit.[11–13]

In recent years, in order to interpret the laboratory test results on unsaturated soils and precisely assess the coupled unsaturated hydro-mechanical responses of the soils, implementing an effective stress approach by considering unsaturated state variables has been vastly taken into consideration by researchers.[4,14–20] It has been shown that by considering an effective stress approach, the changes in unsaturated stress state variables can be directly related to water retention behavior of the soil under validity of the continuum mechanics concepts. Implementing the concept of effective stress in unsaturated soils has also been reported to be very efficient in demonstrating the collapsible behavior of loessial soils.[3,4,6,7,21] Garakani,[3] Haeri et al.,[4,21,22] and Garakani et al.[23] have conducted a series of unsaturated tests, using suction-control triaxial and oedometer devices, on intact, reconstituted, and lime-stabilized loessial specimens and by considering an effective stress approach under unsaturated conditions developed constitutive models to predict different aspects of the hydro-mechanical behaviors of collapsible loessial soils (e.g., shear strength, isotropic compression, soil behavior at critical state, load-collapse behavior, and stress–strain behavior in k_o condition). Considering previous studies, it is revealed that using an effective stress concept can lead to an efficient way of demonstrating the stress state variables for the unsaturated state of the soils.

DOI: 10.1201/9781003362081-8

Among different approaches implemented to model the mechanical behavior of different types of materials, the disturbed state concept (DSC) has been vastly used by the researchers, so far, to predict the mechanical behavior of different materials such as structured soils. The DSC, well known as a simple and comprehensive approach, was first proposed by Desai[24–26] and developed afterward as a unified framework for predicting the behavior of different materials including soils.[6,23,27–38] Taking into account the factors affecting material behavior (e.g., inherent and induced discontinuities, initial or in situ stress or strain, isotropic and anisotropic hardening, stress (load) path dependency, degradation and softening in material structure), DSC has been used to explain the deformation behavior of a variety of engineering materials, namely, soil, concrete, ceramics, asphalt, and masonry.[35,39–43] It also has been utilized to interpret the unsaturated soils' behavior,[6,22,23,38,44–46] to study liquefaction in loose sandy soils,[29,47] to investigate the soil behavior under applying cyclic loads,[48,49] to study the behavior of structured soils,[50,51] and to numerically model the geotechnical problems.[52,53]

According to DSC, the mechanical response of a soil to changes in state variables (e.g., hydromechanical parameters such as stress level, matric suction) can be simply and accurately predicted by having the soil responses for two extreme structural reference conditions, namely, "minimum soil structure disturbance" and "maximum soil structure disturbance."

In this chapter, the functionality of the DSC for prediction of the triaxial shear behavior of intact and reconstituted unsaturated loessial specimens and the oedometer load-deformation characteristics of the lime-stabilized unsaturated loessial specimens is investigated by considering two research works conducted by Haeri et al.[6,22] and Garakani et al.[23]

6.2 SHEAR BEHAVIOR OF UNSATURATED COLLAPSIBLE LOESSIAL SOILS

Haeri et al.[6] studied the effect of initial soil structure and specimen disturbance on the hydromechanical behavior of a highly collapsible Aeolian loessial soil by conducting a total of 16 wetting-induced collapse and shear tests on "intact" and "reconstituted" loess specimens at three different net stresses of 50, 200, and 400 kPa and various matric suctions.

6.2.1 TESTED MATERIAL AND EXPERIMENTAL PROGRAM

The loessial soil examined by Haeri et al.[6] was a clayey silt soil (i.e., ML, in accordance with the Unified Soil Classification System[54]), which was obtained from the "Hezar-pich" hills in the city of Gorgan northeast Iran. The loess sequence in this area is part of the Eurasian loess belt extending from Northwest Europe to Central Asia and China, which was created by severe storms blowing in from the north of Iran, forming a wide plain of suspended silt particles in the form of hills and mounds in the Gorgan region.[55–57]

Haeri et al.[6] used a suction-controlled unsaturated triaxial testing device for their experimental investigations on the shear strength of the examined collapsible loess. Details of the implemented device and also the sampling procedure can be found in Garakani[3] and Haeri et al.[4] Index properties of the tested material are summarized in Table 6.1. Based on the results of the filter paper tests presented by Garakani,[3] Haeri et al.,[4] and Ghazizadeh[58] for reconstituted and intact specimens with an

---

TABLE 6.1

Index Properties of the Loess Tested by Haeri et al.[6]

Initial Void Ratio	Initial Dry Unit Weight	Initial Moisture Content	Initial Degree of Saturation	Pass. No. 4 Sieve	Pass. No. 200 Sieve	Clay Frac. (<2 μm)	Atterberg Limits	
e_0	γ_{do} (kN/m³)	w_o (%)	S_{ro}	d_4 (%)	d_{200} (%)	d_c (%)	w_L	w_P
0.770	15.07	7.1	0.25	100	96.5	25.9	32	26

TABLE 6.2

Characteristics of the Loessial Specimens and Their Stress State Conditions Tested by Haeri et al.[6]

Specimen State	Initial Mean Net Stress p_{no} (kPa)	Final Matric Suction ψ (kPa)
Intact	50	0, 50, 400
	200	0, 50, 300
	400	0, 50
Reconstituted	50	10, 50, 300, 400
	200	10, 50, 300, 400

initial void ratio of about 0.770, air entry suction values of 8 and 10 kPa, respectively, were obtained and residual degree of saturation, $S_{r, res}$, for reconstituted and intact specimens were reported as 0.05 and 0.14, respectively.

In triaxial tests on unsaturated samples performed by Haeri et al.,[6] four stages of "equalization," "isotropic compression," "wetting," and "shearing" were conducted on intact and reconstituted loessial specimens. During equalization, an initial matric suction of 750 kPa was introduced to the boundaries of each specimen and the soil suction was then allowed to equalize along the specimen height. The specimens were then isotopically compressed to the desired mean net stress by increasing the cell pressure while maintaining constant pore air and water pressures (i.e., isotropic compression stage). After reaching an equilibrium state, the soil specimens were wetted by increasing the applied water pressure to the bottom of the specimen, while keeping the air pressure constant at the top (wetting stage). Then, the specimens were sheared at a displacement rate of 0.002 mm/min until approaching an axial strain, ε_a, of 20% to ensure fully drained and constant suction conditions up to high strain levels. The states of the tested specimens and their stress state conditions are summarized in Table 6.2.

6.2.2 Results of Unsaturated Triaxial Shearing Tests on the Examined Loess

Figures 6.1–6.3 present variations in shear stress, volume, and degree of saturation of intact and reconstituted specimens subjected to different mean net stress and matric suctions, in accordance with Haeri et al.[6] As shown in these figures, in both reconstituted and intact specimens, the absolute values of peak shear strength increase as either the applied mean net stress or the matric suction increase. However, depending on the level of applied unsaturated stress state variables (initial mean net stress and matric suction), the consequent magnitude of wetting-induced collapse experienced by the specimen before shear, and the initial structure (intact or reconstituted) of the tested specimens, different stress-deformation behavior may be observed in the post peak range.

6.2.3 Implementing DSC for Prediction of the Stress–Strain Behavior of Loessial Soils under Triaxial Shearing

In the study performed by Haeri et al.,[6] a general concept in the framework of the DSC, which was developed by Desai[24,25] and extended by Geiser et al.,[44,45] was utilized to characterize the stress–strain behavior of the intact and reconstituted specimens of the tested collapsible soil during shear. DSC is a general approach to express the behavior of a deforming material (e.g., soils, masonry, concrete, and asphalt concrete), with accommodation of different influencing factors.[24–28,30–33,36,38,44,45] Some of the influencing factors are initial stress/strain conditions, deformation and hardening/

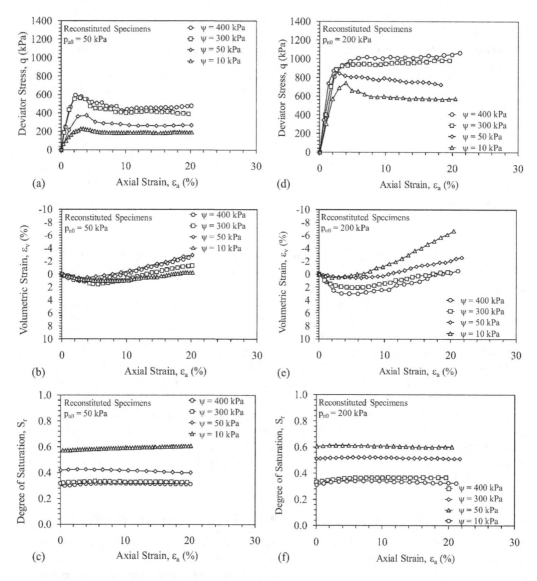

FIGURE 6.1 Variations in (a, d) deviatoric stress, (b, e) volumetric strain, and (c, f) degree of saturation of reconstituted specimens under different mean net stresses and matric suctions.[6]

softening parameters, anisotropy, discontinuity, and loading conditions. In this approach, the deformation behavior of a geologic material during loading is defined in terms of two different constituent parts: the continuum part, which is the part of the material that remains continuous and is called the relative intact (RI) state, and the part which experiences microcracking due to self-adjustment/organization[32] upon loading and is called the fully adjusted (FA) state. The domain of disturbance is the space between RI and FA states among which the observed behavior of the material is located. The FA state is referred to as the reference state or the state of stress with no disturbance. This state is considered by Geiser et al.[44] as the saturated state for unsaturated materials. Based on the DSC, disturbance, D, is a term, which represents the microstructural changes that lead to microcracking, damage, and softening during loading[32] and can be expressed as a "Weibull" function in terms of

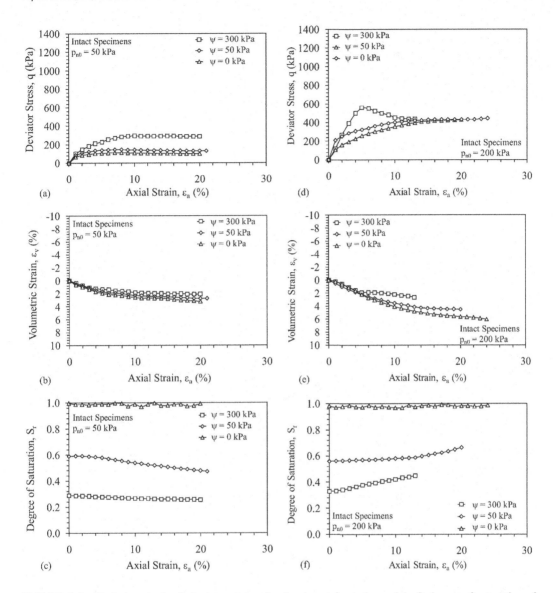

FIGURE 6.2 Variations in (a, d) deviator stress, (b, e) volumetric strain, and (c, f) degree of saturation of intact specimens under initial mean net stresses of 50 and 200 kPa and matric suctions of 10, 50, and 300 kPa.[6]

some internal variables such as accumulated deviator plastic strains, ζ_D, or plastic work, as shown by Eq. 6.1:

$$D = D_u \left[1 - \exp\left(-A.\zeta_D^Z\right) \right] \tag{6.1}$$

where A, Z, and D_u are the material parameters. The deviator stress (q) related to the RI state of the material can be expressed using a hyperbolic function proposed by Konder[59] in terms of the axial strain as:

$$q = \frac{\varepsilon_a}{a + b \cdot \varepsilon_a} \tag{6.2}$$

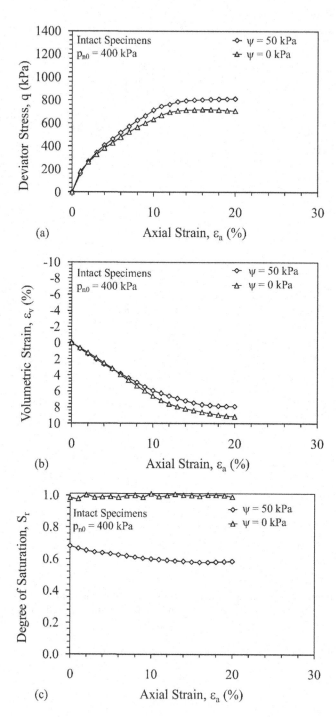

FIGURE 6.3 Variations in (a) deviatoric stress, (b) volumetric strain, and (c) degree of saturation of intact specimens subjected to an initial mean net stress of 400 kPa and matric suctions of matric suctions of 0, 50 and 400 kPa.[6]

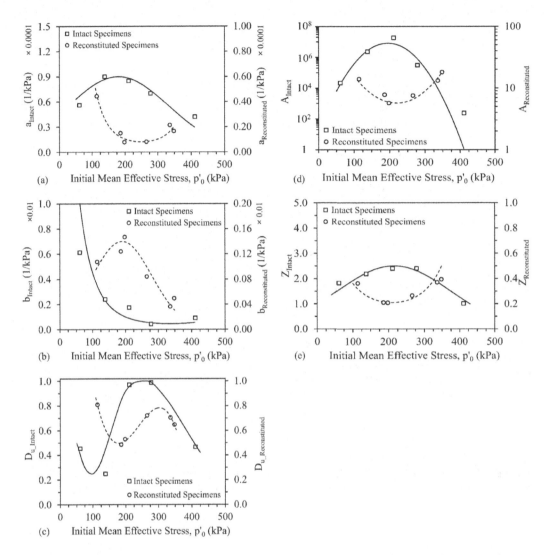

FIGURE 6.4 Variation of the material parameters against initial mean effective stress for intact and reconstituted specimens: (a) a; (b) b; (c) D_u; (d) A; and (e) Z.[6]

in which q and ε_a are deviator stress and axial strain at RI state, respectively, and a and b are the material parameters. With the values of deviator stress at RI (q_i) and FA (q_c), and the value of disturbance (D), the observed deviator stress, q_a, can be obtained as:

$$qa = qi - (qi - qc) \times D \tag{6.3}$$

Haeri et al.[6] used the shear stress–strain data, presented in Figures 6.1–6.3, to obtain the model parameters for the intact and reconstituted specimens under different mean net stresses and matric suctions. They treated the disturbance parameters a, b, A, and Z as fitting parameters and obtained the value of D_u from the ultimate FA state. Figure 6.4 presents the measured model parameters in terms of initial mean effective stress before shearing, p'_o, for both reconstituted and intact specimens. As presented in this figure, the values of model parameters may change depending on the

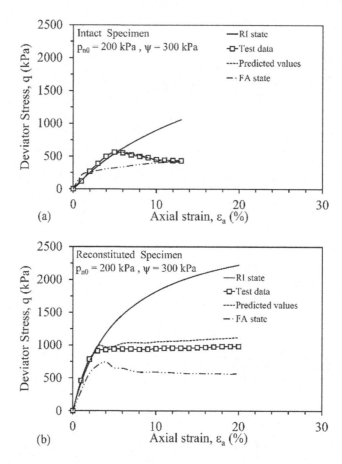

FIGURE 6.5 Relative intact (RI) and fully adjusted (FA) states along with predicted and experimental deviatoric stress values for intact and reconstituted specimens tested under $P_{no} = 200\,\text{kPa}$ and $\psi = 300\,\text{kPa}$.[6]

magnitude of applied effective stress or the structure of the soil specimen (i.e., being intact or reconstituted).

Using the model parameters presented in Figure 6.4 in Eqs. 6.1–6.3, the values of deviator stress for the tests conducted under the initial mean net stress of 300 kPa and the matric suction of 200 kPa were predicted for intact and reconstituted specimens, as shown in Figure 6.5. In this figure, the experimental data are presented as dots, while the predicted values of deviator stress and those at RI and FA states are presented as lines. The values of deviator stress at the FA state are obtained from the shearing tests conducted at saturated conditions (i.e., zero-suction tests). As shown in Figure 6.5, for both intact and reconstituted specimens, the measured experimental data are in an excellent agreement with those predicted using the DSC.

6.3 ONE-DIMENSIONAL STRESS–STRAIN BEHAVIOR OF LIME-STABILIZED UNSATURATED COLLAPSIBLE LOESSIAL SOILS

The influence of lime-stabilization on collapsible soils' behavior (especially loess) has been studied by many researchers. For instance, Roohparvar,[60] Noorzad and Pakniyat,[61] and Haeri et al.[22] have investigated the changes in the collapse potential of lime-stabilized loess specimens by performing single or double oedometer tests, under either saturated or unsaturated conditions. They reported that the addition of lime to the soil as well as increasing the curing time generally reduces the collapse potential upon wetting. Moreover, their research revealed that soil stabilization, even with low amounts of lime, can result in considerable decrease in the volume changes of the specimens before and after inundation, over a wide range of applied stresses.

Haeri et al.[22] conducted a series of oedometer tests on unsaturated, unstabilized, and lime-stabilized specimens to investigate the stress-deformation behavior of the Gorgan loess. Accordingly, they studied the oedometer behavior of unsaturated loess specimens treated with two different lime contents of 1% and 3%, and as for comparison, they also conducted oedometer tests on unsaturated, unstabilized specimens with 0% lime content. During these tests, water content and volume change of the specimens were continuously recorded. Besides, they also performed filter paper tests under both wetting and drying paths and calculated Soil-Water Retention Curve (SWRC) parameters for the examined collapsible loess.

On the basis of the experimental results reported by Haeri et al.,[22] a DSC-based framework was proposed by Garakani et al.[23] for evaluation of the stress–strain behavior of the lime-stabilized, unsaturated collapsible loessial soils.

6.3.1 Tested Material and Experimental Program

The soil examined by Haeri et al.[23] was collapsible Gorgan loess whose index parameters were previously presented in Table 6.1. As a stabilizing agent to treat the tested loessial soil, an ordinary type of calcium oxide (quicklime) with the chemical formula of CaO was utilized. The used lime was in the form of a white powder with a dry unit weight of 33.1 kN/m³.

Oedometer tests were carried out on the unsaturated specimens with 0%, 1%, and 3% lime contents by applying different constant matric suctions (i.e., $\psi=0$, 50, 100, 200, and 400 kPa) and imposing staged increasing net vertical stresses (i.e., $p_n=1$, 100, 200, 400, 800, 1600, and 2400 kPa) to the tested specimens.

To evaluate the soil water retention behavior of the examined lime-stabilized loess, filter paper tests were also carried out by Haeri et al.[22] in accordance with ASTM (American Society for Testing and Materials) D 5298-16.[62]

Details of the implemented oedometer device, filter paper tests, and the method of samples preparation can be found in Haeri et al.[22]

6.3.2 Results of Unsaturated Oedometer Tests on the Examined Lime-Stabilized Loess

In order to investigate the effect of addition of lime on the soil water retention behavior, three groups of filter paper tests were conducted by Haeri et al.[22] on unstabilized (i.e., $L=0\%$) and lime-stabilized loess specimens with 1% and 3% of lime content in both wetting and drying paths. Results are presented in Figure 6.6a–c as SWRCs in terms of the degree of saturation, S_r, versus matric suction, ψ. In order to quantitatively evaluate the SWRC curves, a model proposed by van Genuchten[63] was implemented for curve fitting to measured data from laboratory tests. van Genuchten's model is a smooth continuous function, which comprises three fitting parameters related to the soil behavior and is presented in Eq. 6.4:

$$S_r = \frac{S_{r,\psi o}}{\left(1+\left(a\times\psi\right)^n\right)^m}$$

(6.4)

where S_r is the degree of saturation, ψ is matric suction, $S_{r,\psi o}$ is the degree of saturation at zero matric suction, and a, m, and n are fitting parameters. For the tested specimens, fitting parameters in Eq. 6.4 have been determined for different lime contents by implementing the least square method along with applying visual judgment to gain the best rational fits and the results are depicted in Figure 6.6d.

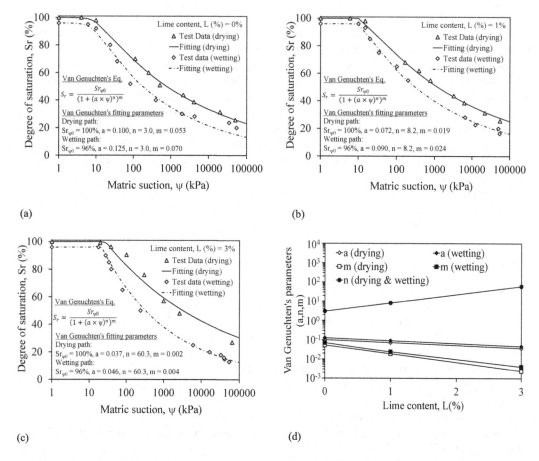

FIGURE 6.6 SWRC curves obtained from filter paper tests in wetting and drying paths for: (a) unstabilized specimens (*L*=0%), (b) lime-stabilized specimens (*L*=1%), (c) lime-stabilized specimens (*L*=3%), and (d) changes in SWRC parameters versus lime content, L.[22]

Figure 6.7 displays the variations of the void ratio, *e*, against the net vertical stress, p_n, for the lime-stabilized collapsible loessial specimens with different lime contents examined during unsaturated oedometer tests.

In preliminary studies on the behavior of unsaturated soils, the researchers used to try providing a unique equation to relate the total stress and the pore air and water pressures to the effective stress variable. In this regard, Bishop[64] suggested the effective stress formula for unsaturated soils by modifying Terzaghi's classical effective stress equation:

$$p' = (p - u_a) + c \cdot (u_a - u_w) \tag{6.5}$$

In Eq. 6.5, p' is effective stress, p is total stress, u_a is pore air pressure, u_w is pore water pressure, and χ is the effective stress parameter. Considering the difference between total stress and pore air pressure as net stress, p_n, and the difference between pore air and pore water pressures as matric suction, ψ, Eq. 6.5 can be rewritten in the form of Eq. 6.6:

$$p' = p_n + \chi \cdot \psi \tag{6.6}$$

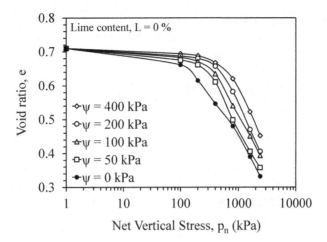

(a)

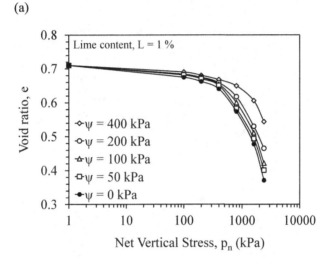

(b)

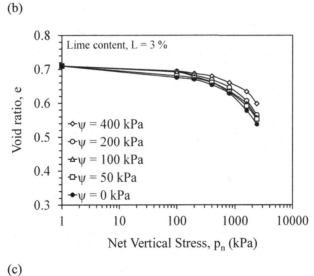

(c)

FIGURE 6.7 Changes in void ratio versus net vertical stress for: (a) unstabilized specimens (L=0%), (b) lime-stabilized specimens (L=1%), and (c) lime-stabilized specimens (L=3%).[22]

Several methods have been proposed so far to determine the value of χ. In this regard, taking the value of χ equal to the value of the effective degree of saturation, S_e, for loessial soils is suggested by several researchers,[4,7,8,20] as shown by Eq. 6.7:

$$S_e = \frac{S_r - S_{r,\text{res}}}{S_{r,\psi o} - S_{r,\text{res}}} \tag{6.7}$$

in which S_r is the degree of saturation, $S_{r,\psi o}$ is the degree of saturation at zero matric suction, and $S_{r,\text{res}}$ is the residual degree of saturation.

By considering data presented in Figures 6.6 and 6.7 and using Eqs. 6.6 and 6.7, Haeri et al.[22] calculated the variations of the effective vertical stress, p', against the vertical strain, ε, for the lime-stabilized collapsible loessial specimens with different lime contents, as shown in Figure 6.8.

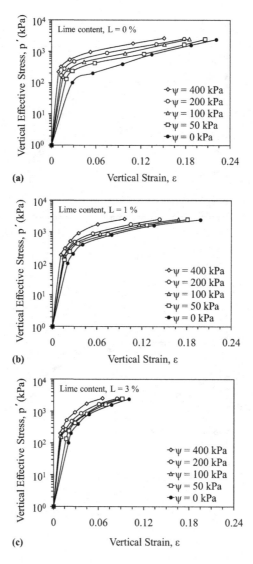

FIGURE 6.8 Changes in vertical effective stress versus vertical strain for: (a) unstabilized specimens ($L=0\%$), (b) lime-stabilized specimens ($L=1\%$), and (c) lime-stabilized specimens ($L=3\%$).[22]

6.3.3 IMPLEMENTING DSC FOR PREDICTION OF THE 1D STRESS–STRAIN BEHAVIOR OF LIME-STABILIZED LOESSIAL SOILS

Considering data reported by Haeri et al.,[22] a DSC-based framework was proposed by Garakani et al.[23] to demonstrate the stress–strain behavior of the unsaturated lime-stabilized loessial soils under oedometer loading conditions.

In accordance with the DSC, given the specific responses of the soil in the two reference states as P_{RI} and P_{FA}, the soil response in any arbitrary condition (observed state, P_a) can be obtained using Eq. 6.8:

$$P_a = P_{RI} - (P_{RI} - P_{FA}) \times D \qquad (6.8)$$

where the parameter D is the disturbance function. The disturbance function, D, in the DSC model, defines the coupling between the RI and FA states and represents the changes in the microstructure, which includes formation of microcracks, damage, and softening during the loading process. Actually, the disturbance parameter indicates the deviation of observed response from the two reference states. As previously shown by Eq. 6.1, D can be defined in the form of an exponential function based on some material response variables such as accumulated deviator plastic strain, ξ_D, and some fitting parameters, namely, A and Z, and the maximum disturbance parameter, D_u.

Garakani et al.[23] considered vertical effective stress versus vertical strain data provided in Figure 6.8 for unstabilized and lime-stabilized specimens to analyze hydro-mechanical behavior of the lime-stabilized loessial soils using DSC framework. Accordingly, by fitting Eq. 6.1 on the data provided in Figure 6.8, they calculated the constitutive parameters (i.e., D_u, A, Z) in terms of lime content for matric suctions of 50, 100, and 200 kPa, as illustrated in Figure 6.9. It is noted that for each lime content, the tests conducted in saturated (i.e., $\psi = 0$ kPa) and in the highest applied matric suction (i.e., $\psi = 400$ kPa) conditions are considered as the reference states of FA and RI, respectively.

As shown in Figure 6.9, by increasing the lime content, deviations in the constitutive parameters of the DSC model decrease and reach asymptotic states; hence, one can postulate that increasing lime content beyond a specific value (i.e., $L = 3\%$) would have no further significant effect on the deformation behavior of the tested collapsible soil.

According to the SWRC of the tested specimens (as displayed by Figure 6.6) and the D_u values shown in Figure 6.9a, Garakani et al.[23] developed a semi-empirical relationship for predicting the values of D_u as a function of matric suction, ψ, and degree of saturation, S_r, in the wetting path, as presented by Eq. 6.9:

$$D_u(\psi) = 1 - \left(\frac{S_{r\_RI}}{S_r(\psi)} \times \frac{\log(1+\psi)}{\log(1+\psi_{RI})} \right)^k \qquad (6.9)$$

where ψ (kPa) is matric suction, S_{rRI} and ψ_{RI} (kPa) are degree of saturation and corresponding matric suction at the RI state, respectively, and k is a dimensionless parameter whose value depends on the material properties. One feature of the proposed model by Garakani et al.[23] is that the obtained values of $D_u(\psi)$ for RI and FA states of the soil become zero (the least possible value) and unity (the maximum possible value), respectively. Furthermore, in Eq. 6.9, the disturbance (as a mechanical constitutive function) is expressed as a function of the degree of saturation (as a hydraulic behavior index of the soil), which represents a coupled correlation between mechanical and hydraulic (hydro-mechanical) behaviors. The measured and predicted values of disturbance parameter, D_u, in terms of matric suction, ψ, for stabilized loessial specimens with various lime contents are depicted in Figure 6.10a. In addition, variation of fitting parameter k versus lime percentage is shown in

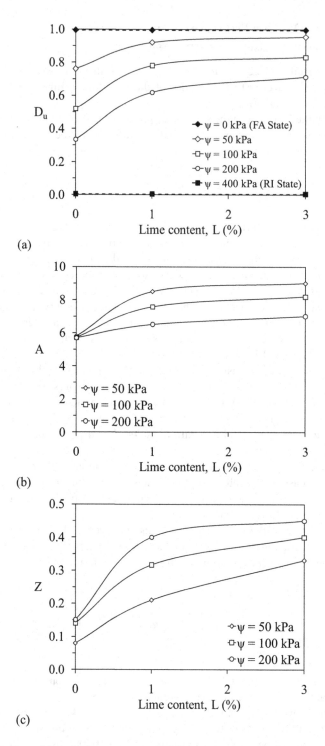

(a)

(b)

(c)

FIGURE 6.9 Variations of disturbed state concept (DSC) parameters versus lime content for different matric suctions: (a) D_u, (b) A, and (c) Z.[23]

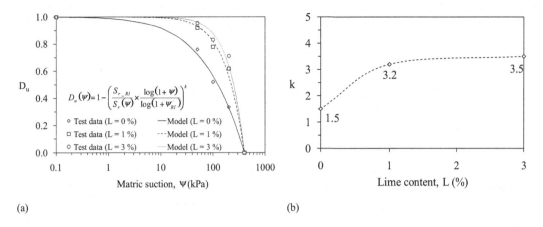

(a)   (b)

FIGURE 6.10 (a) Predicted and measured values of D_u versus matric suction for specimens with different lime contents and (b) variation of parameter k versus lime content.[23]

Figure 6.10b. As shown in Figure 6.10a, the proposed model shows a proper capability in predicting the disturbance parameter versus matric suction. Also, as can be seen in Figure 6.10b, variation in parameter k reduces as lime percentage increases, which means that the effect of adding lime to the soil is substantially reduced as it passes a certain value (i.e., $L=3\%$).

According to the concept of DSC and its constitutive parameters, which are calculated for the tested soil specimens (as shown in Figures 6.9 and 6.10), and considering Eqs. 6.8 and 6.9, the consolidation behavior of the soil specimens in the form of vertical effective stress versus vertical strain was predicted and plotted in Figure 6.11. To this end, nine sets of "observed data" were used for matric suctions of 50, 100, and 200 kPa and three lime contents of 0%, 1%, and 3%. The two reference states of FA and RI (corresponding to the matric suctions of zero and 400 kPa, respectively) alongside the obtained consolidation curves from tests and predicted results are also shown in Figure 6.11. As shown in Figure 6.11, for all specimens, the predicted values correlate very well with the measured ones obtained from the experimental tests.

Using data presented in Figure 6.11, the DSC model was used to calculate the effective yield stress value for soil samples ($p'_{c,\ DSC}$) based on the curves of vertical effective stress versus vertical strain. These values were then compared to the measured values ($p'_{c,\ test}$) obtained from unsaturated oedometer tests, as illustrated in Figure 6.12. The comparison was carried out on the nine sets of "observed data" (i.e., tests conducted under matric suctions of 50, 100, and 200 kPa and three lime contents of 0%, 1%, and 3%). In Figure 6.12, an identity line is also drawn to compare between the predicted and measured values.

Referring data presented in Figure 6.11, the overall normalized root mean square (NRMSE) values between the predicted and observed parameters were calculated as 4.1% implying desirable functionality of the DSC model in predicting the effective yield stress values for the examined unsaturated lime-stabilized loessial soil.

6.4 CONCLUSIONS

Two effective stress-based DSC models were introduced in this chapter for predicting the hydro-mechanical behavior of an unsaturated collapsible loess under triaxial shear and one-dimensional oedometer conditions. Accordingly, experimental data for the intact and the reconstituted specimens examined using triaxial shear tests were used, as well as load-displacement data for lime-stabilized specimens tested under oedometer conditions. In order to fully capture the effect of unsaturated parameters on the behavior of the studied soil, model parameters were defined as

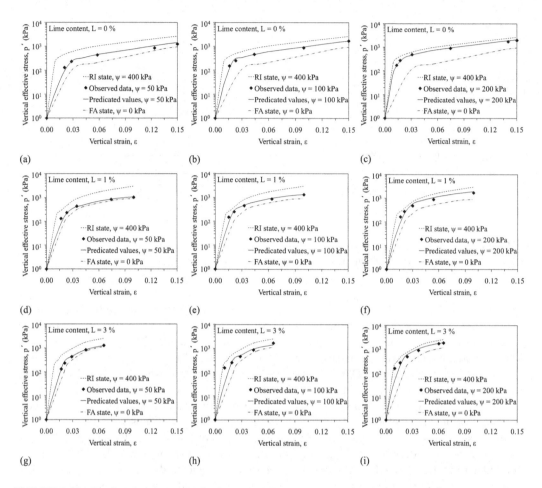

FIGURE 6.11 Predicted and measured values of vertical effective stress versus vertical strain for different values of matric suctions for: (a–c) unstabilized specimens ($L = 0\%$), (d–f) lime-stabilized specimens ($L = 1\%$), and (g–i) lime-stabilized specimens ($L = 3\%$).[23]

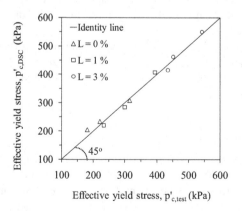

FIGURE 6.12 Comparison between model predicted, $p'_{c,\,DSC}$, and measured, $p'_{c,\,test}$, values of the effective yield stress.[23]

functions of unsaturated material and state variables, such as lime content and matric suction of the soil. Comparisons were made between the corresponding model predicted and experimental data, which reveal the excellent functionality of the proposed DSC models in predicting the deviator stress versus axial strain in triaxial tests, as well as the effective stress versus vertical strain and also the effective yield stress values in oedometer tests, for the examined collapsible loessial specimens.

REFERENCES

1. Peck, R. B., Hanson, W. E., and Thornburn, T. H. *Foundation Engineering*. 2nd Ed., John Wiley & Sons, New York, 1974.
2. Estatiev, D. "Loess improvement methods." *Eng. Geol.*, 25, 1988, 341–366. https://doi.org/10.1016/0013-7952(88)90036-1.
3. Garakani, A. A. "Laboratory assessment of the hydro-mechanical behavior of unsaturated undisturbed collapsible soils—Case study: Gorgan loess." Dissertation for Doctoral Degree, Sharif University of Technology, Tehran, Iran, 2013.
4. Haeri, S. M., Garakani, A. A., Khosravi, A., and Meehan, C. L. "Assessing the hydro-mechanical behavior of collapsible soils using a modified triaxial test device." *ASTM, Geotech. Test. J.*, 37, 2, 2014, 190–204. https://doi.org/10.1520/GTJ20130034.
5. Haeri, S. M., Khosravi, A., Ghazizadeh, S., Garakani, A. A., and Meehan, C. "Characterization of the effect of disturbance on the hydro-mechanical behavior of a highly collapsible loessial soil." In *Proceedings of 6th International Conference on Unsaturated Soils*, Taylor & Francis Group, London, 2014.
6. Haeri, S. M., Khosravi, A., Garakani, A. A., and Ghazizadeh, S. "Effect of soil structure and disturbance on hydromechanical behavior of collapsible loessial soils." *ASCE, Int. J. Geomech.*, 17, 1, 2017, 04016021. https://doi.org/10.1061/(ASCE)GM.1943-5622.0000656.
7. Garakani, A. A., Haeri, S. M., Khosravi, A., and Habibagahi, G. "Hydro-mechanical behavior of undisturbed collapsible loessial soils under different stress state conditions." *Eng. Geol.*, 195, 2015, 28–41. https://doi.org/10.1016/j.enggeo.2015.05.026.
8. Garakani, A. A., Haeri, S. M., Cherati, D. Y., Givi, F. A., Tadi, M. K., Hashemi, A. H., Chiti, N., and Qahremani, F. "Effect of road salts on the hydro-mechanical behavior of unsaturated collapsible soils." *Transp. Geotech.*, 17, 2018, 77–90. https://doi.org/10.1016/j.trgeo.2018.09.005.
9. Ziyaee, A., Pashaei, A., Khormali, F., and Roshani, M. R. "Some physico-chemical, clay mineralogical and micro morphological characteristics of loess-paleosols sequences indicators of climate change in south of Gorgan." *J. Water Soil Cons.*, 20, 1, 2013, 1–27. https://www.sid.ir/en/journal/ViewPaper.aspx?id=306736.
10. Lawton, E. C., Fragaszy, R. J., and Hetherington, M. D. "Review of wetting-induced collapse in compacted soil." *Int. J. Geotech. Eng.*, 118, 9, 1992, 1376–1393. https://doi.org/10.1061/(ASCE)0733-9410(1992)118:9(1376).
11. Dudley, J. H. "Review of collapsing soils." *Soil Mech. Found. Eng. Div.*, 97, SM1, 1970, 925–947. https://doi.org/10.1061/JSFEAQ.0001426.
12. Barden, L., Madedor, A. O., and Sides, G. R. "Volume change characteristics of compacted clays." *Soil Mech. Found. Eng. Div.*, 96, SM4, 1969, 33–39. https://doi.org/10.1061/JSFEAQ.0001226.
13. Mitchell, J. K. *Fundamentals of Soil Behavior*. Wiley, New York, 1976.
14. Khalili, N., and Khabbaz, M. H. "A unique relationship for the determination of the shear strength of unsaturated soils." *Géotechnique*, 48, 5, 1998, 681–687. https://doi.org/10.1680/geot.1998.48.5.681.
15. Wheeler, S. J., Sharma, R. S., and Buisson, M. S. R. "Coupling of hydraulic hysteresis and stress–strain behaviour in unsaturated soils." *Géotechnique*, 53, 1, 2003, 41–54. https://doi.org/10.1680/geot.2003.53.1.41.
16. Gallipoli, D., Gens, A., Sharma, R., and Vaunat, J. "An elasto-plastic model for unsaturated soil incorporating the effects of suction and degree of saturation on mechanical behaviour." *Géotechnique*, 53, 1, 2003, 123–136. https://doi.org/10.1680/geot.2003.53.1.123.
17. Lu, N., and Likos, W. J. "Suction stress characteristic curve for unsaturated soil." *ASCE, J. Geotech. Geoenviron.*, 132, 2, 2006, 131–142. https://doi.org/10.1061/(ASCE)1090-0241(2006)132:2(131).
18. Nuth, M., and Laloui, L. "Effective stress concept in unsaturated soils: Clarification and validation of a unified framework." *Int. J. Numer. Anal. Met.*, 32, 7, 2008, 771–801. https://doi.org/10.1002/nag.645.

19. Lu, N., Godt, J., and Wu, D. "A closed-form equation for effective stress in unsaturated soil." *Water Resour.*, 46, 2010, W05515. https://doi.org/10.1029/2009WR008646.

20. Haeri, S. M., and Garakani, A. A. "Hardening behavior of a hydro collapsible loessial soil." *Jpn. Geotech. Soc. Spec. Publ.*, 2(4), 2016, 253–257. https://doi.org/10.3208/jgssp.IRN-02.

21. Haeri, S. M., Garakani, A. A., and Beigi, M. "A hydromechanical model for the unsaturated behavior of lime-stabilized collapsible soils." In *19th International Conference on Soil Mechanics and Geotechnical Engineering (19th ICSMGE)*, Seoul, Korea, 2017.

22. Haeri, S. M., Garakani, A. A., Roohparvar, H. R., Desai, C. S., Seyed Ghafouri, S. M. H., and Kouchesfahani, K. S. "Testing and constitutive modeling of lime-stabilized collapsible loess. I: Experimental investigations." *ASCE, Int. J. Geomech.*, 19, 4, 2019, 4019006. https://doi.org/10.1061/(ASCE)GM.1943-5622.0001364.

23. Garakani A. A., Haeri. S. M., Desai, C. S., Seyed Ghafouri, S. M. H., Sadollahzadeh, B., and Hashemi Senejani, H. "Testing and constitutive modeling of lime-stabilized collapsible loess. II: Modeling and validations." *ASCE, Int. J. Geomech.*, 19, 4, 2019, 04019007. https://doi.org/10.1061/(ASCE)GM.1943-5622.0001386.

24. Desai, C. S. "A consistent finite element technique for work-softening behavior of geologic media." In *Proceedings of the International Conference on Computational Methods in Nonlinear Mechanics.* Edited by Oden J.T., Univ. of Texas, Austin, Texas, 1974.

25. Desai, C. S., "A consistent numerical technique for work softening behavior of geologic media." In *Proceedings of the International Conference on Computational Methods in Nonlinear Mechanics.* Edited by Oden J.T., Univ. of Texas, Austin, Texas, 1974.

26. Desai, C. S. "Finite element residual schemes for unconfined flow." *Int. J. Numer. Methods. Eng.*, 10, 6, 1976, 1415–1418. https://doi.org/10.1002/nme.1620100622.

27. Desai, C. S. "Hierarchical single surface and the disturbed state constitutive models with emphasis on geotechnical application." *In Geotechnical Engineering: Emerging Trend in Design and Practice*, Edited by Saxeng K.R. Oxford IBH Publishing Company, New Delhi, 32, 3, 1994, 115–154. https://doi.org/10.1016/0148-9062(95)90125-o.

28. Desai, C. S. "Constitutive modeling using the disturbed state as a microstructure self - adjustment concept." *In* Chapter 8, *Continuum Models for Material with Microstructure*, Edited by Mühlhaus H.B. John Wiley and Sons, London, 1995.

29. Desai, C. S. "Evaluation of liquefaction using disturbed state and energy approaches." *J. Geotech, Geoenviron.*, 126, 7, 2000, 618–631. https://doi.org/10.1061/(ASCE)1090-0241(2000)126:7(618).

30. Desai, C. S. *Mechanics of Materials and Interfaces: The Disturbed State Concept*, CRC Press, Boca Raton, FL, 2001. https://doi.org/10.1201/9781420041910.ch2.

31. Desai, C. S. "Disturbed state concept (DSC) for constitutive modeling of geologic materials and beyond." In *Proceedings of the Symposium on Constitutive Modeling by Geomaterials*, Tsinghua Univ., Beijing, 2013, 27–45.

32. Desai, C. S. "Constitutive modeling of materials and contacts using the disturbed state concept. Part 1: Background and analysis." *J. Comput. Struct.*, 146, 2015, 214–233. https://doi.org/10.1016/j.jrmge.2016.01.003.

33. Desai, C. S. "Constitutive modeling of materials and contacts using the disturbed state concept. Part 2: Validations at specimen and boundary value problem levels." *J. Comput. Struct.*, 146, 2015, 234–251. https://doi.org/10.1016/j.jrmge.2016.01.003.

34. Desai, C. S. "Disturbed state concept as unified constitutive modeling approach." *J. Rock Mech. Geotech. Eng.*, 8, 3, 2016, 277–293. https://doi.org/10.1016/j.jrmge.2016.01.003.

35. Desai, C. S., and Toth, J. "Disturbed state constitutive modeling based on stress-strain and nondestructive behavior." *J. Solids. Struct.*, 33, 11, 1996, 1619–1650. https://doi.org/10.1016/0020-7683(95)00115-8.

36. Desai, C. S., and Zhang, W. "Computational aspects of disturbed state constitutive models." *Comput. Methods. Appl. Mech. Eng.*, 151, 3–4, 1998, 361–376. https://doi.org/10.1016/S0045-7825(97)00159-X.

37. Desai, C. S., Basaran, C., and Zhang, W. "Numerical algorithms and mesh dependence in the disturbed state concept." *Int. J. Numer. Methods. Eng.*, 40, 1997. https://doi.org/10.1002/(SICI)1097-0207(19970830)40:16<3059::AID-NME182>3.0.CO;2-S.

38. Garakani, A. A., Pirjalili, A. and Desai, C. S. "An effective stress-based DSC model for predicting the coefficient of lateral soil pressure in unsaturated soils." *Acta Geotechnica*, 2021, 1–18. https://doi.org/10.1007/s11440-021-01376-6.

39. Armaleh, S. H., and Desai, C. S. "Modelling and testing of a cohesionless material using the disturbed state concept." *J. Mech. Behav. Mater.*, 5, 3, 1994, 279–296. https://doi.org/10.1515/JMBM.1994.5.3.279.

40. Loizos, A., Partl, M. N., Scarpas, T., and Al-Qadi, I. L. "3D finite element modeling of polymer modified asphalt base course mixes." In *Advanced Testing and Characterization of Bituminous Materials, Two Volume Set*, Edited by Loizos A., Partl M.N., Scarpas, T., Al-Qadi I.L. CRC Press, Boca Raton, FL, 2009, 971–980.

41. Vallego, M. J., and Tarefdar, R. A. "Predicting failure behavior of polymeric composites using a unified constitutive model." *J. Mech.*, 27, 3, 2011, 379–88. https://doi.org/10.1017/jmech.2011.40.

42. Akhaveissy, A. H., Desai, C. S., Mostofinejad, D., and Vafai, A. "FE analysis of RC structure using DSC model with yield surfaces for tension and compression." *Comput. Concr.*, 11, 2, 2013, 123–48. https://doi.org/10.12989/cac.2013.11.2.123.

43. Ouria, A. "Disturbed state concept–based constitutive model for structured soils." *ASCE, Int. J. Geomech.*, 17, 7, 2017, 04017008. https://doi.org/10.1061/(ASCE)GM.1943-5622.0000883

44. Geiser, F., Laloui, L., and Vulliet, L. "Constitutive modelling of unsaturated sandy silt." In *IACMAG, 9th International Conference on Computer Methods and Advances in Geomechanics*, Wuhan, China, 2, LMS-CONF-2006-023, 1997, 899–904.

45. Geiser, F., Laloui, L., Vulliet, L., and Desai, C. S. "Disturbed state concept for constitutive modeling of partially saturated porous materials." *In Proceedings of the 6th International Symposium on Numerical Models in Geomechanics*, Balkema, Rotterdam, Netherlands, 1997, 129–134.

46. Zhou, C., Shen, Z. J., Chen, S. S., and Chen, T. L. "A hypoplasticity disturbed state model for structured soils." *Chinese J. Geotech. Eng.*, 26, 4, 2004, 435–439.

47. Park, I. J., and Desai, C. S. "Cyclic behavior and liquefaction of sand using disturbed state concept." *ASCE, J. Geotech. Geoenv. Eng.*, 126, 9, 2000, 834–846. https://doi.org/10.1061/(ASCE)1090-0241(2000)126:9(834).

48. Katti, D. R., and Desai, C. S. "Modeling and testing of cohesive soil using disturbed-state concept." *J. Eng. Mech.*, 121, 5, 1995, 648–658. https://doi.org/10.1061/(ASCE)0733-9399(1995)121:5(648).

49. Ouria, A., Desai, C. S., and Toufigh, V. "Disturbed state concept–based solution for consolidation of plastic clays under cyclic loading." *ASCE, Int. J. Geomech.*, 15, 1, 2013, 04014039. https://doi.org/10.1061/(ASCE)GM.1943-5622.0000336.

50. Liu, M. D., Carter, J. P., and Desai, C. S. "Modeling compression behavior of structured geomaterials." *ASCE, Int. J. Geomech.*, 3, 2, 2003, 191–204. https://doi.org/10.1061/(ASCE)1532-3641(2003)3:2(191).

51. Liu, W. Z., Shi, M. L., and Miao, L. C. "Analysis of compressibility of structural soils based on disturbed state concept." *Rock Soil Mech.*, 31, 11, 2010, 3475–3480.

52. Minh N. H., Suzuki K., Oda M., Tobita, T., and Desai, C. S. "Numerical simulation using disturbed state concept (DSC) model for softening behavior of sand." *J. Southeast Asian Geotech. Soc.*, 34, 1, 2008, 25–35.

53. Kalantary, F., Yazdi, J. S., Bazazzadeh, H. "Validation and application of evolutionary computational techniques on disturbed state constitutive model." *Int. J. Civ. Eng.*, 12, 3, 2014, 216–224. http://ijce.iust.ac.ir/article-1-845-en.html.

54. ASTM D2488. "Standard Practice for Description and Identification of Soils (visual-manual procedure." ASTM, West Conshohocken, 2000.

55. Pashaei, A., "Study of physical and chemical characteristics and the source of loess deposits in gorgan and plain region." *Earth Sci. J.*, 23, 1997, 67–78. Iranian Geology Organization.

56. Feiznia, S., Ghauomian, J., and Khajeh, M. "The study of the effect of physical, chemical and climate factors on surface erosion sediment yield of loess soils (case study in Golestan Province)." *J. Pajouhesh Sazandegi.*, 66, 2005, 14–24. https://www.sid.ir/en/Journal/ViewPaper.aspx?ID=67819.

57. Frechen, M., Kehl, M., Rolf, C., Sarvati, R., and Skowronek, A. "Loess chronology of the Caspian lowland in northern Iran." *Quat. Int.*, 198, (1–2), 2009, 220–233. https://doi.org/10.1016/j.quaint.2008.12.012.

58. Ghazizadeh, Sh. "Assessment of the Hydro-Mechanical and Shear Behavior of Collapsible Soils by Using Suction-Controlled Triaxial Apparatus – Case Study: Loess of Gorgan." Dissertation for the MSc Degree, Sharif University of Technology, Tehran, Iran, 2014.

59. Konder, R. L. "Hyperbolic stress-strain response: Cohesive soils." *J. Soil Mech. Found. Div.*, 89, SM1, 1963, 115–143. https://doi.org/10.1061/JSFEAQ.0000479.

60. Roohparvar, H. R. "Evaluation of improvement of collapsible soils with adding lime in conventional and suction controlled oedometer tests - case study of Gorgan loess". Dissertation for the MSc Degree, Sharif University of Technology, Tehran, Iran, 2012.

61. Noorzad, R., and Pakniat, H. "Investigating the effect of sample disturbance, compaction and stabilization on the collapse index of soils." *Env. Earth Sci.*, 75, 18, 2016, 1262. https://doi.org/10.1007/s12665-016-6073-8.

62. ASTM D5298-16. "Standard Test Method for Measurement of Soil Potential (Suction) Using Filter Paper." In *ASTM International*, West Conshohocken, PA, 2016.

63. van Genuchten, M. T. "A closed-form equation for predicting the hydraulic conductivity of unsaturated soils." *Soil Sci. Soc. Am. J.*, 44, 5, 1980, 892–898. https://doi.org/10.2136/sssaj1980.0361599500440005 0002x.

64. Bishop, A. W. "The principle of effective stress." *Teknisk Ukeblad*, 39, 1959, 859–863.

7 Applications of DSC in Modelling Strain-Softening Stress–Strain Relationship of Unsaturated Soils

Xiuhan Yang and Sai K. Vanapalli
Department of Civil Engineering, University of Ottawa

7.1 INTRODUCTION

Soil shear stress can increase until a peak state is reached and then decrease gradually to a lower ultimate value with increasing shear deformation. This behaviour is referred to as strain-softening behaviour of soils [1]. The strain-softening behaviour of saturated soils has been well recognized in overconsolidated clays and dense sands. During the last 25 years, several laboratory investigations (e.g., [2–8]) were undertaken to investigate the strain-softening behaviour of various types of unsaturated soils. These studies highlight the significant influence of suction on the strain-softening behaviour of unsaturated soils (as shown in Figure 7.1). The soils that are in a state of unsaturated condition exhibit significant strain-softening behaviour in comparison to the saturated soils. This phenomenon can be attributed to the destruction of water menisci among soil particles or aggregates under continuous shearing beyond the peak state [7]. As a result, the contribution of suction to the shear strength at the critical state is often lower than that at the peak state. In compacted fine-grained soils, such behaviour can also be explained based on the aggregated fabric [8]. When fine-grained soils are compacted at the dry side of optimum, the fine particles in the soils can be aggregated to form soil packets. In an unsaturated condition, suction can prevent the breakdown of the soil packets even during shearing. In other words, the soil packets during the shearing stage can still retain their aggregated form and act as large individual particles. This phenomenon makes the soils behave in a more granular way (i.e., greater brittleness and dilatancy) than would be justified by its grading.

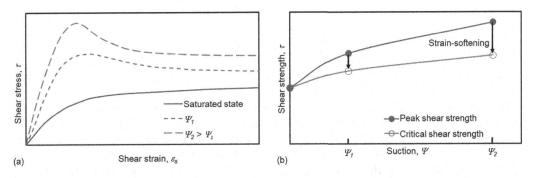

FIGURE 7.1 Typical shearing behaviours of unsaturated soils: (a) stress–strain curve and (b) shear strength envelops.

DOI: 10.1201/9781003362081-9

The strain-softening behaviour plays an important role in the rational interpretation of the mechanical behaviour of slopes in unsaturated soils. In slopes that are in a state of unsaturated condition, several environmental factors can contribute to relatively large shear deformation in local zones; for example, the rainfall infiltration on slope surface [9,10], the seasonal wetting–drying cycles of soils in slope [11] and the variation of the water level in nearby river [12]. Once relatively large local deformations are generated, the shear strength of unsaturated soils in the corresponding zones will be reduced due to the strain-softening behaviour. As a result, the shear stress exceeding the post-peak shear strength will be redistributed to the surrounding zones. Such behaviour contributes to a progressive expansion of the large shear deformation in the slope until the failure condition is reached. The soils in the slip zone will experience a reduction in shear strength and, consequently, the factor of safety of the slope will decrease. Therefore, the strain-softening behaviour and its associated progressive failure should be taken into account in the rigorous analyses and design of slopes in unsaturated soils.

The strain-softening behaviour of unsaturated soils can be modelled extending different types of approaches; for example, the Cam-clay-type model [13] and the bounding surface plasticity model [14]. In addition to these approaches, the disturbed state concept (DSC)-based model is another approach that is receiving attention because of its simplicity in modelling complex behaviours of a wide range of engineering materials. The strain-softening behaviour of saturated soils and interfaces has been simulated by various DSC-based models [15–24]. However, its application in modelling the behaviours of unsaturated soils is still emerging [25–28].

7.2   DSC-BASED MODEL

According to DSC, the soil under an applied load can be assumed to gradually deform simultaneously undergoing microstructural changes and reaches a "disturbed" state. Consequently, the deforming soil transforms gradually from the relative intact (RI) state to the fully adjusted (FA) state (Figure 7.2a). The RI state is defined as the initial condition where the soil is not influenced by the disturbance, and the FA state is defined as the final condition where the soil is fully disturbed [21]. At any intermediate stage during the deforming process, a representative soil element can be considered to be in a combined random mixture of zones in RI and FA states. Therefore, the apparent response of the soil element under an applied load can be expressed as a weighted average of the responses of the zones in RI and FA states as follows:

$$\sigma_{ij}^{a} = (1-D)\sigma_{ij}^{RI} + D\sigma_{ij}^{FA} \tag{7.1a}$$

$$\sigma_{ij}^{a} = (1-D)C_{ijkl}^{RI}\varepsilon_{kl}^{RI} + DC_{ijkl}^{FA}\varepsilon_{kl}^{FA} \tag{7.1b}$$

where σ_{ij}^{a} is the apparent stress tensor; σ_{ij}^{RI} and σ_{ij}^{FA} are the stress tensors of soils in RI and FA states, respectively; ε_{kl}^{RI} and ε_{kl}^{FA} are the strain tensors of soils in RI and FA states, respectively; C_{ijkl}^{RI} and C_{ijkl}^{FA} are the constitutive matrices of soils in RI and FA states, respectively; and D is the disturbance function.

For calculating the apparent response of the soil element (i.e., solid line in Figure 7.2a), the responses of soils in two reference (RI and FA) states (i.e., long and short dash line in Figure 7.2a) should be determined based on the definition of the disturbance using suitable constitutive models for the two reference states to determine C_{ijkl}^{RI} and C_{ijkl}^{FA} in Eq. (7.1b). Different DSC-based models can be formulated by using different constitutive models for the two reference states for capturing various behaviours of materials. D is defined as the ratio of the volume of the zones in FA states to the total volume of the soil element. It is used to represent the degree of disturbance that occurs in

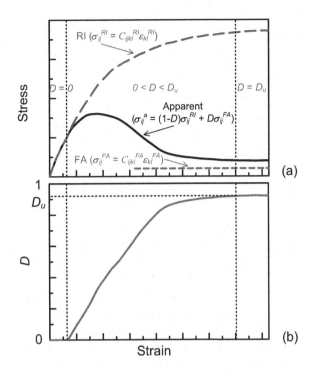

FIGURE 7.2 Schematic of disturbed state concept (DSC)-based model: (a) stress–strain curve and (b) disturbance function curve.

the soil element. A greater value of D means a greater percentage of the zones in the FA state in the soil element (i.e., greater disturbance). During the deforming process, the value of D increases from 0 when the entire element is in the RI state to a value equal to 1 when the entire element transforms to the FA state (Figure 7.2b). Desai [21] suggested two functions to relate D to an internal variable that can reflect the influence of microstructural changes resulting in the disturbance.

$$D = D_u \left[1 - \left\{ 1 + \left(\frac{\xi_D}{h} \right)^w \right\}^{-s} \right] \tag{7.2a}$$

$$D = D_u \left[1 - \exp\left(-A \xi_D^Z \right) \right] \tag{7.2b}$$

where $\xi_D = \int \left(dE_{ij}^p dE_{ij}^p \right)^{1/2}$ is the deviatoric plastic strain trajectory; E_{ij}^p is the deviatoric part of the plastic strain tensor; h, w, s and A, Z are parameters in the disturbance function; D_u is the ultimate value of D in practice, which is less than 1 since FA can only be attained in an idealized condition.

A commonly used DSC-based approach is the disturbed state concept-hierarchical single surface (DSC-HISS) model developed by Desai [21]. In this model, the RI state behaviour is assumed to be the elastoplastic hardening response, which can be simulated by using the HISS model. Different versions of the HISS model were proposed by Desai [21] that can capture the isotropic and anisotropic hardening and associated and non-associated plasticity characterizations of the soil behaviour. For example, the yield surface, flowing rule and hardening law of the non-associated HISS-δ1 model are presented as follows:

$$F = \frac{J_{2D}}{p_a^2} - \left[-\alpha \left(\frac{J_1 + R}{p_a} \right)^n + \gamma \left(\frac{J_1 + R}{p_a} \right)^2 \right] \left(1 - \beta \frac{\sqrt{27} J_{3D}}{2 J_{2D}^{1.5}} \right)^{-0.5} = 0 \qquad (7.3a)$$

$$Q = \frac{J_{2D}}{p_a^2} - \left[-\alpha_Q \left(\frac{J_1 + R}{p_a} \right)^n + \gamma \left(\frac{J_1 + R}{p_a} \right)^2 \right] \left(1 - \beta \frac{\sqrt{27} J_{3D}}{2 J_{2D}^{1.5}} \right)^{-0.5} = 0 \qquad (7.3b)$$

$$\alpha = \frac{a_1}{\xi^{\eta_1}}, \quad \alpha_Q = \alpha + \kappa (\alpha_0 - \alpha)(1 - \xi_v / \xi) \qquad (7.3c)$$

where J_1 is the first invariant of the total stress tensor; J_{2D} and J_{3D} are the second and third invariants of the deviatoric stress tensor, respectively; $\xi = \int \left(d\varepsilon_{ij}^p d\varepsilon_{ij}^p \right)^{1/2}$ is the plastic strain trajectory; $\xi_v = |\varepsilon_{ii}^p| / \sqrt{3}$ is the volumetric plastic strain trajectory; p_a is the atmospheric pressure; γ, β, n, R, a_1, η_1, and κ are material parameters; and α_0 is the value of α at the end of initial (hydrostatic) loading.

In the DSC-HISS model, the FA state behaviour is assumed to be the response of soils at the critical state where the original structure of the soils is considered to be fully destroyed. The critical state response of the soils can be simulated by using Eq. (7.4).

$$\sqrt{J_{2D}} = \bar{m} J_1 \qquad (7.4a)$$

$$e = e_0^c - \lambda \ln \left(J_1 / 3 p_a \right) \qquad (7.4b)$$

where e is the void ratio; e_0^c is the value of e on the critical state line corresponding to $J_1 = 3p_a$; and $\bar{m}$ and λ are parameters associated with the critical state line.

Therefore, the constitutive matrices, C_{ijkl}^{RI} and C_{ijkl}^{FA}, in Eq. (7.1) can be derived using Eqs. (7.3) and (7.4), respectively. Either Eq. (7.2a) or Eq. (7.2b) can be used to describe the disturbance function, D, in Eq. (7.1). Eventually, the DSC-HISS model can be formulated by combining Eq. (7.1)–(7.4).

7.3 APPLICATIONS OF DSC IN MODELLING STRAIN-SOFTENING BEHAVIOUR OF SATURATED SOILS

The DSC-based model is relatively simple because it avoids using the advanced soil plasticity concepts and its approach for explaining the strain-softening behaviour is based on limited and well-defined parameters. The DSC can be conveniently accommodated into different existing reliable constitutive models to capture more complex behaviours of the engineering materials. The simplicity and flexibility of the DSC has encouraged various researchers to develop DSC-based models to capture several behaviours (e.g., creep, strain-softening, collapse, dynamics, fracture, and healing) of a wide range of materials and interfaces. These studies have been well reviewed by Desai [29–31]. In this section, the applications of DSC-based models in simulating the strain-softening behaviour of saturated soils will be reviewed briefly.

The DSC-HISS model discussed in the last section has been successfully applied by Desai and co-workers to model the strain-softening behaviour of saturated sandy soils [15–17] and saturated interfaces [18,19]. In addition, Xiao and Desai [20] proposed a DSC-based model to capture the strain-softening behaviour of saturated, overconsolidated clays. This model was formulated by incorporating a new disturbance function (Eq. 7.5) that takes account of the effects of overconsolidation ratio into the potential failure and dilatancy surfaces at the current and referenced states.

$$D_{OCR} = 1 - \left(\frac{1}{O_\delta} \right)^\zeta \tag{7.5}$$

where D_{OCR} is the disturbance function for overconsolidated clays; O_δ is the ratio of the preconsolidated stress to the current consolidated stress; and ζ is a material parameter.

In addition to these models, there are several other types of DSC-based models that can capture the strain-softening behaviour of different saturated materials [22–24] using different constitutive models for the two reference states. Hu and Pu [22] developed a DSC-based model for saturated soil–structure interface. In this model, the RI state behaviour is assumed to be hyperbolic elastoplastic hardening response. The FA state behaviour is assumed to be the critical state response of interfaces. Eq. (7.2b) is used as the disturbance function. The Hu and Pu model [22] was successfully applied to predict the experimental results of direct and simple shear tests on saturated soil–structure interfaces. The model was further incorporated in the finite element method (FEM) and satisfactorily modelled two soil–structure interaction problems that are of interest in practice in a simple way [22].

Seo et al. [23] developed a DSC-based model for saturated geosynthetic interfaces that exhibit strain-softening behaviour under large shear displacement. This model used elastic-perfectly plastic response as RI state behaviour, critical state response as FA state behaviour and Eq. (7.2b) as disturbance function. The shear strength behaviour of geosynthetic interfaces over a large displacement of 100 mm was successfully simulated using the model [23]. The other combined model was also proposed [23] for capturing the strain-softening behaviour of saturated geosynthetic interfaces by using two hyperbolic equations to model the pre-peak and post-peak stress–strain curve, respectively. However, comparisons between the two approaches (i.e., DSC-based model and combined model) suggest that DSC-based model is more suitable for modelling large-displacement shearing behaviour of saturated geosynthetic interfaces because it can be implemented into FEM more conveniently.

Besides the models assuming an elastoplastic response in the RI state and critical state response in the FA state, Veiskarami et al. [24] developed another DSC-based model for granular materials based on different assumptions. In this model, the RI state behaviour is assumed to be nonlinear elastic response described by hyperbolic stress–strain relationship and the FA state behaviour is assumed to be elastoplastic hardening response. The model [24] can be used in simulating the behaviour of granular materials that is influenced by stress level and shear stress ratio.

7.4 APPLICATIONS OF DSC IN MODELLING STRAIN-SOFTENING BEHAVIOUR OF UNSATURATED SOILS

7.4.1 EXTENDING DSC TO ELASTOPLASTIC MODELS OF UNSATURATED SOILS

Unsaturated soils typically exhibit significant strain-softening and dilative behaviour during shearing process. Due to this reason, simple and flexible DSC-based approaches that can capture the complex behaviour of various engineering materials will be of significant interest in engineering practice applications.

Geiser et al. [25] extended the HISS-δ1 model (i.e., Eq. 7.3) to develop a constitutive model (HISS-δ1$_{unsat}$) for unsaturated soils taking account of the influence of suction. The main modifications include: (i) the suction, which is defined as the difference between pore air pressure and pore water pressure ($u_a - u_w$), was considered as an independent variable and J_1 in Eq. (7.3) was replaced by the first invariant of the saturated effective stress tensor, J_1'; (ii) a mechanical yield surface (Eq. 7.6a) was developed from the saturated HISS-δ1 model (Eq. 7.3) to describe the soil yield in the $p' - q$ plane at constant suction by assuming two material parameters in Eq. (7.3) (i.e., a_1 and R) as non-linear functions of suction (Eq. 7.6d); and (iii) a hydric yield surface (Eq. 7.6e) was proposed to describe the change of ξ due to the change of suction under constant p' when suction is less than the air entry value. Thus, the constitutive model (HISS-δ1$_{unsat}$) for unsaturated soils can be derived based on Eq. (7.6a)–(7.6e). However, the prediction results of the stress–strain curves of an unsaturated silt using the HISS-δ1$_{unsat}$ model show that the strain-softening behaviour cannot be captured [25]. Due to this reason, the HISS-δ1$_{unsat}$ model was extended based on the DSC approach. In the DSC-HISS-δ1$_{unsat}$ model, the RI state behaviour is assumed to be elastoplastic hardening response, which can be simulated using the HISS-δ1$_{unsat}$ model. The FA state corresponds to a saturated state with a modified initial saturated effective mean stress, $p'_{0,m}=(p'_0 - u_a)+S(u_a - u_w)$, where S is the degree of saturation. The disturbance function (Eq. 7.6f) was modified from Eq. (7.2b) by assuming the parameters, A and Z, in Eq. (7.2b) as functions of suction. The corresponding functions, $A(s)$ and $Z(s)$, in Eq. (7.6f) were determined by fitting experimental results.

$$F_1 = \frac{J_{2D}}{p_a^2} - \left[-\alpha(s)\left(\frac{J_1' + R(s)}{p_a}\right)^n + \gamma\left(\frac{J_1' + R(s)}{p_a}\right)^2\right]\left(1-\beta\sqrt{\frac{27}{2}}\frac{J_{3D}}{J_{2D}^{1.5}}\right) = 0 \qquad (7.6a)$$

$$Q_1 = \frac{J_{2D}}{p_a^2} - \left[-\alpha_Q(s)\left(\frac{J_1' + R(s)}{p_a}\right)^n + \gamma\left(\frac{J_1' + R(s)}{p_a}\right)^2\right]\left(1-\beta\sqrt{\frac{27}{2}}\frac{J_{3D}}{J_{2D}^{1.5}}\right) = 0 \qquad (7.6b)$$

$$\alpha(s) = \frac{a_1(s)}{\xi^m}, \alpha_Q = \alpha + \kappa(\alpha_0 - \alpha)(1 - \xi_v/\xi) \qquad (7.6c)$$

$$a_1(s) = \begin{cases} a_1(0), & s < s_e \\ a_1(0)\left[0.9\exp\left[-a_2(s - s_e)\right]+0.1\right], & s \geq s_e \end{cases}, R(s) = \begin{cases} R(0), & s < s_e \\ R(0)+r\sqrt{s}, & s \geq s_e \end{cases} \qquad (7.6d)$$

$$\begin{cases} Q_2 = F_2 = -\left[-\frac{a_3}{\xi^m}\left(\frac{3s}{p_a}\right)^n + \gamma\left(\frac{3s}{p_a}\right)^2\right]\left(1-\beta\sqrt{\frac{27}{2}}\frac{J_{3D}}{J_{2D}^{1.5}}\right), & s < s_e \\ Q_2 = F_2 < 0, & s \geq s_e \end{cases} \qquad (7.6e)$$

$$D = D_u\left[1 - \exp\left(-A(s)\xi_D^{Z(s)}\right)\right] \qquad (7.6f)$$

where $J_1' = 3p'$ is the first invariant of saturated effective stress tensor; p' is the effective mean stress; $s = (u_a - u_w)$ is the suction; s_e is the air entry suction; a_2, a_3, and r are material parameters; and $A(s)$ and $Z(s)$ are the parameters of disturbance function that are functions of suction.

Extending similar logic used in the DSC-HISS-δ1$_{unsat}$ model proposed for the unsaturated soils [25], Hamid and Miller [26] proposed a DSC-based model for unsaturated soil interfaces. In this model [26], a yield surface was developed in the shear stress – net normal stress space (i.e., $\tau - \sigma_{net}$ space) for the unsaturated soil–steel interface. The function of the yield surface is proposed in analogy to that in the DSC-HISS-δ1$_{unsat}$ model [25]. Based on this yield surface, a non-associated elastoplastic model was further developed [26] for the RI state with several parameters that are expressed as functions of suction. In the model [26], the shear stress of interfaces in the FA state is assumed to be constant at zero despite net normal stress. In addition, Eq. (7.2b) was modified to a piecewise function to describe the disturbance. It is assumed that no disturbance occurs ($D=0$) before the peak point is reached ($\xi_D < \xi_D^*$, where ξ_D^* is the value of ξ_D at peak point, which can be expressed as a function of suction). After that, when $\xi_D \geq \xi_D^*$, the disturbance function was $D = \left[\left(\tau_p - \tau_r \right) / \tau_p \right] \times \left[1 - \exp\left(-\left(\xi_D - \xi_D^* \right)^2 \right) \right]$, where τ_p and τ_r are peak and residual shear stress that are functions of stress state variables.

The two models [25,26] have been successfully validated in modelling the strain-softening behaviour of unsaturated soils and unsaturated interfaces, respectively. These models provide valuable information for researchers and practitioners for extending the DSC-based approaches to solve complex problems associated with unsaturated soils.

7.4.2 Extending DSC to Prediction Model of Stress– Strain Relationship of Unsaturated Soils

In addition to the elastoplastic models, there are also some simpler models developed based on the DSC to simulate the stress–strain relationship of unsaturated soils for predicting the variation of shear strength of unsaturated soils during the shear deformation process.

For example, Haeri et al. [27] developed a DSC-based model to reproduce the stress–strain curve of unsaturated collapsible soils. In this model, a hyperbolic curve is used to represent the response of unsaturated soils in the RI state and, in analogy to the DSC-HISS-$\delta 1_{unsat}$ model [25], the stress–strain curve of saturated soil is used as the FA state behaviour of the corresponding unsaturated soils. However, this model did not propose a constitutive model for the FA state and the saturated stress–strain curve is obtained by fitting experimental results. In addition, the variation functions of several parameters with suction were also obtained by fitting experimental results. Due to this reason, there are limitations in extending this model in predicting the stress–strain response of unsaturated soils under various loading conditions.

Yang and Vanapalli [28] developed a simple DSC-based model for predicting the variation of shear stress under consolidated drained triaxial compression condition for a wide range of unsaturated soils. The general equation of DSC, Eq. (7.1a), has been modified as below for formulating a prediction model of stress–strain relationship.

$$q^a = (1 - D)q^{RI} + Dq^{FA} \tag{7.7}$$

where $q^a = \sigma_1^a - \sigma_3^a$ is the apparent deviatoric stress; σ_1^a and σ_3^a are the apparent major and minor principal stresses; and q^{RI} and q^{FA} are the deviatoric stresses sustained by the RI and FA parts, respectively.

Several experimental studies have shown that the strain-softening behaviour of unsaturated soils can be attributed to the generation and development of the shear band where the original soil particle structure is destroyed [32] and local degree of saturation is decreased [33] due to shearing. Thus, the generation and development of the shear band can be treated as the disturbance that causes the strain-softening behaviour of unsaturated soils. As a result, the nonlinear hardening stress–strain relationship at the pre-peak state where the shear band does not develop significantly can be assumed as the response in the RI state where no disturbance has occurred yet. The pre-peak stress–strain relationship can be predicted by a hyperbolic model (Eq. 7.8) in terms of axial strain. The stress–strain relationship at the critical state where the shear band has been well defined in the specimen can be assumed as the response in the FA state where the soil is fully disturbed. The critical state stress–strain relationship can be described by a linear model (Eq. 7.9):

$$q^{RI} = \frac{\varepsilon_a}{\dfrac{1}{E_i} + \dfrac{1}{q_{ult}}\varepsilon_a} \tag{7.8}$$

where ε_a is the axial strain; E_i is the initial tangent modulus (i.e., the initial slope of the stress–strain curve); and q_{ult} is the ultimate deviatoric stress of the hardening stress–strain curve extending the pre-peak curve (i.e., the asymptotic value of the deviatoric stress when ε_a is infinity).

$$q^{FA} = q_c \tag{7.9}$$

where q_c is the critical deviatoric stress.

The disturbance function (Eq. 7.2a) proposed by Desai [21] was modified to describe the evolution of disturbance with shear deformation in unsaturated soils. Firstly, Eq. (7.2a) should be rewritten in terms of axial strain to be consistent with Eqs. (7.8) and (7.9). Secondly, D_u in Eq. (7.2a) is assumed to be 1 considering the critical state is assumed to be the FA state, which can usually be reached at the end of triaxial shear tests. Thirdly, after several fitting trials, Eq. (7.2a) is found to be over-parameterized for modelling the disturbance of unsaturated soils during the strain-softening process; thus, an assumption, $s=1$, is introduced to simplify the parameters. Thus, Eq. (7.2a) can be modified as:

$$D = 1 - \left[1 + \left(\frac{\varepsilon_a}{h} \right)^w \right]^{-1} \tag{7.10}$$

The parameters h and w in Eq. (7.10) should be expressed as functions of suction for extending DSC-based models to capture the unsaturated behaviour [25,26]. For this purpose, a typical apparent stress–strain curve of unsaturated soils with strain-softening behaviour measured in experiment is analysed extending a semi-logarithmic relationship (i.e., solid line with hollow circles in Figure 7.3a). As references, the stress–strain curves of RI and FA states (i.e., long dash lines in Figure 7.3a) are also plotted in this figure that are obtained by fitting the pre-peak part and the critical state part of the apparent stress–strain curve using Eqs. (7.8) and (7.9), respectively. In addition, a typical disturbance function curve (i.e., solid line in Figure 7.3b) described by Eq. (7.10) is also plotted extending a semi-logarithmic relationship for the stress–strain curve. In Figure 7.3b, three reference lines (short dash line) are plotted including two horizontal lines at $D=0$ and $D=1$ and a tangent line of the $D - \log(\varepsilon_a)$ curve passing through its inflection point. Thus, two intersection points, A and B, can be determined whose positions are controlled by h and w. As shown in Figure 7.3b, the position of point A can represent the axial strain where disturbance just begins, and the position of point B can represent the axial strain where most of disturbance has occurred.

Two assumptions are proposed about the positions of A and B for building relations between h and w and suction:

i. Point A corresponds to the peak point (A') on the apparent stress–strain curve, which can be expressed as:

$$h \times 10^{\left(\frac{-20}{\ln 10w} \right)} = \varepsilon_p = \frac{q_p q_{ult}}{E_i \left(q_{ult} - q_p \right)} \tag{7.11}$$

where q_p is the peak deviatoric stress; ε_p is the axial strain at peak deviatoric stress that can be determined by using Eq. (7.8) since A' falls on the RI state curve.

ii. Point B corresponds to the post-peak point of maximum curvature (B') on the apparent stress–strain curve, which can be expressed as:

$$\log \left(h \times 10^{\left(\frac{2}{\ln 10w} \right)} \right) = \log \varepsilon_c' = \log \varepsilon_p + \frac{q_p - q_c}{M} \tag{7.12}$$

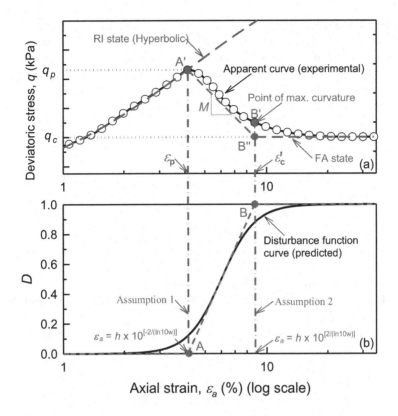

FIGURE 7.3 Derivation of the disturbance function for unsaturated soils. (Modified after Yang and Vanapalli [28].)

where ε'_c is the axial strain at the post-peak point of maximum curvature on apparent stress–strain curve; $M = (q_p - q_c)/(\log \varepsilon'_c - \log\varepsilon_p)$ is defined as post-peak modulus (i.e., the slope of A'B'' in Figure 7.3a).

The equations for determining h and w can be derived from Eqs. (7.11) to (7.12) and then can be substituted into Eq. (7.10) to obtain the disturbance function (Eq. 7.13b) for modelling strain-softening behaviour of unsaturated soils. Combining Eqs. (7.7)– (7.12), the DSC-based approach can be extended to formulate a model (Eq. 7.13) that can be used in prediction of the variation of shear stress in unsaturated soils under consolidated drained triaxial compression condition

$$q^a = (1-D) \times \frac{\varepsilon_a}{\dfrac{1}{E_i} + \dfrac{1}{q_{\text{ult}}}\varepsilon_a} + D \times q_c \tag{7.13a}$$

$$D = 1 - \left[1 + \left(\varepsilon_a \frac{E_i\left(q_{\text{ult}} - q_p\right)}{q_p q_{\text{ult}}} 10^{\frac{q_c - q_p}{2M}} \right)^{\frac{4M}{\ln 10\left(q_p - q_c\right)}} \right]^{-1} \tag{7.13b}$$

According to the definition, M and $(q_p - q_c)$ should equal zero for hardening response; however, this will make Eq. (7.13b) offer no solution. Therefore, M and $(q_p - q_c)$ are assumed equal one for hardening response.

In Eq. (7.13), five mechanical parameters (i.e., E_i, M, q_p, q_{ult}, and q_c) of unsaturated soils are involved, which need to be expressed as functions of suction to consider its influence on the stress–strain relationship of unsaturated soils. The variation functions of those parameters with suction are given as Eqs. (7.14)–(7.18) and are described as follows.

The linear relationship between $\log(E_i/p_a)$ and $\log(\sigma'_3/p_a)$ (σ'_3 is effective minor principal stress) proposed by Duncan and Chang [34] can be extended for interpreting the initial tangent modulus of unsaturated soils by taking account of the independent contribution of suction and the influence of degree of saturation. Thus, the prediction model for E_i of unsaturated soils can be given as:

$$\log\left(\frac{E_i}{p_a}\right) = \log K_E + \left[n_E + \alpha_E(u_a - u_w)\right]\log\left(\frac{\sigma_3 - u_a}{p_a}\right)$$
$$+ \beta_E \log\left[\left(\frac{u_a - u_w}{p_a}\right) \times \left(\frac{S - S_r}{1 - S_r}\right) + 1\right] \tag{7.14}$$

where S is the degree of saturation; S_r is the residual degree of saturation; K_E and n_E are initial tangent modulus parameters of saturated soil; and α_E and β_E are fitting parameters for the initial tangent modulus of unsaturated soil.

The prediction model for M of unsaturated soils is analogous to the model for E_i as:

$$\log\left(\frac{M}{p_a}\right) = \log K_M + \left[n_M + \alpha_M(u_a - u_w)\right]\log\left(\frac{\sigma_3 - u_a}{p_a}\right)$$
$$+ \beta_M \log\left[\left(\frac{u_a - u_w}{p_a}\right) \times \left(\frac{S - S_r}{1 - S_r}\right) + 1\right] \tag{7.15}$$

where K_M and n_M are post-peak modulus parameters of saturated soil; α_M and β_M are fitting parameters for the post-peak modulus of unsaturated soil.

The model proposed by Vanapalli et al. [35] is modified to predict the peak deviatoric stress, q_p, of unsaturated soils in terms of $(\sigma_3 - u_a)$ and $(u_a - u_w)$ as:

$$q_p = \frac{3}{3 - M_p}\left[m_p + (\sigma_3 - u_a)M_p + (u_a - u_w)M_p S^{\kappa_p}\right] \tag{7.16}$$

where m_p and M_p are peak shear strength parameters of saturated soil (i.e., the intercept and slope of saturated peak failure envelope in the deviatoric stress – effective mean stress $(q-p')$ space); κ_p is the fitting parameter for peak shear strength of unsaturated soil.

The prediction model for the critical deviatoric stress, q_c, of unsaturated soils can be represented using the relationship below, which is similar to the model for q_p as:

$$q_c = \frac{3}{3 - M_c}\left[m_c + (\sigma_3 - u_a)M_c + (u_a - u_w)M_c S^{\kappa_c}\right] \tag{7.17}$$

where m_c and M_c are critical shear strength parameters of saturated soil (i.e., the intercept and slope of saturated critical failure envelope in the $q-p'$ space); κ_c is the fitting parameter for critical shear strength of unsaturated soil.

For saturated soils, the ultimate deviatoric stress of a hardening stress–strain curve when ε_a is infinity, q_{ult}, can be predicted using $q_{ult} = R_f \times q_f$, where R_f is the failure ratio and q_f is the deviatoric

TABLE 7.1

Summary of the DSC-Based Model for Unsaturated Soils [28]

DSC Model	Mechanical Parameters in Eq. (7.13)	Parameters Required	Determination Methods of Parameters
Eq. (7.13)	E_i (Eq. 7.14)	K_E and n_E	Fitting experimental results of saturated soils (E_i and σ'_3) using Eq. (7.14) with $(\sigma_3 - u_a) = \sigma'_3$ and $(u_a - u_w) = 0$
		α_E and β_E	Fitting experimental results of unsaturated soils (E_i, $(\sigma_3 - u_a)$ and $(u_a - u_w)$) using Eq. (7.14) with determined K_E and n_E
	M (Eq. 7.15)	K_M and n_M	Fitting experimental results of saturated soils (M and σ'_3) using Eq. (7.15) with $(\sigma_3 - u_a) = \sigma'_3$ and $(u_a - u_w) = 0$
		α_M and β_M	Fitting experimental results of unsaturated soils (M, $(\sigma_3 - u_a)$ and $(u_a - u_w)$) using Eq. (7.15) with determined K_M and n_M
	q_p (Eq. 7.16)	m_p and M_p	Fitting experimental results of saturated soils (q_p and σ'_3) using Eq. (7.16) with $(\sigma_3 - u_a) = \sigma'_3$ and $(u_a - u_w) = 0$
		κ_p	Fitting experimental results of unsaturated soils (q_p, $(\sigma_3 - u_a)$ and $(u_a - u_w)$) using Eq. (7–16) with determined m_p and M_p
	q_c (Eq. 7.17)	m_c and M_c	Fitting experimental results of saturated soils (q_c and σ''_3) using Eq. (7.17) with $(\sigma_3 - u_a) = \sigma''_3$ and $(u_a - u_w) = 0$
		κ_c	Fitting experimental results of unsaturated soils (q_c, $(\sigma_3 - u_a)$ and $(u_a - u_w)$) using Eq. (7.17) with determined m_c and M_c
	q_{ult} (Eq. 7.18)	R_f	Fitting experimental results of unsaturated soils (q_{ult}, $(\sigma_3 - u_a)$ and $(u_a - u_w)$) using Eq. (7.18) with determined m_p, M_p, and κ_p
	-	a, n, m and S_r	Fitting results of conventional pressure plate tests using SWCC equation [36]

stress beyond which the specimen will fail [34]. Extending the same philosophy of this approach, q_{ult} of unsaturated soils can be predicted by:

$$q_{\text{ult}} = R_f q_p = \frac{3R_f}{3 - M_p} \left[m_p + (\sigma_3 - u_a) M_p + (u_a - u_w) M_p S^{\kappa_p} \right] \qquad (7\text{–}18)$$

Eventually, a simple prediction model (Eqs. 7.13–7.18) for stress–strain relationship of unsaturated soils can be formulated by extending the DSC-based approach based on two stress state variables (i.e., net minor principal stress and suction). All the parameters of this model can be divided into three categories: (i) the soil-water characteristic curve (SWCC) parameters (a, n, m, and S_r), (ii) the saturated soil parameters (i.e., K_E, n_E, K_M, n_M, m_p, M_p, m_c, and M_c), and (iii) the fitting parameters (i.e., α_E, β_E, α_M, β_M, κ_p, κ_c, and R_f). All these parameters and corresponding determination methods are summarized in Table 7.1. The mechanical parameters in Table 7.1 can be determined from several $q - \varepsilon_a$ curves measured during consolidated drained triaxial shear tests of saturated and unsaturated soils.

7.4.3 Modelling Applications to Stress–Strain Relationships of Unsaturated Soils

Geiser et al. [25,37] reported several triaxial shear tests on unsaturated sandy silt (CL-ML). The information of the soil and triaxial shear tests have been summarized in Table 7.2. HISS-δ1$_{\text{unsat}}$ model [25] and DSC-HISS-δ1$_{\text{unsat}}$ model [25] introduced in Section 7.4.1 were used to reproduce several stress–strain curves of the consolidated drained triaxial shear tests on unsaturated specimens (Figure 7.4). It can be found that the HISS-δ1$_{\text{unsat}}$ model cannot capture the strain-softening behaviour of unsaturated specimens in spite of modelling most of the features of unsaturated soils

TABLE 7.2

Information of the Studied Soils and Corresponding Triaxial Shear Tests. (Modified after Yang and Vanapalli [28].)

	CL-ML (25, 37)	SM (7)	ML (6)	Expansive clay (3)
Gradation	Sand: Silt: Clay=20: 72: 8	Sand: Silt: Clay=55: 37: 8	Sand : Silt : Clay=5: 90: 5	Clay: 24.8%
OMC (%)	-	12.2	14.5	21.4
$\rho_{d,\,max}$ (g/cm³)	-	1.87	1.74	1.63
Atterberg Limits (%)	LL=25.4, PL=16.7	-	LL=29, PL=10	LL=58.3, PL=26.5
w_{ini} (%)	-	14.2	10	17
Specimen preparation technique	K_0 consolidation (100 kPa vertical stress)	Static compaction (1600 kPa final vertical stress)	Static compaction (1600 kPa final vertical stress)	Static compaction
Saturated tests	σ'_3=400, 600 kPa	σ'_3=100, 200, 300 kPa	σ'_3=50, 100, 200, 300, 400 kPa	σ'_3=50, 100, 150 kPa
Unsaturated tests (back-prediction)	(u_a-u_w)=100, 280 kPa	(u_a-u_w)=250, 750 kPa	(u_a-u_w)=100, 300 kPa	(u_a-u_w)=80, 200 kPa
Unsaturated tests (prediction)	(u_a-u_w)=50, 200 kPa	(u_a-u_w)=50, 500 kPa	(u_a-u_w)=200 kPa	(u_a-u_w)=50, 120 kPa
Saturated parameters in model [28]	K_E=0.72, n_E=1.08, K_M=0.057, n_M=2.69, m_p=0, M_p=1.36, m_c=0, M_c=1.26,	K_E=1.58, n_E=0.92, K_M=0.01, n_M=0, m_p=0, M_p=1.4, m_c=0, M_c=1.4,	K_E=1.55, n_E=0.26, K_M=0.01, n_M=0, m_p=45.84, M_p=1.18, m_c=45.84, M_c=1.18,	K_E=4.14, n_E=0.94, K_M=2.13, n_M=0.57, m_p=82.7, M_p=0.84, m_c=47.44, M_c=0.57,
Fitting parameters in model [28]	α_E=0.0055, β_E=-1.09, α_M=0.0063, β_M=-0.019, κ_p=0.62, κ_c=0.5, R_f=1.27,	α_E=-0.0006, β_E=3.16, α_M=0.0009, β_M=10.1, κ_p=0.71, κ_c=1.24, R_f=1.26	α_E=-0.0015, β_E=4.65, α_M=-0.0088, β_M=8.72, κ_p=1.23, κ_c=1.45, R_f=1.09	α_E=-0.0042, β_E=0.48, α_M=-0.0011, β_M=0.37, κ_p=1.25, κ_c=-0.67, R_f=1.34
SWCC parameters in model [28]	a=83.78, n=2.03, m=1.25, S_r=0.047	a=20.7, n=0.95, m=1.1, S_r=0.125	a=13.13, n=0.91, m=0.93, S_r=0.178	a=324.01, n=1.17, m=1.12, S_r=0.107

Notes: OMC, optimum moisture content; $\rho_{d,\,max}$, maximum dry density; LL, liquid limit; PL, plastic limit; w_{ini}, initial water content of compacted specimens.

(short dash lines in Figure 7.4). However, this can be improved by the introduction of DSC (long dash lines in Figure 7.4).

The DSC-based model [28] introduced in Section 7.4.2 was also used to predict the stress–strain relationships of unsaturated CL-ML specimens. The saturated parameters (i.e., K_E, n_E, K_M, n_M, m_p, M_p, m_c, and M_c) and SWCC parameters (i.e., a, n, m, and S_r) in Table 7.1 were determined from the experimental results of saturated triaxial shear tests and conventional pressure plate tests, respectively. However, only two drained triaxial shear tests were reported for saturated CL-ML specimens. Three more consolidated undrained triaxial shear tests were performed with pore water pressure measurements. Therefore, the saturated shear strength parameters of CL-ML were determined combining the drained and consolidated undrained triaxial shear test results, while the saturated modulus parameters were determined only using the drained triaxial shear tests results. The fitting parameters (i.e., α_E, β_E, α_M, β_M, κ_p, κ_c, and R_f) in Table 7.1 were determined from the experimental results of a portion of unsaturated triaxial shear tests. The values of the saturated SWCC and fitting

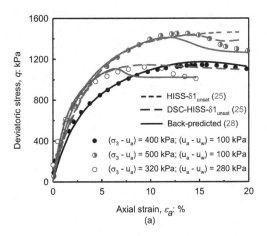
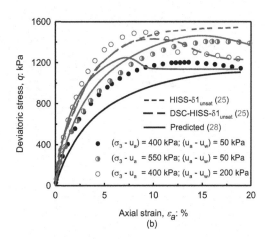

FIGURE 7.4 Comparisons between experimental and predicted results for stress–strain curves of CL-ML (25): (a) Back-prediction [28] and (b) Prediction [28]. (Modified after Geiser et al. [25].)

parameters are summarized in Table 7.2. The model [28] with these parameters in Table 7.2 was used to back-predict the stress–strain curves of the unsaturated triaxial shear tests that were used to determine the fitting parameters. Then, it was used to predict the stress–strain curves of the unsaturated triaxial shear tests that were not used to determine the fitting parameters. The experimental results and the back-predicted and predicted results are compared in Figure 7.4a and b, respectively. The summarized results suggest that the DSC-based model [28] can perform well to reproduce the strain-softening stress–strain curves of unsaturated CL-ML specimens.

The stress–strain curves of other three types of unsaturated soils (i.e., SM, ML, and expansive clay) obtained by consolidated drained triaxial shear tests are collected from literature [3,6,7]. The information of the soils and corresponding triaxial shear tests in these studies has been summarized in Table 7.2. The DSC-based model [28] was used to reproduce the stress–strain curves of these unsaturated soils. The calculation procedures are similar to that used for the unsaturated CL-ML specimens described above. The information of SWCC of ML [6] was not provided; thus, the values of the degree of saturation of compacted specimens that reaches the suction equilibrium state under different suctions before the beginning of triaxial shear tests were used to determine the SWCC parameters for ML. The experimental results and the back-predicted and predicted results of the three unsaturated soils are compared in Figures 7.5–7.7, respectively. It can be found that the back-predicted and predicted results are reasonable for most of stress–strain curves in Figures 7.5–7.7. This means the DSC-based model [28] can well capture the strain-softening behaviour of different types of unsaturated soils varying from coarse- to fine-grained soils.

The strain-softening behaviour of soils can be quantified by several feature parameters, which include: (i) peak and critical shear strength (q_p and q_c); (ii) brittleness index (I_B); (iii) post-peak modulus (M); and (iv) axial strain at peak deviatoric stress (ε_p). The brittleness index, I_B, is defined by Bishop [38] as $(q_p - q_r)/q_p$, where q_r is the residual deviatoric stress. In this chapter, q_c is used as q_r. The definitions of q_p, q_c, M, and ε_p have been shown in Figure 7.3 in Section 7.4.2. I_B, M, and ε_p can be used to represent the magnitude, rate, and starting position of strain-softening, respectively.

The applicability of DSC in modelling strain-softening behaviour of unsaturated soils is studied quantitatively using the feature parameters (q_p, q_c, I_B, M, and ε_p) of unsaturated SM [7], CL-ML [25], ML [6], and expansive clay [3] specimens. The experimental results of these feature parameters and the corresponding predicted results obtained by the DSC-based model [28] are compared in Figure 7.8. It is difficult to determine the values of ε_p of hardening stress–strain relationship; thus, only the information of ε_p of strain-softening stress–strain relationships is included in Figure 7.8d. The comparisons in Figure 7.8 suggest that these feature parameters of strain-softening behaviour

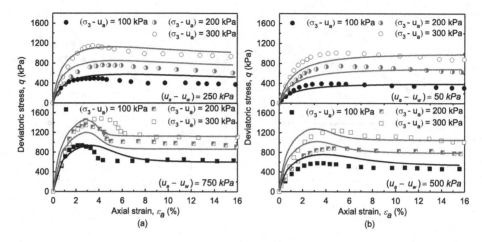

FIGURE 7.5 Comparisons between experimental and predicted results for stress–strain curves of SM (7): (a) Back-prediction [28] and (b) Prediction [28]. (Modified after Yang and Vanapalli [28].)

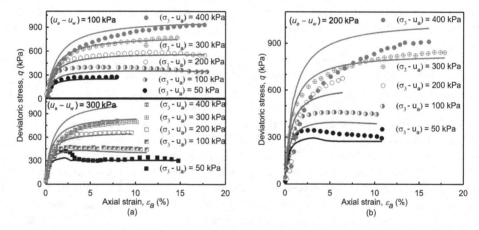

FIGURE 7.6 Comparisons between experimental and predicted results for stress–strain curves of ML [6]: (a) Back-prediction [28] and (b) Prediction [28]. (Modified after Yang and Vanapalli [28].)

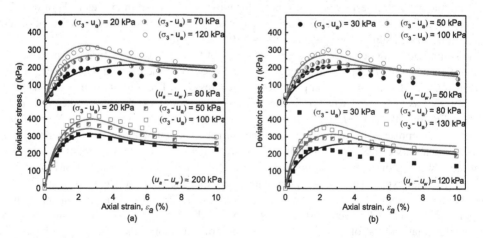

FIGURE 7.7 Comparisons between experimental and predicted results for stress–strain curves of expansive clay [3]: (a) Back-prediction [28] and (b) Prediction [28]. (Modified after Yang and Vanapalli [28].)

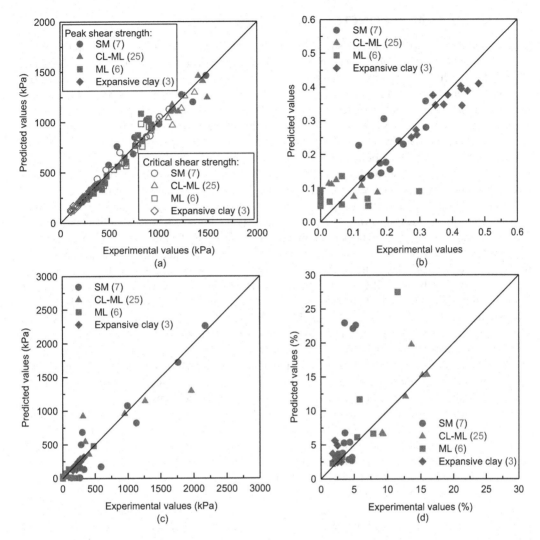

FIGURE 7.8 Comparisons between experimental and predicted results for feature parameters of strain-softening behaviour of unsaturated soils: (a) peak and critical shear strength; (b) brittleness index; (c) post-peak modulus; and (d) axial strain at peak deviatoric stress. (Modified after Yang and Vanapalli [28].)

of unsaturated soils can be predicted reasonably by the DSC-based model [28] except several significant outliers in predictions of ε_p. This further validates the capability of the DSC-based approach in modelling the strain-softening behaviour of a wide range of unsaturated soils.

The DSC-based elastoplastic models [25,26] can be implemented in FEM for solving boundary value problems associated with unsaturated soils as that has been conducted for saturated soils [30,31]. The simple DSC-based model for stress–strain relationship of unsaturated soils [28] did not consider the volume change behaviour of the soils; thus, it can only be used for the stress paths where the values of $(u_a - u_w)$ and $(\sigma_3 - u_a)$ are known. However, this model [28] can be incorporated into an elastoplastic constitutive model of unsaturated soils as a shear strength criterion to describe the variation of shear strength with shear deformation. In other words, it is feasible to extend this model [28] for any stress path and to implement it in FEM to solve the boundary value problems associated with unsaturated soils in engineering practice.

7.5   SUMMARY AND CONCLUSIONS

Unsaturated soils exhibit significant strain-softening behaviour in comparison to the saturated soils due to the influence of suction. Such behaviour plays an important role in rigorous analyses and design of geotechnical structures undergoing progressive failure. For example, an unsaturated soil slope can experience a progressive failure due to the influence of several environmental factors (e.g., the rainfall infiltration on slope surface, the seasonal wetting–drying cycles of soils in slope and the variation of the water level in nearby river). During this process, the shear strength of unsaturated soils in the slip zone drops with increasing shear deformation from the peak to the critical value due to the strain-softening behaviour. As a result, the factor of safety of the slope is reduced gradually with the development of shear deformation. Therefore, the strain-softening behaviour of unsaturated soils should be modelled reliably in engineering practice associated with unsaturated soils.

The DSC is a unified modelling approach that can capture various complex behaviours of a wide range of engineering materials. In DSC, a soil element at any intermediate stage during the deforming process can be considered to be in a combined random mixture of zones in RI and FA states; thus, the apparent response of the soil element can be expressed as a weighted average of the responses of the zones in RI and FA states with an assumed disturbance function as the weight. In this chapter, for example, a commonly used DSC-based model, DSC-HISS, developed by Desai [21] is introduced. Then, the applications of the DSC-based approach in modelling the strain-softening behaviour of saturated soils and interfaces are summarized. It is highlighted that the DSC-based model is relatively simple because it avoids using the advanced soil plasticity concepts and its approach for explaining the strain-softening behaviour is based on limited and well-defined parameters. The DSC can be conveniently accommodated into different existing reliable constitutive models to capture more complex behaviours of a wide range of engineering materials. In addition, the DSC-based model can be implemented into FEM conveniently for solving boundary value problems.

The simple and flexible DSC-based approach can be of significant interest in engineering practice for simulating the shearing behaviour of unsaturated soils. Due to this reason, the DSC-based approach was extended to two elastoplastic models [25,26] for unsaturated soils and interfaces, respectively. The extension methods were reviewed in this chapter, which can provide valuable information for other researchers and practitioners for extending the DSC-based approach to solve complex problems associated with unsaturated soils.

Furthermore, the DSC-based approach was extended to formulate a simple prediction model (Eq. 7.13) for the stress–strain relationship of unsaturated soils taking account of the influence of two stress state variables (i.e., net minor principal stress and suction). In this model, the hyperbolic hardening stress–strain relationship at pre-peak state of unsaturated soils is used as the RI state response and the stress–strain relationship at the critical state of unsaturated soils is used as the FA state response. A function with two parameters is used as the disturbance function and the two parameters are related to mechanical parameters of unsaturated soils based on two assumptions (Figure 7.3). Eventually, a simple prediction model (Eq. 7.13) of stress–strain relationship of unsaturated soils can be formulated by extending the DSC-based approach with five mechanical parameters (i.e., E_i, M, q_p, q_{ult}, and q_c) of unsaturated soils. The five mechanical parameters are expressed as functions (Eqs. 7.14–7.18) of stress state variables. All the parameters in this model (Table 7.1) can be obtained from conventional saturated and unsaturated triaxial shearing tests and pressure plate tests.

Finally, the applications of the DSC-based models in reproducing the stress–strain curves of different types of unsaturated soils are presented. The HISS-$\delta 1_{unsat}$ model [25] and DSC-HISS- $\delta 1_{unsat}$ model [25] were used to predict stress–strain curves of unsaturated CL-ML specimens, respectively. The results suggest that the HISS-$\delta 1_{unsat}$ model cannot capture the strain-softening behaviour of unsaturated specimens, and this can be improved by the introduction of DSC. The DSC-based model [28] was used to reproduce stress–strain curves of four different types of unsaturated soils (i.e., SM, CL-ML, ML, and expansive clay). The results suggest that the DSC-based model [28] can reasonably capture the strain-softening behaviour of different types of unsaturated soils varying from coarse- to fine-grained soils. In addition, the DSC-based model is a promising approach in solving boundary value problems associated with unsaturated soils for use in engineering practice applications extending finite element methods.

REFERENCES

1. Skempton, A. W. Long-term stability of clay slopes. *Géotechnique*, 14, 2, 1964, 77–102.
2. Cui, Y. J., and Delage, P. "Yielding and plastic behaviour of an unsaturated compacted silt." *Géotechnique*, 46, 2, 1996, 291–311.
3. Miao, L., Liu, S., and Lai, Y. "Research of soil–water characteristics and shear strength features of Nanyang expansive soil." *Eng. Geol.*, 65, 4, 2002, 261–267.
4. Rahardjo, H., Heng, O. B., and Choon, L. E. "Shear strength of a compacted residual soil from consolidated drained and constant water content triaxial tests." *Can. Geotech. J.*, 41, 3, 2004, 421–436.
5. Kayadelen, C., Tekinsoy, M. A., and Taşkıran, T. "Influence of matric suction on shear strength behaviour of a residual clayey soil." *Eng. Geol.*, 53, 4, 2007, 891–901.
6. Estabragh, A. R., and Javadi, A. A. "Critical state for overconsolidated unsaturated silty soil." *Can. Geotech. J.*, 45, 3, 2008, 408–420.
7. Patil, U. D. *Response of unsaturated silty sand over a wider range of suction states using a novel double-walled triaxial testing system.* Ph.D. thesis, University of Texas at Arlington, Arlington, Tex; 2014.
8. Toll, D. G. "A framework for unsaturated soil behaviour." *Géotechnique*, 40, 1, 1990, 31–44.
9. Widger, R. A., and Fredlund, D. G. "Stability of swelling clay embankments." *Can. Geotech. J.*, 16, 1, 1979, 140–151.
10. Ng, C. W. W., Zhan, L. T., Bao, C. G., Fredlund, D. G., and Gong, B. W. "Performance of an unsaturated expansive soil slope subjected to artificial rainfall infiltration." *Géotechnique*, 53, 2, 2003, 143–157.
11. Take, W. A., and Bolton, M. D. "Seasonal ratcheting and softening in clay slopes, leading to first-time failure." *Géotechnique*, 61, 9, 2011, 757–769.
12. Yang, X., and Vanapalli, S. K. "Slope stability analyses of outang landslide based on the peak and residual shear strength behavior." *Adv. Eng. Sci.*, 51, 4, 2019, 55–68.
13. Alonso, E. E., Gens, A., and Josa, A. "A constitutive model for partially saturated soils." *Géotechnique*, 40, 3, 1990, 405–430.
14. Russell, A. R., and Khalili, N. "A unified bounding surface plasticity model for unsaturated soils." *Int. J. Numer. Analyt. Methods Geomech.*, 30, 3, 2006, 181–212.
15. Frantziskonis, G., and Desai, C. S., "Analysis of a strain softening constitutive law." *Int. J. Solids Struct.*, 23, 6, 1987, 733–750.
16. Armaleh, S. H., and Desai, C. S. "Modelling and testing of a cohesionless material using the disturbed state concept." *J. Mech. Behav. Mater.*, 5, 3, 1994, 279–296.
17. Minh, N. H., Suzuki, K., Oda, M., Tobita, T., and Desai, C. S. "Numerical simulation using disturbed state concept (DSC) model for softening behavior of sand." *J. Southeast Asian Geotech. Soc.*, 39, 1, 2008, 25–35.
18. Desai, C. S., and Ma, Y. "Modelling of joints and interfaces using the disturbed-state concept." *Int. J. Numer. Analyt. Methods Geomech.*, 16, 9, 1992, 623–653.
19. Navayogarajah, N., Desai, C. S., and Kiousis, P. D., "Hierarchical single surface model for static and cyclic behavior of interfaces." *J. Eng. Mech.*, 118, 5, 1992, 990–1011.
20. Xiao, Y., and Desai, C. S., "Constitutive modeling for overconsolidated clays based on disturbed state concept: I- Theory, II- parameter analysis and validations." *Int. J. Geomech.*, 19, 9, 2019, 04019101–04019102.
21. Desai, C. S. *Mechanics of Materials and Interfaces: The Disturbed State Concept.* Boca Raton, FL: CRC; 2001.
22. Hu, L., and Pu, J. "Testing and modeling of soil-structure interface." *J. Geotech. Geoenviron. Eng.*, 130, 8, 2004, 851–860.
23. Seo, M. W., Park, I. J., and Park, J. B. "Development of displacement-softening model for interface shear behaviour between geosynthetics." *Soils Found.*, 44, 6, 2004, 27–38.
24. Veiskarami, M., Ghorbani, A., and Alavipour, M. "Development of a constitutive model for rockfills and similar granular materials based on the disturbed state concept." *Frontiers Struct. Civil Eng.*, 6, 4, 2012, 365–378.

25. Geiser, F., Laloui, L., and Vulliet, L. "Modelling the behaviour of unsaturated silt." In *Experimental Evidence and Theoretical Approaches in Unsaturated Soils*, Edited by Tarantino A. and Mancuso C. CRC Press, Boca Raton, FL, 2000, 163–184.

26. Hamid, T. B., and Miller, G. A. "A constitutive model for unsaturated soil interfaces." *Int. J. Numer. Analyt. Methods Geomech.*, 32, 13, 2008, 1693–1714.

27. Haeri, S. M., Khosravi, A., Garakani, A. A., and Ghazizadeh, S. "Effect of soil structure and disturbance on hydromechanical behaviour of collapsible loessial soils." *Int. J. Geomech.*, 17, 1, 2017, 04016021.

28. Yang, X., and Vanapalli, S. K. "Model for predicting the variation of shear stress in unsaturated soils during strain-softening." *Can. Geotech. J.*, 58, 2020. https://doi.org/10.1139/cgj-2020-0312.

29. Desai, C. S. "Constitutive modeling of materials and contacts using the disturbed state concept: Part 1–Background and analysis." *Comput. Struct.*, 146, 2015, 214–233.

30. Desai, C. S. "Constitutive modeling of materials and contacts using the disturbed state concept: Part 2–Validations at specimen and boundary value problem levels." *Comput. Struct.*, 146, 2015, 234–251.

31. Desai, C. S. "Disturbed state concept as unified constitutive modeling approach." *J. Rock Mech. Geotech. Eng.*, 8, 3, 2016, 277–293.

32. Higo, Y., Oka, F., Kimoto, S., Sanagawa, T., and Matsushima, Y. "Study of strain localization and microstructural changes in partially saturated sand during triaxial tests using microfocus X-ray CT." *Soils Found.*, 51, 1, 2011, 95–111.

33. Kido, R., and Higo, Y. "Distribution changes of grain contacts and menisci in shear band during triaxial compression test for unsaturated sand." *JGS Spec. Publ.*, 7, 2, 2019, 627–635.

34. Duncan, J. M., and Chang, C.-Y. "Nonlinear analysis of stress and strain in soils." *J. Soil Mech. Found. Div.*, 96, 5, 1970, 1629–1653.

35. Vanapalli, S. K., Fredlund, D. G., Pufahl, D. E., and Clifton, A. W. "Model for the prediction of shear strength with respect to soil suction." *Can. Geotech. J.*, 33, 3, 1996, 379–392.

36. Fredlund, D. G., and Xing, A. "Equations for the soil-water characteristic curve." *Can. Geotech. J.*, 31, 4, 1994, 521–532.

37. Geiser, F., Laloui, L., and Vulliet, L. "Elasto-plasticity of unsaturated soils: Laboratory test results on a remoulded silt." *Soils Found.*, 46, 5, 2006, 545–556.

38. Bishop, A. W. "The influence of progressive failure on the choice of stability analysis." *Géotechnique*, 21, 2, 1971, 168–172.

8 Coefficient of Lateral Soil Pressure in Unsaturated Soils

Amir Akbari Garakani
Power Industry Structures Research Dept.,
Niroo Research Institute (NRI)

8.1 INTRODUCTION

The coefficient of lateral soil pressure, k, is a key factor in the analysis of underground structures, slope stability, retaining walls, etc. Due to the dependence of this parameter on several state variables or material parameters, its accurate determination is very challenging, especially in unsaturated soils.

Many studies have been conducted to examine the value(s) of k for saturated soils.[1-9] However, there are very few investigations on the coefficient of lateral soil pressure and its relevant state variables (or parameters) for *unsaturated* soils, either experimentally, analytically, or numerically. As previous studies have shown, the lateral soil pressure in unsaturated soils is remarkably affected by the soil type,[10,11] the initial soil saturation, S_r, the matric suction, ψ,[10,12-20] the stress state within the soil,[12,18-20] the climate parameters (e.g., infiltration or evaporation),[17,21,22] and the geometrical aspects of the retaining structure that supports the soil against lateral deformations.[11,23] In general, it has been indicated that the at-rest and the active soil pressure coefficients (k_{ou} and k_{au}, respectively) increase by increasing the degree of saturation of the soil and the principal stress level within the soil body, or by decreasing the matric suction of the soil. In addition, it was found that the higher the soil density, the less the lateral soil pressure in unsaturated soil deposits.

Among the limited studies on assessing lateral soil pressure in unsaturated soils, only the at-rest or the active coefficients of lateral soil pressure were studied, and the *continuous* variation of the unsaturated lateral soil pressure from the *at-rest* to the *active* state of the examined soils has not been investigated.

Recently, Garakani et al.[24] have developed a suction-dependent effective stress-based analytical framework on the basis of the disturbed state concept (DSC) to predict the continuous variation of unsaturated lateral soil pressure coefficient from the at-rest to the active state of the soil. DSC, well-known as a comprehensive method for predicting the mechanical behavior of geomaterials, was first proposed by Desai[25-28] and developed afterward by many researchers.[29-34] According to the DSC, the mechanical response of soil due to changing state variables can be defined by considering the soil response in two extreme soil structural reference conditions, namely, *the relative intact (RI) soil structure* and *the fully adjusted (FA) soil structure*.

The functionality of the proposed analytical model by Garakani et al.[24] is verified quantitatively against experimental data reported by Pirjalili et al.[10] The experimental data used for verification included a series of suction-controlled drained tests' results for two different unsaturated soils that were examined using a suction-controlled ring device under six different matric suctions. The ring device, implemented by Pirjalili et al.,[10] is capable of continuously measuring the unsaturated lateral soil pressure and corresponding lateral (radial) strain, ε_r, as well as controlling the matric suction and measuring the water content of the soil specimens during the tests.

The DSC-based analytical solution proposed by Garakani et al.[24] can easily predict the continuous variation of the lateral soil pressure in unsaturated soils, from the at-rest to the active conditions. In this chapter, details of this method are presented.

DOI: 10.1201/9781003362081-10

8.2 DEVELOPMENT OF THE ANALYTICAL FORMULATIONS

8.2.1 LATERAL SOIL PRESSURE IN UNSATURATED SOILS AS A FUNCTION OF EFFECTIVE STRESS

Generally, the coefficient of lateral soil pressure, k, should be calculated from the effective vertical stress and the effective horizontal stress, as shown by Eq. 8.1:

$$k = \frac{\sigma'_h}{\sigma'_v} \tag{8.1}$$

where σ'_h and σ'_v are the effective horizontal and the effective vertical stresses, respectively. The single-phase effective stress relationship in unsaturated soils was first formulated by Bishop,[35] as:

$$\sigma' = (\sigma - u_a) + \chi \times (u_a - u_w) = \sigma_{net} + \chi \times \psi = \sigma_{net} + \sigma_s \tag{8.2}$$

where σ', σ, σ_{net}, ψ, u_a, u_w, and σ_s are effective stress, total stress, net stress, matric suction, pore air pressure, pore water pressure, and suction-stress, respectively. In Eq. 8.2, χ is the effective stress parameter that is a function of soil saturation[36–45] and reflects the contribution of matric suction to effective stress. Typically, χ varies between zero (in fully dry soils) and unity (in fully saturated soils), and it is suggested that χ can be simply taken as the degree of saturation (S_r)[37,46,47] or as the effective saturation (S_e).[42–45]

In accordance with Eq. 8.2, effective horizontal and vertical stresses can be attained by Eqs. 8.3a and 8.3b, respectively:

$$\sigma'_h = \sigma_h + \chi \times (u_a - u_w) = \sigma_h + \sigma_s \tag{8.3a}$$

$$\sigma'_v = \sigma_v + \chi \times (u_a - u_w) = \sigma_v + \sigma_s \tag{8.3a}$$

where σ_h and σ_v are the net horizontal and net vertical stresses, respectively, that are conventionally controlled or measured during unsaturated tests. Therefore, by knowing the soil water retention behavior (i.e., soil water characteristic curve, SWRC) and considering a suitable relationship for changing the effective stress parameter (χ) against the matric suction (ψ), it is possible to calculate the coefficient of lateral soil pressure (k) for unsaturated soils by implementing Eqs. 8.1 and 8.3.

8.2.2 IMPLEMENTING THE DISTURBED STATE CONCEPT (DSC) FOR CALCULATING K

The DSC was first introduced by Desai[25–27] and extended by Geiser et al.[29] for constitutive modeling of unsaturated soils. Details of the DSC are presented by Desai.[28] As a constitutive framework, DSC defines the overall behavior of a deforming material in terms of the behavior of component materials with regard to continuum and disturbed states. The former is often denoted as *Relative Intact* (*RI*) and the latter as *Fully Adjusted* (*FA*).

Accordingly, the mechanical response of the soil in an *intermediate state*, P_{Int}, can be determined by knowing the two corresponding values in *RI* and *FA* states, namely, P_{RI} and P_{FA}. Hence, in accordance with DSC, the expression for calculating P_{Int} is:

$$P_{Int} = P_{RI} - (P_{RI} - P_{FA}) \times D \tag{8.4}$$

Eq. 8.4 provides a continuous expression to define P_{Int} in terms of P_{RI} and P_{FA}, in which P_{RI} and P_{FA} are soil responses at *RI* and *FA* states, respectively, and D is a disturbance parameter.

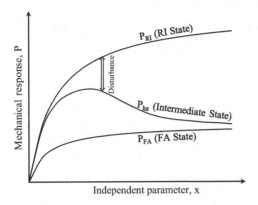

FIGURE 8.1 Schematic of P_{Int}, P_{RI}, and P_{FA}.

The disturbance parameter, D, defines the progression of degradation (softening) or healing and coupling between the *RI* and *FA* responses, and varies between zero and unity during the *RI* toward the ultimate *FA* state. D is typically defined in the form of an exponential function in terms of accumulative plastic volumetric strain, ζ_v, as shown by Eq. 8.5:

$$D = D_u\left(1 - e^{\left(-a\zeta_v^{\bar{z}}\right)}\right) \tag{8.5}$$

In Eq. 8.5, D_u, a, and z are model parameters that depend on the type of material and the loading conditions. In Figure 8.1, variations of P_{Int}, P_{RI}, and P_{FA} versus typical independent variable, x (e.g., a stress state parameter), are shown schematically.

According to the basic concepts of DSC, the soil experiences the most structural disturbance in the *FA* state, in which the soil has the least stiffness. On the other hand, in the *RI* state, the soil experiences a gradual structural disturbance.

Unsaturated soils can be considered to be in the *FA* condition when the matric suction of the soil is zero and the soil is fully saturated. Moreover, since the soil stiffness is *relatively* the highest (compared to other cases) under maximum applied matric suction, unsaturated soils can be considered to be in the *RI* condition when maximum matric suction is applied to the soil.

The conceptual framework of DSC can be implemented to demonstrate the variations of the unsaturated coefficient of lateral soil pressure, k, in terms of the effective vertical stress, σ'_v. Accordingly, by considering the coefficient of the lateral soil pressure at a given matric suction, k_ψ, as P_{int} and considering k values corresponding to the 0-kPa and the maximum applied matric suctions as k_{FA} and k_{RI} in Eq. 8.4, respectively, a general formulation for calculating k at a given matric suction, ψ, is obtained, as shown by Eq. 8.6.

$$k_\psi = k_{\text{RI}} - \left(k_{\text{RI}} - k_{\text{FA}}\right) \times D \tag{8.6}$$

In Eq. 8.6, k_{FA} and k_{RI} are two input parameters, whose variations against the effective vertical stress, σ'_v, must be known. Also, the disturbance parameter, D, is a function of three material parameters (namely, D_u, a, and z), as previously denoted in Eq. 8.5. Now, D_u, a, and z can be defined in terms of the applied matric suction to the soil, as shown by Eqs. 8.7–8.9:

$$D_u(\psi) = 1 - \left(\frac{S_{r(\text{RI})}}{S_r(\psi)} \times \frac{\log(1+\psi)}{\log(1+\psi_{\text{RI}})}\right)^{C_{Du}} \tag{8.7}$$

$$a(\psi) = a_o \times \left(\frac{S_{r(\mathrm{RI})}}{S_r(\psi)} \times \frac{\log(1+\psi)}{\log(1+\psi_{\mathrm{RI}})} \right)^{C_o} \tag{8.8}$$

$$z(\psi) = z_o \times \left(\frac{S_{r(\mathrm{RI})}}{S_r(\psi)} \times \frac{\log(1+\psi)}{\log(1+\psi_{\mathrm{RI}})} \right)^{C_z} \tag{8.9}$$

In Eqs. 8.7–8.9, $S_{r(\mathrm{RI})}$ is the degree of saturation of the soil at a matric suction corresponding to RI condition (ψ_{RI}) and $S_r(\psi)$ is the degree of saturation of the soil at a given matric suction, ψ. In addition, a_o, z_o, C_{Du}, C_a, and C_z are fitting parameters. The form of the relationships presented in Eqs. 8.7–8.9 for calculating D_u, a, and z guarantees the consistency of the stress-deformation behavior of the examined soils with the SWRC behavior of unsaturated soils and also satisfies the fundamentals of the DSC model.[33] Actually, these equations show a continuous behavior at a matric suction range from zero to the air entry suction value of the soil (i.e., fully saturated soil conditions) and also result in the maximum disturbance of the soil (with $D=1$), which is proportional to the FA state of the soil. Also, suction values higher than ψ_{RI} in Eqs. 8.7–8.9 result in a minimal disturbance in the soil structure and lead to $D=0$ in a continuous manner.

8.3   EXPERIMENTAL DATA

The constitutive parameters of the proposed DSC model are calculated in accordance with the experimental data from Pirjalili et al.[10] who performed laboratory tests on two different unsaturated soils (namely, Firouzkouh Clay [CL] and Sand-Kaolin mixture [SC]) using a developed unsaturated suction-controlled ring device. Accordingly, reconstituted specimens of a Sand-Kaolin mixture (with two different initial void ratios, e_o, of 0.52 and 0.72) and a Firouzkouh Clay (with two different initial void ratios of 0.71 and 0.92) were tested under six different matric suctions ($\psi = 0$, 10, 30, 50, 70, and 90 kPa) to examine the dependency of the unsaturated coefficient of lateral soil pressure on the void ratio and matric suction.

SWRCs of the examined soil materials under wetting paths are shown in Figure 8.2 along with the fitted curves and corresponding parameters obtained from van Genuchten's suggested model,[48] as presented by Eq. 8.10:

$$\psi = \frac{S_{ro}}{\left(1+(\alpha\psi)^n\right)^m} \tag{8.10}$$

In Eq. 8.10, S_{ro} is the degree of saturation of the soil at zero matric suction, and α, m, and n are fitting parameters that are related to the pore size and pore size distribution of the soil.

Figure 8.3 plots the variations of the net horizontal stress, σ_h, versus the net vertical stress, σ_v, for different soil specimens. In addition, the variations of the void ratio of the soil, e, versus net vertical stress, σ_v, are shown in Figure 8.4.

As previously mentioned, the two upper and lower structural disturbance states of unsaturated soils, FA and RI states, can be taken as the soil states at the minimum and the maximum applied matric suction conditions, respectively. Accordingly, data presented for $\psi=0$ kPa and $\psi=90$ kPa are considered as FA and RI data sets, respectively. Variations of σ'_h vs. σ'_v are used in Eq. 8.1 to calculate the experimental values of k_{FA} and k_{RI}. Figure 8.5 illustrates variations of k_{FA}, and k_{RI} vs. σ'_v for the examined soils.

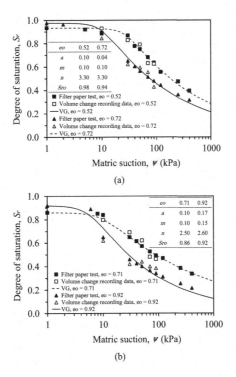

FIGURE 8.2 Soil water characteristic curve (SWRC) of the examined soils: (a) Sand-Kaolin mixture and (b) Firouzkouh Clay.[10]

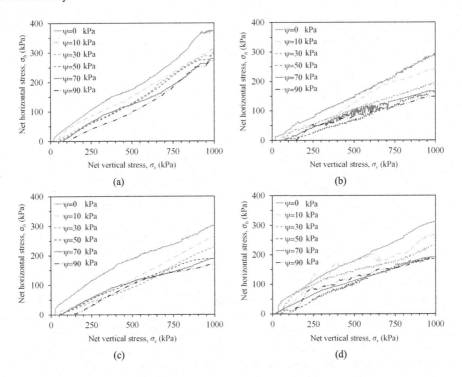

FIGURE 8.3 Variations of the net horizontal stress, σ_h, vs. the net vertical stresses, σ_v, for: (a) Sand-Kaolin Mixture with e_o=0.52, (b) Sand-Kaolin Mixture, e_o=0.72, (c) Firouzkouh Clay, e_o=0.71, and (d) Firouzkouh Clay, e_o=0.92.[10]

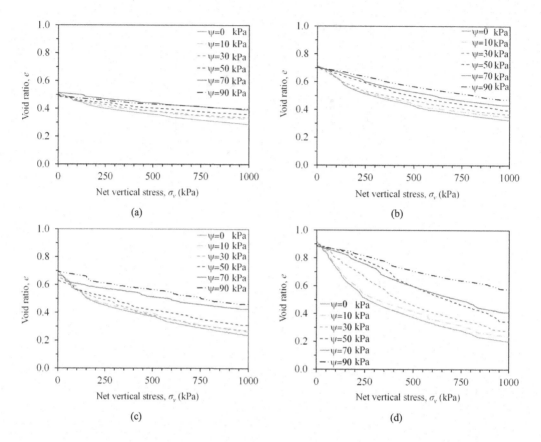

FIGURE 8.4 Variation of the void ratio, e, vs. net vertical stress, σ_v, for: (a) Sand-Kaolin Mixture, $e_o=0.52$, (b) Sand-Kaolin Mixture, $e_o=0.72$, (c) Firouzkouh Clay, $e_o=0.71$, and (d) Firouzkouh Clay, $e_o=0.92$.[10]

8.4 CALCULATION OF MODEL PARAMETERS

Suction-dependent DSC parameters (*i.e.*, D_u, a, and z) are calculated for different soils by considering the constitutive relationships mentioned in the previous sections, using the experimental data reported by Pirjalili et al.,[10] and the results are plotted in Figure 8.6. In addition, values of parameters a_0 and z_0 and variations of C_{Du}, C_a, and C_z against matric suction were calculated using the least square error estimation method for different soil conditions, and the results are presented in Table 8.1 and Figure 8.7, respectively.

As depicted in Figure 8.6, similar trends for each DSC parameter are observed against matric suction for all studied soils. Accordingly, D_u decreases as matric suction values increase so that it has a value of unity in the *FA* state (when $\psi=0$) and then gradually decreases to zero in the *RI* state (when $\psi=90\,$kPa). Similar trends are observed for variations of a and z versus matric suction, as these parameters have zero values in the *FA* state and then gradually reach their maximum values (a_o and z_o, respectively) in the *RI* state. Figure 8.6 also shows that for a given soil type, larger a and z parameters are obtained for specimens with lower initial void ratios in comparison with looser samples. In contrast, larger D_u values were obtained for specific soil-type specimens with higher initial void ratios than dense specimens, which implies a higher structural disturbance potential in looser soil specimens. Moreover, Table 8.1 suggests that the values of a_o and z_o depend on the soil type and the initial void ratio of the soil. Therefore, we deduce that for each type of soil, a_o and z_o have greater values at lower initial void ratios in comparison with the loose samples.

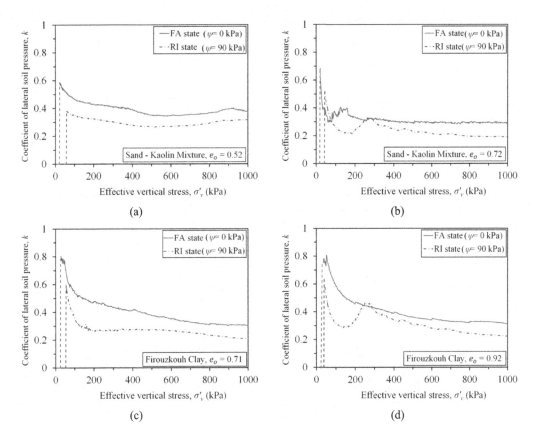

FIGURE 8.5   Variation of k_{FA} and k_{RI} vs. σ'_v for: (a) Sand-Kaolin Mixture, e_o=0.52; (b) Sand-Kaolin Mixture, e_o=0.72; (c) Firouzkouh Clay, e_o=0.71; and (d) Firouzkouh Clay, e_o=0.92.[10]

Data presented in Figure 8.7 show that parameters C_{Du}, C_a, and C_z decrease as the matric suction increases. In C_a and C_z, this reduction continues until their values reach zero at high matric suctions. In addition, it is observed that C_{Du}, C_a, and C_z have greater values for the specimens with higher initial void ratios in comparison with dense soil specimens.

The variation of the coefficient of lateral soil pressure of unsaturated soil, k, is calculated by considering experimental data reported by Pirjalili et al.,[10] implementing the calculated values of k_{FA} and k_{RI} as shown in Figure 8.5 and considering values of C_{Du}, C_a, C_z and D_u, z, a, as presented by Figures 8.6 and 8.7; it is illustrated in terms of the effective vertical stress, σ'_v, in Figures 8.8 and 8.9 for Sand-Kaolin Mixture and Firouzkouh Clay specimens, respectively. Corresponding experimental values of k against σ'_v are also plotted for comparison in Figures 8.8 and 8.9. Accordingly, very good agreement is observed between the model predictions and the experimental data.[i]

Data illustrated in Figures 8.8 and 8.9 indicate that k increases rapidly after increasing the effective vertical stress up to a maximum in the early stages of loading and then decreases to an asymptotic value. In Figures 8.8 and 8.9, the maximum value of k represents the unsaturated at-rest coefficient of the lateral soil pressure, k_o. By further increasing the effective vertical stress and surpassing the at-rest condition, the lateral deformations within the soil mass increase, and k continually decreases until the soil reaches its active limit state of failure and k asymptotically approaches its corresponding active value, k_a.

The validity of the proposed DSC model is assessed by making a quantitative comparison between the experimental and model-predicted k values. To this end, the root mean square error

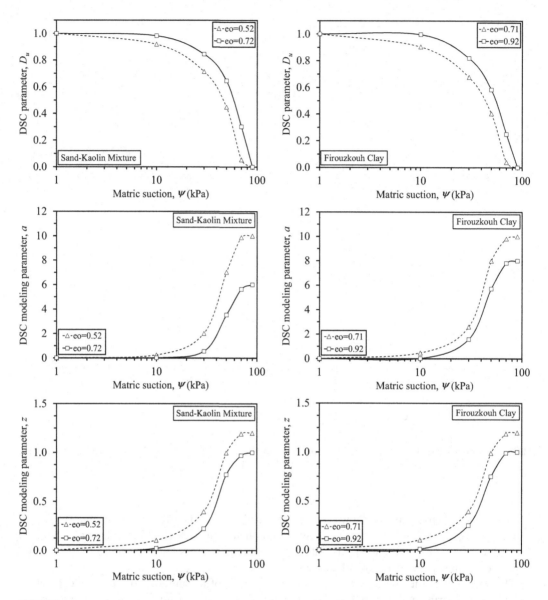

FIGURE 8.6 Variations of D_u, a, and z vs. matric suction, ψ, for the examined soils.

TABLE 8.1
Suggested Values for a_o and z_o for the Examined Soils

Parameter	Sand-Kaolin Mixture		Firouzkouh Clay	
	$e_o = 0.52$	$e_o = 0.72$	$e_o = 0.71$	$e_o = 0.92$
a_o	10	6	10	8
z_o	1.2	1	1.2	1

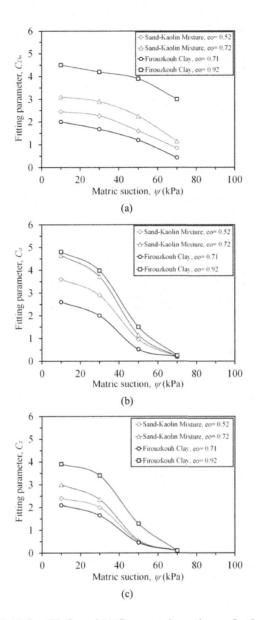

FIGURE 8.7 Variations of: (a) C_{Du}, (b) C_a, and (c) C_z vs. matric suction, ψ, for the examined soils.

(*RMSE*) and the normalized root mean square error (*NRMSE*) parameters are calculated for comparative data sets by implementing Eqs. 8.11 and 8.12:

$$\text{RMSE} = \sqrt{\frac{\sum_{i=1}^{N}\left(k_{\text{experimental}} - k_{\text{predicted}}\right)^2}{N}} \tag{8.11}$$

$$\text{NRMSE} = \frac{\text{RMSE}}{\text{Max}\left(k_{\text{experimental}}\right) - \text{Min}\left(k_{\text{experimental}}\right)} \tag{8.12}$$

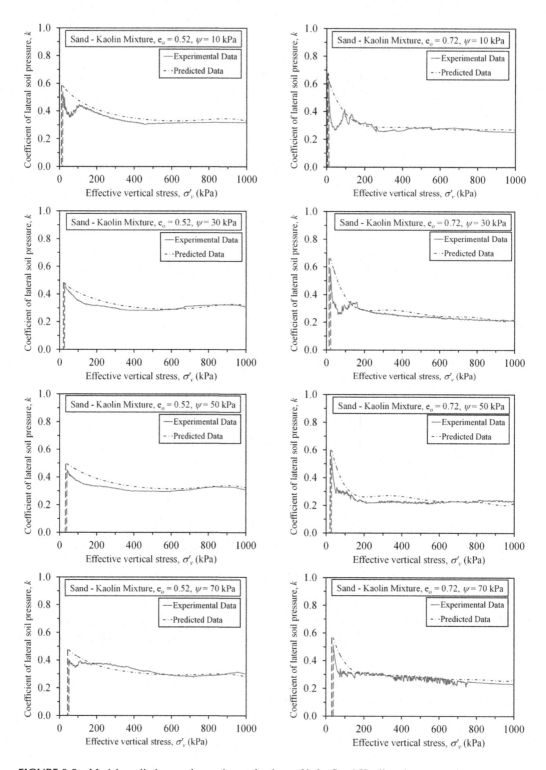

FIGURE 8.8 Model predictions and experimental values of k for Sand-Kaolin mixture specimens.

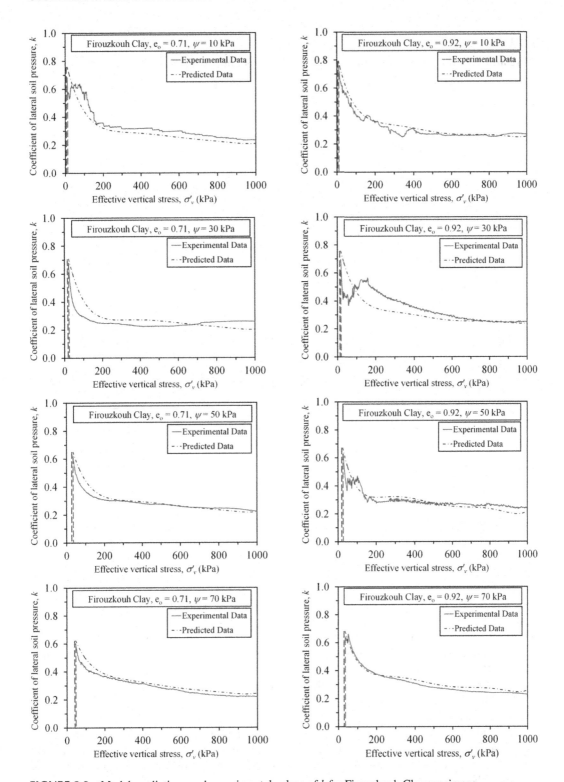

FIGURE 8.9 Model predictions and experimental values of k for Firouzkouh Clay specimens.

TABLE 8.2

RMSE and NRMSE Parameters for Model Predictions and Experimental k Values

Soil Type	Matric Suction ψ (kPa)	RMSE	NRMSE (%)	RMSE	NRMSE (%)
Sand-Kaolin		$e_o = 0.52$		$e_o = 0.72$	
Mixture	10	0.041	8.0	0.052	8.4
	30	0.026	5.4	0.050	8.1
	50	0.035	7.8	0.053	8.5
	70	0.038	7.9	0.049	7.9
Firouzkouh		$e_o = 0.71$		$e_o = 0.92$	
Clay	10	0.051	7.4	0.031	4.5
	30	0.06	8.2	0.065	10.2
	50	0.029	4.6	0.035	5.2
	70	0.035	5.4	0.021	3.2

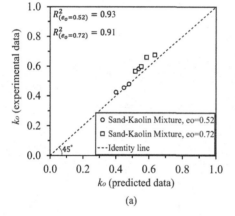
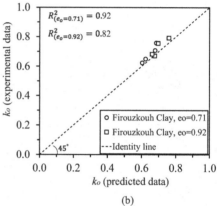

(a)   (b)

FIGURE 8.10   Comparison between the model-predicted and experimental values of k_o for: (a) Sand-Kaolin Mixture and (b) Firouzkouh Clay.

In Eq. 8.11, $k_{\text{experimental}}$ and $k_{\text{predicted}}$ are the experimental and model-predicted values of k, respectively, N is number of data in each data set, and i is the numeral. In addition, in Eq. 8.12, $\text{Max}(k_{\text{experimental}})$ and $\text{Min}(k_{\text{experimental}})$ are the maximum and minimum values of the experimental values of k, respectively. Calculated values of RMSE and NRMSE parameters are shown in Table 8.2 for the comparative data sets.

Data presented in Table 8.2 indicate that the maximum relative and normalized errors are (0.053% and 8.5%) and (0.065% and 10.2%) for the Sand-Kaolin mixture and the Firouzkouh Clay specimens, respectively.

By considering data summarized in Table 8.2 and taking into account Figures 8.8 and 8.9, we see excellent conformance is obtained between the model predictions and experimental variations of k versus effective vertical stress, σ'_v. In addition, the three distinct phases in variation of k versus σ'_v, namely the at-rest, transition, and active phases, are shown to be well captured by the proposed effective stress-based DSC model. To display the accuracy and reliability of the proposed DSC model, experimental and model-predicted values of k_o and k_a are compared in Figures 8.10 and 8.11, respectively, along with the identity lines and corresponding R-squared values. Figures 8.10 and 8.11

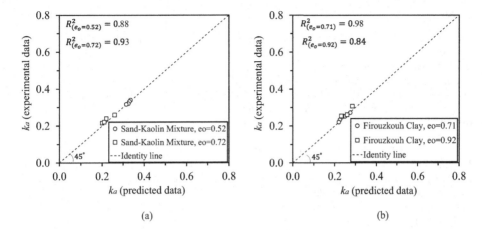

FIGURE 8.11 Comparison between the model-predicted and experimental values of k_a for: (a) Sand-Kaolin Mixture and (b) Firouzkouh Clay.

clearly show that predicted and experimental data are mostly aligned with the identity lines, which implies that the proposed model can predict the at-rest and active k values for the examined soils appropriately.

8.5 CONCLUSIONS

An effective stress-based analytical approach has been introduced for predicting the coefficient of lateral pressure in unsaturated soils, k, based on the DSC. The parameters of the proposed model have been defined as SWRC-compatible functions. The proposed analytical framework has been validated by considering experimental data for a Sand-Kaolin Mixture and a Firouzkouh Clay; each had two different initial void ratios and was tested under six applied matric suctions. Accordingly, k values measured under the minimum and maximum applied matric suction were used to calculate the unsaturated coefficient of the lateral soil pressures at the lower and upper structural disturbance boundaries, respectively. It has been found that the disturbance parameter, D, of the examined unsaturated soils is considerably affected by the soil matric suction and initial void ratio of the examined soils.

Quantitative comparisons made between the model predictions and experimental values of k revealed that *NRMSE* values vary in the range of 3.2%–10.2%. This indicates that the proposed model is capable of properly predicting the unsaturated coefficient of lateral soil pressure values under increasing effective vertical stress and for different state variables and soil parameters, such as the matric suction and initial void ratio.

NOTE

1 *Note:* Effective stress values recorded by implemented sensors have shown irregular fluctuations and jumps during the experimental work and at the beginning of the loading stages, and this might be due to instrument errors at very low stress values, inter-particle rearrangement within the soil texture, and initiation of stress-strain mobilization. These fluctuations were also observed in variations of the lateral soil pressure versus effective vertical stress before the soil reached its maximum lateral soil pressure (*i.e.,* its at-rest state). Since the mathematical description and conceptual interpretation of these irregularities were not possible using the proposed DSC framework, the mentioned jumps and fluctuations are replaced with vertical dashed-lines according to their general trends in Figures 8.8 and 8.9.

REFERENCES

1. Terzaghi, K. "Old earth-pressure theories and new test results." *Eng. News-Record*, 85, 14, 1920, 632–637.
2. Jacky, J. "The coefficient of earth pressure at rest." In Hungarian (A nyugalmi nyomas tenyezoje), *J. Soc. Hung. Eng. Arch.* (Magyar Mernok es Epitesz-Egylet Konzlonye), 1944, 355–358.
3. Jacky, J. "Pressure in Soils." In *Proceeding of the 2nd International Conference on Soil Mechanic and Foundation Engineering*, Rotterdam, Netherlands, 1948, 107–130.
4. Bishop, A.W., and Henkel, D. *The Measurement of Soil Properties in the Triaxial Test.* E. Arnold, London, UK, 1957.
5. Okochi, Y., and Tatsuoka, F. "Some factors affecting ko values of sand measured in triaxial cell." *Soils Found.*, 24, 3, 1984, 52–68. https://doi.org/10.3208/sandf1972.24.3_52.
6. Mesri, G., and Hayat, T. "The coefficient of earth pressure at rest." *Can. Geotech. J.*, 30, 4, 1993, 647–666. https://doi.org/10.1139/t93-056.
7. Chu, J., and Gan, C. "Effect of void ratio on ko of loose sand." *Géotechnique*, 54, 4, 2004, 285–288. https://doi.org/10.1680/geot.2004.54.4.285.
8. Northcutt, S., and Wijewickreme, D. "Effect of particle fabric on the coefficient of lateral soil pressure observed during one-dimensional compression of sand." *Can. Geotech. J.*, 50, 5, 2013, 457–466. https://doi.org/10.1139/cgj-2012-0162.
9. Mayne, P. W., and Kulhawy, F. H. "Ko-OCR relationships in soil." *J. Geotech. Eng. Div.*, 108, 6, 1982, 851–872. https://doi.org/10.1061/ajgeb6.0001306.
10. Pirjalili, A., Garakani, A. A., Golshani, A., and Mirzaii, A. "A suction-controlled ring device to measure the coefficient of lateral soil pressure in unsaturated soils." *Geotech. Test. J.*, 43, 6, 2020, 1379–1396. https://doi.org/10.1520/GTJ20190099.
11. Li, Z. W., and Yang, X. L. "Active earth pressure for soils with tension cracks under steady unsaturated flow conditions." *Can. Geotech. J.*, 55, 12, 2018, 1850–1859. https://doi.org/10.1139/cgj-2017-0713.
12. Zhang, R., Zheng, J. L., and Yang, H. P. "Experimental study on K0 consolidation behavior of recompacted unsaturated expansive soil." In *Recent Advancement in Soil Behavior, in Situ Test Methods, Pile Foundations, and Tunneling: Selected Papers, from the 2009 GeoHunan International Conference*, 2009, 27–32. https://doi.org/10.1061/41044(351)5.
13. Yang, H. P., Xiao, J., Zhang, G. F., and Zhang, R. "Experimental study on ko coefficient of Ningming unsaturated expansive soil." In *Proceedings of the 5th International Conference on Unsaturated Soils*, 1, 2011, 397–401.
14. Oh, S., Lu, N., Kim, T., and Lee, Y. "Experimental validation of suction stress characteristic curve from nonfailure triaxial ko consolidation tests." ASCE, *J. Geotech. Geoenviron.*, 139, 9, 2013, 1490–1503. https://doi.org/10.1061/(ASCE)GT.1943-5606.0000880.
15. Li, J., Xingand, T., and Hou, Y. *Centrifugal Model Test on the at-Rest Coefficient of Lateral Soil Pressure in Unsaturated Soils.* CRC Press, Perth, AU, 2014, 931–936.
16. Monroy, R., Zdravkovic, L., and Ridley, A. "Evaluation of an active system to measure lateral stresses in unsaturated soils." *ASTM, Geotech. Test. J.*, 37, 1, 2014, 1–14. https://doi.org/10.1520/GTJ20130062.
17. Silva, T. A., Delcourt, R. T., and de Campos, T. M. "Study of the Coefficient of At-Rest Earth Pressure for Unsaturated Residual Soils with Different Weathering Degrees." *In PanAm Unsaturated Soils*, 2017, 451–460. https://doi.org/10.1061/9780784481707.045.
18. Abrantes, L. G., and Pereira de Campos, T. M. "Evaluation of the coefficient of earth pressure at rest (K0) of a saturated-unsaturated colluvium soil." A. Tarantino and E. Ibraim, eds. *E3S Web of Conferences*, 92, 2019 p.07006. https://doi.org/10.1051/e3sconf/20199207006.
19. Liang, W. B., Zhao, J. H., Li, Y., Zhang, C. G., and Wang, S. "Unified solution of Coulomb's active earth pressure for unsaturated soils without crack." In *Appl. Mech. Mater.* 170, 2012, 755–761. Trans Tech Publications Ltd. https://doi.org/10.4028/www.scientific.net/AMM.170-173.755.
20. Fathipour, H., Siahmazgi, A. S., Payan, M., and Chenari, R. J. "Evaluation of the lateral earth pressure in unsaturated soils with finite element limit analysis using second-order cone programming." *Comput. Geotech.*, 125, 2020, p.103587. https://doi.org/10.1016/j.compgeo.2020.103587.
21. Vahedifard, F., Leshchinsky, B. A., Mortezaei, K., and Lu, N. "Active earth pressures for unsaturated retaining structures." *ASCE, J. Geotech. Geoenviron.*, 141, 11, 2015, p.04015048. https://doi.org/10.1061/(ASCE)GT.1943-5606.0001356.
22. Shahrokhabadi, S., Vahedifard, F., Ghazanfari, E., and Foroutan, M. "Earth pressure profiles in unsaturated soils under transient flow." *Eng. Geol.*, 260, 2019, p.105218. https://doi.org/10.1016/j.enggeo.2019.105218.

23. Li, Z., and Yang, X. "Three-dimensional active earth pressure for retaining structures in soils subjected to steady unsaturated seepage effects." *Acta Geotech.*, 2019, 1–13. https://doi.org/10.1007/s11440-019-00870.2.

24. Garakani, A. A., Pirjalili, A. and Desai, C. S. "An effective stress-based DSC model for predicting the coefficient of lateral soil pressure in unsaturated soils." *Acta Geotech.*, 2021, 1–18. https://doi.org/10.1007/s11440-021-01376-6.

25. Desai, C. S. "A consistent finite element technique for work-softening behavior." In *Proceedings of the International Conference on Computational Methods in Nonlinear Mechanics.* Edited by Oden J.T. Univ. of Texas, Austin, TX, 1974.

26. Desai, C. S. "Hierarchical single surface and the disturbed state constitutive models with emphasis on geotechnical application." In *Geotechnical Engineering: Emerging Trend in Design and Practice*, Edited by Saxeng K.R. Oxford IBH Publishing Company, New Delhi, 32, 3, 1994, 115–154. https://doi.org/10.1016/0148-9062(95)90125-o.

27. Desai, C. S. "Constitutive modeling using the disturbed state as a microstructure self - adjustment concept." In Chapter 8, *Continuum Models for Material with Microstructure*, Edited by Mühlhaus H.B. John Wiley and Sons, London, 1995.

28. Desai, C. S. *Mechanics of Materials and Interfaces: The Disturbed State Concept.* CRC Press, Boca Raton, FL, 2001. https://doi.org/10.1201/9781420041910.ch2.

29. Geiser, F., Laloui, L., Vulliet, L., and Desai, C. S. "Disturbed state concept for constitutive modeling of partially saturated porous materials." In *Proceedings of the 6th International Symposium on Numerical Models in Geomechanics*, Balkema, Rotterdam, Netherlands, 1997, 129–134.

30. Desai, C. S. "Constitutive modeling of materials and contacts using the disturbed state concept. Part 1: Background and analysis." *J. Comput. Struct.*, 146, 2015a, 214–233. https://doi.org/10.1016/j.jrmge.2016.01.003.

31. Desai, C. S. "Constitutive modeling of materials and contacts using the disturbed state concept. Part 2: Validations at specimen and boundary value problem levels." *J. Comput. Struct.*, 146, 2015b, 234–251. https://doi.org/10.1016/j.jrmge.2016.01.003.

32. Desai, C. S. "Disturbed state concept as unified constitutive modeling approach." *J. Rock Mech. Geotech. Eng.*, 8, 3, 2016, 277–293. https://doi.org/10.1016/j.jrmge.2016.01.003.

33. Garakani A. A., Haeri. S. M., Desai, C. S., Seyed Ghafouri, S. M. H., Sadollahzadeh, B., and Hashemi Senejani, H. "Testing and constitutive modeling of lime-stabilized collapsible loess. II: Modeling and validations." *ASCE, Int. J. Geomech.*, 19, 4, 2019, 04019007. https://doi.org/10.1061/(ASCE)GM.1943-5622.0001386.

34. Haeri, S. M., Garakani, A. A., Roohparvar, H. R., Desai, C. S., Seyed Ghafouri, S. M. H., and Kouchesfahani, K. S. "Testing and constitutive modeling of lime-stabilized collapsible loess. I: Experimental investigations." *ASCE, Int. J. Geomech.*, 19, 4, 2019, 4019006. https://doi.org/10.1061/(ASCE)GM.1943-5622.0001364.

35. Bishop, A. W. "The principle of effective stress." *Teknisk ukeblad*, 39, 1959, 859–863.

36. Lu, N., and Likos, W. J. "Suction stress characteristic curve for unsaturated soil." *ASCE, J. Geotech. Geoenviron.*, 132, 2, 2006, 131–142. https://doi.org/10.1061/(ASCE)1090-0241(2006)132:2(131).

37. Khalili, N., and Khabbaz, M. H. "A unique relationship for the determination of the shear strength of unsaturated soils." *Géotechnique*, 48, 5, 1998, 681–687. https://doi.org/10.1680/geot.1998.48.5.681.

38. Wheeler, S. J., Sharma, R. S., and Buisson, M. S. R. "Coupling of hydraulic hysteresis and stress–strain behaviour in unsaturated soils." *Géotechnique*, 53, 1, 2003, 41–54. https://doi.org/10.1680/geot.2003.53.1.41.

39. Nuth, M., and Laloui, L. "Effective stress concept in unsaturated soils: Clarification and validation of a unified framework." *Int. J. Numer. Anal. Met.*, 32, 7, 2008, 771–801. https://doi.org/10.1002/nag.645.

40. Lu, N., Godt, J., and Wu, D. "A closed-form equation for effective stress in unsaturated soil." *Water Resour.*, 46, 2010, W05515. https://doi.org/10.1029/2009WR008646.

41. Khalili, N., and Zargarbashi, S. "Influence of hydraulic hysteresis on effective stress in unsaturated soils." *Géotechnique*, 60, 9, 2010, 729–734. https://doi.org/10.1680/geot.09.T.009.

42. Garakani, A. A. "Laboratory assessment of the hydro-mechanical behavior of unsaturated undisturbed collapsible soils—Case study: Gorgan loess." Dissertation for Doctoral Degree, Sharif University of Technology, Tehran, Iran, 2013.

43. Haeri, S. M., Garakani, A. A., Khosravi, A., and Meehan, C. L. "Assessing the hydro-mechanical behavior of collapsible soils using a modified triaxial test device." *ASTM, Geotech. Test. J.*, 37, 2, 2014, 190–204. https://doi.org/10.1520/GTJ20130034.

44. Haeri, S. M., Khosravi, A., Garakani, A. A., and Ghazizadeh, S. "Effect of soil structure and disturbance on hydromechanical behavior of collapsible loessial soils." *ASCE, Int. J. Geomech.*, 17, 1, 2017,
04016021. https://doi.org/10.1061/(ASCE)GM.1943-5622.0000656.

45. Garakani, A. A., Haeri, S. M., Khosravi, A., and Habibagahi, G. "Hydro-mechanical behavior of undisturbed collapsible loessial soils under different stress state conditions." *Eng. Geol.*, 195, 2015, 28–41.
https://doi.org/10.1016/j.enggeo.2015.05.026.

46. Schrefler, B. A. The finite element method in soil consolidation. Doctoral dissertation, University
College of Swansea, UK. 1984.

47. Borja, R. I. "Cam-Clay plasticity. Part V: A mathematical framework for three-phase deformation and
strain localization analyses of partially saturated porous media." *Comp. Met. Appl. Mech. Eng.*, 193,
48–51, 2004, 5301–5338. https://doi.org/10.1016/j.cma.2003.12.067.

48. van Genuchten, M. T. "A closed-form equation for predicting the hydraulic conductivity of unsaturated
soils." *Soil Sci. Soc. Am. J.*, 44, 5, 1980, 892–898. https://doi.org/10.2136/sssaj1980.0361599500440005
0002x.

Section 3

Structured and Bio- or Chem-Treated Soils

9 Compression of Structured Geomaterials

Martin D. Liu
School of Civil, Mining and Environmental Engineering,
University of Wollongong

John P. Carter
School of Engineering, University of Newcastle

9.1 INTRODUCTION

The compression behavior of geomaterials is often investigated because valuable information can be obtained, such as important parameters for deformation analyses, e.g., Chai and Carter [1]. Theoretical modeling of the compression behavior of geomaterials can also be employed as a basis for the generation of constitutive models that may apply to more general stress conditions, e.g., Liu et al. [2,3]. However, predicting the compression behavior of geomaterials, which may appear to be simple because of the inherent stress and strain conditions, is surprisingly complicated. Furthermore, theoretical modeling has traditionally been carried out by classifying geomaterials into several groups and treating them separately. The major groups include clays, sands and gravels, calcareous soils, rock, and geomaterials artificially treated by the addition of mechanical reinforcement or by adding chemical agents.

Two major aspects of geomaterials are usually recognized. One is the variation in material composition, i.e., grain size and mineralogy. The other is the variation in the structure of the geomaterial, i.e., the arrangement and bonding of the constituents. Structure is defined here in a very broad sense in that it encompasses all mechanical behavioral features of a geomaterial that are different from those of the material with the same mineralogy at a selected reference state. Therefore, the geomaterial structure and its influence, as defined in this chapter, are relative quantities and depend on the selected reference state. The removal of geomaterial structure is commonly referred to as "destructuring" or "destructuration" and is usually due to stress excursions. Destructuring is generally an irrecoverable and progressive process. The presence and variation of material structure during loading constitutes a major challenge for reliable predictions of the structured geomaterial's mechanical response. Previous research in this area may be found in papers by Burland [4], Liu and Carter [5], and Mesri and Vardhanabhuti [6].

The DSC/HISS, proposed by Desai [7–9] and Desai and Toth [10], describes the response of any material by relating it to the responses of that material at two reference states. It thus provides a powerful theoretical framework for studying the behavior of a structured geomaterial by using knowledge of the properties of the material at these reference states. A key to the successful use of this approach is identification of an appropriate 'disturbance function', which defines the transition from one reference state to the other states. Liu et al. [11,12] conducted studies involving modeling the compression behavior of geomaterials with various structures. They demonstrated that the disturbed state concept (DSC) framework provides a useful method for describing the transition in behavior of structured clay from that of the material in its undisturbed state to that at its corresponding reference (or reconstituted) state.

Ouria [13] suggested a compression model for structured soils with a variable slope of the compression curve. This model was developed using the distributed state concept and was implemented in the modified Cam-clay constitutive model. It was verified using experimental data available in the literature for oedometer and triaxial tests on structured soils.

Xu et al. [14] investigated the triaxial compression behavior of loess soil and developed a constitutive model accounting for the structural effect of loess based on the DSC. By analyzing triaxial compression test results, they established a disturbance function with respect to the volumetric and shear moduli parameters. They verified their constitutive model using laboratory test results and indicated that a double-parameter disturbance function evolves in an exponential form, capturing well the effect of moisture content and confining pressure on the behavior of loess.

The previously cited studies and others, including Zhou et al. [15], Liu et al. [16], Qian et al. [17], and Liu et al. [18], all indicate the importance of identifying an appropriate disturbance function in order to capture accurately the destructuring of initially structured soils. In this chapter, a general disturbance function first proposed by Liu et al. [12] is introduced. The general compression model thus formed simulates successfully the behavior of a very wide range of structured geomaterials such as clay, sand, gravel, calcareous soil, soft rock, and reinforced, cemented, and chemically treated soils. The powerful potential of the DSC for modeling the influence of material structures on the mechanical responses is demonstrated.

9.2 SUMMARY OF THE DISTURBED STATE CONCEPT

9.2.1 DISTURBED STATE CONCEPT (DSC)

The DSC was first proposed by Desai [8]. In the theory, the deformation of a material to stress changes is written in terms of the deformation of the material at two reference states, namely, the "relatively intact" (RI) state and the "fully adjusted" (FA) state, e.g., Desai [8] and Desai and Toth [10]. The observed response of the material is then related to the responses at the reference states through the disturbance function, which provides a coupling and interpolation mechanism between the response of the material in the RI and FA states. The basic constitutive equation is generally written in incremental forms and as follows (Chapter 1):

$$d\varepsilon_{ij}^a = \left(1 - D_\varepsilon\right)d\varepsilon_{ij}^i + D_\varepsilon d\varepsilon_{ij}^c + dD_\varepsilon\left(\varepsilon_{ij}^c - \varepsilon_{ij}^i\right) \tag{9.1}$$

where ε_{ij} is the Cartesian component of the strain tensor, the superscript 'a' indicates quantities associated with the observed or the actual material response, the superscript 'c' indicates quantities associated with the response of the FA material, and the superscript 'i' indicates quantities associated with the response of the RI material. D is a scalar disturbance function relating the actual material response to the responses of the material at the two reference states. D_ε indicates specifically the disturbance function for strain quantities.

9.2.2 GENERAL STRESS–STRAIN EQUATIONS BASED ON THE DSC

For modeling the influence of the geomaterial structure, the FA state is selected. This state is the selected reference state. Different reference states can be selected based on the particular problem. For clay, the FA state is usually selected as its response to loading in the reconstituted state; thus, there is no influence of soil structure on its deformation. The RI state is chosen to be the 'zero state', i.e., the state with no response to stress (a perfectly rigid material). In this case, $\varepsilon_{ij}^i \equiv 0$, based on Eq. (9.1), the following general equations for geomaterials are obtained:

$$d\varepsilon_v^a = \bar{D}_{\varepsilon v}d\varepsilon_v^c \tag{9.2}$$

$$d\varepsilon_d^a = \bar{D}_{\varepsilon d}d\varepsilon_d^c \tag{9.3}$$

where $\bar{D}_{\varepsilon v}$ and $\bar{D}_{\varepsilon d}$ are the separate disturbance functions for volumetric deformation ε_v and distortional deformation ε_d. For modeling the compression behavior of geomaterials with various structures, only the volumetric deformation is considered. Therefore, Eq. (9.2) is employed to simulate the compression behavior of structured geomaterials, and the most important work is to define the disturbance function for volumetric deformation.

9.2.3 REFERENCE STATES FOR STRUCTURED GEOMATERIALS

The framework of DSC is hierarchical, and thus possesses the possibility and convenience to select the reference state behavior according to the practical problem of interest and the knowledge and data available. As previously indicated, the influence of geomaterial structure is defined as the difference in the mechanical behavior between the structured material and the material at a selected reference state. Theoretically, there is no restriction on the choice of selected reference state, but because it is employed as a basis for the phenomenological modeling of the behavior of a material with structure, it is rational that the state chosen for the reference behavior should be one that has been relatively well studied, and preferably simple.

The following suggestions are made for geomaterials based on our research, e.g., Liu et al. [11,12,22]. For cohesive soils such as clay and some calcareous soils, the reference state is normally selected as the behavior in the reconstituted state. However, for noncohesive materials such as sands and gravels, a standard method to reconstitute the material is yet to be defined, and the terms 'reconstituted sand' and 'reconstituted gravel' have little meaning in geotechnical engineering. For sands, the behavior of materials with minimum or maximum voids ratio are contenders for adoption as the reference state behavior. For geomaterials artificially treated, such as by mechanical reinforcement or by the addition of chemical agents, the mechanical behavior of the material without artificial treatment may be selected as the reference state. For artificially treated geomaterials, it should be pointed out that differences exist in material compositions between the material at the reference state and the material with structures. For clarity, the properties of a geomaterial at the reference state are called the reference properties and are often denoted by the symbol * attached to the relevant mathematical symbols.

9.3 DSC COMPRESSION MODEL FOR STRUCTURED GEOMATERIALS

The general DSC compression model proposed by Liu et al. [12] is introduced here. The model is based on three assumptions.

Assumption 1: The compression behavior of a geomaterial is divided into two regions in stress space by the current yield stress, i.e., an elastic region and a virgin yielding region. The initial yield stress before loading is determined by the initial structure of the geomaterial. During virgin yielding, the current yield stress increases, and its value is equal to the current mean effective stress. The current yield stress remains constant for deformations within the elastic region.

The material idealization of the compression behavior of a structured geomaterial is consequently formed, shown in Figure 9.1. For structured geomaterials, there usually exists an initial yield stress, denoted by $p'_{y,i}$. For initial loading with $p' \leq p'_{y,i}$, purely elastic deformation is produced. For loading with $p' > p'_{y,i}$, virgin yielding occurs with both elastic and plastic deformation, and the current stress is the current yield stress, denoted by p'_y. Therefore, for initial loading AB and unloading CD and reloading DC, the geomaterial behaves as an elastic material. For loading along stress paths BC and CE, virgin yielding behavior is predicted. The following equation for the current yield stress p'_y is thus proposed:

$$\begin{cases} p'_y = p'_{y,i} & \text{before initial yielding} \\ p'_y = p'_{\max} & \text{after initial yielding} \end{cases} \tag{9.4}$$

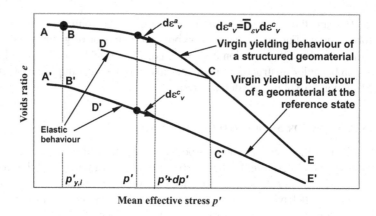

FIGURE 9.1   Compression behavior of geomaterials at reference and structured states.

Assumption 2: During purely elastic behavior, the deformation of a geomaterial is independent of the material structure.

In the constitutive equation for elastic volumetric deformation suggested previously by the authors [12], it was proposed that the volumetric strain increment at the structured state is equal to that at the reference state. Because the voids ratios at the two states are different (as seen in Figure 9.1), the voids ratio change in the two states will generally be different, and the assumption that "the elastic deformation of a geomaterial is independent of the material structure" is not strictly satisfied. In the version of the general compression model for structured geomaterials presented here, a new constitutive equation is proposed:

$$d\varepsilon_v^a = \begin{cases} \dfrac{(1+e^*)}{(1+e)} d\varepsilon_v^c & \text{for } p' < p_y' \\[2ex] \bar{D}_{\varepsilon v} d\varepsilon_v^c & \text{for } p' > p_y' \end{cases} \tag{9.5}$$

A consequence of this assumption is that the elastic behavior of a structured geomaterial is the same as that of the material at the reference state. Based on the available experimental evidence, e.g., Burland [4], Hight et al. [19], and Lorenzo and Bergado [20], it is seen that material structure can have some influence on its elastic deformation, and Assumption 2 is an approximation for the purpose of simplification. The approximation of this assumption will be examined in Section 9.4 on model performance.

Assumption 3: During compression loading, the disturbance function $\bar{D}_{\varepsilon v}$ for a geomaterial with a given structure varies with the current yield stress only and can be expressed by the following equation:

$$\bar{D}_{\varepsilon v} = 1 + b \left(\frac{p_{y,i}'}{p_y'} \right)^r \tag{9.6}$$

Parameters b and r quantify the effect of compression disturbance on geomaterial behavior. For a geomaterial with a given structure, the parameters b, r, and $p_{y,i}'$ are assumed to be constant. A DSC compression model is thus formulated. The general constitutive equation is expressed as follows:

$$
d\varepsilon_v^a =
\begin{cases}
\dfrac{\left(1+e^*\right)}{\left(1+e\right)} d\varepsilon_v^c & \text{for } p' < p_y' \\[4mm]
\left[1 + b\left(\dfrac{p_{y,i}'}{p_y'}\right)^r\right] d\varepsilon_v^c & \text{for } p' > p_y'
\end{cases}
\tag{9.7}
$$

The general compression equation for structured geomaterials, Eq. (9.7), has a very simple form with only two parameters. This is because of the hierarchical nature of the DSC framework, where the deformation of the geomaterial at the selected reference state (or with no structure) is directly input for computation.

In geotechnical engineering, the vertical effective stress σ_v' is usually substituted for the mean effective stress p' when describing one-dimensional compression behavior, e.g., Burland [4] and Liu and Carter [5]. For modeling soil behavior in situations like this, the general Eq. (9.7) is modified and re-written in terms of vertical effective stress as follows:

$$
d\varepsilon_v^a =
\begin{cases}
\dfrac{\left(1+e^*\right)}{\left(1+e\right)} d\varepsilon_v^c & \text{for } \sigma_v' < \sigma_{vy}' \\[4mm]
\left[1 + b_v\left(\dfrac{\sigma_{vy,i}'}{\sigma_{vy}'}\right)^r\right] d\varepsilon_v^c & \text{for } \sigma_v' > \sigma_{vy}'
\end{cases}
\tag{9.8}
$$

Similarly, $\sigma_{vy,i}'$ is the initial vertical yield stress associated with initial material structure, σ_{vy}' is the current vertical yield stress, and b_v is the disturbance parameter corresponding to one-dimensional compression conditions. It is assumed that parameter r is the same as in Eq. (9.7), where soil compression behavior is described in terms of mean effective stress. A study on the model parameters can be found in the paper by Liu et al. [11,12].

The compression behavior of clay exposed to metal-rich liquids ("metal-rich clay") has been extensively studied recently because metal contamination of clay is found worldwide and increasingly poses environmental risks, e.g., Sharma and Reddy [21]. However, studies predicting the compression behavior of clays contaminated with various metal concentrations are very limited, but the highly nonlinear nature of the behavior of these soils is well known. Fan et al. [22] performed a study of this topic using the DSC compression model and found that a special and simple version of the general compression model predicts satisfactorily the highly nonlinear compression of contaminated metal-rich clay. The suggested compression constitutive equation is given as:

$$
d\varepsilon_v^a =
\begin{cases}
\dfrac{\left(1+e^*\right)}{\left(1+e\right)} d\varepsilon_v^c & \text{for } p' < p_y' \\[4mm]
\left[1-b\right] d\varepsilon_v^c & \text{for } p' > p_y'
\end{cases}
\tag{9.9}
$$

It is seen here that the constitutive equation is a very special and simple case of Eq. (9.7) with $r = 0$. It is also noted that the value of b is normally positive when the uncontaminated soil with the highest voids ratio is selected as the reference state.

9.4 PERFORMANCE OF THE GENERAL DSC COMPRESSION MODEL

The DSC compression models, represented by Eqs. (9.7), (9.8), and (9.9), are employed in this section to simulate and/or predict the behavior of structured geomaterials. Experimental data used in the analyses are obtained from previous publications. Details of the original papers, the tests, the values of model parameters identified, and the reference states selected are listed in Table 9.1. In this table, NCL represents the normal compression line and ICL represents the isotropic compression line. Detailed comparisons of the compression behavior of these geomaterials at their selected reference states and with various structures are shown in Figures 9.2–9.11.

The initial yield stress, $p'_{y,i}$ (or $\sigma'_{vy,i}$), is determined directly from the original compression curves plotted in e-ln p' (or e-ln σ'_v) coordinates. As seen in the inset on Figure 9.2, the yielding point and the corresponding value of the yield stress is estimated schematically. The values of parameters r and b (or b_v) were determined by finding the best match between the simulations and the test data for virgin compression, because the parameters control different features of the virgin compression behavior. The test data for reinforced soil in Figure 9.11 were presented in terms of volumetric strain ε_v and p' relationships, and so the simulations are presented here in the same form.

The simulations are represented on the figures by dark solid lines, and those at the reference state by dark broken lines. Test data are represented by solid circles or squares, and those at the reference state by open circles. The test data are linked by thin solid lines for some tests in order to illustrate clearly the stress paths in these tests. For most predictions, the selected initial yield point may also be seen on the plots from the sharp change in the theoretical curve.

The geomaterials included are soft and stiff clays with various structures, fissured clays, sands, gravels, rocks, calcareous clays and sands, soft rocks, cemented soils, reinforced soils, and soils artificially treated by chemicals and polluted by heavy metals. The influence of material structure on the mechanical responses and the degradation of that structure during monotonically increasing loading as well as during cyclic loading are clearly illustrated. Some further observations may be made.

1. The simulations cover a wide range of geomaterials with various structures. They are clays from very soft to very stiff states, clay and clay shale with fissures, sands of different mineralogy, crushed rock, calcareous clay and sand, soft rock, clays artificially treated by adding lime, cement, or chemicals, and sand reinforced by geotextiles. For reinforced soils and soils artificially treated, the structured geomaterial and the material at the selected reference state do not possess precisely the same composition. The forms of the structure are naturally formed structures, structures formed artificially in the laboratory without adding new compositions, structures formed with adding lime, cement, or chemicals, sand reinforced by geotextiles, and the structure of geomaterials that have experienced some destructuring.

2. The loadings considered here include both monotonic loading and cyclic loading with a very wide range of stress variation. The applied stress varies from 100 to 830,000 kPa for the test on Cambria sand (Figure 9.5). For the tests shown in Figures 9.2, 9.3, and 9.9, comparisons of purely elastic behavior of geomaterials at reference states and at structured states are seen. Assumption 2 for the compression model is an overall acceptable approximation.

3. For smectite exposed to $Cu(NO_3)_2$ (Figure 9.10), the behavior of the soil with contamination is selected as a reference state. As a result, the soil without contamination and with different magnitudes of contamination is theoretically modeled as structured soils. Overall, the model simulations are satisfactory. It is seen from this example that model parameter b is directly dependent on the selected reference state and parameter r appears to be less dependent on the reference state. The merit of the flexibility for selecting a reference state in the DSC plasticity is clearly seen here.

TABLE 9.1

Details of Compression Tests and Values of Model Parameters

Test No.	Geomaterial (Fig. No.)	Reference	Initial Yield Stress (kPa)	b or b_v	r	General Information and Comments
1a	Bangkok clay,	Lorenzo	Virgin yld.	Recons.	-	Reconstituted, NCL, as reference state.
1b	reconstituted,	et al. [20]	$\sigma'_{vy,\,i}=75$	$b_v=1.4$	1	NCL: $e^*=3.1-0.29\ln(\sigma'_v)$.
1c	naturally		$\sigma'_{vy,\,i}=180$	$b_v=2.5$	1	Unloading: $de^*=-0.05(d\sigma'_v/\sigma'_v)$.
1d	structured,		$\sigma'_{vy,\,i}=430$	$b_v=2.5$	1	1b with natural structure;
1e	cemented		$\sigma'_{vy,\,i}=530$	$b_v=3.2$	1	1c, 1d, and 1e with cementation 5%, 10%, and
	(9.2)					15%, respectively.
2	Stiff Pleistocene	Cotecchia	$\sigma'_{vy,\,i}=2900$	$b_v=1.05$	1	Reconstituted as reference state.
	clay	et al. [23]				Unloading: $de^*=-0.024(d\sigma'_v/\sigma'_v)$.
	(9.3)					For intact soil, cyclic loadings, and σ'_v from 18 to 30,000 kPa.
3	Fissured stiff	Lehane	$\sigma'_{vy,\,i}=1250$	$b_v=-0.35$	1	Reconstituted NCL as reference state.
	lodgement till	et al. [24]				Unloading: $de^*=-0.018(d\sigma'_v/\sigma'_v)$.
	(9.4)					Cyclic loading, and σ'_v from 25 to 9,000 kPa.
4a	Cambria sand	Yamamuro	$\sigma'_{vy,\,i}=2000$	$b_v=4$	1.5	Specimen with $e=0.55$ as reference state.
4b	(9.5)	et al. [25]	$\sigma'_{vy,\,i}=2000$	$b_v=12$	1.5	Unloading, $de^*=-6\times10^{-8}d\sigma'_v$.
						Samples by dry-pluviated.
						Very high stress and σ'_v from 0.1 to 830 MPa.
5a	Crushed syenite	Pestana	$\sigma'_{vy,\,i}=10$	$b_v=2.4$	0.3	Specimen with $e=0.65$ as reference state.
5b	rock	et al. [26]	$\sigma'_{vy,\,i}=10$	$b_v=4$	0.3	NCL: $e^*=0.664-4.8\times10^{-5}\,p'$.
5c	(9.6)		$\sigma'_{vy,\,i}=10$	$b_v=5.6$	0.3	Different uniformity coefficient: $C_u=1.4$ for
5d			$\sigma'_{vy,\,i}=10$	$b_v=6.2$	0.3	test 5b, and $C_u=2.8$ for tests 5a, 5c, 5d.
6	Calcareous soil	Lagioia	$p'_{y,\,i}=2400$	$b=35$	25	Reconstituted as reference state.
	(9.7)	et al. [27]				ICL: $e^*=2.57-0.208\ln(p')$.
						Intact. Unloading: $de^*=-0.0165(dp'/p')$.
7a	Bioclastic	Pestana	$\sigma'_{vy,\,i}=10$	$b_v=1.6$	0.45	Specimen with $e=0.826$ as reference state,
7b	calcareous sand	et al. [26]	$\sigma'_{vy,\,i}=10$	$b_v=4.3$	0.45	Cyclic loading, σ'_v from 20 to 76,000 kPa.
	(9.8)					
8a	Pyroclastic soft	Cecconi	$\sigma'_{vy,\,i}=550$	$b_v=0.55$	0.8	Reconstituted as reference state,
8b	rock	et al. [28]	$\sigma'_{vy,\,i}=700$	$b_v=0.2$	0.8	Unloading: $de^*=-0.009(d\sigma'_v/\sigma'_v)$
	(9.9)					8a: partially destructured
						8b: intact
						Cyclic loading and stress from 20 to 2000 kPa.
9a	Smectite	Yong et al.	Not	$b_v=0.67$	0	Contaminated soil as reference state.
9b	exposed to	[29]	needed.	$b_v=0.55$	0	Different metal concentrations.
9c	$Cu(NO_3)_2$			$b_v=-0.18$	0	
9d	(9.10)			$b_v=-0.28$	0	
10a	Sand reinforced	Varadarajan	$p'_{y,\,i}=1$	$b=0.2$	0.05	No reinforcement as reference state,
10b	by nonwoven	et al. [30]		$b=1.3$	0.1	10a, b, c: 1, 2, and 3 layers of geotextile,
10c	geotextiles			$b=3.9$	0.2	respectively.
	(9.11)					

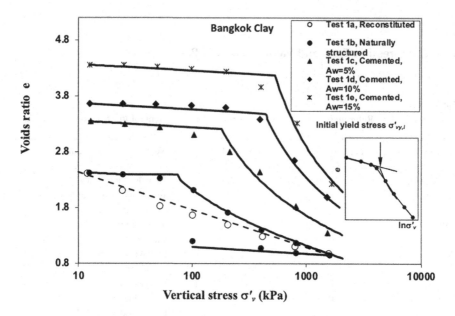

FIGURE 9.2   One-dimensional compression tests on soft Bangkok clay [20].

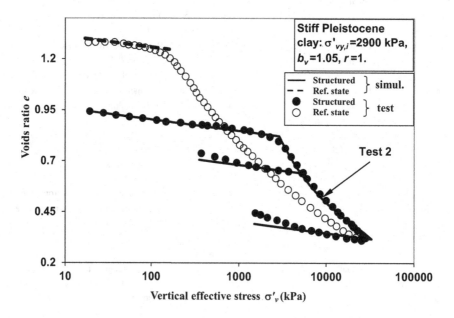

FIGURE 9.3   One-dimensional compression tests on stiff Pleistocene clay [23].

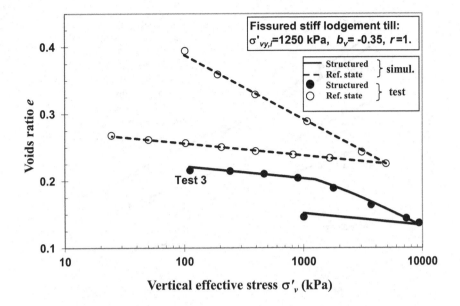

FIGURE 9.4 One-dimensional compression tests on fissured stiff lodgement till [24].

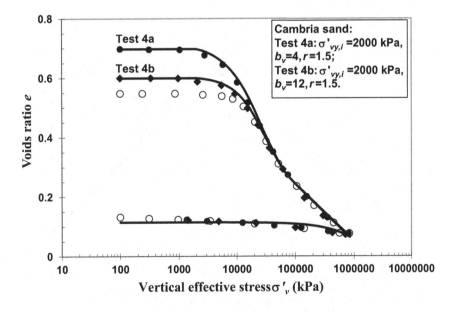

FIGURE 9.5 One-dimensional compression tests on Cambria sand [25].

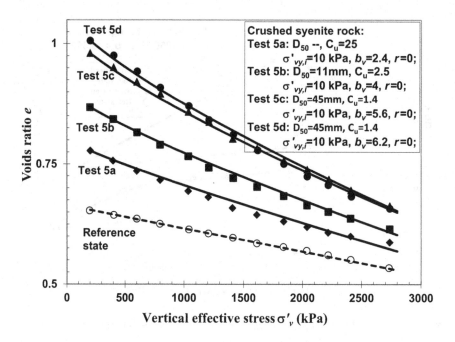

FIGURE 9.6   One-dimensional compression tests on crushed syenite rock [26].

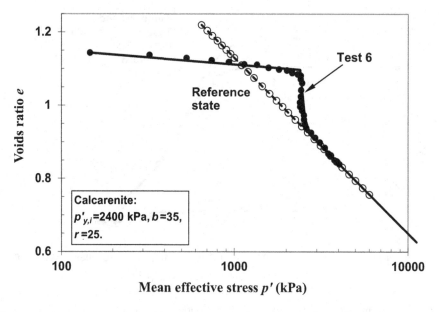

FIGURE 9.7   Isotropic compression tests on a calcarenite [27].

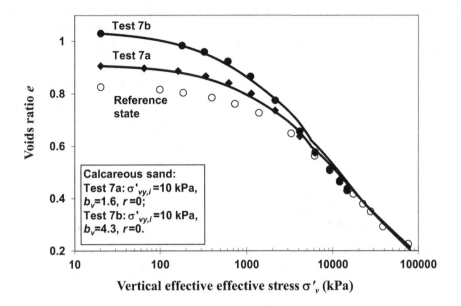

FIGURE 9.8   One-dimensional compression tests on calcareous sand [26].

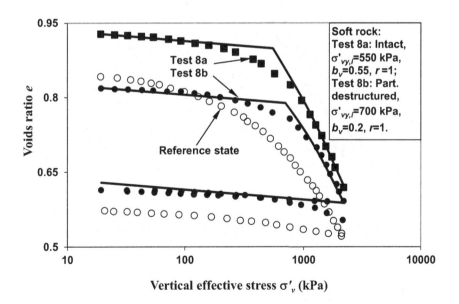

FIGURE 9.9   One-dimensional compression tests on a pyroclastic soft rock [28].

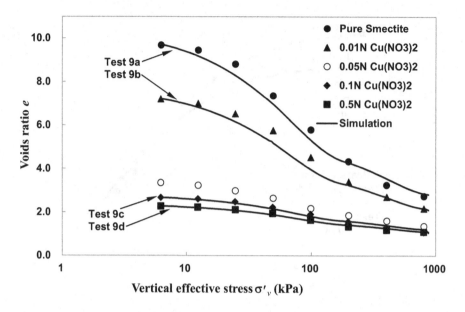

FIGURE 9.10   One-dimensional compression tests on smectite exposed to $Cu(NO_3)_2$ [29]

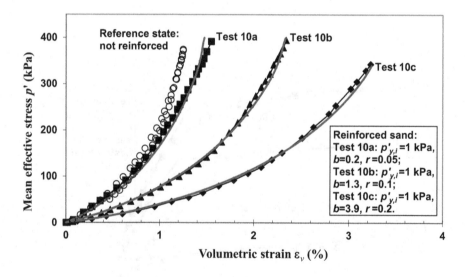

FIGURE 9.11   Isotropic compression tests on sand reinforced by nonwoven geotextile [30].

9.5   CONCLUSIONS

Constitutive modeling of geomaterials has always been a challenge because of the variation in material composition, i.e., grain size and mineralogy, and the variation in the structure of the geomaterial, i.e., the arrangement and bonding of the constituents. Currently, different constitutive models are generally proposed for different categories of geomaterials and even for the same materials with different structures. Modeling the compression behavior of geomaterials within the framework of the DSC is presented in this chapter. In particular, a simple but general compression model for structured geomaterials was introduced in detail within the general theoretical framework of the DSC proposed by Desai [8]. It was demonstrated that the general compression model can successfully simulate the behavior of a very wide range of structured geomaterials. The powerful potential of the DSC for modeling the influence of material structures on the material mechanical responses has been demonstrated.

REFERENCES

1. Chai, J.-C. and Carter, J. P. *Deformation Analysis in Soft Ground Improvement*. Springer, Dordrecht, 2011.
2. Liu, M. D., Carter, J. P. and Airey, D. W. "Sydney Soil model: (I) theoretical formulation", *International Journal of Geomechanics*, ASCE, 11, 3, 2011, 211–224.
3. Liu, M. D. and Carter, J. P. "A structured Cam Clay model", *Canadian Geotechnical Journal*, 39, 6, 2002, 1313–1332.
4. Burland, J. B. "On the compressibility and shear strength of natural soils", *Géotechnique*, 40, 1990, 329–378.
5. Liu, M. D. and Carter, J. P. "Virgin compression of structured soils", *Géotechnique*, 49, 1, 1999, 43–57.
6. Mesri, G. and Vardhanabhuti, B. "Compression of granular materials", *Canadian Geotechnical Journal*, 46, 2009, 369–392.
7. Desai, C. S. "Constitutive modeling using the disturbed state as microstructure self-adjustment concept", *Continuum Models for Materials with Microstructure*, Mühlhaus (ed.), John Wiley, UK, 1995, 239–296.
8. Desai, C. S. *Mechanics of Materials and Interfaces: The Disturbed State Concept*, CRC Press, Boca Raton, FL, 2001.
9. Desai, C. S. "Disturbed state concept as unified constitutive modeling approach". *Journal of Rock Mechanics and Geotechnical Engineering*, 8, 2016, 277–293.
10. Desai, C. S. and Toth, J. "Disturbed state constitutive modeling based on stress-strain and non-destructive behavior", *International Journal of Solids and Structures*, 33, 1996, 1619–1650.
11. Liu, M. D., Carter, J. P., Desai, C. S. and Xu, K. J. "Analysis of the compression of structured soils using the disturbed state concept". *International Journal for Numerical and Analytical Methods in Geomechanics*, 24, 2000, 723–735.
12. Liu, M. D., Carter, J. P. and Desai, C. S. "Modeling compression behaviour of geo-materials". *International Journal of Geomechanics, ASCE*, 3, 3/4, 2003, 191–204.
13. Ouria, A. "Disturbed state concept–based constitutive model for structured soils". *International Journal of Geomechanics*, 17, 7, 2017, 04017008.
14. Xu, Y., Guo, P., Wang, Y., Zhu, C-W., Cheng, K. and Lei, G. "Modelling the triaxial compression behavior of loess using the Disturbed State Concept". *Advances in Civil Engineering*, 2021. https://doi.org/10.1155/2021/6638715.
15. Zhou, C., Shen, Z., Chen, S. and Chen, T. "A hypoplasticity disturbed state model for structured soils". *Chinese Journal of Geotechnical Engineering* 26, 4, 2004, 435–439.
16. Liu, W. Z., Shi, M. L. and Miao, L. C. "Analysis of structural soil compression characteristics based on the concept of disturbance state". *Rock and Soil Mechanics*, 31, 11, 2010 3475–3480. (In Chinese).
17. Qian, S., Shi, J. and Ding, J. W. "Modified Liu-Carter compression model for natural clays with various initial water contents". *Advances in Civil Engineering*, 2016. https://doi.org/10.1155/2016/1691605.
18. Liu, E.-L., Yu, H.-S., Zhou, C., Nie, Q. and Luo, K.-T. "A binary-medium constitutive model for artificially structured soils based on the Disturbed State Concept and Homogenization theory". *International Journal of Geomechanics*, 17, 7, 2017, 04016154.
19. Hight, A. D., Bond, W. J. and Legge, J. D. "Characterisation of the Bothkennar clay: An overview". *Géotechnique*, 42, 1992, 303–347.
20. Lorenzo, G. A. and Bergado, D. T. "Fundamental parameters of cement-admixed clay: new approach", *Journal of the Geotechnical Engineering Division, ASCE*, 130, 10, 2004, 1042–1050.
21. Sharma, H. D. and Reddy, K. R. *Geoenvironmental Engineering: Site Remediation, Waste Containment, and Emerging Waste Management Technologies*. John Wiley & Sons, Inc., New York, 2004.
22. Fan, R., Liu, M. D., Du, Y. and Horpibulsuk, S. "Estimating the compression behaviour of metal-rich clays via a disturbed state concept (DSC) model". *Applied Clay Science*, 132, 2016, 50–58.
23. Cotecchia, F. and Chandler, R. J. "The influence of structure on the pre-failure behaviour of a natural clay". *Géotechnique*, 47, 1997, 523–544.

24. Lehane, B. and Faulkner, A. "Stiffness and strength characteristics of a hard lodgement till". *The Geotechnics of Hard Soil – Soft Rocks*, Evangelista A. and Picarelli L. (eds.), 637–646. CRC Press, Boca Raton, FL, 1998.

25. Yamamuro, J. A., Bopp, P. A. and Lade, P. V. "One-dimensional compression of sands at high pressures". *Journal of Geotechnical Engineering, ASCE*, 122, 1996, 147–154.

26. Pestana, J. M. and Whittle, A. J. "Compression model for cohesionless soils". *Géotechnique*, 45, 1995, 621–631.

27. Lagioia, R. and Nova, R. "An experimental and theoretical study of the behaviour of a calcarenite in triaxial compression", *Géotechnique*, 45, 1995, 633–648.

28. Cecconi, M., Viggiani, G. and Rampello, S. "An experimental investigation of the mechanical behaviour of a pyroclastic soft rock". *The Geotechnics of Hard Soils - Soft Rocks*, Evangelista A. and Picarelli L. (eds.), 473–482. CRC Press, Boca Raton, FL, 1998.

29. Yong, R.N., Ouhadi, V.R., and Goodarzi, A.R., "Effect of Cu^{2+} ions and buffering capacity on smectite microstructure and performance" *Journal of Geotechnical and Geoenvironmental Engineering, ASCE*, 135, 12, 2009, 1981–1985.

30. Varadarajan, A., Sharma, K. G. and Soni, K. M. "Behaviour of a reinforced sand during triaxial loading", *Indian Geotechnical Journal*, 29, 1999, 242–261.

10 Chemically Treated Soils

Buddhima Indraratna
Transport Research Centre, University of Technology Sydney (UTS)

Jayan S. Vinod
Faculty of Engineering and Information Sciences,
University of Wollongong (UOW)

Rasika Athukorala
Golder Associates

10.1 INTRODUCTION

Dispersive and erodible soils are considered the most common problematic soil type found worldwide. The structures (e.g. embankment dams, rail/road embankments, foundation) founded on erodible and dispersive soils are in danger of surface or piping erosion. Therefore, it is imperative to adopt a suitable ground improvement technique to control the erosion of these soils. Chemical (admixture) soil stabilisation is one of the oldest and most widespread techniques among the ground improvement methods. The soils such as erodible, dispersive, soft clay have been stabilised by traditional mixtures of lime, cement, fly ash, milled slag, bitumen and calcium chloride. Cement, gypsum, lime and other alkaline admixtures have been commonly used for the construction of highways, rail tracks and airport runways to enhance bearing capacity, reduce settlement, control shrinking/swelling and reduce erosion, permeability (Balasubramaniam et al., 1989, 1998; Indraratna et al., 1995; Indraratna 1996; Bergado et al., 1996; Rajasekaran and Narasimha Rao, 1997; Uddin et al., 1997; Kamon, 2000; Hausmann, 1990, Chew et al., 2003; Consoli et al., 2011; Mehenni et al., 2016, among others). However, such traditional admixtures (i.e. cement, lime, fly ash, etc.) are not frequently useable because of stringent occupational health and safety concerns apart from various threats to the environment due to an inevitable increase in soil and groundwater alkalinity. Recently, lignosulfonate (LS) has been proven to be effective in enhancing the erosion resistance of highly erodible and dispersive soils overcoming the environmental problems caused by the traditional chemical stabilisers (Indraratna et al., 2008, 2013; Vinod et al., 2010). Lignosulfonate belongs to a family of lignin-based organic polymers derived as a waste by-product from wood and paper processing industry. Karol (2003) had described it as a dark brown liquid whose composition would vary with batch depending on the source and wood process. Compared to highly alkaline and sometimes corrosive chemical admixtures, LS is an environmentally friendly, non-corrosive and non-toxic chemical that does not alter the soil pH upon treatment. Due to the very small amount needed for effective soil treatment, there is also insignificant leaching to adversely affect the groundwater chemistry. The LS is a completely soluble, dark brown liquid having a pH value of approximately 4. It is characterised as an inflammable, non-corrosive and non-hazardous chemical according to the National Occupational Health and Safety Commission (NOHSC) criteria (Chemstab, 2003). Indraratna et al. (2008) conducted a series of internal erosion tests on LS-treated dispersive soils using a Process Simulation Apparatus for Internal Crack Erosion (PSAICE), and they highlighted that LS treatment improves the erosion resistance of the treated soil similar to lime and cement. An analytical model was developed for LS-treated soil to evaluate the reduction in erosion capturing its enhanced tensile behaviour (Indraratna et al., 2009). Lignosulfonate treatment increases the critical shear stress required to initiate soil erosion and decreases the coefficient of soil

DOI: 10.1201/9781003362081-13

149

erosion. The improved performance can be attributed to a reduction of the double layer thickness by the neutralisation of surface charges and the subsequent formation of a stable clay particle cluster (Vinod et al., 2010). Moreover, it has now been revealed that the amount of LS required to make the soil non-erodible is considerably less than that of cement for certain types of soft and erodible soils.

10.1.1 Stress–Strain Behaviour of Lignosulfonate-Treated Soil

Figure 10.1 compares the shear stress–shear strain and volumetric strain–shear strain responses of soil at different levels of sodium-based LS treatments under an effective normal stress (σ'_n) of 5kPa. The details of the soil characteristics and testing methodology can be found elsewhere (e.g. Indraratna et al., 2009; Athukorala et al., 2015). It is evident from Figure 10.1 that the increase in the level of LS treatment enhances the shear stress of the soil. The occurrence of well-defined peak shear stresses is evident, followed by post-peak strain softening under an effective normal stress of 5 kPa for both treated and untreated specimens. The volumetric response exhibits a dilative behaviour with an increase in the percentage of LS. However, the increase in the stiffness, due to stabilisation is slight, and the ductility of the soil remains unchanged upon treatment compared to untreated soil. This is in contrast to most chemical admixtures that make the treated soil exhibit a pronounced brittle behaviour on one hand, while increasing the stiffness significantly on the other (e.g. cement or lime treatment).

10.2 REVIEW OF DSC/HISS MODELLING

The DSC is a unified approach, and it adheres to a notable critical state and various capped yield models as special cases for modelling geomaterials (Desai, 2001). The additional factors such as associated and non-associated flow rules, linear or non-linear elastic responses, isotropic or aniso-tropic hardening behaviour as well as a range of conditions pertaining to elastoplasticity, visco-elas-ticity, thermoplasticity and thermo-viscoplasticity can be captured in various hierarchical models compatible with the DSC framework.

In DSC, the observed response of a material is expressed using two reference states, i.e., relative intact (RI) state and fully adjusted (FA) state. The material is usually considered to be at the RI state, initially. However, upon disturbance from external loading, the material may undergo micro-structure adjustments and transform from the RI to the FA state. These states are then combined through the disturbance function (D) to obtain the actual response of the material. The disturbance function, D, can be defined based on the stress–strain responses and the volumetric strain responses

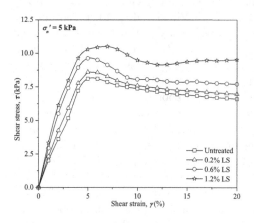
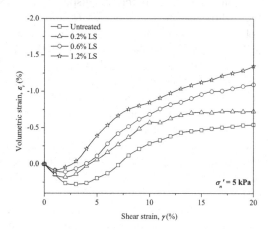

FIGURE 10.1 Shear stress and volumetric strain versus shear strain for treated and untreated soils. (Reproduced from Athukorala et al., 2015 with permission from ASCE.)

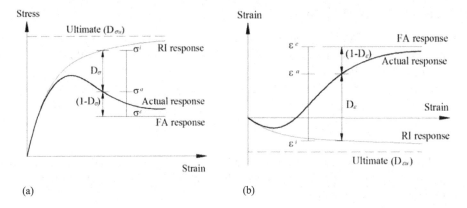

FIGURE 10.2 Model for disturbance (a) stress–strain and (b) volumetric strain–strain. (Reproduced from Athukorala et al., 2015 with permission from ASCE.)

(Figure 10.2). Equations 10.1 and 10.2 present the disturbance function in terms of stress–strain and volumetric strain, respectively (Desai, 2001):

$$D_\sigma = \frac{\sigma^i - \sigma^a}{\sigma^i - \sigma^c} \tag{10.1}$$

$$D_\varepsilon = \frac{\varepsilon^i - \varepsilon^a}{\varepsilon^i - \varepsilon^c} \tag{10.2}$$

The superscripts i, a and c are used to indicate the RI response, the observed (actual) response and the FA response, respectively.

10.2.1 RELATIVELY INTACT (RI) RESPONSE

The RI behaviour of untreated soil was modelled using the δ_0-version of HiSS plasticity models, which is based on associative plasticity and isotropic hardening. In this approach, the yield function F is given in the two-dimensional form as (Desai, 2001):

$$F = \left(\frac{\tau}{P_a}\right)^2 + \alpha\left(\frac{\sigma'_n}{P_a}\right)^n - \gamma\left(\frac{\sigma'_n}{P_a}\right)^q = 0 \tag{10.3}$$

where σ'_n is the effective normal stress, τ is the shear stress, P_a is the atmospheric pressure, n is the phase change parameter, γ is the ultimate parameter, q is a parameter that depends on the linearity of the ultimate envelope and α is the hardening function given by:

$$\alpha = \frac{a}{\xi_D^{\ b}} \tag{10.4}$$

In Eq. (10.4), a and b are the hardening parameters and ξ_D is the trajectory of the deviatoric plastic strains.

10.2.2 FULLY ADJUSTED (FA) RESPONSE

The untreated soil at the FA state behaves in a manner similar to the shear behaviour at 20% of the shear strain. Based on the experimental shear test of the untreated soil, the following relationship could be proposed for the FA state:

$$\left(\frac{\tau^c}{P_a} \right) = c_1 \left(\frac{\sigma_n^c}{P_a} \right)^{c2} \tag{10.5}$$

where c_1 and c_2 are the FA material parameters and σ_n^c is the effective normal stress corresponding to the FA state. The relationship between the FA value of normal strain (ε_n^c) and the effective normal stress can be given as (Desai, 2001):

$$\varepsilon_n^c = c_3 \left(\frac{\sigma_n^c}{P_a} \right) - \varepsilon_0 \tag{10.6}$$

where c_3 is a material parameter and ε_0 is the FA normal strain when $\sigma_n^c = 0$. The incremental shear stress and normal strain equations for the FA can be derived from Eqs. (10.5) and (10.6) as:

$$d\tau^c = c_1 c_2 \left(\frac{\sigma_n^c}{P_a} \right)^{(c_2 - 1)} d\sigma_n^c \tag{10.7}$$

$$d\varepsilon_n^c = \frac{c_3}{P_a} d\sigma_n^c \tag{10.8}$$

13.2.3 ACTUAL RESPONSE OF THE UNTREATED SOIL

The incremental stress–strain relationships for the actual (observed) behaviour can be evaluated combining the disturbance function (Desai, 2001):

$$d\underset{\sim}{\sigma}^a = (1 - D) \underset{\sim}{C}^{epi} d\underset{\sim}{\varepsilon}^i + D \underset{\sim}{C}^c \underset{\sim}{C}^{epi} d\underset{\sim}{\varepsilon}^i + \underset{\sim}{\sigma}_R dD \tag{10.9}$$

where $\underset{\sim}{C}^{epi}$ is the elasto-plastic constitutive matrix for the RI behaviour, $d\underset{\sim}{\varepsilon}^i$ is the vector of incremental strains of RI response $\underset{\sim}{\sigma}_R = \left(\underset{\sim}{\sigma}^c - \underset{\sim}{\sigma}^i \right)$ and $\underset{\sim}{C}^c$ is a matrix corresponding to the FA state:

$$\underset{\sim}{C}^c = \left\{ \begin{array}{cc} 0 & c_1 c_2 \left(\dfrac{\sigma_n^i}{P_a} \right)^{(c_2 - 1)} \\[2ex] 0 & 1 \end{array} \right\}$$

The disturbance function, D, in Eq. (10.9) can be expressed in terms of the deviatoric plastic strain trajectory, ξ_D, as (Desai, 2001):

$$D = D_u \left[1 - e^{-A\left(\xi_D - \xi_D^* \right)^Z} \right] \tag{10.10}$$

where D_u is the ultimate value of D at the residual, ξ_D^* is the deviatoric plastic strain trajectory below which no disturbance occurs, and A and Z are material parameters. An expression for dD in Eq. (10.9) can be derived from Eq. (10.10) as:

$$dD = \underset{\sim}{C}^R d\underset{\sim}{\varepsilon}^i \tag{10.11}$$

where $\underset{\sim}{C}^R$ is given by:

$$\underset{\sim}{C}^R = \frac{\left[D_u AZ\left(\xi_D - \xi_D^*\right)^{(Z-1)} e^{-A\left(\xi_D - \xi_D^*\right)^Z}\right]\left[\left(\frac{\partial F}{\partial \underset{\sim}{\sigma}}\right)^T \cdot \frac{\partial F}{\partial \underset{\sim}{\sigma}}\right]^{1/2}}{\left(\frac{\partial F}{\partial \underset{\sim}{\sigma}}\right)^T \cdot \underset{\sim}{C}^e \cdot \frac{\partial F}{\partial \underset{\sim}{\sigma}} - \frac{\partial F}{\partial \xi_D} \cdot \left[\left(\frac{\partial F}{\partial \underset{\sim}{\sigma}}\right)^T \cdot \frac{\partial F}{\partial \underset{\sim}{\sigma}}\right]^{1/2}} \left(\frac{\partial F}{\partial \underset{\sim}{\sigma}}\right)^T \cdot \underset{\sim}{C}^e$$

The stress–strain relationship for the actual response, Eq. (10.9) becomes:

$$d\underset{\sim}{\sigma}^a = \underset{\sim}{C}^{DSC} d\underset{\sim}{\varepsilon}^i \tag{10.12}$$

where $\underset{\sim}{C}^{DSC}$ is the DSC constitutive matrix given by:

$$\underset{\sim}{C}^{DSC} = (1-D)\underset{\sim}{C}^{epi} + D\underset{\sim}{C}^c \underset{\sim}{C}^{epi} + \sigma_R \underset{\sim}{C}^R$$

10.2.4 DSC Modelling of Lignosulfonate Bonds

The observed behaviour of the LS bonds was modelled by considering a linear elastic RI response and a zero stress FA response as the two reference states (Figure 10.3). The disturbance function, D^b, for LS bonds:

$$D^b = \frac{\sigma^{bi} - \sigma^{ba}}{\sigma^{bi}} \tag{10.13}$$

where σ^{bi} is the shear stress of RI response and σ^{ba} is the shear stress of actual response.

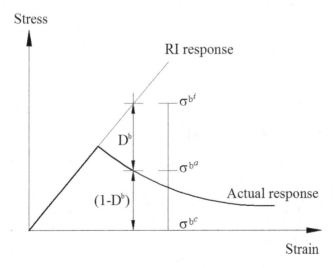

FIGURE 10.3 Disturbed state concept (DSC) model for lignosulfonate (LS) bonds. (Reproduced from Athukorala et al., 2015 with permission from ASCE.)

10.2.5   Relative Intact (RI) Response of Lignosulfonate Bonds

The stress–strain behaviour of LS bonds; the RI response of LS was taken as linear elastic and, therefore, the incremental stress–strain relationship of the RI response can be written as:

$$d\underset{\sim}{\sigma}^{bi} = \underset{\sim}{C}^{b(e)i} d\underset{\sim}{\varepsilon}^{bi} \tag{10.14}$$

where $d\underset{\sim}{\sigma}^{bi}$ and $d\underset{\sim}{\varepsilon}^{bi}$ are the vectors of incremental LS bond stresses and strains, respectively, and $\underset{\sim}{C}^{b(e)i}$ is the elastic constitutive matrix for the LS bonds at the RI state.

10.2.6   Actual Response of Lignosulfonate Bonds

Similar to the untreated soil, from the definition of the disturbance function (Eq. 10.13), the incremental stress–strain relationships for the actual response of the LS bonds can be evaluated by:

$$d\underset{\sim}{\sigma}^{ba} = \underset{\sim}{C}^{bDSC} d\underset{\sim}{\varepsilon}^{bi} \tag{10.15}$$

where $\underset{\sim}{C}^{bDSC}$ is the DSC constitutive matrix for LS bonds given by:

$$\underset{\sim}{C}^{bDSC} = \left(1 - D^b\right)\underset{\sim}{C}^{b(e)i} - \sigma^{bi}\underset{\sim}{C}^{bR} \tag{10.16}$$

The disturbance function for LS bonds D^b in Eq. (10.16) can be expressed in terms of the shear strain γ as:

$$D^b = D_u^b \left[1 - e^{-A^b\left(\gamma - \gamma^*\right)^{Z^b}}\right] \tag{10.17}$$

where D_u^b is the ultimate value of D^b at the residual, γ^* is the shear strain below which the disturbance is zero, and A^b and Z^b are model parameters for LS bonds. $\underset{\sim}{C}^{bR}$ in Eq. (10.16) is given by a 1×2 matrix as:

$$\underset{\sim}{C}^{bR} = \left[\left(D_u^b A^b Z^b \left(\gamma - \gamma^*\right)^{\left(Z^b - 1\right)} e^{-A^b\left(\gamma - \gamma^*\right)^{Z^b}} \right) \quad 0 \right]$$

10.3   DSC MODELLING OF LIGNOSULFONATE-TREATED SOIL

The stress–strain response of LS-treated soil using the DSC; the response of the FA state was considered as the response of the LS bonds obtained using Eq. (10.15). The reference states and the disturbance for LS-treated soils are shown in Figure 10.4. Based on Eq. (10.1), the disturbance for LS-treated soil can be given as:

$$D^t = \frac{\sigma^{ti} - \sigma^{ta}}{\sigma^{ti} - \sigma^{ba}} \tag{10.18}$$

The superscript "t" used in Eq. (10.18) represents the LS treatment.

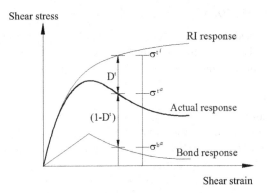

FIGURE 10.4 Disturbance for lignosulfonate (LS)-treated soil. (Reproduced from Athukorala et al., 2015 with permission from ASCE.)

Similar to the behaviour of an untreated soil, the RI response of LS-treated soil was simulated using the δ_0-version of HiSS plasticity models. Then, the incremental stress–strain relationship for the actual response of LS-treated soil can be evaluated by:

$$d\underset{\sim}{\sigma}^{ta} = \underset{\sim}{C}^{t\mathrm{DSC}} d\underset{\sim}{\varepsilon}^{ti} \tag{10.19}$$

where $d\underset{\sim}{\varepsilon}^{ti}$ is the vector of incremental strains of LS-treated soil in the RI state and $\underset{\sim}{C}^{t\mathrm{DSC}}$ is the DSC constitutive matrix for LS-treated soil given by:

$$\underset{\sim}{C}^{t\mathrm{DSC}} = \left(1 - D^t\right)\underset{\sim}{C}^{t(ep)i} + D^t\underset{\sim}{C}^{b\mathrm{DSC}} + \sigma_R^t\underset{\sim}{C}^{tR} \tag{10.20}$$

In Eq. (10.20), $\sigma_R^t = \left(\underset{\sim}{\sigma}^{ba} - \underset{\sim}{\sigma}^{ti}\right)$, $\underset{\sim}{C}^{t(ep)i}$ is the elasto-plastic constitutive matrix for the RI behaviour of LS-treated soil, and the disturbance function D^t can be expressed in terms of the plastic strain trajectory, ξ, as follows:

$$D^t = D_u^t \left[1 - e^{-A^t\left(\xi_D^t - \xi_D^{t*}\right)^{Z^t}} \right] \tag{10.21}$$

where D_u^t is the ultimate value of D^t at the residual, ξ_D^{t*} is the deviatoric plastic strain trajectory below which the disturbance is zero and A^t and Z^t are the model parameters for LS-treated soil. $\underset{\sim}{C}^{tR}$ is given by:

$$\underset{\sim}{C}^{tR} = \frac{\left[D_u^t A^t Z^t \left(\xi_D^t - \xi_D^{t*}\right)^{\left(Z^t - 1\right)} e^{-A^t\left(\xi_D^t - \xi_D^{t*}\right)^{Z^t}} \right] \left[\left(\dfrac{\partial F^t}{\partial \underset{\sim}{\sigma}}\right)^T \cdot \dfrac{\partial F^t}{\partial \underset{\sim}{\sigma}} \right]^{1/2}}{\left(\dfrac{\partial F^t}{\partial \underset{\sim}{\sigma}}\right)^T \cdot \underset{\sim}{C}^{t(e)i} \cdot \dfrac{\partial F^t}{\partial \underset{\sim}{\sigma}} - \dfrac{\partial F^t}{\partial \xi_D^t} \left[\left(\dfrac{\partial F^t}{\partial \underset{\sim}{\sigma}}\right)^T \cdot \dfrac{\partial F^t}{\partial \underset{\sim}{\sigma}} \right]^{1/2}} \left(\dfrac{\partial F^t}{\partial \underset{\sim}{\sigma}}\right)^T \cdot \underset{\sim}{C}^{t(e)i}$$

10.3.1 Model Parameters for Untreated Soil

The model parameters and validation of the proposed model was conducted using the drained from direct shear tests conducted on LS-treated and untreated soil. The effective stress–strain curves and volume change–shear strain behaviour were obtained and quantified and presented in Figure 10.1.

10.3.1.1 Elastic Parameters

The values of the shear and normal moduli (E_s and E_n, respectively) of the soil are usually determined from the unloading slopes of the shear and normal stress curves. However, E_n can be assumed to be zero when the laboratory direct shear tests are conducted under a constant effective normal stress condition. The value of E_s was determined by the initial slope of the shear stress versus shear strain curves.

10.3.1.2 Relative Intact (RI) Parameters

The RI response was obtained by fitting the experimental results to a hyperbolic curve as proposed by Clough and Duncan (1971):

$$\tau = \frac{\gamma}{\dfrac{1}{E_s} + \dfrac{\gamma}{\tau_{ult}}} \tag{10.22}$$

where τ is the shear resistance, γ is the shear strain and τ_{ult} is the ultimate shear strength (asymptotic value of the shear at infinite strain of the hyperbolic curve). Subsequently, the model parameters corresponding to that hyperbolic curve were determined.

The ultimate failure envelopes were envisaged as non-linear, and therefore, the value of q could be determined from the experimental results. At the ultimate state where the shear strength reaches an asymptotic value at infinite strain, the hardening parameter (α) becomes zero. Therefore, the values of parameters γ and q can be determined by plotting the yield function in a logarithmic form at $\alpha=0$. The plot of $\ln(\tau_{ult}/P_a)$ versus $\ln(\sigma'_n/P_a)$, shown in Figure 10.5a, gives the values of γ from the intercept and q from the slope of the average straight line. The phase change parameter, n, was calculated by considering the transition point where the normal stress changes from compression to dilation. Equation (10.3) was solved for α and then substituted in $\partial F/\partial \sigma'_n = 0$ to obtain the value of n. To obtain the hardening parameters a and b, Eq. (10.4) was written in the logarithmic form and the plot of $\ln \alpha$ versus $\ln \xi_D$ was obtained (Figure 10.5b). Then, the values of a and b can be calculated from the intercept and the slope of the fitted straight line, respectively. The value of plastic strain trajectory ξ_D was determined from the stress–strain curves for stress increment i as:

$$\xi_D = \int \left(d\gamma^p \cdot d\gamma^p \right)^{1/2} \tag{10.23}$$

where $d\gamma^p$ is the increment of plastic shear strain. The value of the hardening function α corresponding to the total stress after each stress increment was calculated using Eq. (10.3).

10.3.2 Fully Adjusted (FA) Parameters

The plot of $\ln(\tau^c/P_a)$ versus $\ln(\sigma_n^c/P_a)$ gives the value of c_2 from the slope of the approximated linear fit (Figure 10.5c), and the value of c_1 can be calculated from the vertical intercept. After taking the logarithms of Eq. (10.6), the plot of ε_n^c versus σ_n^c/P_a (Figure 10.5d) gives the values of c_3 and ε_0 from the slope and the intercept of the linear fit of the experimental results, respectively.

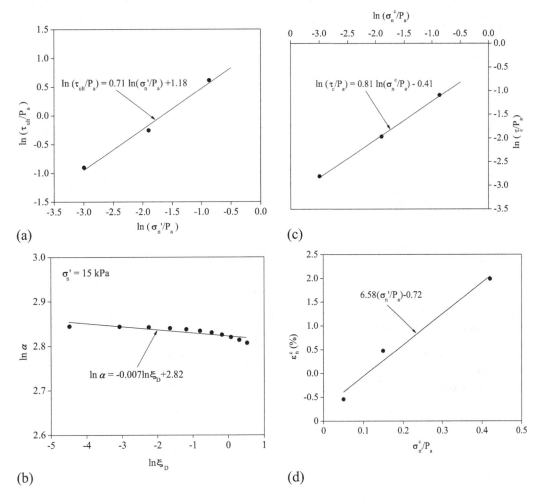

FIGURE 10.5 Determination of model parameters, relative intact (RI) and fully adjusted (FA) state, for untreated soil. (Reproduced from Athukorala et al., 2015 with permission from ASCE.)

10.3.2.1 Disturbance Function (D) Parameters

The values of disturbance for stress and strain were determined from experimental results using Eqs. 10.1 and 10.2, respectively, for the untreated soil. The disturbance parameters were then determined by fitting the calculated disturbance values to corroborate with Eq. (10.10), as illustrated in Figure 10.6. The calculated model parameters for untreated soil used in this study are summarised in Table 10.1.

10.3.3 Model Parameters for Lignosulfonate Bonds and Treated Soil

10.3.3.1 Relative Intact (RI) Parameters for Lignosulfonate bonds

The values of shear modulus $\left(E_s^b\right)$ were taken as the initial slope of the LS bond strength $\left(\tau^b\right)$ versus shear strain curves (Figure 10.7). The bond strength was calculated from the laboratory stress–strain curves as the difference between the shear stresses of LS-treated and untreated responses at a given shear strain. Then, the initial slopes of these stress–strain curves for different amounts of LS were plotted against the corresponding effective normal stress (Figure 10.8). It can be observed from Figure 10.8 that the E_s^b increased initially and then dropped beyond an effective normal stress

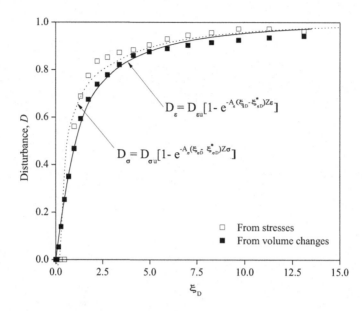

FIGURE 10.6 Disturbance parameters for soil at 5 kPa effective normal stress. (Reproduced from Athukorala et al., 2015 with permission from ASCE.)

TABLE 10.1
The Model Parameters for Untreated Soil

Relative Intact and Fully Adjusted		Disturbance Function	
Parameter	Value	Parameter	Value /Relationship
q	1.44	$\xi^*_{\sigma D}$	0.3
n	1.71	$D_{\sigma u}$	1
g	10.3	Z_σ	$Z_\sigma = -0.011\sigma'_n + 0.53$
a	16.78	A_σ	$A_\sigma = 0.026\sigma'_n + 0.095$
b	0.007	$\xi^*_{\varepsilon D}$	0
c_1	0.66	$D_{\varepsilon u}$	1
c_2	0.81	A_ε	$A_\varepsilon = 0.16\ln\sigma'_n + 0.11$
c_3	6.58	Z_ε	$Z_\varepsilon = -0.19\ln\sigma'_n + 1.32$
ε_0	0.72		

Modified after Athukorala et al. (2015).

of 10 kPa. The initial increase in E_s^b may be attributed to the contribution to the bond strength by the effective normal stress. At higher levels of effective normal stress, the bond strength is reduced, and hence, a reduction in E_s^b occurs. Moreover, it can be concluded that the effect of LS bonds on E_s^b would be negligible at higher effective normal stresses. Consequently, E_s^b and the shear strength of LS bonds can be expected to become zero at very high effective normal stress levels. Based on the above observations, the following equation is proposed to determine E_s^b.

$$E_s^b = \left(E_s\right)_p^{ut} \times \left(\frac{E_s^b}{\left(E_s\right)_p^{ut}}\right)_{(\sigma'_n=0)} \times \left(1 - \frac{\sigma'_n}{\sigma'_{nc}}\right)^{d_1} \quad (10.24)$$

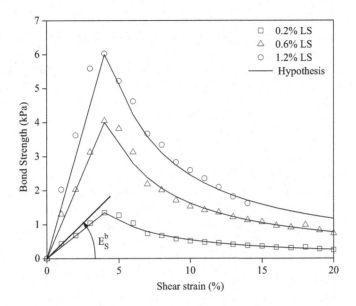

FIGURE 10.7 Stress–strain curves of lignosulfonate (LS) bonds at 10 kPa of effective normal stress. (Reproduced from Athukorala et al., 2015 with permission from ASCE.)

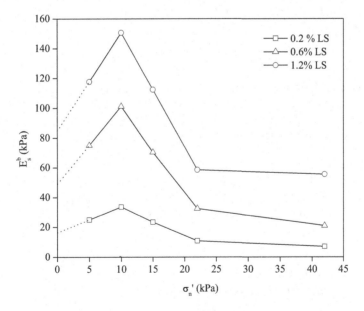

FIGURE 10.8 Variation of E_s^b with the effective normal stress. (Reproduced from Athukorala et al., 2015 with permission from ASCE.)

In Eq. (10.24), $(E_s)_p^{ut}$ is the secant shear modulus of the untreated soil at peak stress; σ'_{nc} is the effective normal stress at which the bond strength becomes zero; and d_1 is a model parameter. A similar approach has been used by Haeri and Hamidi (2009) to model the stress–strain behaviour of cemented bonds. The values of $\left(\dfrac{E_s^b}{(E_s)_p^{ut}} \right)_{(\sigma'_n=0)}$ in Eq. (10.24) were found by plotting the experimental ratio of $\left(\dfrac{E_s^b}{(E_s)_p^{ut}}\right)$ against the corresponding effective normal stress for different amounts of LS and

extrapolating the curves to $\sigma_n' = 0$. The model parameter d_1 was found by fitting the experimental values of E_s^b to Eq. (10.24) and can be given by Eq. (10.25) as a function of the effective normal stress:

$$d_1 = -0.0089\left(\sigma_n'\right)^2 + 0.573\sigma_n' - 6.97 \tag{10.25}$$

10.3.3.1.1   Disturbance Function (D^b) Parameters

The values of disturbance for LS bonds at different effective normal stresses using Eq. (10.13); the actual responses of LS bonds were modelled first. For this purpose, the bond strength was considered to be linearly elastic up to the peak bond strength, and then decreased as a decaying power function, as shown in Figure 10.7. Equation (10.26) is proposed to calculate the peak strength $\left(\tau_p^b\right)$ of LS bonds.

$$\tau_p^b = \tau_p^{ut} \times \left(\dfrac{\tau_p^b}{\tau_p^{ut}}\right)_{(\sigma_n'=0)} \times \left(1 - \dfrac{\sigma_n'}{\sigma_{nc}'}\right)^{d_2} \tag{10.26}$$

In Eq. (10.26), τ_p^{ut} is the peak shear stress of the untreated soil under the considered effective normal stress. By fitting the experimental peak bond strengths to Eq. (10.26), the model parameter d_2 was found to be a function of the effective normal stress as given by:

$$d_2 = -0.0064\left(\sigma_n'\right)^2 + 0.453\sigma_n' - 6.9 \tag{10.27}$$

Post-peak behaviour of LS bonds was modelled by fitting the experimental results to a power function. The disturbance was then quantified using Eq. (10.13) for LS bonds and the corresponding disturbance parameters were then determined by fitting calculated disturbance values to corroborate with Eq. (10.17). A summary of these values is given in Table 10.2.

10.3.3.1.2   Model Parameters for Lignosulfonate-Treated Soil

Similar to the RI behaviour of untreated soil, the RI response of LS-treated soil was simulated using δo version of the HiSS plasticity model. The LS bond response, which becomes the FA response for the LS-treated soil (as conceptually illustrated in Figure 10.4), could be quantitatively

---

TABLE 10.2

Model Parameters for Disturbance Functions of Lignosulfonate Bonds

Disturbance function	Value/Relationship
D_u^b	1
Z^b	$Z^b = 0.24\ln\sigma_n' + 0.19$
γ^*	$\gamma^* = 0.24\sigma_n' + 2.35$
A^b	$A^b = -0.3\ln\sigma_n' + 1.12$

Modified after Athukorala et al. (2015).

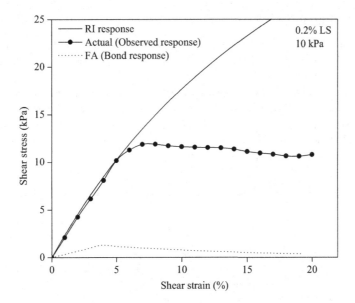

FIGURE 10.9 Relative intact (RI), fully adjusted (FA) and actual responses used to calculate disturbance corresponding to 0.2% lignosulfonate (LS)-treated soil at 10 kPa effective normal stress. (Reproduced from Athukorala et al., 2015 with permission from ASCE.)

TABLE 10.3
The Calculated Model Parameters for Lignosulfonate-Treated Soil

Relative Intact		Disturbance Function	
Parameter	Value	Parameter	Value/Relationship
a^I	20.25	$\xi_{\sigma D}^{*I}$	0
q^I	1.36	$D_{\sigma u}^I$	0.8
n^I	1.70	Z_σ^I	$Z_\sigma^I = 0.35e^{-0.085\sigma_n'}$
b^I	0.037	A_σ^I	$A_\sigma^I = -0.22\ln\sigma_n' + 0.88$
γ^I	11.71	$D_{\varepsilon u}^I$	0.8
$\xi_{\varepsilon D}^{*I}$		A_ε^I	0.6
Z_ε^I			0
		$Z_\varepsilon^I = -0.36\ln\sigma_n' + 1.40$	

Modified after Athukorala et al. (2015).

plotted using the above-mentioned model parameters (Eqs. 10.24–10.27). The RI, FA and the actual responses corresponding to 0.2% LS-treated soil at 10 kPa effective normal stress are shown in Figure 10.9. These responses were then used to calculate the values of experimental disturbances for LS-treated soil using Eq. (10.18). Disturbance parameters were then determined using the same approach used for untreated soil. Table 10.3 gives the calculated model parameters for LS-treated soil used in this study.

10.4   MODEL PREDICTIONS

In predicting the stress–strain and volumetric behaviour of LS-treated and untreated soil, the model parameters were determined using the experimental results corresponding only to the effective normal stresses of 10 and 22 kPa.

The stress–strain and volumetric behaviours of 0.2%, 0.6% and 1.2% LS-treated soil at different effective normal stresses considered are shown in Figure 10.10. The comparisons of the predicted stress–strain curves with the experimental data clearly show that the increase in shear strength as well as the initial slopes of the stress–strain curves, at various effective normal stresses, can be predicted reasonably well by the proposed model.

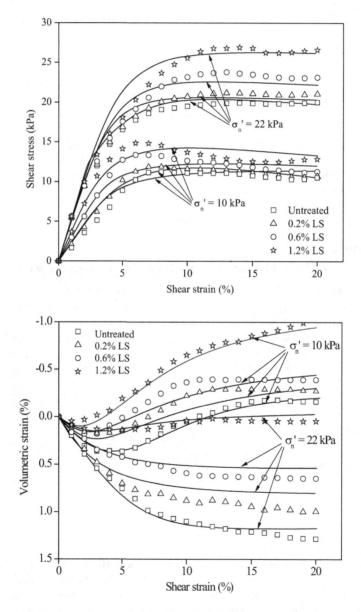

FIGURE 10.10   Predictions of shear stress–shear strain and volume change–shear strain behaviour of untreated and lignosulfonate-treated soil for $\sigma_n = 10$ and 22 kPa. (Reproduced from Athukorala et al., 2015 with permission from ASCE.)

It should be noted that the results presented in Figure 10.10 provide an independent prediction that can be used to validate the proposed model, because the experimental results under these two effective normal stress levels have not been used to determine the model parameters. Figure 10.10 confirms that the experimentally observed stress–strain behaviour is well captured by the DSC inspired model. The model predictions also confirm that the amount of compression of the untreated soil at these effective normal stresses was reduced due to LS treatment, and in agreement with the experimental observations.

10.5   CONCLUSIONS

In this chapter, a constitutive model for chemically treated soil in relation to LS was proposed, incorporating the original DSC. Specifically, the δ_0-version of the HiSS plasticity models was used to simulate the RI behaviour, and the LS bond strength was considered as the FA state for this LS-stabilised soil. The bond strength was modelled separately using a linear elastic RI response and zero strength state as the reference states. A series of drained direct shear tests were carried out for a LS-treated soil and the test results were used to calculate the relevant model parameters.

The proposed model has captured the key features of the LS-treated soil similar to the laboratory experiments. In particular, the proposed DSC model has captured accurately the shear strength, stiffness and compressibility with LS treatment. The DSC model has also captured the ductility of the LS-treated soil, different to soil stabilisation using traditional admixtures (e.g. cement), which often exhibit brittle behaviour.

The validated DSC model for bonded soil can be reliably used as supplementary aid in designing geotechnical structures (e.g. unpaved roads and transport embankments).

ACKNOWLEDGEMENTS

The work described in this chapter is a result of a past Australian Research Council's (ARC) Industry Linkage grant. The authors wish to express their gratitude to the ARC, to the Queensland Department of Transport and Main Roads (Brisbane), to Robert Armstrong (Chemstab Consulting Pvt. Ltd., Wollongong) and Coffey Geotechnics (Sydney) for providing financial support for this research.

The authors also acknowledge the PhD thesis of the third author, which contains greater details of this research, and this doctoral thesis is available to the readership through the University of Wollongong Library. Much of the content including the artwork has been reproduced from a past journal article of the authors with kind permission obtained from ASCE, while some modification to the artwork has been introduced during re-drawing.

REFERENCES

Athukorala, R., Indraratna, B. and Vinod, J. S. "A disturbed state concept inspired constitutive model for ligno-sulfonate treated silty sand". *International Journal of Geomechanics, ASCE*, 15, 6, 2015, 1–10.

Balasubramaniam, A. S., Phicrrwej, N., Lin, D. G., Karuzzaman, A. H. M., Uddin, K. and Bergado, D. T. "Chemical stabilisation of Bangkok Clay with cement, lime and fly ash additives", *Proceedings of 13th –Southeast Asian Geotechnical Conference*, Taipei, Taiwan, 1998, 253–258.

Balasubramaniam, A. S., Bergado, D. T., Buensuceso, B. R. and Yang, W. C. "Strength and deformation characteristics of lime-treated soft clays". *Geotechnical Engineering*, 20, 1, 1989, 49–65.

Bergado, D. T., Anderson, L. R., Miura, N. and Balasubramaniam, A. S. *Soft Ground Improvement in Lowland and Other Environments*, Published by ASCE Press, Reston, VA, 1996.

Chemstab. *Technical Manual*, CHEMSTAB Consulting Pty Ltd, Horsley, NSW Australia, 2003.

Chew, S. H., Kamruzzaman, A. H. M. and Lee, F. H. "Physicochemical and engineering behavior of cement treated clays", *Journal of Geotechnical and Geoenvironmental Engineering*, 130, 7, 2003, 696–706.

Clough, G. W. and Duncan, J. M. "Finite element analysis of retaining wall behaviour", *Journal of Soil Mechanics and Foundation Engineering Division, ASCE*, 97, 12, 1971, 1657–1673.

Consoli, N. C., Lopes, L. D. S., Prietto, P. D. M., Festugato, L. and Cruz, R. C. "Variables controlling stiffness and strength of lime-stabilized soils." *Journal of Geotechnical and Geoenvironmental Engineering*, 2011. doi: 10.1061/ (ASCE)GT.1943-5606.0000470.

Desai, C. S. *Mechanics of* Materials *and* Interfaces*: The* Disturbed State Concept, CRC Press, Boca Raton, FL, 2001.

Hausmann, M. R. *Engineering Principles of Ground Modification*, McGraw-Hill Publishing Company, New York, 1990.

Indraratna, B. "Utilization of lime, slag and fly ash for improvement of a colluvial soil in New South Wales", *Australia, Journal of Geotechnical & Geological Engineering*, 14, 1996, 169–191.

Indraratna, B., Athukorala, R. and Vinod, J. S. "Estimating the rate of erosion of a silty sand treated with lignosulfonate". *Journal of Geotechnical and Geoenvironmental Engineering, ASCE*, 139, 5, 2013, 701–714.

Indraratna, B., Balasubramaniam, A. S. and Khan, M. J. "Effect of fly ash with lime and cement on the behaviour of a soft clay", *Quarterly Journal of Engineering Geology & Hydrogeology*, 28, 2, 1995, 131–142.

Indraratna, B., Muttuvel, T. and Khabbaz, H. "Modelling the erosion rate of chemically stabilized soil incorporating tensile force-deformation characteristics". *Canadian Geotechnical Journal*, 46, 2009, 57–68.

Indraratna, B., Muttuvel, T., Khabaaz, H. and Armstrong, B. "Predicting the erosion rate of chemically treated soil using a process simulation apparatus for internal crack erosion". *Journal of Geotechnical and Geoenvironmental Engineering, ASCE*, 134, 6, 2008, 837–844.

Kamon, M. "Remediation techniques by use of ground improvement", *Proceedings of Conference on Soft Ground Technology*, ASCE Geotechnical Special Publication Number, 112, 2000, 374–387.

Karol, R. H. *Chemical Grouting and Soil Stabilization*, Marcel Decker, INC., New York, Basel, 2003.

Mehenni, A., Cuisinier, O. and Masrouri, F. " Impact of lime, cement, and clay treatments on the internal erosion of compacted clays", *Journal of Materials in Civil Engineering, ASCE*, 28, 9, 2016, 01016071.

Rajasekaran, G. and Narasimha Rao, S. "The microstructure of lime-stabilized marine clay", *Ocean Engineering*, 24, 9, 1997, 867–878.

Uddin, K., Balasubramaniam, A. S. and Bergado, D. T. "Engineering behaviour of cement treated Bangkok soft clay". *Geotechnical Engineering Journal*, 28, 1, 1997, 89–119.

Vinod, J. S., Indraratna, B. and Mahamud, M. A. A. "Stabilisation of an erodible soil using a chemical admixture". *Proceedings of the Institution of Civil Engineers Ground Improvement*, 163, GII, 2010, 43–51.

11 Binary-Medium Constitutive Model (BMCM) for Structured Soils

Enlong Liu
College of Water Resource and Hydropower, Sichuan University

11.1 INTRODUCTION

Natural soils, usually called structured soils, have bonding among soil grains and big pores in the process of sedimentation, which leads to them behaving differently from the reconstituted soils. The stress–strain relationships for artificially structured soils are shown in Figure 11.1.[1] For triaxial tests, under lower confining pressures, the structured soil samples exhibit strain-softening behavior and initially contract, then dilate (the lower the confining pressure, the more the sample dilates), and under higher confining pressures, the samples exhibit strain-hardening behavior and contract at all times (the larger the confining pressure, the more the sample contracts). All the remolded soil samples having the same density of the structured ones (as shown in Figure 11.2)[1] exhibit stress-hardening under confining pressures ranging from 25 to 400 kPa, and have similar tendency of volume change as structured soil samples, among which the difference is that the structured soil samples dilate more at lower confining pressures.

Furthermore, for a structured soil element tested under triaxial compression conditions, the strain distribution within it is not uniform, especially when the shear band appears, companied by strain localization, and thus the stresses within the soil element are also distributed inhomogeneously. This is due to the fact that: upon loading, the higher local stress that equals the strength of the bonds between soil particles within structured soil samples will result in their breakup, which leads to the nonuniform distribution of stress and strain within structured soil samples.

Many constitutive models have been formulated to describe the stress–strain properties of soils, of which the widely used one is the Cam-clay model for normally consolidated or reconstituted clay,[2,3] displaying strain hardening and volumetric contraction. However, many soils in nature have

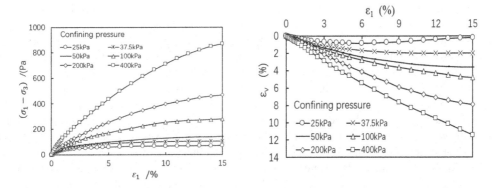

FIGURE 11.1 Stress–strain relationship for structured soils: (a) deviatoric stress– axial strain and (b) volumetric strain–axial strain.[1]

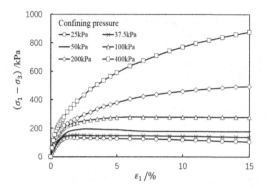 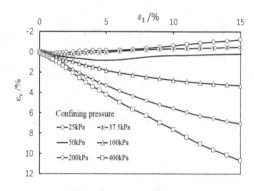

FIGURE 11.2 Stress–strain relationship for remolded soils: (a) deviatoric stress–axial strain and (b) volumetric strain–axial strain.[1]

cementation effects between soil particles, which usually display strain softening, initially volumetric contraction followed by dilatancy. Therefore, it is necessary to formulate a constitutive model for structured soils to consider the nonuniform stress and strain in the soil element and reflect on the macroscopic strain softening by use of the parameters considering the micro-deformation mechanism, which is the aim of binary-medium model (BMM), and this will be elucidated in detail in the subsequent sections. In the following, the breakage mechanism of structured soils and the concept of BMM is introduced first, followed by the generalized stress–strain relationships for BMCM, then the model is compared with the test results.

There are many constitutive models for structured soils, which can be found in references[4–7] and will not be reviewed here. This chapter is mainly a review on the development of BMM and some results have been published,[1,8–24] which can be found for detail in the corresponding references.

11.3   CONCEPT OF BINARY-MEDIUM MODEL (BMM)

In this section, the breakage mechanism of structured soils is analyzed first, followed by introducing the concept of BMM.[1,10] For natural soils, the bearing capacity of the soil element comes from both the cohesive resistance and frictional resistance, and they are not mobilized simultaneously at different deformation or strain levels,[25] with the former reaching a peak value within a relatively small strain and the latter making a full contribution within a relative large deformation or strain. The cohesion essentially comes from the cementation bonding between particles, whose distribution is not uniform among geological materials. In the process of sedimentation of geological materials, the bonded blocks are formed where the cementation bonding strength is stronger, and the weakened bands are formed where the cementation bonding is weaker, so the heterogeneous structured soils are gradually developed. During the loading process, the brittle bonded blocks gradually break up, transforming to the elasto-plastic weakened bands, so the two components bear the loading collectively. With the development of the breakage process, the bearing capacity of the bonded blocks will decrease, and that of the weakened bands will increase; however, the structured soil wholly exhibits strain-hardening or strain-softening behavior, depending on the increase of the bearing capacity of the weakened bands and the decrease of the bearing capacity of the bonded blocks. In view of the understanding of the breakage mechanism of structured soils mentioned previously, the structured soil can be conceptualized as a binary-medium material consisting of the bonded blocks and weakened bands bearing the capacity collectively.[10] In the following, the bonding blocks are called the bonded elements, and the weakened bands are called frictional elements. Figure 11.3 presents the sketch of BMM, where the bonded element is composed of a spring (E_b) and a brittle bond (q) and the frictional element is composed of a spring (E_f) and a plastic slider (f). For

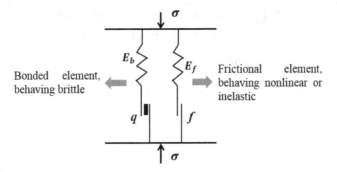

FIGURE 11.3 Sketch of binary-medium material.

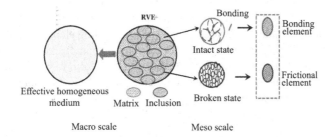

FIGURE 11.4 Schematic of binary-medium model (BMM).[26]

the brittle bond, it does not deform when the stress is less than the bond strength q and fails once the stress reaches the value of q. In a continuum of structured soil sample, there are many bonded elements and frictional elements. Upon loading, some bonded elements may break up and transfer to frictional elements and bear external loads collectively.

Upon loading, the bonding elements within the soil element will break up gradually and transform to be the frictional elements, when the breakage criterion is satisfied for the bonding element. For the soil element (RVE, Representative Volume Element), initially the bonding elements mainly bear the external loading, and with the increasingly breaking process of the bonding among soil particles, the frictional elements increasingly bear more external loads. And thus, the frictional elements/bonding elements can be assumed to be matrix or inclusions within the soil element, in which their interactions should be considered. From Figure 11.4,[26] we see that the distribution of stress/strain within the soil element will be nonuniform due to the different stiffness of the bonding elements and frictional elements. Before the bonding element breaks up, it may behave elastic, linear or nonlinear, which is mainly resulted from the soil solid matrix properties and cementation/bonding effects among soil particles. The frictional elements, however, may exhibit elasto-plastic properties, which is mainly the result of the slip and rotation among soil particles. There is a similar concept of disturbed state concept (DSC) proposed by Desai and coworkers,[4,27] in which the continuum element is assumed to be composed of intact (RI) and adjusted (FA) states and has been used for soils (sands and clays), rocks, rockfill, asphalt, concrete, silicon, polymers, and interfaces and joints. In DSC, a deforming material is a mixture of (RI and FA states and similar in binary-medium materials) components that interact with each other to lead to the observed behavior. The material mixture can undergo degradation or softening and stiffening or healing. However, the basis in the damage approach is different; it starts from the assumption that a part of the material is damaged or cracked. The observed behavior is then defined based essentially on behavior of the undamaged part, and both do not interact because the damaged part is assumed to possess no strength.

11.4　GENERALIZED STRESS–STRAIN RELATIONSHIPS FOR BINARY-MEDIUM CONSTITUTIVE MODEL (BMCM)

11.4.1　ONE PARAMETER-BASED BMCM

For the RVE as shown in Figure 11.4, the stress and strain for BMCM can be expressed as follows[8–10] by homogenization theory:

$$\sigma_{ij} = (1-\lambda)\sigma_{ij}^b + \lambda\sigma_{ij}^f \tag{11.1a}$$

$$\varepsilon_{ij} = (1-\lambda)\varepsilon_{ij}^b + \lambda\varepsilon_{ij}^f \tag{11.1b}$$

in which σ_{ij}, σ_{ij}^b, and σ_{ij}^f are the stresses of RVE, bonding element, and frictional element, respectively; ε_{ij}, ε_{ij}^b, and ε_{ij}^f are the strains of RVE, bonding element, and frictional element, respectively; λ is the breakage ratio and equal to the volume ratio of the bonding elements to frictional elements within the RVE. At the initial loading stage, the RVE is mainly composed of the bonding elements, which break up gradually with the process of loading and transform to be the frictional elements, and at failure, the frictional elements dominate the behavior of RVE. The nonuniform distribution of strain/stress can be realized by introducing the local strain/stress coefficient, C_{ijkl} or A_{ijkl}, linking the strain/stress of the bonding elements with that of RVE, expressed as follows:

$$\varepsilon_{ij}^b = C_{ijkl}\varepsilon_{kl} \text{ or } \sigma_{ij}^b = A_{ijkl}\sigma_{kl} \tag{11.2}$$

The tangential stiffness matrixes of the bonded elements and frictional elements are represented by D_{ijkl}^b and D_{ijkl}^f, respectively, and we have:

$$d\sigma_{ij}^b = D_{ijkl}^b d\varepsilon_{kl}^b \tag{11.3a}$$

$$d\sigma_{ij}^f = D_{ijkl}^f d\varepsilon_{kl}^b \tag{11.3b}$$

Expressing (11.1a) and (11.1b) in the incremental form, and combining (11.2) and (11.3a), (11.3b), we can have the generalized stress–strain relationship for BMCM, expressed as follows:

$$d\sigma_{ij} = \left[\left(1-\lambda^0\right)\left(D_{ijkl}^b - D_{ijkl}^f\right)C_{klmn}^0 + D_{ijmn}^f\right]d\varepsilon_{mn} + \left(1-\lambda^0\right)\left(D_{ijkl}^b - D_{ijkl}^f\right)dC_{klmn}\varepsilon_{mn}^0 \tag{11.4}$$

in which λ^0 is the current breakage ratio; C_{klmn}^0 is the current local strain coefficient.

11.4.2　TWO PARAMETERS-BASED BMCM

The stress and strain can be divided into the spherical and deviatoric parts, and thus we have:

$$\sigma_m = (1-\lambda_v)\sigma_m^b + \lambda_v\sigma_m^f \tag{11.5a}$$

$$\varepsilon_v = (1-\lambda_v)\varepsilon_v^b + \lambda_v\varepsilon_v^f \tag{11.5b}$$

$$s_{ij} = (1 - \lambda_s) s_{ij}^b + \lambda_s s_{ij}^f \qquad (11.5c)$$

$$e_{ij} = (1 - \lambda_s) e_{ij}^b + \lambda_s e_{ij}^f \qquad (11.5d)$$

in which $\sigma_m = \sigma_{kk}/3$, $\varepsilon_v = \varepsilon_{kk}$, $s_{ij} = \sigma_{ij} - \delta_{ij}\sigma_m$, $e_{ij} = \varepsilon_{ij} - \delta_{ij}\varepsilon_m/3$, and δ_{ij} is the Kronecker symbol; λ_v and λ_s are the volumetric breakage ratio and area breakage ratio, respectively, which are introduced to account for the breakage of the bonding elements and slip along the shear band, commonly companying strain softening. Similarly, the stress and strain for both the bonding element and frictional element can be divided into the spherical and deviatoric parts, so we have:

$$d\sigma_m^b = K^b d\varepsilon_v^b \qquad (11.6a)$$

$$ds_{ij}^b = G_{ijkl}^b de_{kl}^b \qquad (11.6b)$$

$$d\sigma_m^f = K^f d\varepsilon_v^f \qquad (11.6c)$$

$$ds_{ij}^f = G_{ijkl}^f de_{kl}^f \qquad (11.6d)$$

in which K^b, G_{ijkl}^b and K^f, G_{ijkl}^f are the moduli in spherical and deviatoric parts for both bonded and frictional elements, respectively.

The strains of the bonding element and RVE have the following relationship:

$$\varepsilon_v^b = C_v \varepsilon_v \text{ and } e_{ij}^b = C_{ijkl} e_{ij} \qquad (11.7)$$

in which C_v and C_{ijkl} are strain local coefficients for volumetric and deviatoric strains, respectively.

In the same way for deriving (11.4), we can obtain the two parameters-based BMCM in the following general form:

$$d\sigma_{ij} = \left[K^f + (1 - \lambda_v^0)(K^b - K^f) B^v \right] d\varepsilon_v \delta_{ij} + \frac{\varepsilon_v^0}{\lambda_v^0} \left[(C_v^0 - 1) K^f + D_m^0 - K^{b0} C_v^0 \right] d\lambda_v \delta_{ij}$$

$$+ \left\{ G_{ijkl}^f + (1 - \lambda_s^0)(G_{ijmn}^b - G_{ijmn}^f) B_{mnkl}^s \right\} de_{kl} \qquad (11.8)$$

$$+ \left\{ \left[(C_{ijmn}^{s0} - 1) G_{mnkl}^f + G_{ijkl}^{s0} - G_{ijmn}^{s0} C_{mnkl}^{s0} \right] \right\} \frac{e_{kl}^0}{\lambda_s^0} d\lambda_s$$

in which $B^v = C_v^0 + \frac{\partial C_v}{\partial \varepsilon_v} \varepsilon_v^0$; $D_m^0 = \sigma_m^0 / \varepsilon_v^0$; $K^{b0} = \sigma_m^{b0} / \varepsilon_v^{b0}$; $B_{ijkl}^s = C_{ijkl}^{s0} + \frac{\partial C_{ijkl}}{\partial e_{mn}} e_{mn}^0$; $s_{ij}^0 = G_{ijkl}^{s0} e_{kl}^0$; $s_{ij}^{b0} = G_{ijkl}^{sb0} e_{kl}^{b0}$; λ_v^0, C_v^0, λ_s^0, C_{ijmn}^{s0} are the current values.

When deriving the generalized stress–strain relationships (11.4) and (11.8), we used the local strain coefficients to link the strain of the bonding element and that of RVE. In addition, we can also use the local stress coefficient to derive the generalized stress–strain relationship for BMCM, referring to references[9,23].

11.4.3 Approach for Determining Model Parameters

There are four sets of model parameters in BMCM, including the parameters of the bonding elements, the frictional elements, the breakage ratios, and the local strain/stress coefficient. From the definition of the bonding element, we know that its model parameters can be determined for the sample at the initial loading stage, and during this stage the behavior of the sample is dominated by the bonding elements and can be described by an elastic stress–strain relationship. For structured soils, the model parameters are determined within 0.25% of the axial strain.[1,24] From the definition of the frictional element, we know that its model parameters can be determined when the sample fails, which behaves in an elasto-plastic manner. For structured soils, the behavior of the frictional elements can be determined on the reconstituted sample with the same density of the intact sample. According to the test results on reconstituted soils, the frictional elements can be described by the Cam-clay model, Lade-Duncan model, double hardening model, or elastic-perfectly plastic constitutive model.

In the BMCM, both the breakage ratios and the local strain/stress coefficient are the structural parameters, which can be determined by the following approaches. The breakage ratio is an internal variable, having similar meaning as the hardening parameter in plasticity and the damage variable in damage mechanics, which denotes the breakage of the bonding element in some degree. When the breakage ratio is expressed as the function of stress or strain of RVE, it has the evolving rule of Weibull distribution. In addition, the energy conservation in meso- and macroscales can be employed to solve the breakage ratio, which is implied in the constitutive model.[22]

The local strain/stress coefficient bridges the strain/stress of the bonding element to that of RVE, which reflects the nonuniform distribution of strain/stress in REV. There are two approaches that can be used to determine it, one of which is to express it as the function of the breakage ratio, and the other is to use the mesomechanics-based method (such as Mori-Tanaka method), and thus the bonding element/frictional elements can be idealized to be matrix/inclusions transforming to inclusions/matrix.[17,20,22] The mesomechanical-based method used to determine the local strain/stress coefficient can consider the interactions between the bonding element and frictional element, and the local strain/stress coefficient can be implicitly solved.

11.5 BINARY-MEDIUM CONSTITUTIVE MODEL (BMCM) FOR STRUCTURED SOILS

11.5.1 Constitutive Relationship of the Bonded Elements

The bonded elements have bonding and big pores within them, whose behaviors are similar to those of artificially structured soils in the initial loading within very small strain having almost intact structures. Natural soils are formed in layers by sedimentation, whose mechanical properties are isotropic in horizontal planes and different in horizontal and vertical directions. Therefore, we assume here that the bonded elements are cross-anisotropic elastic-brittle materials. When setting the symmetry axis along the z-direction and x-axis and y-axis in the horizontal plane, the stress–strain relationship Eq. (11.15) of the bonded elements can be rewritten as follows:

$$\left\{ \begin{array}{c} d\sigma_x \\ d\sigma_y \\ d\sigma_z \\ d\tau_{yz} \\ d\tau_{zx} \\ d\tau_{xy} \end{array} \right\}_b = \left[\begin{array}{cccccc} D_{11} & D_{12} & D_{13} & 0 & 0 & 0 \\ D_{12} & D_{11} & D_{13} & 0 & 0 & 0 \\ D_{13} & D_{13} & D_{33} & 0 & 0 & 0 \\ 0 & 0 & 0 & D_{44} & 0 & 0 \\ 0 & 0 & 0 & 0 & D_{44} & 0 \\ 0 & 0 & 0 & 0 & 0 & \dfrac{D_{11}-D_{12}}{2} \end{array} \right]_b \left\{ \begin{array}{c} d\varepsilon_x \\ d\varepsilon_y \\ d\varepsilon_z \\ d\varepsilon_{yz} \\ d\varepsilon_{zx} \\ d\varepsilon_{xy} \end{array} \right\}_b$$

(11.9)

where five material constants, D_{11}, D_{12}, D_{13}, D_{33}, and D_{44}, can be determined by the stress–strain curves at the initial loading stage of the tested samples, during which the structured samples are hardly damaged and could be regarded as the bonded elements. When $D_{11} = D_{33}$, $D_{12} = D_{13}$ and $D_{44} = (D_{11} - D_{12})/2$, Eq. (11.9) can be reduced to the stress–strain relationship of isotropic materials with two constants.

11.5.2 Constitutive Relationship of the Frictional Elements

The frictional elements are transformed from bonded elements when the bonds between soil particles are broken completely, whose mechanical properties could be assumed as those of remolded soils. From the test results of remolded soils shown in Figure 11.2, we know that the stress–strain relationship of frictional elements can be described by the Lade-Duncan model.[28,29] For the Lade-Duncan model, the incremental strain of soils consists of elastic and plastic parts as follows:

$$\{d\varepsilon\}_f = \{d\varepsilon^e\}_f + \{d\varepsilon^p\}_f \tag{11.10}$$

in which $\{d\varepsilon^e\}_f$ is the incremental elastic strain and $\{d\varepsilon^p\}_f$ is the incremental plastic strain.

According to the Lade-Duncan model, the elastic strain can be expressed as follows:

$$
\begin{Bmatrix}
d\varepsilon_x^e \\
d\varepsilon_y^e \\
d\varepsilon_z^e \\
d\varepsilon_{yz}^e \\
d\varepsilon_{zx}^e \\
d\varepsilon_{xy}^e
\end{Bmatrix}_f
= \frac{1}{E_f}
\begin{Bmatrix}
d\sigma_x - v_f(d\sigma_y + d\sigma_z) \\
d\sigma_y - v_f(d\sigma_z + d\sigma_x) \\
d\sigma_z - v_f(d\sigma_x + d\sigma_y) \\
2(1+v_f)d\tau_{yz} \\
2(1+v_f)d\tau_{zx} \\
2(1+v_f)d\tau_{xy}
\end{Bmatrix}_f
\tag{11.11}
$$

where E_f and v_f are the tangential deformational modulus and tangential Poisson's ratio of the remolded samples, respectively. In the Lade-Duncan model, the failure criterion is $f_1 = \dfrac{I_1^3}{I_3} = K_f$, yielding function is $f = \dfrac{I_1^3}{I_3} = K_0$ and plastic potential $g = I_1^3 - K_2 I_3$, where I_1 and I_3 are the first and third invariants of stress, respectively, and $K_f = K_0$ at failure. Therefore, according to hardening elasto-plastic theory, we can obtain the incremental plastic strain as follows:

$$
\begin{Bmatrix}
d\varepsilon_x^p \\
d\varepsilon_y^p \\
d\varepsilon_z^p \\
d\varepsilon_{yz}^p \\
d\varepsilon_{zx}^p \\
d\varepsilon_{xy}^p
\end{Bmatrix}_r
= d\vartheta \cdot K_2
\begin{Bmatrix}
\dfrac{3I_1^2}{K_2} - \sigma_y\sigma_z + \tau_{yz}^2 \\
\dfrac{3I_1^2}{K_2} - \sigma_z\sigma_x + \tau_{zx}^2 \\
\dfrac{3I_1^2}{K_2} - \sigma_x\sigma_y + \tau_{xy}^2 \\
2\sigma_x\tau_{yz} - 2\tau_{xy}\tau_{zx} \\
2\sigma_y\tau_{zx} - 2\tau_{xy}\tau_{yz} \\
2\sigma_z\tau_{xy} - 2\tau_{yz}\tau_{zx}
\end{Bmatrix}_f
\tag{11.12}
$$

where $d\vartheta$ is the plastic multiplier and K_2 is model constant. The detail description of the Lade-Duncan model can be found in references[28,29].

11.5.3 STRUCTURAL PARAMETERS OF BREAKAGE RATIO AND LOCAL STRAIN COEFFICIENT

The breakage ratio λ is a structural parameter, whose evolving rules are closely related to soil type, stress and strain level, and stress path and history. At the initial stage of loading, λ has a very small value close to zero for the external loads are mainly born by the bonded elements; with the process of loading, λ increases gradually accompanied by the bonded elements transferring to the frictional elements, with both of them bearing the external loading; when the strain is very large, λ tends to be 1.0 and the external loads are mainly borne by the frictional elements at the moment. In view of the determination method of damage factor and hardening parameters, we assume that breakage ratio λ is the function of volumetric strain and generalized shear strain with the following expression:

$$\lambda = 1 - exp\left[-\beta\left(\alpha\varepsilon_z + \varepsilon_x + \varepsilon_y\right)^\psi - \left(\xi\varepsilon_s\right)^\theta \right] \tag{11.13}$$

where $\varepsilon_s = \sqrt{2e_{ij}e_{ij}/3}$, $e_{ij} = \varepsilon_{ij} - \varepsilon_{kk}\delta_{ij}/3$, and α, β, ξ, ψ and θ are materials parameters, with the symmetry axis along the z-direction and x-axis and y-axis in the horizontal plane.

The local strain coefficient matrix bridges the strains of bonded elements and RVE, varying with loading history and strain level. We assume here that the elements in the local strain coefficient matrix are the same and are represented by C of function of the generalized shear strain as follows:

$$C = exp\left[-\left(t_c \times \varepsilon_s\right)^{r_c} \right] \tag{11.14}$$

where t_c and r_c are model parameters.

The breakage ratio and local strain coefficient are both internal variables, which should be determined by meso-mechanics at the meso-scale.[20,22] In this section, another method is used. We establish their evolving relationships using the similar determination methods of hardening parameters in plasticity or damage factor in damage mechanics. Based on analysis of the breakage mechanism of artificially structured soils from meso-scale to macro-scale, we formulate their expressions in which those model parameters could be determined by test results.

11.5.4 DETERMINATION OF MODEL PARAMETERS UNDER TRIAXIAL STRESS CONDITIONS

Under conventional triaxial stress conditions in which three kinds of soil samples including initially isotropic structured, initially stress-induced anisotropic structured, and remolded samples previously mentioned are tested, the vertical direction is set as z-axial direction, along which the maximal principal stress applies, and the other two principal stresses apply in the horizontal plane.[1,24] Combining the test results provided above, we present the determination method of model parameters under triaxial stress conditions in the following sections:

a. Parameter determination for the bonded elements

Under conventional triaxial stress conditions, the stress–strain relationship of the bonded elements, Eq. (11.9), can be simplified as follows:

$$\left\{ \begin{array}{c} d\sigma_1 \\ d\sigma_3 \end{array} \right\}_b = \frac{E_{vb}}{(1-v_{hhb})E_{vb} - 2v_{vhb}^2 E_{hb}} \left[\begin{array}{cc} (1-v_{hhb})E_{vb} & 2v_{vhb}E_{hb} \\ v_{vhb}E_{hb} & E_{hb} \end{array} \right] \left\{ \begin{array}{c} d\varepsilon_1 \\ d\varepsilon_3 \end{array} \right\}_b \tag{11.15}$$

where there are four material parameters of E_{vb}, E_{hb}, v_{vhb} and v_{hhb}, in which E_{vb} and E_{hb} represent the elastic moduli of the boned elements in vertical and horizontal directions, respectively, and v_{vhb} and v_{hhb} the Poisson ratios of boned elements in vertical and horizontal directions, respectively.

Within a small strain range upon initial loading, there are mainly bonded elements in RVE to bear the external loads, so the stress–strain curve of the structured samples can be very similar to that of bonded elements. Here, we use the strain of 0.25% of artificially structured samples tested to determine E_{vb}, E_{hb}, v_{vhb} and v_{hhb}. Using Eq. (11.15), we can only solve the values of E_{vb} and v_{vhb}. For initially stress-induced anisotropic structured samples, when $E_{vb} = E_{hb}$ and $v_{vhb} = v_{hhb}$, they become initially isotropic structured samples, and thus we assume the values of E_{hb} and v_{hhb} of initially stress-induced anisotropic structured samples are the same as those of initially isotropic structured samples. E_{vb}, E_{hb}, v_{vhb} and v_{hhb} are functions of confining pressure σ_3 expressed as $E_{vb}\left(or\ E_{vb} \right) = b_1 ln\left(\dfrac{\sigma_3}{p_a} \right) + b_2$ and $v_{hhb}\left(or\ v_{vhb} \right) = b_3 \left(\dfrac{\sigma_3}{p_a} \right)^{b_4}$, in which b_1, b_2, b_3, b_4 are material constants, and p_a is atmospheric pressure of 0.1014 MPa.

b. Parameter determination for the frictional elements

Frictional elements are transferred from the bonded elements and bonding between soil particles breaking up fully, whose mechanical properties are similar to those of remolded soils and can be described by the Lade-Duncan model as abovementioned. Based on the test results of remolded soils, we give the parameters of the Lade-Duncan model here.

Under conventional triaxial stress conditions, Eq. (11.10) can be rewritten as follows:

$$\left\{ \begin{array}{c} d\sigma_1 \\ d\sigma_3 \end{array} \right\}_f = [D]_f^{ep} \left\{ \begin{array}{c} d\varepsilon_1 \\ d\varepsilon_3 \end{array} \right\}_f \tag{11.16}$$

in which $[D]_f^{ep}$ is the stiffness matrix of frictional elements. According to the Lade-Duncan model, the elastic parameters of E_f and v_f can be determined by the nonlinear elastic model of the Duncan-Chang hyperbolic model[28,29] as follows:

$$E_f = K p_a \left(\frac{\sigma_3}{p_a} \right)^n \left[1 - \frac{R_f (\sigma_1 - \sigma_3)(1 - \sin\varphi)}{2(c\cos\varphi + \sigma_3 \sin\varphi)} \right]^2 \tag{11.17}$$

$$v_f = \frac{G - F \lg(\sigma_3 / p_a)}{\left[1 - \dfrac{D(\sigma_1 - \sigma_3)}{K p_a \left(\dfrac{\sigma_3}{p_a} \right)^n \left[1 - \dfrac{R_f(\sigma_1 - \sigma_3)(1 - \sin\varphi)}{2(c\cos\varphi + \sigma_3 \sin\varphi)} \right]^2} \right]^2} \tag{11.18}$$

where K, n, R_f, G, F, and D are material constants, c and φ are cohesion and internal frictional angle of remolded soils, respectively; and the stress is that of frictional elements.

Setting $m_1 = \dfrac{E_f(1-v_f)}{(1+v_f)(1-2v_f)}$, we can present $[D]_f^{ep}$ as follows:

$$[D]_f^{ep} = \begin{bmatrix} m_1 - \dfrac{n_3}{n_9} & \dfrac{2m_1 v_f}{1-v_f} - \dfrac{n_4}{n_9} \\[3mm] \dfrac{m_1 v_f}{1-v_f} - \dfrac{n_5}{n_9} & \dfrac{m_1}{1-v_f} - \dfrac{n_6}{n_9} \end{bmatrix}, \tag{11.19}$$

in which $n_1 = m_1\left[3I_1^2 - K_2\sigma_3^2 + \dfrac{2v_f}{1-v_f}\left(3I_1^2 - K_2\sigma_1\sigma_3\right)\right]$,

$$n_2 = m_1\left[\dfrac{v_f}{1-v_f}\left(3I_1^2 - K_2\sigma_3^2\right) + \dfrac{1}{1-v_f}\left(3I_1^2 - K_2\sigma_1\sigma_3\right)\right],$$

$$n_3 = \dfrac{m_1^2 \times n_1}{I_3^2}\left[\left(3I_1^2 I_3 - \sigma_3^2 I_1^3\right) + \dfrac{v_f}{1-v_f}\left(3I_1^2 I_3 - \sigma_1\sigma_3 I_1^3\right)\right],$$

$$n_4 = \dfrac{m_1^2 \times n_1}{I_3^2}\left[\dfrac{2v_f}{1-v_f}\left(3I_1^2 I_3 - \sigma_3^2 I_1^3\right) + \dfrac{1}{1-v_f}\left(3I_1^2 I_3 - \sigma_1\sigma_3 I_1^3\right)\right],$$

$$n_5 = \dfrac{m_1^2 \times n_2}{I_3^2}\left[\left(3I_1^2 I_3 - \sigma_3^2 I_1^3\right) + \dfrac{v_f}{1-v_f}\left(3I_1^2 I_3 - \sigma_1\sigma_3 I_1^3\right)\right],$$

$$n_6 = \dfrac{m_1^2 \times n_2}{I_3^2}\left[\dfrac{2v_f}{1-v_f}\left(3I_1^2 I_3 - \sigma_3^2 I_1^3\right) + \dfrac{1}{1-v_f}\left(3I_1^2 I_3 - \sigma_1\sigma_3 I_1^3\right)\right],$$

$$n_7 = \dfrac{m_1}{I_3^2}\left[\left(3I_1^2 I_3 - \sigma_3^2 I_1^3\right) + \dfrac{v_f}{1-v_f}\left(3I_1^2 I_3 - \sigma_1\sigma_3 I_1^3\right)\right],$$

$$n_8 = \dfrac{m_1}{I_3^2}\left[\dfrac{2v_f}{1-v_f}\left(3I_1^2 I_3 - \sigma_3^2 I_1^3\right) + \dfrac{1}{1-v_f}\left(3I_1^2 I_3 - \sigma_1\sigma_3 I_1^3\right)\right],$$

$n_9 = \dfrac{\left[1-\beta'(f-f_t)\right]\sigma_3}{\alpha'}\dfrac{m_1}{I_3^2}\left[n_7\left(3I_1^2 - K_2\sigma_3^2\right) + n_8\left(3I_1^2 - K_2\sigma_1\sigma_3\right)\right]$, and the stresses in these expressions are those of frictional elements. K_2 and stress level f have the following relationship:

$$K_2 = Af + 27(1-A) \tag{11.20}$$

where A is materials constant and f has the relationship with plastic work, which can be expressed as follows:

$$f - f_t = \dfrac{W_p}{\alpha' + \beta' W_p} \tag{11.21}$$

in which $f_t = 27$ for the remolded soils tested, and α', β' are model parameters. Substituting f, f_t, and W_p into Eq. (11.21), we can obtain $\beta' = 0.01$ and α' varying with the confining pressure as $\alpha' = r_1\left(\sigma_3 / p_a\right) + r_2$, in which r_1 and r_2 are materials constants.

c. Parameter determination for structural parameters

Under conventional triaxial stress conditions, the breakage ratio λ of Eq. (11.13) can be written as:

$$\lambda = 1 - \exp\left[-\beta(\alpha\varepsilon_1 + 2\varepsilon_3)^\psi - \left(\frac{2}{3}\xi(\varepsilon_1 - \varepsilon_3)\right)^\theta \right] \qquad (11.22)$$

For the artificially structured soils, the parameters of β and ψ are constants; α, ξ, and θ vary with the confining pressure as follows: $\alpha\left(\text{or } \xi, \theta\right) = e_1\left(\sigma_3 / p_a\right)^{e_2}$, in which e_1 and e_2 are constants.

The local strain coefficient of C in Eq. (11.14) can be expressed under triaxial stress conditions as follows:

$$C = \exp\left[-\left(\frac{2}{3}t_c(\varepsilon_1 - \varepsilon_3)\right)^{r_c} \right] \qquad (11.23)$$

in which $t_c = s_1\left(\sigma_3 / p_a\right) + s_2$, in which s_1 and s_2 are constants; for initially isotropic structured soils $r_c = 1.0$, and for initially stress-induced anisotropic structured soils $r_c = 1.5$.

11.5.5 Comparisons between Model Predicted and Tested Results

There are four sets of parameters, including those of bonded elements, frictional elements, and structural parameters of breakage ratio and local strain coefficient, required to be provided in the proposed BMCM for artificially structured soils. For the bonded elements, the parameters are obtained as follows: $b_1 = 9.8383$ and $b_2 = 30.37$ for E_{vb}, $b_1 = 9.1511$ and $b_2 = 28.61$ for E_{hb}, $b_3 = 0.2134$ and $b_4 = -0.41$ for ν_{hhb}, and $b_3 = 0.1389$ and $b_4 = -0.668$ for ν_{vhb}. For the frictional elements, the parameters are obtained as follows: $K = 88.797$, $n = 0.3425$, $R_f = 0.95$, $G = 0.242$, $F = 0.313$, $D = 0.0113$, $c = 0$, $\varphi = 32.062°$, and A = 0.3535; $r_1 = -14.0$, $r_2 = \_10.0$ when $\sigma_3 < 100$ kPa and $r_1 = -155.0$, $r_2 = -66.67$ when $\sigma_3 \geq 100$ kPa. For the structural parameters, when determining α, $e_1 = 100.55$, $e_2 = 0.1135$ for initially isotropic structured soils, $e_1 = 105.58$, $e_2 = 0.1081$ for initially stress-induced structured soils; when determining ξ, $e_1 = 40.56$, $e_2 = 40.0$ at $\sigma_3 < 100$ kPa and $e_1 = 2.535$, $e_2 = 100.0$ at $\sigma_3 \geq 100$ kPa; when determining θ, $e_1 = 0.0$, $e_2 = 0.15$ at $\sigma_3 < 100$ kPa and $e_1 = 0.0435$, $e_2 = 0.325$ at $\sigma_3 \geq 100$ kPa; and $s_1 = 11.859$, $s_2 = 30.854$.

The curves of the deviatoric stress–axial strain and volumetric strain–axial strain of artificially structured soils computed and tested are shown in Figures 11.5 and 11.6 for initially isotropic structured samples and initially stress-induced anisotropic structured samples, respectively. From the deviatoric stress–axial strain curves shown in Figures 11.5a and 11.6a, although there are some slight differences in values computed and tested, the proposed constitutive model can reflect the deformational features of artificially structured soils. At low confining pressures of 25, 37.5, and 50 kPa, the computed results give strain-softening behavior, which are in agreement with the tested ones and whose peak values are very close to those of the tested results; at 100 kPa confining pressure, both the computed and tested deviatoric stresses reach plastic flow state simultaneously; at high confining pressures of 200 and 400 kPa, the computed results show strain-hardening behavior, which is also in agreement with the tested ones.

From the volumetric strain–axial strain curves shown in Figures 11.5b and 11.6b, the computed results have similar properties as the tested ones. At low confining pressures of 25 kPa, 37.5, and 50 kPa, the computed volumetric strains first contract and then dilate, with slightly bigger values of contraction than those of the tested ones and very close dilatancy at failure; at high confining pressures of 100, 200, and 400 kPa, both results computed contract continuously till failure,

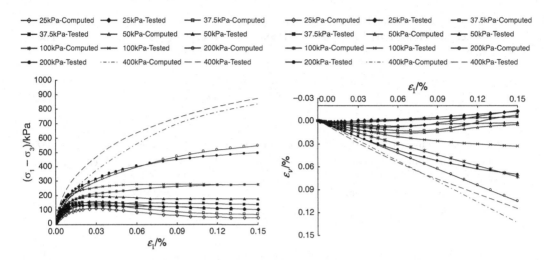

FIGURE 11.5 Stress–strain relationship for initially isotropic structured soils[1]: (a) deviatoric stress–axial strain and (b) volumetric strain–axial strain.

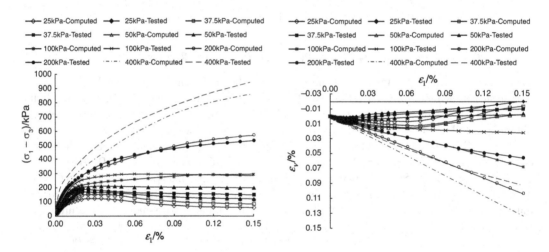

FIGURE 11.6 Stress–strain relationship for initially stress-induced structured soils[24]: (a) deviatoric stress–axial strain and (b) volumetric strain–axial strain.

which agree with tested results with slight differences in values. The deviatoric stresses of initially stress-induced structured soils are bigger than those of initially isotropic structured soils, and the volumetric strains of initially stress-induced structured soils contract more than those of initially isotropic structured soils, which are in agreement with the tested results. The reason for this is that the vertical loads applied during the process of curing the artificially structured soils make the bonding between soil particles along the vertical direction stronger than that in the horizontal plane.

11.6 DISCUSSIONS AND CONCLUSIONS

The new constitutive model, BMCM proposed here for artificially structured soils, idealizes the structured samples as composition of the bonded elements described by elastic-brittle materials and frictional elements described by the Lade-Duncan model, whose distribution of stress and strain can be considered by introducing the local strain coefficient and breakage ratio. The computed results compared with the tested ones demonstrate that the new model can grasp the main mechanical properties of artificially structured soils including strain softening and contraction first followed by dilatancy at low confining pressures and strain-hardening and continuous contraction at high confining pressures. Furthermore, the model also can describe the stress-induced anisotropy of artificially structured soil samples in some degree.

The performance of the model for un-breakage states and completely broken states is discussed here. For un-breakage states, the bonded elements are assumed to be of elastic state in the paper and bear the external loading. Therefore, the structured soil sample can be represented by the bonded elements for un-breakage states. When determining the parameters of the bonded elements, the artificially structured soils at the initial loading within very small strain (e.g. 0.25% axial strain) are used to assure that the bonds between soil particles are in an elastic state and not broken. For completely broken states, the bonded elements are wholly broken and transformed into frictional elements. Therefore, the structured soil sample can be represented by the frictional elements for completely broken states, which bear the external loading. When determining the parameters of the frictional elements, the remolded soil sample prepared by remolding the artificially structured sample tested are used to assure that the bonds between soil particles are completely broken. For the constitutive model for structured soil proposed here, the structured soil sample usually consists of two components or binary media of the bonded elements and frictional elements, and at failure the frictional elements dominate.

There are some differences between BMM and the existing constitutive models. In BMM, the breakage of the bonding element is the study focus, in which the breakage criterion and the evolutions of structural parameters, the breakage ratio and the local strain/stress coefficient, are new concepts to describe the transformation of the bonding elements to the frictional elements and consider the nonuniform distribution of stress/strain in RVE. When determining the model parameters, the mesomechanics-based approaches can be used, which can reduce the model parameters and let the parameters have clear meanings, and thus multi-scale BMCM have been formulated,[26] which enrich the constitutive models for geological materials.

Besides structured soil,[1] the BMCM for structured soils reviewed here has been used to model certain kinds of geomaterials, including frozen soil,[16,18,20,22,30] rock material,[13,19,31] cemented mixed soil,[23] and tailing soil subjected to freeze-thaw cycles.[17] The model parameters and validation results can be found in the corresponding articles, in which both the strain softening, initial contraction followed by dilatancy at low stress level, and the strain hardening and volumetric contraction at high stress level, can be modeled relatively well compared with the test results. It demonstrates that BMCM can grasp the salient features of cemented/bonding geomaterials.

REFERENCES

1. Liu, E.L., Yu, H.-S., Zhou, C., Luo, K.T., and Nie, Q. "A binary-medium constitutive model for artificially structured soils based on the disturbed state concept (dsc) and homogenization theory." *International Journal of Geomechanics*, 17, 7, 2017, 04016154.
2. Schofield, A.N., and Wroth, P. *Critical State Soil Mechanics*. McGraw-Hill, New York, 1968.
3. Roscoe, K. H., Schofield, A., and Wroth, C. P. "On yielding of soils." *Géotechnique*, 8, 1, 1958, 2–53.
4. Liu, M., Carter, J., and Desai, C. "Modeling compression behavior of structured geomaterials." *Int. J. Geomech.*, 3, 2003, 191–204.
5. Baudet, B., and Stallebrass, S. "A constitutive model for structured clays." *Géotechnique*, 54, 2004, 269–278.
6. Belokas, G., and Kavvadas, M. "An anisotropic model for structured soils." *Comput. Geotech.*, 37, 2010, 737–747.
7. Leroueil, S., and Vaughan, P.R. "The important and congruent effects of structure in natural soils and weak rocks." *Géotechnique*, 40, 1990, 467–488.
8. Shen, Z.J. "Breakage mechanics and double-medium model for geological materials." *Hydro-science and Engineering*, 4, 2002, 1–6. (in Chinese).

9. Shen, Z.J., Liu, E.L., and Chen, T.L. "Generalized stress-strain relationship of binary medium model for geological materials." *Chniese Journal of Geotechncal Engineering*, 27, 5, 2005, 489–494. (in Chinese).

10. Shen, Z.J. "Progress in binary medium modeling of geological materials." *Modern Trends in Geomechancis*, Wu, W., and Yu, H.S. (eds.) Springer, Berlin, 2006, 77–99.

11. Liu, E.L., and Shen, Z.J. "Research on the constitutive model of structured soils." *Proceedings of The International Conferences on Problematic Soils*, Famagusta, North Cyprus, 2005, 73–380.

12. Liu, E.L., Nie, Q., and Zhang, J.H. "A new strength criterion for structured soils." *Journal of Rock Mechanics and Geotechnical Engineering*, 5, 2013, 156–161.

13. Liu, E.L., and Zhang, J.H. "Binary medium model for rock sample." Q. Yang et al. (Eds.): Constitutive Modeling of Geomaterials: Advances and New Applications, SSGG, 2013, 341–347. Berlin, Heidelberg: Springer-Verlag, 2013.

14. Luo, F., Liu, E.L., and Zhu, Z.Y. "A strength criterion for frozen moraine soils." *Cold Regions Science and Technology*, 164, 2019, 102786.

15. Wang, P., Liu, E.L., Song, B.T., Liu, X.Y., Zhang, G., and Zhang, D. "Binary medium creep constitutive model for frozen soils based on homogenization theory." *Cold Regions Science and Technology*, 162, 2019, 35–42.

16. Zhang, D., and Liu, E.L. "Binary-medium-based constitutive model of frozen soils subjected to triaxial loading." *Results in Physics*, 12, 2019, 1999–2008.

17. Liu, Y.N., Liu, E.L., and Yin, Z.Y. "Constitutive model for tailing soils subjected to freeze-thaw cycles based on meso-mechanics and homogenization theory." *Acta Geotechnica*, 15, 2020, 2433–2450.

18. Wang, P., Liu, E.L., Zhang, D., Liu, X.Y., Zhang, G., and Song, B.T. "An elastoplastic binary medium constitutive model for saturated frozen soils." *Cold Regions Science and Technology*, 174, 2020, 103055.

19. Yu, D., Liu, E.L., Sun, P., Xiang, B., and Zheng, Q. "Mechanical properties and binary-medium constitutive model for semi-through jointed mudstone samples." *International Journal of Rock Mechanics and Mining Science*, 132, 2020, 104376.

20. Zhang, D., Liu, E.L., and Yu, D. "A micromechanics-based elastoplastic constitutive model for frozen sands based on homogenization theory." *International Journal of Damage Mechanics*, 29, 5, 2020, 689–714.

21. Zhang, D., Liu, E.L., and Huang, J. "Elastoplastic constitutive model for frozen sands based on framework of homogenization theory." *Acta Geotechnica*, 15, 2020, 1831–1845.

22. Wang, P., Liu, E.L., and Zhi, B. "An elastic-plastic model for frozen soil from micro to macro scale." *Applied Mathematical Modelling*, 91, 2021, 125–148.

23. Yu, H.J., and Liu, E.L. "A binary-medium constitutive model for artificially cemented gravel-silty clay mixed soils." *European Journal of Environmental and CivilEngineering*, 26, 12, 2022, 5773–5792.

24. He, C., Liu, E.L., and Nie, Q. "Mechanical properties and constitutive model for artificially structured soils with an initial stress-induced anisotropy." *Acta Geotechnica Slovenica*, 17, 2, 2020, 46–55.

25. Lambe, T.W. "A mechanical picture of shear strength in clay." In *Research Conference on Shear Strength of Cohesive Soils*. University of Colorado, Colorado, 1960, 555–580.

26. Liu, E.L. "Binary-medium constitutive model for geological materials: A multi-scale approach." In *Proceedings of the 20th International Conference on Soil Mechanics and Geotechnical Engineering*. Australian Geomechanics Society, Sydney, 2022, 1325–1328.

27. Desai, C. S. *Mechanics of Materials and Interfaces: The Disturbed State Concept*. Boca Raton, FL: CRC, 2001.

28. Lade, P.V., and Duncan, J.M. "Elasto-plastic stress-strain theory for cohesionless soil." *Journal of the Geotechnical Engineering Division*, 101, 1975, 1037–1053.

29. Lade, P.V. "Elasto-plastic stress-strain theory for cohesionless soil with curved yield surfaces." *International Journal of Solids and Structures*, 13, 1977, 1019–1035.

30. Wang, D., Liu, E.L., Zhang, D., Yue, P., Wang, P., Kang, J., and Yu, Qihao. "An elasto-plastic constitutive model for frozen soil subjected to cyclic loading." *Cold Regions Science and Technology*, 189, 2021, 103341.

31. Yu, D., Liu, E.L., Xiang, B., He, Y.Y., Luo, F., and He, C. "A micro-macro constitutive model for rock considering breakage effect." *International Journal of Mining Science and Technology*, 33, 2023, 173–184.

12 Collapsing of Sandstone
Acoustic Emission Study Based on the DSC Model

Meng Chen and Lei Wang
School of Civil Engineering, Chongqing University

Xiang Jiang
School of Materials Science and Engineering,
Chongqing Jiaotong University
School of Civil Engineering, Chongqing University

Yang Xiao
School of Civil Engineering, Chongqing University

12.1 INTRODUCTION

Quasi-brittle rock materials generally have the following mechanical properties: inhomogeneity, anisotropy, structural discreteness, and nonlinearity. Because of the complexity of its nonlinear evolution process, the triggering mechanism and prediction method of this kind of catastrophic damage have always been a difficult problem for engineering and scientific fields. Although the mechanical properties of rock under loading have been studied in detail, such as compressive force [1], cyclic loading [2], etc., the prediction of catastrophic failure in this situation is still a challenge. Statistical physics has been introduced to study the nonlinear and chaotic systems, such as earthquake [3–6], plastic behavior, and microfracture of solids or collapse of brittle solids [6–11]. Moreover, the concept of crackling noise (where systems under perturbation respond through discrete events with a variety of sizes and energies) has been introduced into material science [12–14]. In the system of brittle solid collapse, acoustic emission (AE) usually was adopted as crackling noise because of the sensitive and fast calculation ability of the AE signal. In the field of rock mechanics and rock engineering, AE has been used for many years [6–8,15]. AE records provide important information to understand the failure mode of solids: the spatial evolution of damage as well as the source mechanisms can be followed using this technique [16]. Analysis of AE induced during brittle and porous rock fracturing at various loading conditions has also been studied, [17] and nucleation of compaction bands could be clearly identified by the appearance of AE clusters inside the samples. Microstructural analysis of fractured samples shows excellent agreement between location of AE hypocenters and faults or the positions of compaction bands [18]. In addition, a series of laboratory experiments has been conducted to record three-dimensional (3-D) locations of AEs and to analyze the compaction bands development of sandstones [19].

Acoustic warning has been investigated widely in solid collapse like natural rock in coal mine disaster [20], porous 316L stainless steel in petrochemical industries, wastewater treatment, and aerospace [21], and even fault activity in earthquake [12]. This predictability is based on the crackling noise, which corresponds to the intermittent response in the deformation of materials [13,22]. Crackling noise corresponds to the intermittent response to a driven external field, and this kind of intermittent response shows a typical avalanche behavior. Under compression, the weak points in solids, such as dislocation [23] and microcracks [24], release elastic waves that could be captured by piezoelectric sensors as AE signals. These AE signals contain abundant information about the development of internal microcracks, and AE measurements are very sensitive and can detect even weak signals with an AE absolute energy near 1 aJ (= 10^{-18} joules) [25].

The disturbed state concept (DSC) [26–30] is a unified framework for modeling the response of different physical systems. It has been used in structured clay [31,32], cemented sand [33], concrete frames [34], and rock and rockfill materials [35–37]. So, in this study, AE has been adopted as a continuous and noninvasive real-time way to monitor the process of sandstone collapsing. And, the DSC model has been introduced to combine with AE analysis in the compression process of sandstone.

12.2   EXPERIMENTS

The sandstone samples are selected from the Triassic period, from southwest of China. Rock sample preparation is done according to the International Society of Rock Mechanics (ISRM) testing guidelines. The samples were drilled with a high-speed rotary saw from larger blocks, and their shapes were cylindrical with 25 mm diameter and 50 mm length (Figure 12.1a shows the sandstone sample after compression). The sides of the specimen were smooth, and the ends of the specimens were flat within ±0.02 mm. The bulk density is 2.2 g/cm³. Sandstone samples are mainly composed of quartz, feldspar, muscovite and these proportion values were determined by X-ray diffraction (see Figure 12.1b). Figure 12.1c and d shows the microbially induced calcite precipitation (MICP) sample before tests and the elements contained in it. Uniaxial compressive test was conducted by Shimadzu I-250 (Japan) testing machine (Figure 12.2a). The AE signals were captured by the AE workstation (DISP from American Physical Acoustics Company) with 16 bit, 10 MSPS digital signal processing rate,

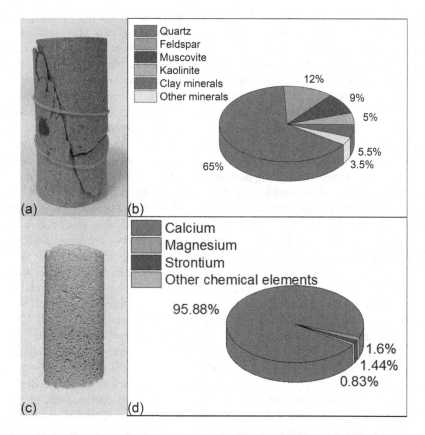

FIGURE 12.1   (a) Sandstone sample after uniaxial compression; (b) the mineral analysis of sandstone samples; (c) MICP sample before uniaxial compression; and (d) the elements contained in MICP samples.

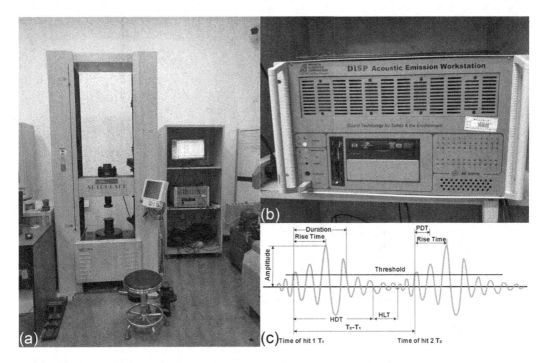

FIGURE 12.2 (a) Photograph of the loading setup; (b) acoustic emission workstation; and (c) typical acoustic emission wave and parameters.

and 1 kHz–1.2 MHz Bandwidth, see Figure 12.2b. The AE sensors are also from the American Physical Acoustics Company (Nano30T). The operating frequency range of this sensor is from 125 to 750 kHz. Two piezoelectric sensors were fixed on the specimen's side surface by plastic bands and were acoustically coupled to the sample by a thin layer of grease. Figure 12.2c shows typical AE wave signals, and for detailed meaning of AE parameters and experimental settings, previous related studies can be referred to [38].

12.3 TEST RESULTS

Figures 12.3–12.5 show the AE energy jerks in dry and saturated sandstone samples undergoing compression, while the jerks of an MICP sample are shown in Figure 12.6. We find several super-jerks in this compression process. The superjerks are defined as bursts with energies greater than any previous signals, and we ignored the influence of initial signals that were caused by the friction effect between the sample and loading equipment. The superjerks were marked in black color. We define normalization timescale as ranging from 0 to 1. Here, 0 indicates the initial state, and 1 represents the final collapsing. Figures 12.7–12.10 show the relation between normalization times-cale and superjerks' energy. According to the DSC theory, the relatively intact (RI) state and fully adjusted (FA) state are the two states of the physical system. For the compression process, the initial stage is considered as the RI state, and the final collapsing state can be represented as the FA state. Hence, the compression process can be predicted as the following:

$$\kappa = D\left(1 - e^{\left(-CE^T\right)}\right) \tag{12.1}$$

where the ultimate disturbance function is obtained from the ultimate FA state, namely, $D = 1$ for sandstone samples and $D = 0.6$ for the MICP sample.

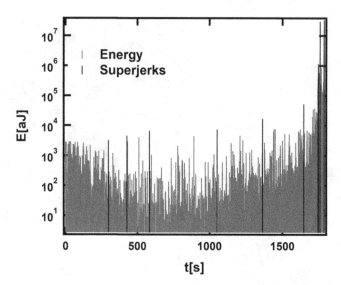

FIGURE 12.3 Acoustic emission absolute energy spectrum of a dry sandstone, and black lines indicate the superjerks' energy signals.

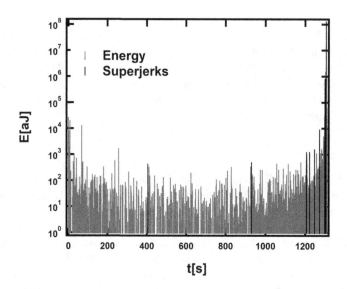

FIGURE 12.4 Acoustic emission absolute energy spectrum of a saturated sandstone sample w25, and black lines indicate the superjerks' energy signals.

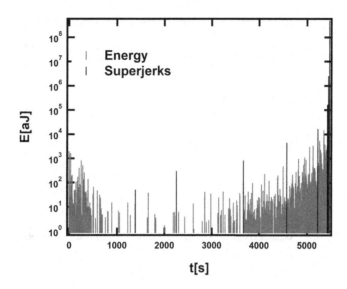

FIGURE 12.5 Acoustic emission absolute energy spectrum of a saturated sandstone sample w50, and black lines indicate the superjerks' energy signals.

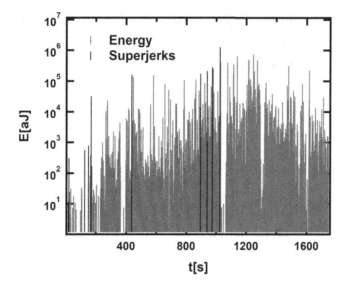

FIGURE 12.6 Acoustic emission absolute energy spectrum of an MICP sample, and black lines indicate the superjerks' energy signals.

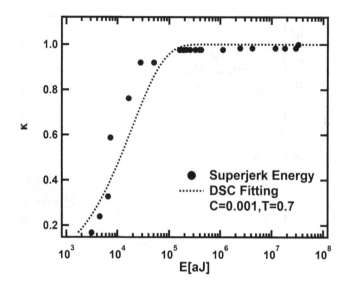

FIGURE 12.7   The relation between normalization timescale and superjerks' energy of a dry sandstone, and the dashed curve is the fitting based on the disturbed state concept (DSC) theory.

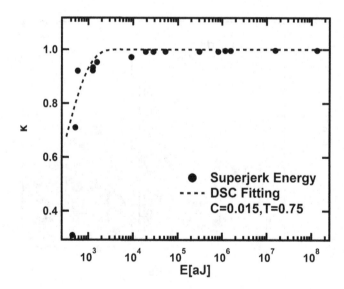

FIGURE 12.8   The relation between normalization timescale and superjerks' energy of a saturated sandstone sample w25, and the dashed curve is the fitting based on the disturbed state concept (DSC) theory.

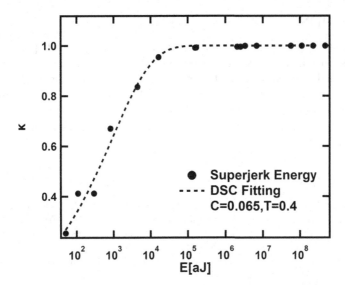

FIGURE 12.9   The relation between normalization timescale and superjerks' energy of a saturated sandstone sample w50, and the dashed curve is the fitting based on the disturbed state concept (DSC) theory.

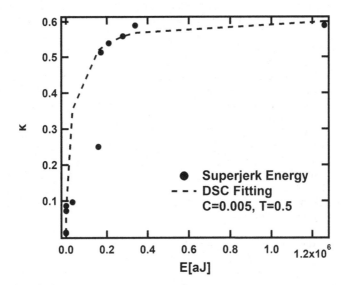

FIGURE 12.10   The relation between normalization timescale and superjerks' energy of an MICP sample, and the dashed curve is the fitting based on the disturbed state concept (DSC) theory.

12.4 CONCLUSIONS

In this study, AE has been used to detect the failure process during the collapse of sandstone under uniaxial compression. The AE spectrum shows typical avalanche dynamics process with irregular intermittent signals, which are usually defined as jerks. Several superjerks have been found in this compression process, and the superjerks were defined as jerks with energies greater than any previous acoustic signals. The saturation of superjerks is related to the final rupture. The DSC theory could be used to model the evolution of superjerk energy and as a reference warning for an approaching disaster.

REFERENCES

1. Xiao, Y., Meng, M., Daouadji, A., Chen, Q., Wu, Z., and Jiang, X. "Effects of particle size on crushing and deformation behaviors of rockfill materials." *Geosci. Front.*, 11, 2, 2020, 375–388.
2. Liu, P., Meng, M., Xiao, Y., Liu, H., and Yang, G. "Dynamic properties of polyurethane foam adhesive-reinforced gravels." *Sci. China. Tech. Sci.*, 64, 3, 2021, 535–547.
3. Bak, P., and Tang, C. "Earthquakes as a self-organized critical phenomenon." *J. Geophys. Res-Sol. Earth. Planet*, 94, B11, 1989, 15635–15637.
4. Gutenberg, B., and Richter, C. F. *Seismicity of the Earth.* Princeton University Press, Princeton, NJ, 1949.
5. Johnston, A. C., and Nava, S. J. "Recurrence rates and probability estimates for the new madrid seismic zone." *J. Geophys. Res-Sol. Earth. Planet*, 90, NB8, 1985, 6737–6753.
6. Desai, C. S., and Sherif, S. A. "Mechanics of materials and interfaces: The disturbed state concept." *Appl. Mech. Rev.*, 54, 4, 2001, B57–B58.
7. Tang, C. A., and Kaiser, P. K. "Numerical simulation of cumulative damage and seismic energy release during brittle rock failure - Part I: Fundamentals." *Int. J. Rock Mech. Min. Sci.*, 35, 2, 1998, 113–121.
8. Kaiser, P. K., and Tang, C. A. "Numerical simulation of damage accumulation and seismic energy release during brittle rock failure - Part II: Rib pillar collapse." *Int. J. Rock Mech. Min. Sci.*, 35, 2, 1998, 123–134.
9. Zapperi, S., Vespignani, A., and Stanley, H. E. "Plasticity and avalanche behaviour in microfracturing phenomena." *Nature*, 388, 6643, 1997, 658–660.
10. Petri, A., Paparo, G., Vespignani, A., Alippi, A., and Costantini, M. "Experimental-evidence for critical-dynamics in microfracturing processes." *Phys. Rev. Lett.*, 73, 25, 1994, 3423–3426.
11. Petri, A. "Acoustic emission and microcrack correlation." *Philos. Mag. B-Phys. Condens. Matter Stat. Mech. Electron. Opt. Magn. Prop.*, 77, 2, 1998, 491–498.
12. Baró, J., Corral, Á., Illa, X., Planes, A., Salje, E. K. H., Schranz, W., Soto-Parra, D. E., and Vives, E. "Statistical similarity between the compression of a porous material and earthquakes." *Phys. Rev. Lett.*, 110, 8, 2013, 088702.
13. Salje, E. K. H., and Dahmen, K. A. "Crackling noise in disordered materials." *Annu. Rev. Condens. Matter Phys.*, 5, 1, 2014, 233–254.
14. Sethna, J. P., Dahmen, K., Kartha, S., Krumhansl, J. A., Roberts, B. W., and Shore, J. D. "Hysteresis and hierarchies - dynamics of disorder-driven 1st-order phase-transformations." *Phys. Rev. Lett.*, 70, 21, 1993, 3347–3350.
15. Cox, S. J. D., and Meredith, P. G. "Microcrack formation and material softening in rock measured by monitoring acoustic emissions." *Int. J. Rock Mech. Min. Sci. Geomech. Abs.*, 30, 1, 1993, 11–24.
16. Xiao, Y., Wang, L., Jiang, X., Evans, T. M., Stuedlein, A. W., and Liu, H. "Acoustic emission and force drop in grain crushing of carbonate sands." *J. Geotech. Geoenviron.*, 145, 9, 2019, 04019057.
17. Stanchits, S., and Dresen, G. "Advanced acoustic emission analysis of brittle and porous rock fracturing." *ICEM 14: 14th International Conference on Experimental Mechanics,* vol. 6, F. Bremand, ed. 2010.
18. Stanchits, S., Fortin, J., Gueguen, Y., and Dresen, G. "Initiation and propagation of compaction bands in dry and wet bentheim sandstone." *Pure. Appl. Geophys.*, 166, 5–7, 2009, 843–868.
19. Fortin, J., Stanchits, S., Dresen, G., and Gueguen, Y. "Acoustic emission and velocities associated with the formation of compaction bands in sandstone." *J. Geophys. Res-Sol. Ea.*, 111, B10, 2006. doi:10.1029/2005JB003854.

20. Jiang, X., Jiang, D. Y., Chen, J., and Salje, E. K. H. "Collapsing minerals: Crackling noise of sandstone and coal, and the predictability of mining accidents." *Am. Mineral.*, 101, 12, 2016, 2751–2758.

21. Chen, Y., Wang, Q., Ding, X., Sun, J., and Salje, E. K. H. "Avalanches and mixing behavior of porous 316L stainless steel under tension." *Appl. Phys. Lett.*, 116, 11, 2020, 111901.

22. Sethna, J. P., Dahmen, K. A., and Myers, C. R. "Crackling noise." *Nature.*, 410, 6825, 2001, 242–250.

23. Chen, Y., Ding, X., Fang, D., Sun, J., and Salje, E. K. H. "Acoustic emission from porous collapse and moving dislocations in granular mg-ho alloys under compression and tension." *Sci. Rep.*, 9, 2019, 1330.

24. Soto-Parra, D., Vives, E., Botello-Zubiate, M. E., Matutes-Aquino, J. A., and Planes, A. "Acoustic emission avalanches during compression of granular manganites." *Appl. Phys. Lett.*, 112, 25, 2018, 251906.

25. Navas-Portella, V., Corral, Á., and Vives, E. "Avalanches and force drops in displacement-driven compression of porous glasses." *Phys. Rev. E*, 94, 3, 2016, 033005.

26. Desai, C. S., and Faruque, M. O. "Constitutive model for (geological) materials." *J. Eng. Mech.*, 110, 9, 1984, 1391–1408.

27. Desai, C. S., Sharma, K. G., Wathugala, G. W., and Rigby, D. B. "Implementation of hierarchical single surface delta-0 and delta-1 models in finite-element procedure." *Int. J. Numer. Anal. Met.*, 15, 9, 1991, 649–680.

28. Desai, C. S., Pradhan, S. K., and Cohen, D. "Cyclic testing and constitutive modeling of saturated sand–concrete interfaces using the disturbed state concept." *Int. J. Geomech.*, 5, 4, 2005, 286–294.

29. Desai, C. S. "Constitutive modeling of materials and contacts using the disturbed state concept: Part 1-Background and analysis." *Comput. Struct.*, 146, 2015, 214–233.

30. Desai, C. S. "Constitutive modeling of materials and contacts using the disturbed state concept: Part 2-Validations at specimen and boundary value problem levels." *Comput. Struct.*, 146, 2015, 234–251.

31. Liu, M. D., Carter, J. P., Desai, C. S., and Xu, K. J. "Analysis of the compression of structured soils using the disturbed state concept." *Int. J. Numer. Anal. Met.*, 24, 8, 2000, 723–735.

32. Liu, M. D., Carter, J. P., and Desai, C. S. "Modeling compression behavior of structured geomaterials." *Int. J. Geomech.*, 3, 2, 2003, 191–204.

33. Desai, C. S., and Xiao, Y. "Constitutive modeling for mechanical behaviors of geomaterials, new designs and techniques in geotechnical engineering." *J. Rock Mech. Geotech.*, 8, 3, 2016, 275.

34. Akhaveissy, A. H., and Desai, C. S. "Application of the DSC model for nonlinear analysis of reinforced concrete frames." *Finite Elem. Anal. Des.*, 50, 1, 2012, 98–107.

35. Desai, C. S., and Salami, M. R. "Constitutive model for rocks." *J. Geotech. Eng.*, 113, 5, 1987, 407–423.

36. Wu, G., and Zhang, L. "Studying unloading failure characteristics of a rock mass using the disturbed state concept." *Int. J. Rock Mech. Min.*, 41, 3, 2004, 437–437.

37. Abbas, S. M. *Constitutive Modeling of Rockfill Materials*. LAP Lambert Academic Publishing AG & Co KG, Riga, 2011.

38. Salje, E. K. H., Liu, H., Jin, L., Jiang, D., Xiao, Y., and Jiang, X. "Intermittent flow under constant forcing: Acoustic emission from creep avalanches." *Appl. Phys. Lett.*, 112, 5, 2018, 054101.

Section 4

Fracturing and Crushing Grains

13 Particle Crushing
DSC/HISS Model Considering the Particle Crushing of Soil

Yufeng Jia and Mengjie Tang
School of Hydraulic Engineering, Dalian University of Technology

13.1 INTRODUCTION

During the soil deformation under loading, the soil microstructure is continuously transformed from the relative intact (RI) state into the fully adjusted (FA) state,[1] and the work performed by loading is consumed by friction among the particles and the particles' roll over, crushing, and restructuring. The soil constitutive characteristics are controlled by the particles' friction, roll rover, breakage, and rearrangement. During these processes, particle breakage changes the contact status among particles, affecting the particle friction, reducing the particle roll rover, and causing particle restructuring.[2] The particle breakage is affected by particle material, particle weathering, particle shape, particle size, stress state, and loading history. It directly modifies the soil structure, influencing its strength, dilatancy, friction angle, and permeability, as well as generating creep deformation, wetting deformation, and residual strain under a seismic load.[2–6] In a freight railway in Queensland, Australia, the ballast was broken by the load of the train. The crushed particles are filled into ballast pores, blocking tracks' drainage and affecting their performance.[7] The creep deformation generated by rockfill particle breakage caused the Tianshengqiao No. 1 concrete face rockfill dam (CFRD) in China, which has a height of 175 m, to undergo a maximum settlement of 100 cm within 1 year after completion. Thus, 52% of the concrete face was separated from the cushion zone.[2] The slaking of weathered basalt rockfill in the Ataturk Dam in Turkey, which was constructed in 1990 and was considered the fourth largest clay-cored rockfill dam at that time, generated additional settlement and landsides in the clay core, which led to longitudinal cracks appearing along the crest.[4] The plastic deformation of the rockfill generated by particle crushing and restructuring in the Wenchuan earthquake ($M_L=8.0$) caused the Zipingpu CFRD in China to undergo a maximum settlement of 810.3 mm and horizontal displacement exceeded 300 mm. Therefore, numerous researches in recent years focused on the particle breakage regulation of soil and the development of the soil constitutive model considering the influence of particle crushing.

13.2 REVIEW

The soil particle crushing regulation was investigated using the single-grain crushing tests,[8,9] point loading tests,[10,11] one-dimensional compression tests,[12,13] ring shear tests,[14,15] triaxial tests,[2,4,16–19] impact tests,[20,21] cyclic tests,[22,23] discrete-element numerical tests,[7,9] and so on. The single-grain crushing tests and point loading tests investigate the single particle crushing stress, breakage modes, and failure processes. The one-dimensional compression test is easy to implement and the soil particle crushing is obvious. It is often used to investigate the crushing law of the soil particle aggregates under long-term load and complex environmental load. The ring shear tests can provide large shear strain, which is mainly used to analyze particle breakage of fine-grained soil in a critical state. The triaxial test can not only simulate complex stress path and natural environment but also apply to soil

DOI: 10.1201/9781003362081-17

with different particle sizes, which is used to analyze the particle crushing under complex stress, wetting deformation, and creep deformation. The impact test is similar to the one-dimensional compression test, except that the impact load is borne by the specimen under lateral constraints. This test focuses on the soil particle breakage distribution in different areas of the specimen. The cyclic test investigates the soil particle breakage under seismic loading such as earthquake, waves, and vehicles. The discrete-element numerical test can not only simulate the crushing process of the soil particles in the specimen but also record the stress and displacement of soil particles in detail. These soil particle crushing tests show that the soil particle tensile strength could be simulated using the Weibull statistics. It decreases with the increase in particle size, additional acidity, and soaking time of the acidic erosion. The soil particle breakage is both influenced by the stress and strain. Under a low confining pressure, the particle sliding and roll over are loosely constrained; the contact force among the particles is small, angular crushing and abrasion of particles appeared in sliding, but negligible particle breakage occurs during deformation. In addition, significant dilatancy will occur, and the shear stress is generated by the sliding friction and dilatancy. The energy generated by the loading is transformed into elastic deformation energy of soil particles, the friction energy consumption among particles and the work done by the dilatancy against volumetric stress. Under a high confining pressure, the particle sliding and roll over are more constrained, the contact force significantly increases, and significant particle breakage occurs, which increased with increasing soil deformation. The shear stress is supported by the sliding friction, particle interlock, and dilatancy. Moreover, the shear stress generated by the particle interlock decreases with an increase in particle breakage, and the crushed small particles fill the pores among particles, which increases the sliding friction and reduces the dilatancy. The plastic work generated by loading is consumed by the friction energy, particle breakage energy and work done by the dilatancy. Therefore, the particle breakage increases with stress and strain, which reduces the particle size and improves particle crushing strength. Once the particle crushing strengths are large enough to endure their contact forces, the particle breakage does not develop, and the soil structure becomes stable and reaches critical state, in which the plastic work is totally consumed by friction among particles. Therefore, the complex constitutive relationship of soil is essentially the external expression of different energy conversion forms of soil structure in the process of the deformation under loading. Particle crushing not only consumes the energy itself but also impacts friction energy consumption and work done by dilatancy, which is a key factor to accurately simulate the constitutive characters of soil.

According to the disturbed state concept (DSC),[1] in the process of soil deformation under loading, the soil microstructure is continuously transformed from RI state into the FA state, which could be simulated by the disturbance function. Particle crushing directly modifies the soil microstructure, which not only influences the disturbance function but also affects the FA state of soil.

In the initial research of particle crushing, the effect of particle breakage on soil shear strength has attracted researcher's attention, and the empirical relationship between particle breakage and soil shear strength is constructed, as follows[5]:

$$\frac{\sigma_1}{\sigma_3} = \frac{1 + B_g}{A_d B_g + B_d} \tag{13.1}$$

where σ_1 is the major principal stress, σ_3 is the minor principal stress, and B_g is the particle breakage factor; Figure 13.1 shows the definition of B_g. A_d and B_d are fitting parameters, which could be determined using the values of (σ_1/σ_3) at failure and B_g under various confining pressures. According to DSC, at the RI state, no particle crushing occurs and B_g is zero. Thus, at the RI state, the value of (σ_1/σ_3) at failure could be determined using Eq. (13.1) as[5]:

$$\frac{\sigma_1}{\sigma_3} = \frac{1}{B_d} \tag{13.2}$$

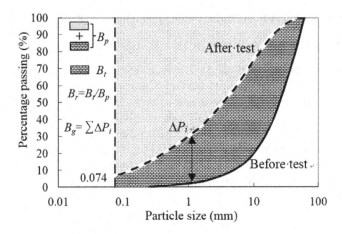

FIGURE 13.1 Definition of particle breakage factor.

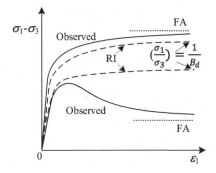

FIGURE 13.2 One relative intact (RI) considers particle breakage.

Figure 13.2 shows the relationship between the RI and B_g. The stress–strain curves at the RI state in Figure 13.2 are assumed to be simulated by the hyperbolic relationship by which the parameters of the hierarchical single surface (HISS) could be determined.

With the development of particle breakage tests, the effect of particle breakage on the critical state line has been found. The critical state line considering particle breakage can be expressed as[18]:

$$e_{cs} = e_{cs0} - \lambda_s \left\{ \frac{3}{3 - M_{cs}} 10^{\left[\frac{1}{\chi_a} \left(\frac{3 B_r (a_u - a_0) + a_0 (3 - a_u)}{B_r (a_u - a_0) + (3 - a_u)} - a_{e0} + \chi_a e_0 \right) \right]} \right\}^{\zeta} \quad (13.3)$$

where e_{cs} is the critical void ratio; e_{cs0}, λ_s, a_{e0}, χ_a, a_{e0}, a_0, a_u, and e_0 are material constants; M_{cs} is the slope of the critical state line in p-q space; and B_r is the relative particle breakage factor. Figure 13.1 shows the definition of B_r. The critical state line simulates the FA state in DSC. Hence, the present DSC constitutive model considering particle crushing focuses on the effect of particle breakage on the RI and FA states. However, particle crushing directly modifies soil microstructure and promotes the transformation from the RI state to the FA state, which is simulated by the disturb function. Therefore, the DSC constitutive model considering particle crushing in the disturb function is introduced next.

13.3 DSC CONSTITUTIVE MODEL CONSIDERING PARTICLE CRUSHING

13.3.1 PARTICLE CRUSHING REGULATION DURING SOIL DEFORMATION

The particle breakage tests show that the particle breakage factor increases with axial strain in the consolidated drain triaxial compression tests. The B_r at the same axial strain increases with confining pressures.[2] No particle breakage occurred for some axial strain values (marked as ε_{1i}) at the beginning of the test. Within this initial axial strain ε_{1i}, all the basalt rockfill particles remained in an elastic-plastic state, and the rockfill deformation consisted of elastic deformation, plastic deformation, sliding, and rolling over.[2] When the confining pressure is not large enough to limit particle roll over, ε_{1i} will cover the whole compression, and no obvious particle breakage occurred in the test. The B_r under different confining pressures could be simulated as:

$$B_r = \frac{B_{r(\text{ult})} \cdot (\varepsilon_1 - \varepsilon_{1i})}{A_B + (\varepsilon_1 - \varepsilon_{1i})} \tag{13.4a}$$

where A_B is the fitting parameter, $B_{r(\text{ult})}$ is the B_r in the critical state, which can be expressed as:

$$A_B = a_0 \ln(\sigma_3 / P_a) + b_0 \tag{13.4b}$$

$$B_{r(\text{ult})} = \frac{a_2 \cdot \sigma_3}{R_B + \sigma_3^{a_2}} \tag{13.4c}$$

where a_0, b_0, a_2, and R_B are fitting parameters.

13.3.2 THE FA STATE CONSIDERING PARTICLE CRUSHING

Figure 13.3 shows the $\sqrt{J_{2D}} / J_1$ (J_{2D} and J_1 are defined by Eq. 13.1) values of the basalt rockfill under different confining pressures. Figure 13.3 shows that the $\sqrt{J_{2D}} / J_1$ values of different confining

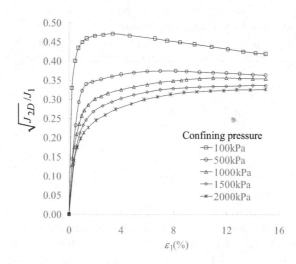

FIGURE 13.3   Strain–stress curves and $\sqrt{J_{2D}} / J_1$ values under different confining pressures.

pressures at failures are obviously different under the influence of particle crushing. Moreover, as discussed in Chapter 1, in the FA state,

$$\sqrt{J_{2D}^c} = \bar{m} J_1^c \tag{13.5}$$

where the superscript "c" denotes the critical state, material constant $\bar{m}$ is the function of M_{cs} and is influenced by the particle breakage, which can be simulated as:

$$\bar{m} = A_m \cdot B_{r(\text{ult})} + B_m \tag{13.6}$$

where A_m and B_m are material constants.

13.3.3 Disturb Function Considering Particle Crushing

The RI state is simulated using the HISS model introduced in Section 2.4, which is expressed as

$$F = \bar{J}_{2D} - \left(-\alpha \bar{J}_1^n + \gamma \bar{J}_1^2\right)\left(1 - \beta S_r\right)^{-0.5} = 0 \tag{13.7a}$$

$$n = A_n \cdot \ln(\sigma_3 / P_a) + B_n \tag{13.7b}$$

where A_n and B_n are fitting parameters. The HISS model parameters can be determined according to the initial test data as shown in Figure 13.4. The hardening function α is given by:

$$\alpha = \frac{a_1}{\xi^m} \tag{13.8a}$$

$$a_1 = \exp\left[a' \cdot \ln\left(\frac{\sigma_3 + R}{p_a}\right) + b'\right] \tag{13.8b}$$

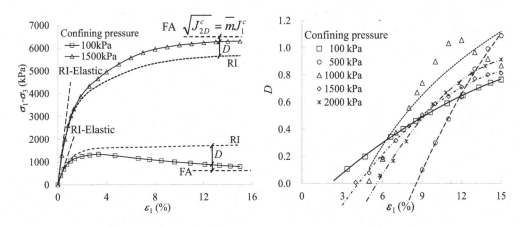

FIGURE 13.4 Disturbances calculated from test data (test data from Jia et al.[2]).

where a', b', and η_1 are the hardening parameters, and ξ is the trajectory of plastic strains $\xi = \int \left(d\varepsilon_1^p \cdot d\varepsilon_1^p + d\varepsilon_2^p \cdot d\varepsilon_2^p + d\varepsilon_3^p \cdot d\varepsilon_3^p \right)^{1/2}$. The elastic modulus E and Poisson's ratio μ are determined according to the unloading stress–strain curves of the basalt. Figure 13.4 shows the E under 100 and 1500 kPa confining pressures, which can be expressed as:

$$E = K \cdot p_a \cdot \left(\frac{\sigma_3}{p_a} \right)^{n_E} \tag{13.9}$$

where K and n_E are material constants.

According to the stress–strain curves of RI and FA, the disturbance, D, can be calculated using Eq. (13.10).

$$D_\sigma = \frac{\sigma^{RI} - \sigma^a}{\sigma^{RI} - \sigma^{FA}} \quad \left(\text{stress-strain behavior} \right) \tag{13.10}$$

Figure 13.4 shows the D under different confining pressures. Figures 13.3 and 13.4 show that the D values under 100 kPa confining pressure do not contain the effect of particle crushing, which can be simulated using typical function as:

$$D = D_u \left(1 - e^{-A\varepsilon_1^B} \right) \tag{13.11}$$

where A and B are material constants.

Figure 13.4 and Eq. (13.11) show that under the 100 kPa confining pressure, the particle sliding and roll over are not constrained by the small confining pressure. The significant dilatancy not only improves the shear stress but also decreases the compactness and sliding friction stress. Once the shear stress increased by dilatancy is less than the reduced value of friction, the soil enters a softening state, in which the structure transforms from the dense RI state to the loose FA state. On the other hand, under 1500 kPa confining pressure, the particle sliding and rollover are constrained. The interlock, sliding friction, and rollover of particles support the shear stress. With the increase of axial strain, the contact stress among particles exceeds particle breakage strength and particle crushing occurs, which releases the interlock, reduces dilatancy, and increases sliding friction. Therefore, particle crushing not only quickly reduces the interlock stress among particles but also slowly increases the friction among particles, which results in the slow transition of soil test stress from the RI state to the FA state in the initial axial strain. So the D values of 500, 1000, 1500, and 2000 kPa are less than that of 100 kPa in the initial axial strain, which could be expressed as:

$$D = D_u \left(1 - e^{-A \cdot \varepsilon_1^B + C \cdot B_r} \right) \tag{13.12}$$

where A, B, and C are fitting parameters.

13.3.4 PREDICATIONS

The incremental stress equations for the DSC are given by:

$$d\sigma^a = (1 - D) d\sigma^i + D d\sigma^c + dD \left(\sigma^c - \sigma^i \right) \tag{13.13}$$

It is assumed that the observed, RI, and FA strain and mean pressure are equal $\varepsilon^a = \varepsilon^i = \varepsilon^c$, $J_1^a = J_1^i = J_1^c$. In the FA state, $d\sigma^c = 0$ and Eq. (13.13) reduces to:

$$d\sigma^a = (1-D)d\sigma^i + dD(\sigma^c - \sigma^i) \tag{13.14}$$

The observed stress increment at step $(n+1)$ is calculated using Eq. (13.14) as:

$$d\sigma^a_{n+1} = (1-D_n)d\sigma^i_{n+1} + dD_{n+1}(\sigma^c_n - \sigma^i_n) \tag{13.15}$$

$$d\sigma^i_{n+1} = C^{ep}_n d\varepsilon^i_{n+1} \tag{13.16a}$$

$$C^{ep}_n = C^e - \frac{C^e \cdot \dfrac{\partial Q}{\partial \sigma^i_n} \cdot \left(\dfrac{\partial F}{\partial \sigma^i_n}\right)^T \cdot C^e}{\left(\dfrac{\partial F}{\partial \sigma^i_n}\right)^T C^e \cdot \dfrac{\partial Q}{\partial \sigma^i_n} - \dfrac{\partial F}{\partial \xi_n}\left[\left(\dfrac{\partial Q}{\partial \sigma^i_n}\right)^T \cdot \left(\dfrac{\partial Q}{\partial \sigma^i_n}\right)\right]^{1/2}} \tag{13.16b}$$

where C^e is the elastic matrix, and Q is the plastic potential function and is expressed as:

$$C^e = \frac{E}{(1+\mu)(1-2\mu)}\begin{bmatrix} 1-\mu & \mu & \mu \\ \mu & 1-\mu & \mu \\ \mu & \mu & 1-\mu \end{bmatrix} \tag{13.16c}$$

$$Q = \bar{J}_{2D} - \left(-\alpha_Q \bar{J}_1^n + \gamma \bar{J}_1^2\right)\left(1 - \beta S_r\right)^{-0.5} = 0 \tag{13.16d}$$

where α_Q is the hardening function and is given by:

$$\alpha_Q = \alpha + \kappa(\alpha_0 - \alpha)(1 - \gamma_v) \tag{13.16e}$$

where α_0 is the value of α at the end of initial (hydrostatic) loading, $\gamma_v = \xi_v / \xi$, $\xi_v = |d\varepsilon^p_v|$, and κ is the nonassociative parameter, which is expressed as:

$$\kappa = A_k(\sigma_3)^{B_k} \tag{13.16f}$$

The observed stress is calculated at a constant value of strain increment $d\varepsilon^a = d\varepsilon^i = d\varepsilon^c = d\varepsilon$. Therefore, dD_{n+1} in Eq. (13.15) can be calculated using Eq. (13.12)

$$dD_{n+1} = D_{n+1} - D_n \tag{13.17}$$

The RI stress and trajectory of plastic strains at $(n+1)$ step are evaluated as:

$$\sigma^i_{n+1} = \sigma^i_n + d\sigma^i_{n+1} \tag{13.18a}$$

$$\xi_{n+1} = \xi_n + d\xi_{n+1} \tag{13.18b}$$

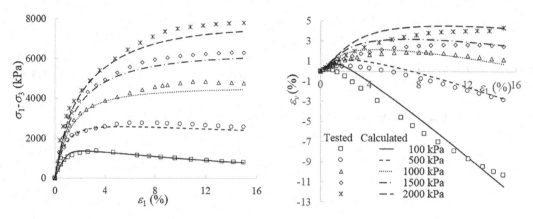

FIGURE 13.5 Predication of stress–strain and volumetric strain curves of basalt rockfill material (test data from Jia et al.[2]).

TABLE 13.1
Rockfill Parameters

	Material Constant
Elastic constant	$K = 1463.3, n_E = 0.365, \mu = 0.165$
Particle crushing	$\varepsilon_{1i} = 0.87, a_0 = 7.6, b_0 = 11.2, R_B = 6418\,\text{kPa}, a_2 = 0.98$
RI state	$A_n = 0.66, B_n = 5.53, \beta = 0.68, \gamma = 0.054, R = 323.8\,\text{kPa},$ $a' = -7.19, b' = -6.98, \eta_1 = 1.21, A_k = 0.195, B_k = -0.515$
FA state	$A_m = -0.229, B_m = 0.382$
Disturbance state	$A = 0.02, B = 1.59, C = 0.066$

σ_{n+1}^i and ξ_{n+1} are evaluated using the drift correction procedure. Figure 13.5 shows the calculated stress–strain and volumetric strain curves. Table 13.1 lists the rockfill parameters.

13.4 USEFULNESS OF THE DSC MODELING PARTICLE CRUSHING

The predication of the stress–strain and volumetric strain curves of the basalt rockfill material show that the DSC model considering particle crushing can simulate the stress soften and dilatancy characters. The influence of particle crushing on the constitutive characters of the soil is mainly reflected by the DSC model in the following aspects:

Firstly, the slope of the critical state line m considering particle breakage shows an obvious non-linear, which impact the disturbance function.

Secondly, the disturbance function considering particle crushing controls the transformation of the soil from the RI state into the FA state, by which the ratio of peak stress and axial strain of the peak stress are simulated. Particle crushing reduces the growth rate of the disturbance function.

Thirdly, the κ in the plastic potential function simulates the dilatancy of the soil and increases with confining pressure, which constrains the particle roll over and increases the particle breakage.

Finally, parameter a_1 in hardening function and parameter n in yield function influence the peak stress and axial strain of the peak stress together, which are influenced by the particle crushing and confining pressure.

13.5 CONCLUSION

Particle crushing directly changes the microstructure of the soil and affects the constitutive characters, which also directly modifies the disturbance function of the DSC model. Under the conventional stress path, the slope of the critical state line decreases with the increase in particle crushing. With the influence of the particle breakage, some typical DSC model parameters, such as a_1, n, and κ, also show obvious confining pressure–related characteristics. This is because under the conventional stress path, particle crushing and dilatancy are significantly influenced by the confining pressure, which covers the actual relationship between DSC model parameters and particle crushing parameters. Therefore, the DSC model considers particle crushing under complex stress path should be developed in the future, by which the influence of particle crushing on the DSC model parameters and disturbance function would be revealed.

REFERENCES

1. C. S. Desai. *Mechanics of Materials and Interfaces: The Disturbed State Concept*. Boca Raton, FL: CRC Press, 2001.
2. Y. Jia, B. Xu, S. Chi, B. Xiang, and Y. Zhou. "Research on the particle breakage of rockfill materials during triaxial tests", *Int. J. Geomech.*, 17, 10, 2017. doi: 1061/(ASCE)GM.1943-5622.0000977.
3. Y. Xiao, H. Liu, C. S. Desai, Y. Sun, and H. Liu. "Effect of intermediate principal-stress ratio on particle breakage of rockfill material", *J. Geotech. Geoenviron. Eng.*, 142, 4, 2016, 06015017.
4. Y. Jia, B. Xu, C. S. Desai, S. Chi, and B. Xiang. "Rockfill particle breakage generated by wetting deformation under the complex stress path", *Int. J. Geomech.*, 20, 10, 2020, 04020166.
5. A. Varadarajan, K. G. Sharma, S. M. Abbas, and A. K. Dhawan. "Constitutive model for rockfill materials and determination of material constants", *Int. J. Geomech.*, 6, 4, 2006, 226–237.
6. Y. Xiao, H. Liu, Y. Chen, and J. Jiang. "Strength and deformation of rockfill material based on large-scale triaxial compression tests. II: Influence of particle breakage", *J. Geotech. Geoenviron. Eng.*, 140, 12, 2014, 04014070.
7. B. Indraratna, N. T. Ngo, C. Rujikiatkamjorn, and J. S. Vinod. "Behavior of fresh and fouled railway ballast subjected to direct shear testing: Discrete element simulation", *Int. J. Geomech.*, 14, 1, 2014, 34–44.
8. Y. Xiao, Z. Sun, C. S. Desai, and M. Meng. "Strength and surviving probability in grain crushing under acidic erosion and compression", *Int. J. Geomech.*, 19, 11, 2019. doi: 10.1061/(ASCE)GM.1943-5622.0001508.
9. J. Wang, S. Chi, X. Shao, and X. Zhou. "Determination of the mechanical parameters of the microstructure of rockfill materials in triaxial compression DEM simulation", *Comput. Geotech.*, 137, 2021, 104265.
10. W. Wang, and M. R. Coop. "An investigation of breakage behaviour of single sand particles using a high-speed microscope camera", *Geotechnique*, 66, 12, 2016, 984–998.
11. W. Wang, and M. R. Coop. "Breakage behaviour of sand particles in point-load compression", *Geotech. Lett.*, 8, 1, 2018, 61–65.
12. Y. Xiao, Z. Yuan, C. S. Desai, M. Zaman, Q. Ma, Q. Chen, and H. Liu. "Effects of load duration and stress level on deformation and particle breakage of carbonate sands", *Int. J. Geomech.*, 20, 7, 2020. doi: 10.1061/(ASCE)GM.1943-5622.0001731.
13. B. Y. Zhang, J. H. Zhang, and G. L. Sun. "Particle breakage of argillaceous siltstone subjected to stresses and weathering", *Eng. Geol.*, 137–138, 2012, 21–28.
14. Z. X. Yang, R. J. Jardine, B. T. Zhu, P. Foray, and C. H. C. Tsuha. "Sand grain crushing and interface shearing during displacement pile installation in sand", *Geotechnique*, 60, 6, 2010, 469–482.
15. M. R. Coop, K. K. Sorensen, T. Bodas Freitas, and G. Georgoutsos. "Particle breakage during shearing of a carbonate sand", *Geotechnique*, 54, 3, 2004, 157–163.
16. Y. Jia, B. Xu, S. Chi, B. Xiang, D. Xiao, and Y. Zhou. "Particle breakage of rockfill material during triaxial tests under complex stress paths", *Int. J. Geomech.*, 19, 12, 2019, 04019124.
17. Y. Xiao, and H. Liu. "Elastoplastic constitutive model for rockfill materials considering particle breakage", *Int. J. Geomech.*, 17, 1, 2017, 1–13.
18. Y. Xiao, H. Liu, X. Ding, Y. Chen, J. Jiang, and W. Zhang. "Influence of particle breakage on critical state line of rockfill material", *Int. J. Geomech.*, 16, 1, 2016, 04015031.

19. A. Varadarajan, K. G. Sharma, K. Venkatachalam, and A. K. Gupta. "Testing and modeling two rockfill materials", *J. Geotech. Geoenviron. Eng.*, 129, 3, 2003, 206–218.
20. Y. Xiao, H. Liu, Q. Chen, L. Long, and J. Xiang. "Evolution of particle breakage and volumetric deformation of binary granular soils under impact load", *Granular Matter*, 19, 4, 2017, 1–10.
21. Y. Xiao, Z. Yuan, Y. Lv, L. Wang, and H. Liu. "Fractal crushing of carbonate and quartz sands along the specimen height under impact loading", *Constr. Build. Mater.*, 182, 2018, 188–199.
22. B. Indraratna, P. K. Thakur, J. S. Vinod, and W. Salim. "Semiempirical cyclic densification model for ballast incorporating particle breakage", *Int. J. Geomech.*, 12, 3, 2012, 260–271.
23. Z. Fu, S. Chen, and C. Peng. "Modeling cyclic behavior of rockfill materials in a framework of generalized plasticity", *Int. J. Geomech.*, 14, 2, 2014, 191–204.

14 Prediction of Strength for the Degradation of Single-Particle Crushing Based on DSC

Chenggui Wang, Zengchun Sun, Hao Cui,
Fang Liang, Huanran Wu, and Yang Xiao
School of Civil Engineering, Chongqing University

14.1 INTRODUCTION

Granular soils are widely used as engineering materials in rockfill dams, railway projects, or foundation engineering [1–4] due to the characteristics of high strength, good permeability, and convenience to obtain. But a key issue is that particle breakage of the granular materials often occurs when the particles endure high stress or large strain, which always threaten the safety, stability, and durability of geotechnical engineering [5–7]. The factors that affected particle crushing mainly include particle size [8,9], particle size distribution [10,11], particle shape [12,13], stress level [14–17], load time [18,19], temperature [20], erosion [21,22], etc. Notably, the effect of particle crushing is a multi-scale and complex problem that the breakage of single particles will lead to a change in the structure of the clusters and then affect the macro response of the soils in engineering.

Hence, the studies on the degradation of single particles due to crushing can give a deep comprehension on the granular materials. Steacy and Sammis [23] found that the crushing of rock particles can be depicted with the fractal model with a fractal dimension of 2.585. McDowell et al. [24] and McDowell [25] stated that the particle breakage is related to the size of the particle and the number of surrounding particles. Xu [26] explained the size effect of single-particle crushing based on the fractal model and linked the particle crushing strength with the single particle size. Nakata et al. [27] conducted 1D compression test and tested the strength of single particles and pointed out that the yield stress of 1D compression test is not only related to particle size but also related to the strength of single particles. Similarly, Yoshimoto et al. [28] established the relationship between the strength of single particle and the shear strength by studying the crushing strength and the shear behaviors of coal ash.

However, the special property of geotechnical engineering materials such as sand or rockfills with nonuniform, defect, and uncertainty, make it difficult and complex to predict the deformation and strength with the elasticity, plasticity, continuum mechanic or fracture mechanics. The disturbed state concept (DSC) proposed by Desai [29] provided a new and unified method [30] to simulate the response of engineering materials for different loading conditions [31–33], the hardening and softening behavior [34], the evaluation of liquefaction [35–37], etc. The main idea of DSC assumes that the microstructure of materials would be disturbed by an external force, which leads to a change from a relative intact (RI) state, with an automatic adjustment, to a fully adjusted (FA) state. The process of self-adjust might lead to a relative movement with micro-crack and damage of particles. Hence, it might be a better method to predict the degradation of the strength of single particles under the influence of particle size, erosion, duration, etc. Hence, in this study, a model based on DSC is established to predict the crushing strength of the limestone particles disturbed with acidic erosion [21].

DOI: 10.1201/9781003362081-18

14.2 MATERIALS AND TEST PROCEDURE

The limestone used in Xiao et al. [21] can be seen as a carbonate mineral, whose main compositions include 97% calcium carbonate, which is prone to be eroded by acid liquid. A series of compression tests were conducted by the Microtest Materials Testing Module on the limestone particles with different sizes (2.5, 5, and 10 mm), different acid environments (pH=2, 4, and 7), and different soaking times (0, 5, 10, and 15 days). Some of the test data of the average crushing stress in different conditions are listed in Table 14.1.

TABLE 14.1

Average Crushing Stress and the Observation Disturbed Values of Limestone with Different Sizes and Acidic Treatments

Immersion State	Soaking Days (t/day)	Particle Size (d/mm)	Average Crushing Stress (σ_{av}/MPa)	Observation Disturbed Value (D)
Original	0	2.5	29.7	0.000
		5	13.39	0.708
		10	7.51	0.963
pH=2 NaCl	5	2.5	27.92	0.077
		5	12.81	0.733
		10	7.24	0.974
	10	2.5	25.69	0.174
		5	11.84	0.775
		10	7.04	0.983
	15	2.5	24.2	0.239
		5	11.43	0.793
		10	6.65	1.000
pH=4 NaCl	5	2.5	28.99	0.031
		5	13.09	0.721
		10	7.26	0.974
	10	2.5	28.03	0.072
		5	12.96	0.726
		10	7.2	0.976
	15	2.5	26.48	0.140
		5	12.34	0.753
		10	7.01	0.984
pH=7 NaCl	5	2.5	29.05	0.028
		5	13.13	0.719
		10	7.38	0.968
	10	2.5	28.66	0.045
		5	13.02	0.724
		10	7.35	0.970
	15	2.5	28.01	0.073
		5	12.9	0.729
		10	7.26	0.974

Data from Xiao et al. [21].

14.3 DISTURBED FUNCTION AND PREDICTIONS

Based on the theory of DSC, it is important to define or determine the RI state and the FA state. As the test stress–strain curves shown in Figures 14.1–14.3, the peak stress σ_f will degrade with the increase of particle size in Figure 14.1, acid concentration in Figure 14.2, and soaking time in Figure 14.3, which means the change in size and acid treatments can be seen as the disturbed factors.

For simplification, the average crushing stress of the particle with smaller size (2.5 mm) and no acid treatments can be seen as the RI state here, and the FA state can be represented by the particle with $d = 10$ mm, pH=2, and soaked for 15 days. Although it seems inaccurate, the main objective

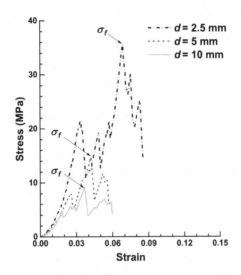

FIGURE 14.1 Stress–strain of limestone particles with no treatment for different sizes. (Data from Xiao et al. [21].)

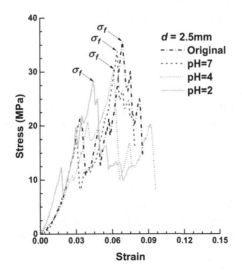

FIGURE 14.2 Stress–strain of limestone particles with acid treatment for 5 days. (Data from Xiao et al. [21].)

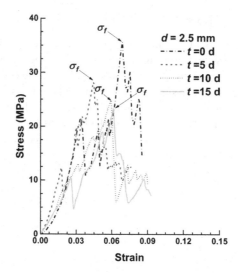

FIGURE 14.3 Stress–strain of limestone particles with pH = 2. (Data from Xiao et al. [21].)

of this study is to make a simple attempt and prediction of the crushing strength of single particles based on DSC. Hence, the crushing strength of single particles can be presented as [38]:

$$\sigma_{cr} = (1 - D)\sigma_{RI} + D\sigma_{FA} \tag{14.1}$$

where σ_{cr} is the crushing stress; σ_{RI} is the crushing stress at the RI state; σ_{FA} is the crushing stress at the FA state; and D is the disturbed function, which can be affected by particle size, pH value, and soaking time, that is, it can be treated as the function of them:

$$D = D(d, pH, t) \tag{14.2}$$

where t represents the soaking time.

To determine the disturbed function and the relationship between particle size, pH, and soaking time, the observation value [38] can be obtained by:

$$D = \frac{\sigma_{av}^{i} - \sigma_{av}^{ob}}{\sigma_{av}^{i} - \sigma_{av}^{c}} \tag{14.3}$$

where σ_{av} is the average crushing stress, and the superscript i, c, and ob represent the initial state without any treatment, critical state with the most serious treatment, and the observation in the current state of the study [21]. And, the calculated observation disturbed values have been listed in Table 14.1. As referred before, the following exponential function is used to fit the disturbed effect of particle size, pH, and soaking time:

$$D = D_u \left[1 - \chi \exp(\xi d) \right] \tag{14.4}$$

where D_u is obtained from the ultimate FA state $D_u = 1$; χ and ξ are material parameters, which can considering the influence of pH and soaking time. It can be rewritten as Eq. (14.5) by fitting the observed disturbed values:

$$D = 1 - (2.99 + 0.058\omega - 0.02t) \exp(-0.493d) \tag{14.5}$$

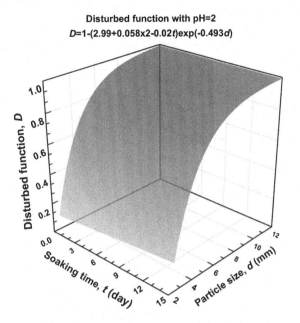

FIGURE 14.4   Evolution of disturbed function with soaking time and particle size at pH=2.

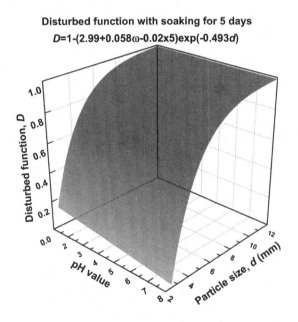

FIGURE 14.5   Evolution of disturbed function with pH and particle size at soaking time of 5 days.

Hence, the evolutions of the disturbed function are presented in Figures 14.4 and 14.5 by controlling the pH=2 in Figure 14.4 and controlling the soaking time for 5 days in Figure 14.5, since it will be more visualized than a 4D plot. It can be found that the disturbed function is mainly affected by the particle size. The influence of soaking time and acid concentration can also be clearly seen in Figure 14.5. And, the disturbed function will approach the ultimate FA state with $D = 1$ when the particle size is large enough, the acid concentration is high enough, and the soaking time is long enough.

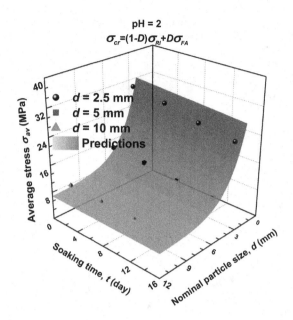

FIGURE 14.6   Test data and predictions of average stress for limestone particles with pH = 2. (Test data from Xiao et al. [21].)

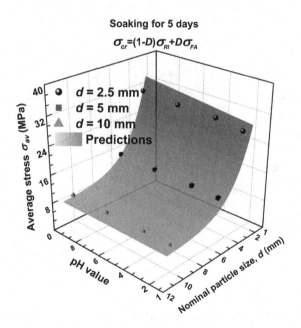

FIGURE 14.7   Test data and predictions of average stress for limestone particles with soaking for 5 days. (Test data from Xiao et al. [21].)

The crushing stress can be predicted with the proposed model based on DSC as Figures 14.6 and 14.7. Similarly, Figure 14.6 gives the prediction of average crushing stress for limestone particles under pH =2 with different soaking times and sizes. And, Figure 14.7 presents the average crushing stress for different particle sizes with soaking for 5 days in different pH liquids. All the test data

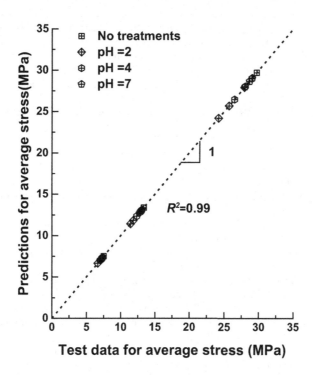

FIGURE 14.8 Comparison between test data and predictions with the disturbed state concept (DSC) for average stress.

and predictions with the proposed DSC model for average crushing stress have been replotted in Figure 14.8 comparing with a line whose slope equals to 1, which shows a good agreement with $R^2 = 0.99$. The comparison verifies that the performance of the proposed crushing model in predicting the strength is good enough and it can simplify the method by considering the size effect in the framework of Weibull statistics [39]. Admittedly, the model can also predict the peak force by combining the size effect proposed in Xiao et al. [21].

14.4 CONCLUSIONS

Single-particle tests for limestone particles have been conducted by Xiao et al. [21] with an analysis on size effect and acid erosion. And, the degradation of the particle strength matches the case of DSC, since the increase of particle size and acid erosion can both be seen as the disturbed factor in the framework of DSC. A disturbed function based on the characteristic of observation is established by considering the impact of particle size, pH, and soaking time. The evolution of disturbed function is analyzed to present the disturbance of particle size and soaking time or pH. The predictions with disturbed function can be consistent with the test data, which means the proposed model based on DSC can consider the size effect and acid erosion.

REFERENCES

1. Indraratna, B., Ionescu, D., and Christie, H. D. "Shear behavior of railway ballast based on large-scale triaxial tests." *J. Geotech. Geoenviron. Eng.*, 124, 5, 1998, 439–449.
2. Indraratna, B., Lackenby, J., and Christie, D. "Effect of confining pressure on the degradation of ballast under cyclic loading." *Geotechnique*, 55, 4, 2005, 325–328.

3. Zhou, W., Li, S., Ma, G., Chang, X., Ma, X., and Zhang, C. "Parameters inversion of high central core rockfill dams based on a novel genetic algorithm." *Sci. China Tech. Sci.*, 59, 5, 2016, 783–794.

4. Xu, B., Zou, D., Kong, X., Zhou, Y., and Liu, X. "Concrete slab dynamic damage analysis of CFRD based on concrete nonuniformity." *Int. J. Geomech.*, 17, 9, 2017, 04017055.

5. Yasufuku, N., and Hyde, A. F. L. "Pile end-bearing capacity in crushable sands." *Geotechnique*, 45, 4, 1995, 663–676.

6. Liu, H., and Zou, D. "Associated generalized plasticity framework for modeling gravelly soils considering particle breakage." *J. Eng. Mech.*, 139, 5, 2013, 606–615.

7. Jia, Y., Xu, B., Chi, S., Xiang, B., and Zhou, Y. "Research on the particle breakage of rockfill materials during triaxial tests." *Int. J. Geomech.*, 17, 10, 2017, 04017085.

8. McDowell, G. R., and Amon, A. "The application of weibull statistics to the fracture of soil particles." *Soils Found.*, 40, 5, 2000, 133–141.

9. Lim, W. L., McDowell, G. R., and Collop, A. C. "The application of Weibull statistics to the strength of railway ballast." *Granul. Matter.*, 6, 4, 2004, 229–237.

10. Huang, J., Xu, S., and Hu, S. "Effects of grain size and gradation on the dynamic responses of quartz sands." *Int. J. Impact. Eng.*, 59, 2013, 1–10.

11. Koohmishi, M., and Palassi, M. "Effect of particle size distribution and subgrade condition on degradation of railway ballast under impact loads." *Granul. Matter*, 19, 3, 2017, 63.

12. Zhang, X., Hu, W., Scaringi, G., Baudet, B. A., and Han, W. "Particle shape factors and fractal dimension after large shear strains in carbonate sand." *Géotech. Lett.*, 8, 1, 2018, 73–79.

13. Yu, J., Shen, C., Liu, S., and Cheng, Y. P. "Exploration of the survival probability and shape evolution of crushable particles during one-dimensional compression using dyed gypsum particles." *J. Geotech. Geoenviron. Eng.*, 146, 11, 2020, 04020121.

14. Miura, N., and Yamamoto, T. "Particle-crushing properties of sands under high stresses." *Technology Reports of the Yamaguchi University*, 1, 4, 1976, 439–447.

15. Mun, W., and McCartney, J. S. "Constitutive model for drained compression of unsaturated clay to high stresses." *J. Geotech. Geoenviron. Eng.*, 143, 6, 2017, 04017014.

16. Wu, Y., Yamamoto, H., Cui, J., and Cheng, H. Y. "Influence of load mode on particle crushing characteristics of silica sand at high stresses." *Int. J. Geomech.*, 20, 3, 2020, 04019194.

17. Xiao, Y., Wang, C. G., Zhang, Z. C., Liu, H. L., and Yin, Z. Y. "Constitutive modeling for two sands under high pressure." *Int. J. Geomech.*, 21, 5, 2021, 04021042.

18. Chen, W.-B., Liu, K., Yin, Z.-Y., and Yin, J.-H. "Crushing and flooding effects on one-dimensional time-dependent behaviors of a granular soil." *Int. J. Geomech.*, 20, 2, 2020, 04019156.

19. Desai, C. S. "Constitutive modeling including creep-and rate-dependent behavior and testing of glacial tills for prediction of motion of glaciers." *Int. J. Geomech.*, 11, 6, 2011, 465–476.

20. He, S.-H., Shan, H.-F., Xia, T.-D., Liu, Z.-J., Ding, Z., and Xia, F. "The effect of temperature on the drained shear behavior of calcareous sand." *Acta Geotechnica*, 16, 2020, 613–633.

21. Xiao, Y., Sun, Z., Desai, C. S., and Meng, M. "Strength and surviving probability in grain crushing under acidic erosion and compression." *Int. J. Geomech.*, 19, 11, 2019, 04019123.

22. Indraratna, B., Nguyen, V. T., and Rujikiatkamjorn, C. "Assessing the potential of internal erosion and suffusion of granular soils." *J. Geotech. Geoenviron. Eng.*, 137, 5, 2011, 550–554.

23. Steacy, S. J., and Sammis, C. G. "An automaton for fractal patterns of fragmentation." *Nature*, 353, 6341, 1991, 250–252.

24. McDowell, G. R., Bolton, M. D., and Robertson, D. "The fractal crushing of granular materials." *J. Mech. Phys. Solids*, 44, 12, 1996, 2079–2101.

25. McDowell, G. R. "On the yielding and plastic compression of sand." *Soils Found.*, 42, 1, 2002, 139–145.

26. Xu, Y. F. "Explanation of scaling phenomenon based on fractal fragmentation." *Mech. Res. Commun.*, 32, 2, 2005, 209–220.

27. Nakata, Y., Kato, Y., Hyodo, M., Hyde, A. F. L., and Murata, H. "One-dimensional compression behaviour of uniformly graded sand related to single particle crushing strength." *Soils Found.*, 41, 2, 2001, 39–51.

28. Yoshimoto, N., Hyodo, M., Nakata, Y., Orense, R. P., Hongo, T., and Ohnaka, A. "Evaluation of shear strength and mechanical properties of granulated coal ash based on single particle strength." *Soils Found.*, 52, 2, 2012, 321–334.

29. Desai, C. "A consistent finite element technique for work-softening behavior." *Proceedings of International Conference on Computational Methods in Nonlinear Mechanics*, University of Texas, Austin.

30. Desai, C. S. "Disturbed state concept as unified constitutive modeling approach." *J. Rock Mech. Geotech. Eng.*, 8, 3, 2016, 277–293.

31. Xu, Y., and Guo, P. "Disturbance evolution behavior of loess soil under triaxial compression." *Adv. Civil Eng.*, 2020, 2020, 1–14.

32. Varadarajan, A., Sharma, K. G., Desai, C. S., and Hashemi, M. "Constitutive modeling of a schistose rock in the Himalaya." *Int. J. Geomech.*, 1, 1, 2001, 83–107.

33. Desai, C. S., and Rigby, D. B. "Cyclic interface and joint shear device including pore pressure effects." *J. Geotech. Geoenviron. Eng.*, 123, 6, 1997, 568–579.

34. Abyaneh, M. J., and Toufigh, V. "Softening behavior and volumetric deformation of rocks." *Int. J. Geomech.*, 18, 8, 2018, 11.

35. Desai, C. S. "Evaluation of liquefaction using disturbed state and energy approaches." *J. Geotech. Geoenviron. Eng.*, 126, 7, 2000, 618–631.

36. Desai, C. S., Park, I., and Shao, C. M. "Fundamental yet simplified model for liquefaction instability." *Int. J. Numer. Anal. Met.*, 22, 9, 1998, 721–748.

37. Park, I. J., and Desai, C. S. "Cyclic behavior and liquefaction of sand using disturbed state concept." *J. Geotech. Geoenviron. Eng.*, 126, 9, 2000, 834–846.

38. Desai, C. S. *Mechanics of Materials and Interfaces the Disturbed State Concept.* CRC Press, Boca Raton, FL, 2001.

39. Weibull, W. "A statistical distribution function of wide applicability." *J. Appl. Mech.*, 18, 3, 1951, 293–297.

Section 5

Geotechnical Structures and Engineering

15 Reinforced Earth Geosynthetic Walls

Chandrakant S. Desai
Dept. of Civil and Architectural Engineering and
Mechanics, University of Arizona

15.1 INTRODUCTION

Reinforced earth material is developed by a combination of frictional earth (soil) and reinforcement [1–10]. The cohesion of reinforced earth occurs due to friction of soil grains against the reinforcing members.

The reinforced earth structures are composed of earth and reinforcing elements composed of strips usually placed in horizontal layers. Reinforcement can be by means of metal strips and rods; geotextile strips, sheets and grids; or wire grids [8,9]. These materials are fastened to the facing unit and extended into the backfill at some distance, Figure 15.1.

Such reinforced soil retaining structures have been widely used because they offer economic benefits compared to conventional retaining systems [1–10]. There are two main approaches for designing reinforced soil structures: limit equilibrium and numerical (finite) element methods. The approaches based on limit equilibrium include two groups: force equilibrium analysis and strain compatibility analysis. Most of the simple design methods based on the use of limit equilibrium do not provide information concerning deformations or stress distributions in either the soil or the reinforcement [2]. Also, boundary conditions as well as stress equilibrium at each point within the reinforced mass and along the slippage surface are not accounted for in the formulation of the conventional limiting equilibrium methods [11]. Furthermore, as reported by Wu [10], limit equilibrium methods underestimated the factors of safety of two large-scale geosynthetic-reinforced soil walls involving granular and cohesive backfill.

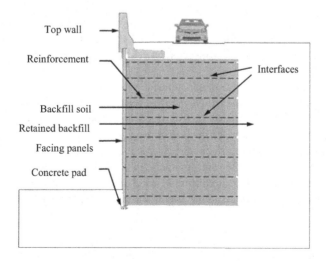

FIGURE 15.1 Schematic of a reinforced earth wall and interfaces.

DOI: 10.1201/9781003362081-20

Reinforced soil walls represent a composite system including the facing, backfill material, reinforcement, and foundation. The performance of the reinforced soil wall relies on the interaction between its components and to a large extent this interaction arises from the relative motions at the interface between the soil and reinforcement. Also, relative strains and deformations occur under working conditions. These requirements can be established by the construction and monitoring of many full-scale test walls. Unfortunately, the cost of performing and suitable monitoring of a sufficiently large number of full-scale walls can be so high that it is often not practical. To overcome this constraint, finite element analysis has been commonly employed for improved analysis and parametric studies. For performing a realistic finite element analysis of a reinforced soil structure, the computer procedure should have the capability of modeling the construction sequences, structural elements (i.e., bars and beams), soil elements, and interface elements that allow nonlinear behavior and relative motions between the soil and reinforcement [12].

Discrete and composite approaches are two different finite element techniques used in analytical modeling of reinforced soil structures. In the discrete approach, the reinforced system is considered as a heterogeneous body in which the soil and reinforcing elements are separately represented using different material properties. The important advantage of this type of model is that detailed information is obtained directly about the behavior of the interface between the soil and the reinforcement, and stress concentration in the soil due to reinforcing members. The reinforced soil is characterized as a "homogeneous" composite structure with the properties of the composite material in the composite approach in which the stiffness of the soil and the reinforcement form the composite element stiffness [13–15].

There are several shortcomings associated with the composite approach, e.g., information about the interaction between the soil and reinforcement such as bond stresses and stress concentration in the soil due to geometric discontinuities because of the presence of reinforcement and edge effects due to the local transfer of stress between the soil and the reinforcement at the boundaries are usually not available. Therefore, in the composite finite element analysis, the strains and forces in the reinforcements at the edge are predicted to be greater than the actual measured values of the structure, which overestimates the effectiveness of the reinforcement [13].

The Goodman-type zero-thickness interface element [16,17] has been often used to model the interface behavior, but it may not provide an accurate mechanism of deformation [18,19]. Different researchers have reported numerical problems when using the zero-thickness interface element. Desai et al. [18] found that the zero-thickness element approach may not provide realistic modeling particularly of the normal stress in soil structure interaction because it is difficult to obtain the appropriate and high value for the normal stiffness. Furthermore, Day and Potts [20] reported that both ill conditioning of the stiffness matrix and high stress gradients cause numerical instability. In this chapter, the thin-layer finite element model with the disturbed state concept (DSC/HISS) [18,21–23] has been used for modeling soil-geosynthetic behavior for the mechanically stabilized earth (MSE) retaining structures.

Realistic constitutive models for soils and interfaces are important and necessary aspects for finite element analyses of reinforced earth structures. Such models should account for factors like elastic, plastic and creep strains; microcracking leading to degradation and softening; relative motions at interface; and environmental effects. The main objective of this chapter is to use unified DSC/HISS modeling for soils and interfaces that yields improved predictions of the field behavior.

The interface behavior between the reinforcement and soil (backfill) is a vital component in the behavior of a reinforced wall. Hence, attention in this chapter is first devoted to interface behavior including thermal and chemical effects. Then, details of an example for field behavior and validations are provided.

15.2 INTERFACE BEHAVIOR

15.2.1 EXAMPLE 1: INTERFACE BEHAVIOR BETWEEN BACKFILL AND TENSAR REINFORCEMENT

Direct shear tests, which can simulate the interface behavior approximately, were performed on interfaces between the Tensar reinforcement and backfill related to the instrumented retaining wall in Tucson, described subsequently. The cyclic multi-degree-of-freedom (CYMDOF) device was used to test interfaces; details of the CYMDOF device have been presented in Desai and Rigby [22]. A view of the device and cross section of the specimen are shown in Figure 15.2.

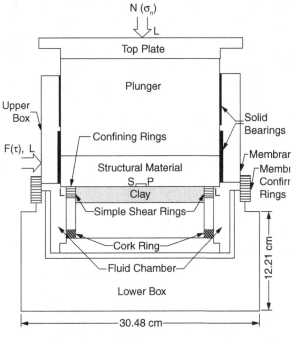

P = To Pore Pressure Transducer
S = Porous Stone
L = LVDT

FIGURE 15.2 Details of the cyclic multi-degree-of-freedom (CYMDOF) test device: (a) general view and (b) a section of the specimen.

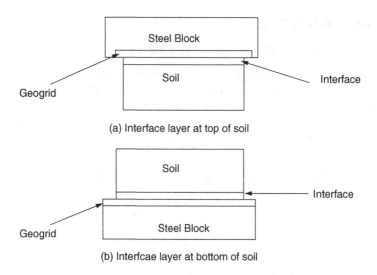

(a) Interface layer at top of soil

(b) Interfcae layer at bottom of soil

FIGURE 15.3 Configuration of the interface layer.

15.2.1.1 Interface Configuration

The interface behavior is affected by complex interaction and interlocking of grains around the grid openings [24]. To identify the effect of openings (apertures) in the reinforcement, two types of tests were performed, using flat geogrid and geogrid with openings.

The diameter of the specimen in the CYMDOF device was 6.5 in (165 mm). The desired initial density of about 18 KN/m³ of the backfill soil was achieved by compacting the specimen in five layers using about 25 blows. The geogrid sample was prepared by cutting the specimen (165 mm) from a large sheet and attaching it to the steel block. It was placed on the soil (dry or bulk with 8% moisture content that was prevalent on the site) in two configurations, Figure 15.3, which were used to study the effect of the location of the interface in the test device.

A normal stress was first applied to the test specimen. Then, the horizontal shear displacement or stress was applied with several loading, unloading, and reloading cycles. Table 15.1 shows the interface friction angles and adhesion for different configurations, and Figure 15.3 shows geogrid with openings and flat geogrid. Because of the interlocking effect, the values for geogrid with openings are higher than those for the flat geogrids. The results for the bulk soil (with moisture content of about 8.0%) with the geogrid at the top of the soil were realistic. They were used for modeling and computer analysis in the subsequent example of a reinforced retaining wall.

TABLE 15.1

Friction and Adhesion for Interface Tests and Configurations

Interface Tested	Interface Friction Angle δ (degree)	Adhesion c_a (kPa)
Dry Soil/Geogrid* at Top, Figure 15.3a	38.5	41
Dry Soil/Geogrid at Bottom, Figure 15.3b	38.2	40
Bulk Soil/Geogrid at Bottom, Figure 15.3b	32.2	58
Bulk Soil/Geogrid at Top, Figure 15.3a	33.8	66
Dry Soil/Flat Geogrid at Top, Figure 15.3a	26.5	29.5
Bulk Soil/Flat Geogrid at Top, Figure 15.3a	27.0	37.0
Bulk Soil/Flat Geogrid at Bottom, Figure 15.3b	25.2	23.0

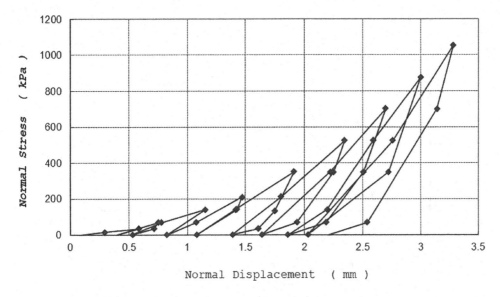

FIGURE 15.4 Normal behavior of bulk soil geogrid with opening at top.

15.2.1.2 Interface Testing

The shear tests were performed under loading, unloading, and reloading at various normal stresses, σ_n =35, 70, 140, 350, and 700 kPa. Typical results for normal stresses, and shear behavior for σ_n =70 and 700 kPa for geogrid at the top of the soil are shown in Figures 15.4 and 15.6. Such test data were used to determine DSC/HISS parameters for interfaces and soil for the subsequent application of Tensar-reinforced wall.

15.2.2 EXAMPLE 2: INTERFACE BEHAVIOR UNDER CHEMICAL AND THERMAL EFFCETS IN LANDFILLS

Behavior of the interfaces between geosynthetics and soil influences the response of geosynthetic liner systems used in waste landfill sites. A schematic of such a site is shown in Figure 15.7. The interface behavior can be affected by external forces, friction, tempertaure, and chemicals. A number of studies have been reported on this subject [25–28]. The study presented by Kwak et al. [25] is included herein.

The main attention in the study by Kwak et al. [25] was devoted to the cyclic shear behavior of geosynthetic–soil interfaces influenced by heat and chemicals at waste landfill sites, depicted in Figure 15.7. Chemical conditins of leachates were introduced with three different pH values: Basic =10, Neutral=7.00, and Acid=4.00. Biodegradation or hydration, and decomposition of organic materials are among the main factors that contribute to the heat production in the waste landfills. Temperarures between 30°C and 85°C have been reported; polymers in geosynthetics are sensitive to such temperatures. The multipurpose interface apparatus (M-PIA), which allows application of temperatures from 20°C to 80°C, was used [25]. Figure 15.8 shows a cross section of the test specimen used in the M-PIA device.

The geosynthetics were geocomposite (geonet plus nonwoven fabric) manufactured by Goldenpow, Korea. The grain-size distribution of the soil used, called Jumunjin sand, from Korea is shown in Figure 15.9. Both the geosynthetics and soil were submerged in basic, neutral, and acidic solutions for 180 days, which represents the short-term behavior of the interface.

The DSC and the associated disturbance function were used to compute the degree of shear stress degradation based on the expertimental results.

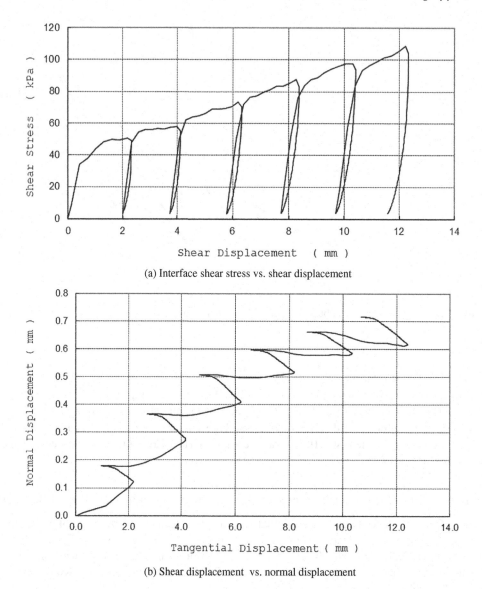

(a) Interface shear stress vs. shear displacement

(b) Shear displacement vs. normal displacement

FIGURE 15.5   Shear test results for dry soil geogrid with opening at top, $\sigma_n = 70$ kP.

15.2.2.1   Testing and Results

A series of cyclic shear tests were peformed under simulation of thermal and chemical conditions. Tests were conducted at room temperature of 20°C with 0.30 MPa normal stress, which corresponds to landfill height of 30m. The unit weight of the soil was 10 kN/m³, which compares to that of typical munucipal waste.

Figures 15.10 and 15.11 show shear stress–shear strain curves under cyclic loading for temperatures of 20°C and 60°C, respectively. Here, the peak shear stress and rate of shear degradation decrease with the number of cycles.

The disturbance function, D, curves were determined based on Figures 15.10 and 15.11, and plotted against the trajectory of deviatoric plastic shear strain, ξ_D, for various chemical conditions, Figure 15.12. The effects of two temperatures for different chemical conditions on the disturbance are shown in Figure 15.13. In these curves, the increase in D represents an increase in the

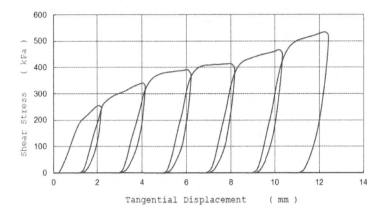

(a) Interface shear stress vs. shear displacement

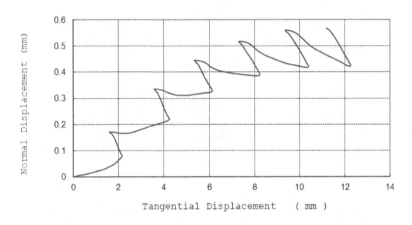

(b) Shear displacement vs. normal displacement

FIGURE 15.6 Normal and shear test results for dry soil geogrid with opening at top, $\sigma_n = 700\,\text{kPa}$.

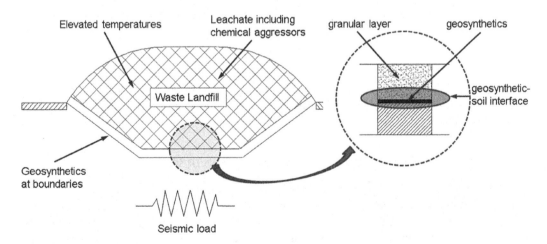

FIGURE 15.7 Schematic of typical landfill liner system. (Reprinted with permission by ASTM from Ref. [25].)

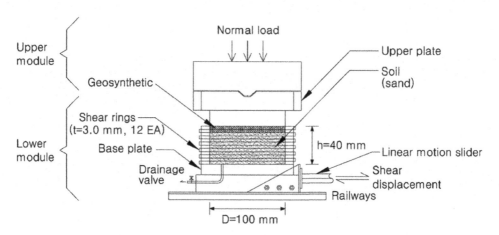

FIGURE 15.8 Details of test box in the multipurpose interface apparatus (M-PIA) simple shear apparatus. (Reprinted with permission by ASTM from Ref. [25].)

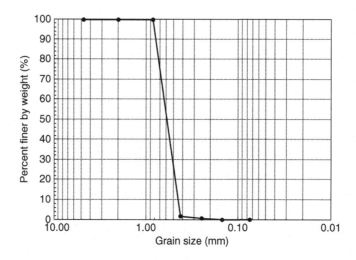

FIGURE 15.9 Grain-size distribution of Jumunjin sand. (Reprinted with permission by ASTM from Ref. [25].)

degradation or damage at the geosyntheic–soil interfaces. The shape of the D function is dependent on the thermal and chemical conditions. In all cases, the greatest disturbance in interfaces occured in the acid condtion.

15.2.3   Example 3: Field Validations for Tensar-Reinforced Wall

15.2.3.1   Description of Wall

Forty-three geogrid-reinforced walls were constructed at grade-separated interchanges at the Tanque Verde-Wrightstown-Pantano site, Tucson, Arizona, USA. This project represents the first use of geogrid reinforcement in MSE retaining walls in a major transportation-related application in North America [29–31]. Here, the behavior of the wall panel No. 26–32 is simulated using the FEM, Figure 15.14.

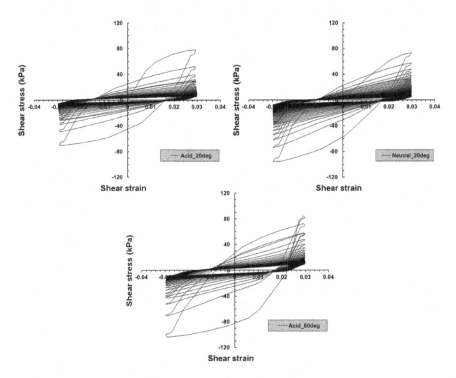

FIGURE 15.10 Shear stress–shear strain responses at 20°C. (Reprinted with permission by ASTM from Ref. [25].)

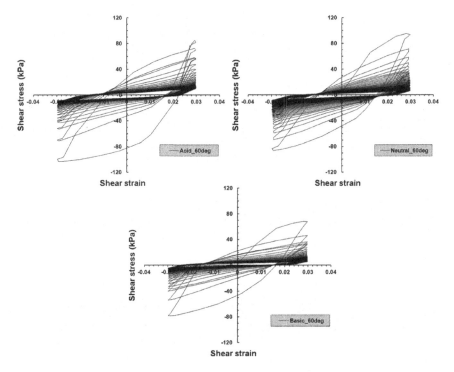

FIGURE 15.11 Shear stress–shear strain responses at 60°. (Reprinted with permission by ASTM from Ref. [25].)

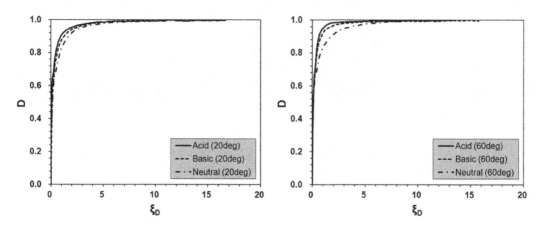

FIGURE 15.12 Disturbance function curves for chemical conditions. (Reprinted with permission by ASTM from Ref. [25].)

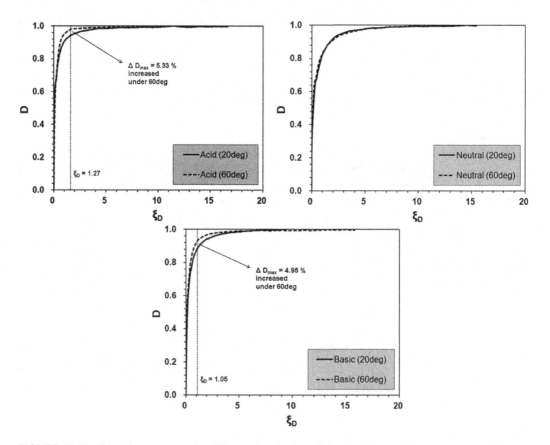

FIGURE 15.13 Disturbance curves for different chemical conditions and temperature. (Reprinted with permission by ASTM from Ref. [25].)

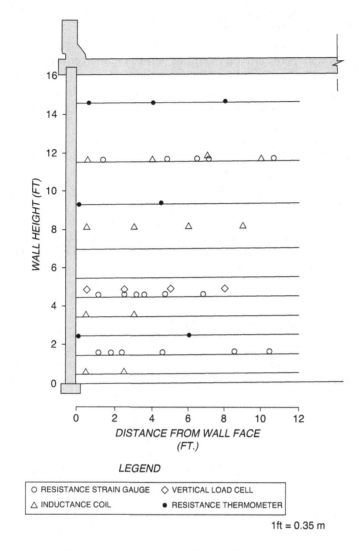

FIGURE 15.14   Locations of instruments for Tensar wall panel No. 26–32 [29,30].

The wall height was 4.88 m (16.0 ft). The reinforced soil mass was faced with 15.24 cm (6.0 in) thick and 3.05 m (10.0 ft) wide precast reinforced concrete panels. Soil-reinforced geogrid was mechanically connected to the concrete facing panels at the elevation shown in Figure 15.14 and extended to a length of 3.66 m (12.0 ft). A pavement structure was constructed on the top of the wall fill; it consisted of 10.16 cm (4.0 in) base course covered by 24.13 cm (9.5 in) of Portland cement concrete. Details of the wall including various geometries are reported in Ref. [32–36]; the Ref. 36 presents a linear finite element analysis for the wall.

The soil reinforcement used was TENSAR SR2 structural geogrid. It is a uniaxial product that is manufactured from high-density polyethylene (HDPE) stabilized with about 2.5% carbon black to provide resistance to attack by ultraviolet (UV) light [29,30]. It is reported to be resistant to chemical substances that normally exist in soils [34]. The geogrids have a maximum tensile strength of 79 kN/m (5,400 lb/ft) and a secant modulus in tension at 2% elongation of 1094 kN/m (75,000 lb/ft). The allowable long-term tensile strength based on creep conditions is reported to be 29 kN/m (1,986 lb/ft) at 10% strain after 120 years. This value was reduced by an overall factor of safety equal to 1.5 to compute a long-term tensile strength equal to 19 kN/m (1,324 lb/ft).

15.2.3.2 Material Parameters and Validations

A comprehensive series of triaxial hydrostatic and shear tests were performed on soils, and CYMDOF shear tests were performed on interfaces between reinforcement and soil; the latter has been presented before under Section 2.1.

Soils: The soil (back fill) samples were collected from the site. The soil was classified as SP poorly graded or gravelly sand. The following are the index properties: specific gravity$=2.64$; D_{10}, D_{30}, $D_{60}=0.48$, 1.00, 1.75 mm; $e_{max}=0.71$, $e_{min}=0.37$; $\gamma_{dmax}=18.84$ kN/m^3; $\gamma_{dmin}=15.35$ kN/mm^{3}02γ_d (field) ≈ 18.0 kN/m^3; optimum moisture content$=8.0\%$.

The soil specimen (71 mm diameter and 142 mm height) was prepared by compacting it in a split model into six layers until the desired initial density (≈ 18 KN/m^3) was obtained. The sample was installed in the triaxial device, and the initial confining stress was applied. Then, the deviatoric stress was applied with the strain rate of about 0.03% per millimeter.

The triaxial shear tests including loading, unloading, and reloading were performed on the samples under different initial normal stresses, $\sigma_3 = 17.5$, 35.0, 52.0, 70.0, 140.0, 210.0, 345.0, and 420.0 kPa [30]. The maximum confining pressure relates to the approximate field pressure of about 480 kPa. Typical test results, hydrostatic and triaxial shear ($\sigma_3 = 35$ and 420 kPa), are shown in Figure 15.15.

15.2.4 Brief Descriptions of DSC/HISS Model

In the DSC/HISS model, it is assumed that a material (soil or interface) element at any stage of loading is composed of the relative intact (RI) and fully adjusted (FA) parts. The basic incremental equation is given in Reference [23]):

$$d\sigma^a = (1-D)C^i \, d\varepsilon^i + DC^c \, d\varepsilon^c + dD\left(\sigma^c - \sigma^i\right) \tag{15.1}$$

where a, i, and c denote observed, RI, and FA responses, respectively, C^i and C^c denote constitutive matrices for RI and FA parts, respectively, σ =stress vector, ε =strain vector, D denotes disturbance, and d denotes increment or rate.

The RI (C^i) behavior can be modeled by using elastic (linear or nonlinear) and plasticity models. Here, it is simulated by using the hierarchical single-surface (HISS) plasticity model in which the yield function for soils is given by Reference [23]:

$$F = \bar{J}_{2D} - \left(-\alpha \, \bar{J}_1^n + \gamma \, \bar{J}_1^2\right)\left(1 - \beta \, S_r\right)^{-0.5} = 0 \tag{15.2}$$

where J_{2D}=second invariant of the deviatoric stress tensor; overbar denotes nondimensional with respect to p_a; the atmospheric pressure, J_1=first invariant of the stress tensor; $S_r = \left(\sqrt{27}/2\right) J_{3D} \cdot J_{2D}^{-1.5}$, J_{3D}=third invariant of deviatoric stress tensor; γ and β are ultimate stress parameters; n is related to the state from compressive to dilative volume change; and α is the hardening or growth function given by:

$$\alpha = \frac{a_1}{\xi^m} \tag{15.3}$$

where a_1 and η_1 are the growth parameters and ξ is the plastic strain trajectory or total plastic strains.

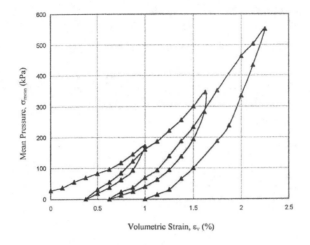

(a) Hydrostatic compression test

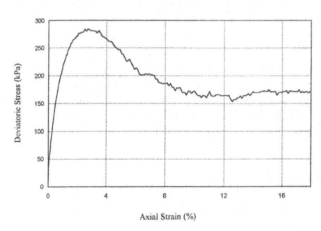

(b) $\sigma_3 = 35$ kPa

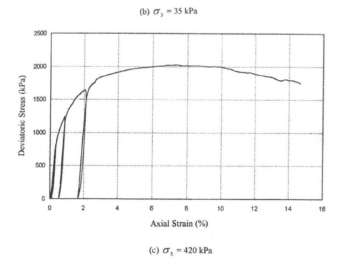

(c) $\sigma_3 = 420$ kPa

FIGURE 15.15   Triaxial test results for backfill soil.

For interfaces, the yield function is given by [23]:

$$F_i = \tau^2 + \alpha_i \sigma_n^n - \gamma \sigma_n^2 = 0 \tag{15.4}$$

where τ and σ_n are the shear and normal stresses, respectively, γ is the ultimate parameter, and ξ is the trajectory of plastic tangential and normal relative displacements.

The disturbance function is given by:

$$D = D_u \left(1 - e^{-A \xi_D^Z} \right) \tag{15.5}$$

where D_u, A, and Z are the disturbance (softening parameters) and ξ_D is the trajectory of deviatoric plastic strain.

Nonassociative Behavior:

The behavior of soils and interfaces were found to be frictional and nonassociative. Hence, modification in the hardening function, α, to account for the nonassociative response introduced an additional parameter, κ [23].

The DSC/HISS model parameters were determined following the procedures published in various publications, e.g., Desai [23]. The soil exhibited softening behavior, Figure 15.15; hence, the disturbance (softening) model (DSC/HISS) was used. The interface did not exhibit softening behavior, Figures 15.5a, and 15.6a; hence, the HISS plasticity model was used. Table 15.2 shows the parameters for the soil and interfaces. The elastic modulus, E, for soils, and shear stiffness, k_s, and normal stiffness, k_n, were found to be functions of confining and normal stress, respectively; relevant equations are shown in Table 15.2 [30].

TABLE 15.2

Material Parameters for Backfill Soil and Interface for Tensar Wall

Material Constant	Symbol	Soil	Interface
Elastic	E or K_n	$f_1 (\sigma_3)^a$	$f_2 (\sigma_n)^b$
	v or K_s	0.3	$f_3 (\sigma_n)$
Plasticity – Ultimate	γ	0.12	2.3
	β	0.45	0.0
Phase Change Parameter	n	2.56	2.8
Growth Parameters	a_1	3.0E-05	0.03
	η_1	0.98	1.0
Nonassociative Constant	κ	0.2	0.4
Disturbance Parameters	D_u	0.93	
	A	0.37	
	Z	1.60	
Angle of Friction and Adhesion	$\phi/\delta/c_a$	$\phi = 40°$	$\delta = 34°$
			$c_a = 66\,\text{kPa}$
Unit Weight (Field)	γ	18.00 kN/m³	
Coefficient of Earth Pressure at Rest	K_o	0.4	

a $E = 62 \times 10^3 \sigma_3^{0.28}$.

b $k_s \left(\text{shear stiffness}\right) = 30 \times 10^3 \sigma_n^{0.28}$.

$k_n \left(\text{normal stiffness}\right) = 18 \times 10^3 \, \sigma_n^{0.29}$.

15.2.4.1 Validations

The parameters (Table 15.2) were used to predict the stress–strain response for typical tests by using the DSC/HISS model (Eqs. 15.1 and 15.4). Satisfactory correlations between the predictions by using the DSC/HISS model were obtained for several tests for the soil and the interface. Figure 15.16a shows typical comparisons for the soil with $\sigma_3 = 52.0\,\text{kPa}$. Similar results for the interface at $\sigma_n = 210\,\text{kPa}$ are shown in Figure 15.16b.

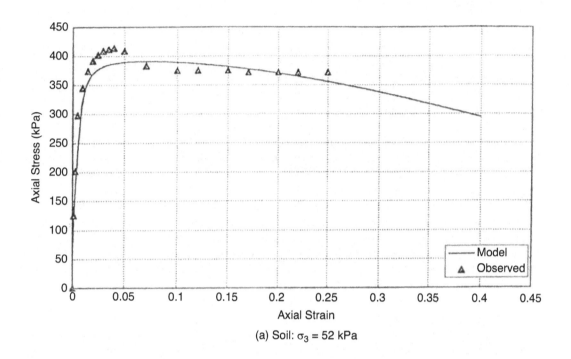

(a) Soil: $\sigma_3 = 52$ kPa

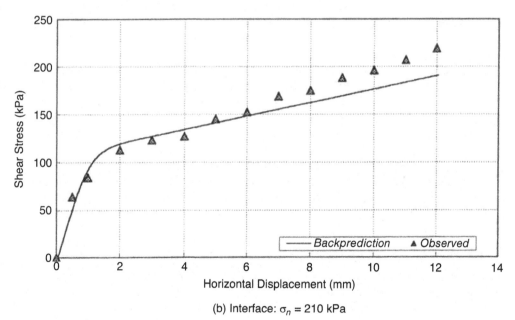

(b) Interface: $\sigma_n = 210$ kPa

FIGURE 15.16 Typical validations for stress–strain and interface behavior.

15.2.4.2 Parameters for Other Materials

The concrete facing with thickness = 152.5 mm was assumed to be linear elastic with the Young's modulus $E = 2.1 \times 10^7$ kPa and Poisson's ratio, $\nu = 0.15$. Geogrid reinforcement with thickness = 1.5 mm was also assumed to be linear elastic with $E = 1.5 \times 10^6$ kPa. The creep in the geogrid was not included. However, as stated before, the tensile strength was reduced to 19 kN/m to account for creep.

15.3 NUMERICAL MODELING

The numerical analysis of the reinforced soil wall was performed using a finite element code called DSC-SST-2D developed by Desai [37]. The code allows for plain strain, plain stress, and axisymmetric idealizations including simulation of construction sequences. Various constitutive models, elastic, elasto-plastic (Von Mises, Drucker–Prager, Mohr–Coulomb, Hoek–Brown, Critical State and Cap, HISS), viscoelastic, plastic and softening (DSC) can be chosen for the analysis. The wall was modeled as a plane strain with two-dimensional idealization. Since the Tensar reinforcement is continuous and normal to the cross section, Figure 15.14, the plane strain idealization is considered appropriate. The code requires material properties to explicitly model the soil, facing panels, reinforcement layers, and the interfaces. It allows incremental fill placement to be simulated (i.e., rows of elements added sequentially as fill placement).

Two finite element meshes, coarse and fine, were used. Figure 15.17a shows the coarse mesh with 184 nodes and 167 elements including 10 wall facing, 18 interfaces between soil and reinforcement, and 9 bar (for reinforcement) elements. In the coarse mesh, only three layers of reinforcement were considered [30]. The fine mesh, Figure 15.17b, contained 1188 nodes and 1370 elements including 480 interfaces, 35 wall facing, and 250 bar elements; it contained 11 layers as in the field. The properties of reinforcement in the coarse and fine mesh were assumed to be the same. The fine mesh was considered to contain a great number of nodes and elements; hence, intermediate and finer meshes were not analyzed. The dimensions for the fine mesh were the same as the coarse mesh; part of the fine mesh near the reinforcement is shown in Figure 15.17b.

As the relative motions between the backfill and reinforcement have a significant effect on the behavior, interface elements were provided between backfill and reinforcement. It was also assumed that the relative motions between wall facing and backfill soil in this problem may not have a significant influence; hence, interface elements were not provided. This is discussed later under Displacements. However, such relative motions can have an influence, and in general, interface elements need to be provided.

The meshes involved four-node quadrilateral elements for soil, wall and interfaces, and one-dimensional elements for reinforcement. As shown in Figure 15.17a, the nodal points at the bottom boundary were fixed, and those on the side boundaries were fixed only in the horizontal direction. The side boundaries were placed at 2.5 times the length of the reinforcement, and the bottom boundary was placed at 3.125 times the height of the wall. Such distances and the assumed boundary conditions are considered to approximately simulate the semi-infinite extent of the problem. These boundary conditions apply to both coarse and fine meshes.

It was found that the fine mesh provided satisfactory and improved predictions compared to those from the coarse mesh. Hence, most of the results presented are for the fine mesh; however, typical comparisons from the coarse mesh are included to show the improvement from the fine mesh.

15.3.1 CONSTRUCTION SIMULATION

The in situ stress was introduced in the foundation soil by adopting coefficient, $K_o = 0.4$. Then, the backfill was constructed into 11 layers, Figure 15.17b, as was done in the field. The soil was compacted in each layer. The reinforcement was placed on a layer before the next layer was installed. The compacted soil in each layer was assigned the material parameters according to the stress state induced after installing the layer. The completion of the sequences of construction is referred to as

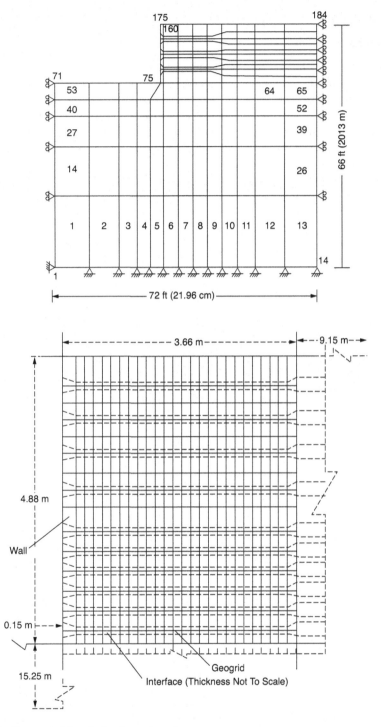

FIGURE 15.17 Coarse and fine finite element meshes.

"end of construction." Then, the surcharge load due to the traffic of 20 kPa was applied uniformly on the top of the mesh, Figure 15.14; this stage is referred to as "after opening to traffic." The concrete pavement was not included in the mesh. However, since it can have an influence on the behavior of the wall, in general, it is desirable to include the pavement.

15.4   COMPARISON BETWEEN PREDICTIONS AND FIELD MEASUREMENTS

15.4.1   Vertical Soil Pressure

Most of the results presented are for the fine mesh; typical results for vertical stress are included to show the improvements from the fine mesh compared to the results from the coarse mesh [30]. Figure 15.18 shows comparisons between computed vertical soil stresses at elevation = 1.53 m at the end of construction, from coarse and fine mesh, respectively. It is evident from these figures that the results from the fine mesh show improved correlation with the field data, compared to those from the coarse mesh. Hence, now onward the results from the fine mesh are presented and analyzed.

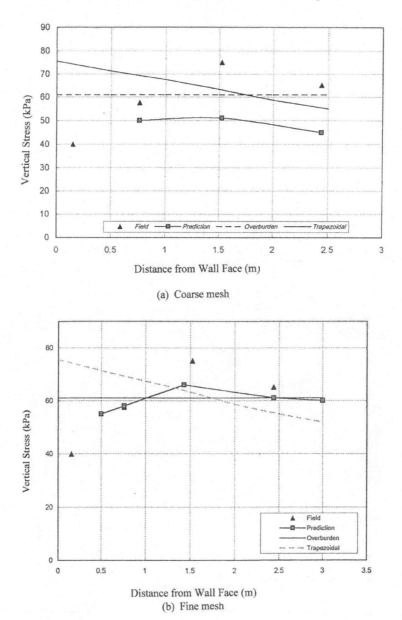

(a) Coarse mesh

(b) Fine mesh

FIGURE 15.18   Comparison between predictions and field measurements for vertical soil stresses at El. 53m at the end of construction.

The measured and predicted vertical soil stress near the wall face is less than the overburden value (i.e., $\sigma_v = \gamma h$, where h = height to elevation = 1.53 m). This can be due to the relative motions between the backfill and reinforcement. It is seen that the vertical stress distribution along the reinforcement layer is nonlinear. The vertical pressure increases in the zone away from the facing panel until reaching a maximum value at about 152 cm from the wall face. Thereafter, it shows a decrease.

15.4.2 Lateral Earth Pressure against the Facing Panel

The distribution of lateral earth pressure on the wall facing was measured based on the four pressure cells located at or near the wall face, about 0.61, 1.22, 2.44, and 3.66 m. from the bottom of the wall. The earth pressure against the facing panel was obtained in the finite element analysis from the horizontal stress in the soil elements near the facing. This pressure distribution is useful for evaluating the magnitude of the stresses exerted on the facing panels and the tension in the geogrid connection. Figure 15.19 shows the typical predicted and measured lateral soil pressure behind the facing panel after opening to traffic, along with the Rankine pressure. Predicted and measured horizontal soil stresses agree very well. The design procedure assumes that no significant lateral earth pressure should be transferred to the reinforcement. Except at the bottom of the wall, the low value of the horizontal soil stress on the wall panel approximately confirms this assumption.

15.4.3 Soil Strains

The predicted values of horizontal soil strains at elevation 2.44 m from the finite element analysis are compared with field data in Figure 15.20. There is very good correlation between predicted and measured values.

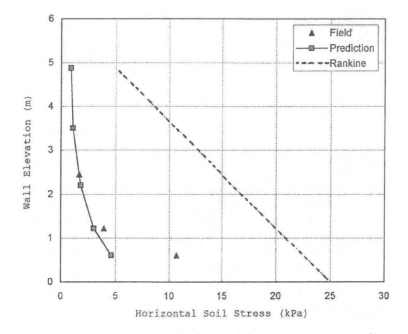

FIGURE 15.19 Comparison between predictions and field measurements for horizontal soil stress after opening to traffic.

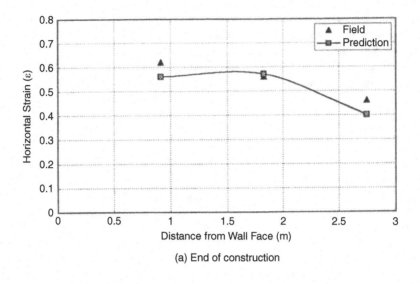

(a) End of construction

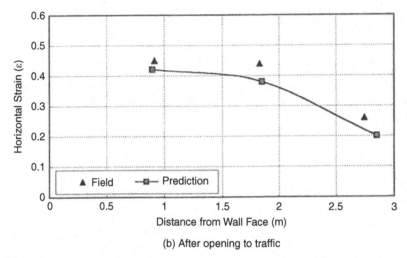

(b) After opening to traffic

FIGURE 15.20 Comparison between predictions and field measurements for horizontal soil strains at El. 2.44 m.

15.4.4 GEOGRID STRAINS

Measured and predicted reinforcement tensile strains at elevations of 1.37 and 4.42 m are shown in Figure 15.21. Agreement between the measured and predicted values is considered very good. The results demonstrate that tensile strains in the geogrids are less than 0.4% corresponding to 4.4-kN/m load in the geogrid. Comparison of this load to the maximum tensile strength of the geogrid, which is 79 kN/m, indicates that the grids are loaded to about 6.0% of the maximum strength.

15.4.5 STRESS CARRIED BY GEOGRID

Figure 15.22 shows comparison between measured and predicted horizontal stresses at different elevations for horizontal stress in the geogrid near the wall face. The measurements are obtained from the strain gauges installed on the geogrid. The predicted geogrid stresses compare well with the measurements.

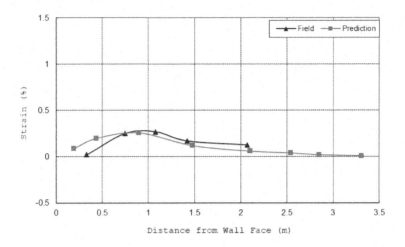

(a) At El. 1.37

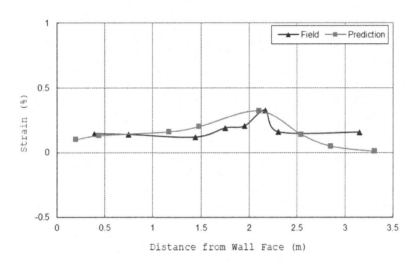

(b) At El. 4.42m

FIGURE 15.21   Comparison between predictions and field measurements for geogrid strains.

15.4.6 DISPLACEMENTS

Figure 15.23 compares predicted and measured wall movements. The correlation is considered satisfactory near the lower heights of the wall; however, it is not satisfactory at other locations. For example, near the top of the wall, the predicted value of about 42 mm is not in good agreement with the measured value of about 76 mm. The finite element analysis using the linear elastic model reported the maximum displacement of about 30 mm [35]. With the present nonlinear soil and interface models, the maximum displacement increased to about 42 mm, Figure 15.23.

The main reason for the discrepancy is the possible errors in the measurements. It is believed that since other measurements compare well with the predictions, the displacements from the finite element prediction can be considered satisfactory.

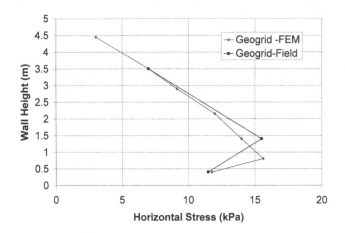

FIGURE 15.22   Comparison of predictions and field measurements for horizontal stress carried by geogrid near wall face.

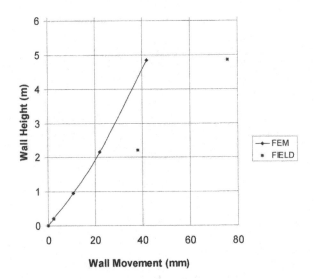

FIGURE 15.23   Comparison between predictions and measurement for wall face movements after opening to traffic.

The magnitude of the maximum wall displacement, δ_{max}, can be estimated from the following equation:

$$\delta_{max} = \delta_r \times H / 75 \tag{15.6}$$

where δ_r=relative displacement found from the chart based on L/H ratio, H=wall height, and L=reinforcement length. According to Eq. (15.6), the $\delta_{max} \approx 60$ mm does not compare well with the measured value of about 78 mm.

The wall rotates about the toe of the wall, Figure 15.23. Also, the displacements of the wall and the soil strains, Figure 15.6, are not high. The maximum displacement is about 1.5% with respect to the wall height. It appears from this behavior that there is no significant relative motion between the wall and soil for this problem. Hence, interface elements between the wall and backfill soil were not included.

15.5 CONCLUSIONS

The importance of the interface behavior between the soil and reinforcement is vital for predicting the behavior of reinforced earth walls. Laboratory test results and DSC/HISS modeling are presented for two sets of interfaces including effects of temperature and chemicals. The nonlinear finite element procedure with the unified and realistic DSC/HISS model is used to predict the field performance of a Tensar-reinforced wall. The DSC model parameters for soil and interfaces are found based on comprehensive soil tests by using triaxial testing and interface tests by using the CYMDOF shear device. The measured behaviors of the wall at the end of construction and after opening to traffic have been compared with the predictions from the finite element analysis with respect to lateral stresses on the wall facing, soil strains, geogrid strains, horizontal and vertical soil stresses, lateral stresses carried by geogrids, and wall displacements. Overall, the DSC/HISS model in nonlinear finite element procedures provides very good correlations between measured and predicted results for all quantities, except the wall displacements; possible errors in the measurement of displacements and computations can be the reasons for the discrepancy.

REFERENCES

1. Christopher, B.R., Gill, S.A., Giroud, J.P., Joran, I., Mitchell, J.K., Schlosser, F., and Dunnicliff, J. (1989). "Reinforced soil structures." *Volume II. Summary of Research and System Information, Report No. FHWA-RD-89-043*, Federal Highway Administration, McLean, VA, 158.
2. Rowe, R.K., and Ho, S.K. (1992). "A review of the behavior of reinforced soil walls." *Preprint Special and Keynote Lecture, International Symposium on Earth Reinforcement Practice*, A.A. Balkema, Rotterdam, 47–76.
3. Holtz, R.D., Christopher, B.R., and Berg, R.R. (1997). *Geosynthetic Engineering*, BiTech Publisher Ltd., Vancouver, BC.
4. Bathurst, R.J., and Hatami, K. (1998). "Seismic response analysis of a geosynthetic-reinforced soil retaining wall," *Geosynth. Int.*, 5(1–2), 127–166.
5. Hatami, K., Bathurst, R.J., and Di Pietro, P. (2001). "Static response of reinforced soil retaining wall with nonuniform reinforcements," *J. Geomech.*, 1(4), 477–506.
6. Tatsuoka, F., Leshchinsky, D., and Ling, H.I. (Eds.) (2003). *Reinforced Soil Engineering: Advances in Research and Practice*, CRC Press, Boca Raton, FL.
7. Saran, S. (2005). *Reinforced Soil and Its Engineering Applications*, I.K. International Publishing House Pvt. Ltd, Delhi.
8. Ehrlich, R.M., and Becker, L. (2019). *Reinforced Soil Walls and Sloes: Design and Construction*, CRC Press, Boca Raton, FL.
9. Wu, J.T.H. (2019). *Geosynthetic Reinforced Soil (GRS) Walls*, Wiley-Blackwell, New York.
10. Wu, J.T.H. (1992). "Predicting performance of the enver walls: general report." In *Geosynthetic-Reinforced Soil Retaining Walls*, A.A. Balkema, 3–20.
11. Leshchinsky, D. (1992). "Discussion: Strain compatibility analysis for geosynthetics reinforced soil walls," *J. Geotech. Eng.*, 188(5), 816–819.
12. Collin, J.G. (1986). "Earth Wall Design." *Ph.D. Dissertation*, University of California, Berkeley, CA.
13. Romstad, K.M., Herrmann, L.R., and Shen, C.K. (1976). "Integrated study of reinforced earth-II: Behavior and design," *J. Geotech Eng. Div.*, ASCE, 102(6), 577–590.
14. Hermann, L.R., Welch, K.R., and Lim, C.K. (1984). "FEM analysis for layered systems," *J. Geotech. Eng. Div.*, ASCE, 110(9), 1284–1302.
15. Wu, J.T.H., and Lin, J.C. (1991). "Analysis and design of geotextile reinforced earth walls," Final Report to Colorado Dept. of Highways, Dept. of Civil Eng., Univ. of Colorado, Denver, CO, USA.
16. Goodman, R.E., Taylor, R.L., and Brekke, T.L., (1968). "A model for the mechanics of jointed rock," *J. Soil Mech. Found. Div.*, ASCE, 94, 637–659.

17. Gens, A., Carol, I., and Alonso, E.E. (1989). "An interface element formulation for the analysis of soil-reinforcement interaction," *Comput. Geotech.*, 7, 133–151.

18. Desai, C.S., Zaman, M., Lightner, J.G., and Siriwardane, H.J. (1984). "Thin-layer element for interfaces and joints," *Int. J. Numer. Anal. Meth. Geomech.*, 18, 19–43.

19. Yi, C.T., Chan, D.H., and Scott, J.D. (1995). "A large slippage finite element model for geosynthetics interface modeling," *Geosynthetics' 95*, Nashville, TN, 1, 93–104.

20. Day, R.A., and Potts, D.M. (1994). "Zero thickness interface elements numerical stability and application," *Int. J. Numer. Anal. Meth. Geomech.*, 18, 689–708.

21. Desai, C.S., and Ma, Y. (1992). "Constitutive modeling of joints and interfaces by using disturbed state concept," *Int. J. Num. Anal. Methods Geomech.*, 16, 623–653.

22. Desai, C.S., and Rigby, D.B. (1997). "Cyclic interface and joint shear device including pore pressure effects," *J. Geotech. Geoenv. Eng., ASCE*, 123(6), 568–579.

23. Desai, C.S. (2001). *Mechanics of Materials and Interfaces: The Disturbed State Concept*, CRC Press LLC, Boca Raton, FL.

24. Jewell, R.A., Milligan, G.W.E., Sarby, R.W. and Dobois, D. (1984). "Interaction between soil and geogrids." *Proceedings, Symposium on Polymer Grid Reinforcement in Civil Engineering*, Science and Engineering Research Council, and Netlon Ltd.

25. Kwak, C.W., Park, I.J. and Park, J.B. (2016). "Development of modified interface apparatus and prototype cyclic simple hear test considering chemical and thermal effects," *Geotech. Testing J., ASTM*, 39(1), 20–34.

26. Gilbert, R.B., Fernandez, F. and Horsfield, D.W. (1996). "Shear strength of reinforced geosynthetics clay liners," *J. Geotech. Geoenviron. Eng.*, 122 (4), 259–266.

27. Seo, M.W., Park, J.B. and Park, I.J. (2007). "Evaluation of interface shear strength between geosynthetics under wet condition," *Soils Found.*, 47(5), 845–856.

28. De, A. and Zimmie, T.F. (1997). "Landfill stability: Static and dynamic geosynthetic interface friction value," *Geosynthetic Asia, '97*, Bangalore, India, 271–278.

29. Tensar Earth-Reinforced Wall Monitoring at Tanque Verde-Wrightstown-Pantano Roads, Tucson, Arizona (1986). *Desert Earth Engineering Preliminary Report*, Pima County Dept. of Transportation and Flood Control District, Tucson, AZ, USA.

30. El-Hoseiny, K.E., and Desai, C.S. (2003). Field Behavior of Reinforced Wall using the Disturbed State Model and Finite Element Analysis, Report, CEEM Dept. Univ. of Arizona, Tucson, AZ, USA.

31. Desai, C.S., and El-Hoseiny, K.E. (2005). "Prediction of field behavior of reinforced soil wall using advanced constitutive model," *J. Geotech. Geoenviron. Eng.*, 131 (6), 729–739.

32. Berg, R.R., Christopher, B.R., and Samatani, N.C. (2007). Design and Construction of Mechanically Stabilized Earth Walls and Reinforced Soil Slopes – Volume I, FHWA-NHI-10–024 Federal Highway Administration FHWA GEC 011, Nov. 2009. 4th Edition.

33. Berg, R.R., Bonaparte, R., Anderson, R.P., and Chouery, V.E. (1986). "Design, construction, and performance of two reinforced soil retaining walls." *Proceedings of the 3rd International Conference on Geotextiles*, Vienna, Austria, 2, 401–406.

34. Fishman, K.L., Desai, C.S., and Berg, R.R. (1991). "Geosynthetic-reinforced soil wall: 4-Year history." In *Behavior of Jointed Rock Masses and Reinforced Soil Structures, Transportation Research Record 1330*, TRB, Washington, DC, USA, 30–39.

35. Fishman, K.L., Desai, C.S., and Sogge, R.L. (1993). "Field behavior of instrumented reinforced wall," *J. Geotech. Eng., ASCE*, 119(8), 1293–1307.

36. Fishman, K.L. and Desai, C.S. (1991). "Response of a geogrid earth reinforced retaining wall with full height precast concrete facing." *Proceedings, Geosynthetics, -91*, Atlanta, GA.

37. Desai, C.S. (1998). "DSC-SST-2D: Computer code for static, dynamic, creep and thermal analysis – solid, structure and soil-structure problems," User's Manuals, Tucson, AZ.

16 Theoretical Solution for Long-Term Settlement of a Large Step-Tapered Hollow Pile in Karst Topography

Song Jiang
College of Civil Engineering, Fujian University of Technology

Ming Huang
College of Civil Engineering, Fuzhou University

An Deng
School of Civil, Environment and Mining Engineering,
University of Adelaide

Dexiang Xu
Urban Planning Design Institute of Ganzhou

16.1 INTRODUCTION

Stepped tapered pile has been increasingly used in recent years as the treatment for bridges and soft ground, and many researchers have studied the bearing characteristics of step-tapered piles under static load or dynamic load (Ismael 2003, 2010; Ghazavi and Lavasan 2006; Wang et al. 2020; Fang and Huang 2019). Considering the complexity of the actual project, step-tapered piles are being increasingly developed.

Karst is a landform formed by the dissolution and sedimentation of soluble rocks by groundwater and surface water, as well as by gravity, collapse, and accumulation, which has brought serious adverse effects on engineering construction (Zhang et al. 2019). When karst appears under bridge foundations, conventional ground treatment methods such as digging, crossing, grouting, and steel pipe retaining walls are often used. However, the use of bored piles suffers from problems, such as mud loss and stuck drilling, which seriously threaten the safety of the project.

For example, in the complex karst formations, as shown in Figure 16.1, a series of bead-shaped karst caves spread into deeper strata. As a result, the piles cannot be embedded in a solid bearing layer, which adversely affects the completion of the proposed bridge projects.

To solve this problem, Jiang et al. (2019) improved the conventional bored hollow piles by taking advantage of compensated foundations, friction group piles, and pier foundations. They significantly shortened the pile length, thus avoiding the penetration of piles in the cave.

As a compensatory measure, they increased the diameter of the pile and designed the diameter stepping depth beyond the length of the pile, leading to a tapered pile solution. The step-tapered setting facilitates the optimization of lateral friction resistance and step resistance, thus allowing the full utilization of the shallow soil layers for load bearing. To reduce the pile weight, hollow space inside the pile was allowed. Three large step-tapered hollow (LSTH) spaces of 16 m in

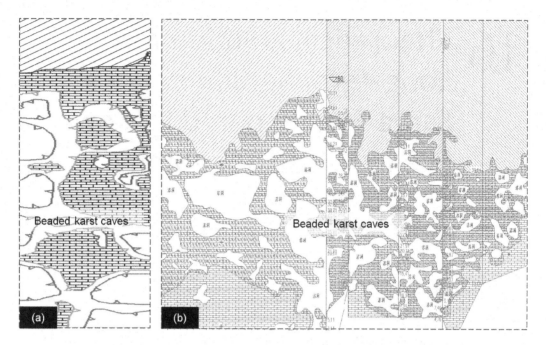

FIGURE 16.1 Complex karst formations: (a) Zengjia bridge project of Yan-Ru Expressway, Hunan, China and (b) Shenzhen Bridge project, Jiangxi, China.

length and 14 and 15 m in diameter were constructed in March 2013 as the foundation solutions for the Shenzhen Bridge in Jiangxi Province, China (Figure 16.2). There are a series of karst caves of more than 80 m below the foundation, and the thickness of the bedrock layer is not more than 4 m. The application of LSTH piles successfully solved the engineering problems caused by the unfavorable geological profile.

The bearing capacity of the LSTH pile is the result of the combined effect of pile friction, step, and pile resistance. Huang et al. (2018) analyzed pile lower shaft resistance. They thought of the additional side pressure generated by soil compression under the variable section (Figure 16.3) and obtained the theoretical solution of the enhanced effect coefficient as follows:

$$\alpha = \frac{\tau_1 + \sigma_{xM} \tan \varphi}{\tau_1} = 1 + \frac{\alpha_{xM} \tan \varphi \sigma_0'}{\tau_1} \tag{16.1}$$

where α is the enhancement effect factor, τ_1 is the shaft resistance in the lower part of the pile, σ_{xM} is the additional lateral pressure at point M, σ_0' is the pressure under the variable cross section, and φ is the angle of internal friction of soil.

Jiang et al. (2019) analyzed the working mechanism of the LSTH pile using the FLAC3D software, considering the effects of pile width, the number of rounded sections on the pile length, and the influence of soil properties around the pile on the vertical bearing capacity of LSTH piles. Huang et al. (2020a) systematically analyzed the load transfer behavior of the LSTH pile by numerical methods and then based on Desai (2001) proposed the disturbance state concept (DSC). A theoretical model was developed with the aim of capturing the load transfer characteristics of the LSTH pile.

Too many settlements can cause serious damage to infrastructures such as railroad or highway bridges, which need to minimize vertical deformation. Many analytical methods have been proposed to assess the load-settling behavior of piles, such as the shear displacement method (Nie and Chen 2005), load transfer method (Tiago and Adam 2018), elastic theoretical method (Ai and Yue 2009; Salgado et al. 2013), and various numerical methods (Yavari et al. 2014; Ju and Francis 2015).

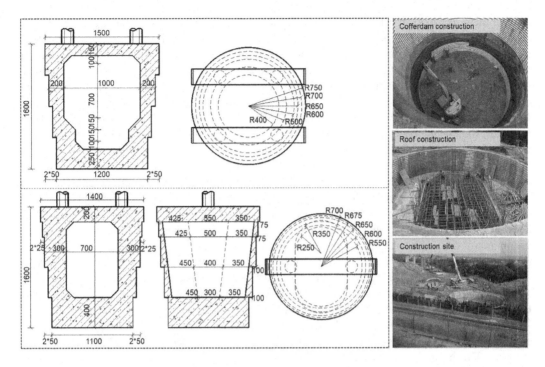

FIGURE 16.2 Sketch of the large step-tapered hollow (LSTH) pile (dimensions in centimeter) and construction photos.

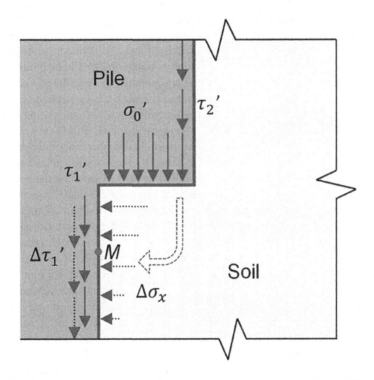

FIGURE 16.3 Enhancement of pile lateral shaft resistance by the lower part of the variable section pile.

Among them, the load transfer is more suitable for predicting the bearing characteristics of piles and has been widely developed due to its simplicity and ability to consider the nonlinearity of soil. Huang et al. (2020a) used the load transfer method to give a theoretical solution for calculating the settlement of LSTH piles and obtained better results for predicting the load settlement. In addition, by modifying the load transfer curve, a variety of complex soil properties and operating conditions can be considered, including soil stress history, nonlinear behavior, and driving effects (Shibata 1963; Randolph and Guo 1999; Mylonakis 2001; Zhu and Chang 2002; Nanda and Patra 2014). However, the evolution of pile–soil interaction is rarely considered in the load transfer method, which calculates the load-settlement response of piles with time (Cui et al. 2019).

After pile installation and loading, the strength and stiffness of the surrounding soil will increase over time due to the dissipation of hyper-pore pressure and subsequent creep behavior of the soil (Mishra and Patra 2018; Cui et al. 2019). This results in changes in the bearing characteristics of the piles, resulting in load-settlement characteristics of the pile over time (Chen et al. 2009; Zhao et al. 2013). Ignoring this phenomenon in pile foundations design will lead to inaccurate estimation of the pile bearing capacity. Unfortunately, the current design criteria mainly focus on the final settlement of piles and neglect the long-term settlement aspect. In recent years, the settlement characteristics of time-dependent piles have attracted much attention in this field. Numerous studies have been conducted on the time effect of single-pile settlement (Booker and Poulos 1976; Edil and Mochtar 1988; Guo 2000; Zeng et al. 2005; Bowman and Soga 2005; Small and Liu 2008; Danno and Kimura 2009; Abbas et al. 2010; Wu et al. 2012; Yang et al. 2014; Li et al. 2015; Feng et al. 2017; Mishra and Patra 2018; Cui et al. 2019). In terms of theoretical methods, Booker and Poulos (1976) proposed a linear viscoelastic model with three parameters to describe the creep properties of the surrounding soil, and obtained the elastic analytical solution of the pile using Mindlin's equation. Guo (2000) used a generalized viscoelastic model and an integration method based on the load transfer method to establish the nonlinear bearing capacity and viscoelastic model of the pile. Wu et al. (2012) used the distributed Voigt model to consider the viscoelasticity of surrounding soil layers, and the virtual soil–pile model to consider the influence of the foundation soil, and gave the analytical solutions of the displacement impedance function and settlement at the top of the monopole. Yang et al. (2014) established the one-dimensional consolidation equation for single-drainage or double-drainage elastic multilayer soils under multilayer loading and derived the effective stress and settlement calculation equations by Laplace numerical inversion transformation. Feng et al. (2014) used the Mesri creep model to describe the soil properties and Mindlin-Geddes method considering pile diameter to calculate the vertical additional stress at the bottom of the pile, and proposed a simple and practical method for calculating post-work settlement of pile foundations of high-speed railway bridges. Li et al. (2015) proposed a time effect analysis method for monopole settlement and obtained the distribution of pile resistance and pile axial force. In the model, dampers were used to consider the viscosity of the soil; a hyperbolic elastoplastic model and an ideal elastoplastic model were used to simulate the nonlinear elasticity of the pile and the soils at the base of the pile, respectively. Feng et al. (2017) proposed a nonlinear method based on a modified Burgers model for predicting the settlement of single and group piles in laminated soils over time. Mishra and Patra (2018) predicted the long-term settlement behavior of group piles using two types of five-parameter viscoelastic soil models. Cui et al. (2019) used the theory of nonstationary flow surface (NSFS) to describe the viscoplastic behavior of surrounding soil and incorporated the evolution of soil aging behavior into the skin friction softening model, and then proposed a theoretical method to estimate the long-term time-dependent load bearing characteristics of piles based on the load transfer method. From the literature review, most of the studies focused on the homogeneous section pile foundations, and there were fewer studies on shaped piles. For step-tapered pile, it is difficult to understand the load transfer mechanism of piles over time due to the multistage nature of soil-deformation process and uncertainty of pile–soil interaction. There are knowledge gaps in the time-dependent load transfer mechanism and time-dependent settlement prediction of the LSTH pile.

Based on the concept of disturbance state, a new time-dependent load transfer model is proposed to describe the time-dependent deformation characteristics (Desai 2001) of the pile–soil interface. The analytical solution of long-term settlement of LSTH piles is further derived using the load transfer method. The accuracy and feasibility of the theoretical method were verified by comparing the data with the numerical simulation calculation results through example analysis.

16.3   TIME-DEPENDENT DSC LOAD TRANSFER MODEL

The load transfer method is a calculation method widely used in single-pile settlement estimation. This approach is based on the idea that the load-settlement response is related to the transfer of reaction forces along the pile and affects local displacements. Mathematically, these relationships are expressed as mobilization function (Zhang and Zhang 2012) at different ends and along the shaft. Using the load transfer function to calculate the long-term settlement of piles can truly reflect the deformation characteristics of the pile–soil interface with time, which is the key to the calculation of the long-term settlement of piles.

16.3.1   TIME-DEPENDENT BEARING CHARACTERISTICS OF AN LSTH PILE

Under long-term loading, the surrounding soil is constantly disturbed and the stress stage of the soil changes, leading to changes in the bearing characteristics of the pile (Chen et al. 2009; Zhao et al. 2013). In order to study the variation law of the bearing characteristics of the LSTH pile with time, a ¼-scale model of a Φ 15 m LSTH pile in one pile was established using FLAC3D software (ICGI 2018), which was applied to the construction of Shenzhen Bridge in Jiangxi Province, China (Figure 16.4). The calculation process includes three stages: the applied geodynamic stress field, the pile installation, and the deformation stage with time. The first two stages refer to the simulation method given by Jiang et al. (2019). After the introduction of the geostatic stress field, the model reaches the pile installation. In the pile installation stage, the model is configured to an equilibrium state after pile installation.

Pile installation is achieved by removing the area where hollow space within the pile exists and replacing the soil model parameters with pile model parameters in the area of the pile. The linear elastic model is used for the pile model and the Mohr–Coulomb model is used for the soil. The material parameters of the corresponding units are shown in Tables 16.1 and 16.2, where the soil parameters are derived from the results of consolidation and drainage triaxial tests in this engineering

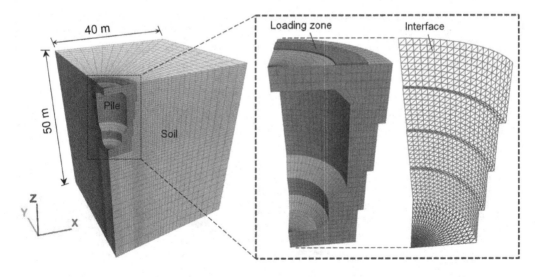

FIGURE 16.4   A numerical model of the Φ15 m large step-tapered hollow (LSTH) pile.

TABLE 16.1

Mechanical Parameters of the Pile and Soil

Element	Density/ (kg·m⁻³)	Poisson's Ratio	Elastic Modulus/ MPa	Bulk Modulus/ MPa	Shear Modulus/ MPa	Cohesion/ kPa	Internal Friction/(°)
Pile	2,500	0.2	30,000	16,700	12,500	–	–
Clay	1,950	0.3	25	20.83	9.62	60	20

TABLE 16.2

Mechanical Parameters of the Pile–Soil Interface

Element	$K_n = K_s$ (MPa)	Cohesion (kPa)	Internal Friction Angle (°)
Interface	1,346	48	16

geological investigation report. In the deformation phase with time, the total vertical constant load in the loading zone is 50 MN. The Burgers model effectively reflects the deformation behavior of the soil with time (Nian et al. 2014; Ma et al. 2014; Huang et al. 2014), which is used to define the soil.

Based on the field data, the soil parameters of the Burgers model were obtained by the back-analysis method (Deng et al. 2012). The details are provided in the appendix. After the pushing construction of the upper structure of the pile top is completed, the JMDL-6210A/HAT hydrostatic leveling system is used to monitor the pile settlement site data. The measuring device of the hydrostatic leveling system consists of a water storage tank, a buoy, a precision liquid level gauge, a protective cover, and a support frame, as shown in Figure 16.5a. Using the measurement system based on the principle of communicating vessels (i.e., liquid level of the liquid reservoirs connected by several connection pipes is always at the same level) and comparing the liquid level height of different liquid reservoirs with the datum point, the relative differential settlement of each measuring point can be obtained.

The monitoring device is installed on the top of the pile, and a reference point is set on the pier on the left side of the abutment. The entire vertical constant load is applied to 50 nm, as shown in Figure 16.5b. A total of 120 days of monitoring data were recorded.

In order to ensure the safety of the project, the settlement monitoring should be carried out 24 h after the load is applied, but the settlement of the pile has actually started before the monitoring. Considering the settlement difference, a reasonable offset value was introduced in the numerical simulation analysis. The comparison of the simulated results and the field average settlement are shown in Figure 16.6. The soil parameters of the Burgers model are shown in Table 16.3.

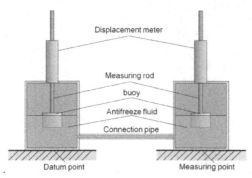

(a) JMDL-6210A/HAT hydrostatic leveling

(b) Field monitoring for the settlement

FIGURE 16.5   The device. (a) JMDL-6210A/HAT hydrostatic leveling. (b) Field monitoring for the settlement.

TABLE 16.3
Burgers Model Parameters

Soil	Viscoelastic Modulus E_M/MPa	Viscoelastic Modulus E_K/MPa	Viscosity/(MPa·h) Maxwell η_M	Kelvin η_K
Clay	663.35	2,536.70	∞	5.38×10^5

It can be seen from Figure 16.7 that the surface friction and axial force of the pile change with time. Under the action of load, the axial force and the surface friction force have great changes in the initial stage, and gradually stabilize with the passage of time. It can be seen from the test results that there is a certain time for the transfer of the bearing capacity of pile under the action of load. During this process, the surrounding soil gradually deforms over time, and the transmitted load between the soil and the pile is dynamically adjusted. Under constant load, the pile–soil interface of LSTH piles has a time-varying disturbance effect.

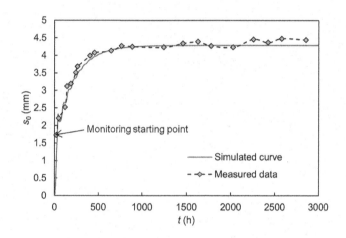

FIGURE 16.6 The *s-t* curve with diameter of 15 m large step-tapered hollow (LSTH) pile under 50 MN load.

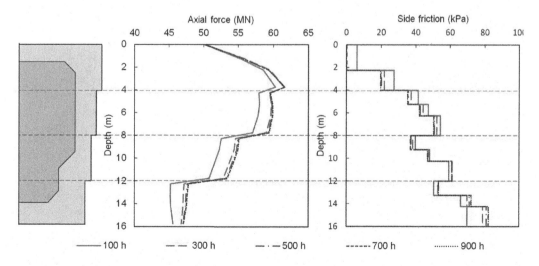

FIGURE 16.7 The disturbance behavior of the soil–pile interface of a large step-tapered hollow (LSTH) pile. (a) Axial force. (b) Skin friction.

16.3.2 Time-Dependent Disturbance at the Pile–Soil Interface of an LSTH Pile

Frantziskonis and Desai (1987) first proposed the DSC of geotechnical materials. Desai and Ma (1992) and Desai and Toth (1996) further developed the idea of static and dynamic constitutive modeling of soils and soil–structure interfaces. With DSC, the material is viewed as a mixture of elements in two states: the relatively intact (RI) state and fully adjusted (FA) state. Under the action of load, the material gradually transitions from the RI state to FA state in its own microstructure through its own adjustment, so that it can effectively bear the load.

Laboratory tests of pile settlement in clay have shown that the settlement of pile growth is mainly due to time-dependent deformation of the pile–soil interface and surrounding soil (Edil and Mochtar 1988). Approximate zone thicknesses can also be obtained through numerical simulations and laboratory experiments.

The concept of "smeared" zones applies to LSTH piles, as shown in Figure 16.8. Under load, the elements of the interface region will effectively carry the load by gradually transitioning from an original intact condition to a final failed state through a continuous transition over time. During this process,

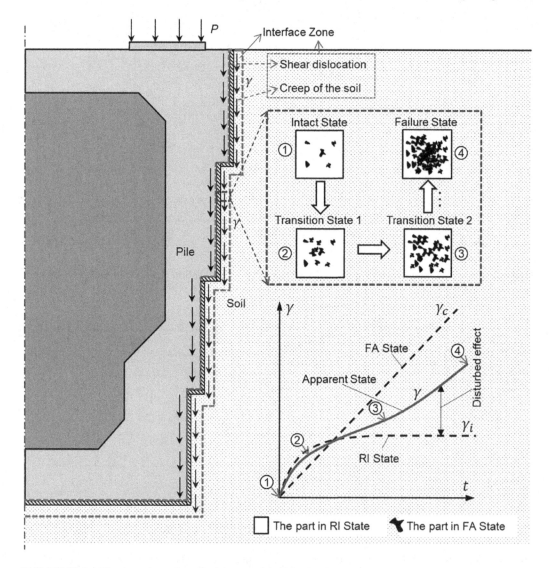

FIGURE 16.8 The time-dependent disturbance of the interface zone.

due to the change of the pile–soil interaction due to the time interference of the interface zone, the stress state of the pile also changes, as shown in Figure 16.7. The stress status of the pile depends on the time-varying disturbance in the boundary zone, but it is hard to judge the extent of the disturbance by the stress condition. Thus, a time-dependent perturbation model must be developed to quantify the behavior of the interfacial zone during the creep phase.

16.3.3 DEVELOPMENT OF THE TIME-DEPENDENT DSC LOAD TRANSFER MODEL

Based on DSC, the shear strain can be obtained as:

$$\gamma = (1 - D)\gamma_i + D\gamma_c \tag{16.2}$$

where γ_i and γ_c represent the shear strains of the soil in the RI and FA states, respectively, and D is the disturbance factor.

The RI state describes the intact or undisturbed part in the interface region that is free from disturbing factors. The material in the FA state is partially surrounded by the RI portion, with certain strength, like a constrained liquid–solid (CLS) (Desai 2001). In the context of creep modeling, the RI part of the interface region can be described by a viscoelastic model, and the FA part of the interface region can be described by a perfect viscoplastic model (Desai 2001; Huang et al. 2020b).

The Kelvin model is a widely adopted viscoelastic model for representing RI states. According to Taylor (1942), the model is expressed in one-dimensional form:

$$\tau_i = G_k\gamma_i + \eta_k\dot{\gamma}_i \tag{16.3}$$

Eq. (16.3) is rearranged as:

$$\gamma_i = \frac{\tau_i}{G_k}\left[1 - \exp\left(-\frac{G_k}{\eta_k}t\right)\right] \tag{16.4}$$

where τ_i is the shear stress in the Kelvin model, G_k is the shear modulus, and η_k is the viscosity coefficient.

For the FA state, this paper uses a complete viscoplastic model (Ling and Cai 2002) composed of viscous elements and plastic elements (St. Venant body) to describe its creep behavior, as follows:

$$\tau_c = \tau_n + \eta_n\dot{\gamma}_c \tag{16.5}$$

From Eq. (16.5), we can obtain:

$$\gamma_c = \frac{\tau_c - \tau_n}{\eta_n}t \tag{16.6}$$

where τ_c is shear stress in the perfectly viscoplastic model, η_n is the viscosity coefficient, and τ_n is the frictional resistance of friction plate.

It should be noted that, according to DSC, a perturbation function D is used to describe the transition of a material from the RI state to the FA state, which is related to dissipated energy, plastic strain, time, entropy, and temperature (Desai 2001). Since the plastic strain of a material is related to the irreversible damage of the material, the evolution of D can be defined in terms of plastic strain when it is subjected to an external load (Bažant 1994). According to the common form of Weibull function, D can be defined as:

$$D = D\left(\xi^p\right) = D_u\left\{1 - \exp\left[\left(-\xi^p / H\right)^N\right]\right\} \tag{16.7}$$

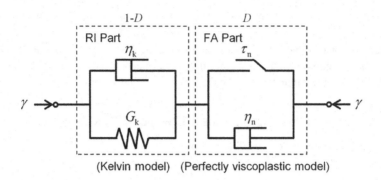

FIGURE 16.9 Load transfer model based on time.

where H and N are the parameters of the disturbance factor D; D_u is the ultimate value of the disturbance factor D, $0 < D_u < 1$, and it is usually taken as 1; and ξ^p is the cumulative plastic strain. In the context of creep, creep strain can be used to estimate D (Huang et al. 2020b), and this study uses the Weibull function of time t to model:

$$D = D\left(\xi^c\right) = D\left(f(t)\right) = 1 - \exp\left[-\left(\frac{t}{A}\right)^B\right] \tag{16.8}$$

where A and B are constants and $\xi^c = f(t)$ is the cumulative creep strain. This relationship essentially shows that there is no creep damage ($D = 0$) initially (the rock components are all in the RI state at $t = 0$) and $D \to 1$ as $t \to \infty$ (the rock components are all in the FA state ultimately).

As shown in Figure 16.9, connect the RI and FA state models in series. Assuming that the stresses are always coordinated during the perturbation, the average stress and stress of the RI state and the FA state are equal:

$$\tau = \tau_i = \tau_c \tag{16.9}$$

Substituting Eqs. (16.4), (16.6), (16.8), and (16.9) into Eq. (16.2), in one dimension, the time-varying load transfer function is shown as:

$$\gamma(t) = \frac{\tau}{G_k}\exp\left[-\left(\frac{t}{A}\right)^B\right]\left[1 - \exp\left(-\frac{G_k}{\eta_k}t\right)\right] + \frac{(\tau - \tau_n)t}{\eta_n}\left\{1 - \exp\left[-\left(\frac{t}{A}\right)^B\right]\right\} \tag{16.10}$$

where $\gamma(t)$ is the total shear strain at time t, G_k is shear modulus, η_k and η_n are viscosity coefficients, τ is initial shear stress, τ_n is frictional resistance of friction plate, and A and B are undetermined constants in disturbance factor D.

16.4 LONG-TERM SETTLEMENT OF THE LSTH PILE

16.4.1 Model Development

Taking into account the homogeneous soil layer, the LSTH pile's long-term settlement is derived. The calculation model is shown in Figure 16.10. A time-varying DSC load transfer model is used to describe the time-varying disturbance of pile–soil interaction. The LSTH pile has n variable cross section steps, and every cross section has a diameter of d_1 to d_n. Divide it into $n + 2$ parts according to the section size of the pile body. Then, those pile elements were named $1, 2, \ldots, n + 1$ and $n + 2$ from pile top to pile tip with the length $l_1, l_2, \ldots, l_{n+1}, l_{n+2}$, the perimeter $U_1, U_2, \ldots, U_{n+1}, U_{n+2}$, and the cross-sectional area $A_{p1}, A_{p2}, \ldots, A_{p(n+1)}, A_{p(n+2)}$. Also, we made the following assumptions:

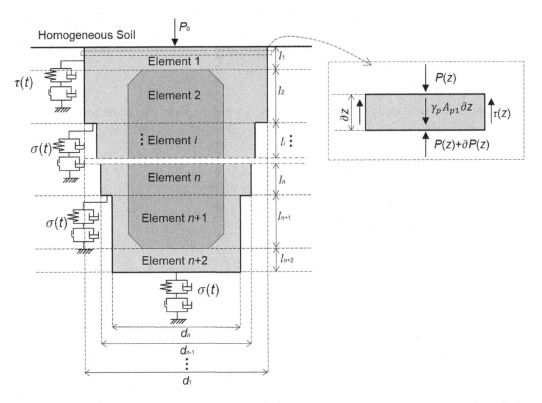

FIGURE 16.10 A calculation model for the settlement of an n-stage variable cross section hollow pile in uniform soil.

i. The pile is divided into several equal units in the direction of its axis, and each element is a uniform isotropic linear elastic body.

ii. On the pile, the displacement at any point is only related to skin friction. Based on this, an independent nonlinear model is developed to simulate the time-varying pile–soil interaction.

iii. The nonlinear model referred to in assumption (ii) may be represented by a functional form.

16.4.2 Formula Derivation

The bearing capacity mechanism of LSTH piles is very complex, and the calculation method of its long-term settlement is very difficult. Therefore, this paper studies the cross section and variable cross section of LSTH piles. In the case of a cylindrical pile section of the LSTH pile, as illustrated in Figure 16.10, a pile element 1 is used. The weight of the pile cannot be neglected, especially in the assessment of settlement. Thus, assuming that the displacement of the pile at z depth is s, its static equilibrium equation can be expressed in the following way:

$$\frac{\partial P(z)}{\partial z} = -U_1 \tau(z) + \gamma_p A_{p1} \tag{16.11}$$

This pile element's elastic compression can be expressed as follows:

$$\frac{\partial s}{\partial z} = -\frac{P(z)}{A_{p1} E_p} \tag{16.12}$$

By combining Eqs. (16.11) and (16.12), the following governing equation can be obtained:

$$\frac{\partial^2 s}{\partial z^2} = \frac{U_1}{A_{p1} E_p} \tau(z) - \frac{\gamma_p}{E_p} \tag{16.13}$$

A time-dependent DSC load transfer model was used to represent the interaction between the pile and soil. Assuming that the pile top load is constant, the creep equation of the pile side can be obtained:

$$\gamma = \frac{(1-D)\tau}{G_k} \times \left(1 - \exp\left(-\frac{G_k}{\eta_k} t\right)\right) + \frac{D(\tau - \tau_n)}{\eta_n} t \tag{16.14}$$

The pile-tip soil's creep equation can be obtained:

$$\varepsilon = \frac{(1-D)\sigma}{E_k} \times \left(1 - \exp\left(-\frac{E_k}{\eta_k} t\right)\right) + \frac{D(\sigma - \sigma_n)}{\eta_n} t \tag{16.15}$$

where ε is the normal strain, σ is the normal stress, $E_k = G_k \times [2(1+v)]$, and v is Poisson's ratio of the soil.

Under the shear stress of the soil at the side of the pile, it can be transformed into:

$$\tau = \frac{\gamma \eta_n + D\tau_n t}{\dfrac{\eta_n(1-D)}{G_k} \times \left(1 - \exp\left(-\dfrac{G_k}{\eta_k} t\right)\right) + Dt} \tag{16.16}$$

Define

$$M_s(t) = \frac{\eta_n}{\dfrac{\eta_n(1-D)}{G_k} \times \left(1 - \exp\left(-\dfrac{G_k}{\eta_k} t\right)\right) + Dt} \tag{16.17}$$

$$N_s(t) = \frac{D\tau_n t}{\dfrac{\eta_n(1-D)}{G_k} \times \left(1 - \exp\left(-\dfrac{G_k}{\eta_k} t\right)\right) + Dt} \tag{16.18}$$

Eq. (16.16) can be reformulated as:

$$\tau = \gamma M_s(t) + N_s(t) = s M_s(t) / l_1 + N_s(t) \tag{16.19}$$

Replacing Eq. (16.13) with Eq. (16.19), we obtain:

$$\frac{\partial^2 s}{\partial z^2} = \frac{U_1}{A_{p1} E_p} [s M_s(t) / l_1 + N_s(t)] - \frac{\gamma_p}{E_p} \tag{16.20}$$

Define

$$\alpha^2(t) = \frac{U_1 M_s(t)}{A_{p1} E_p l_1} \tag{16.21}$$

$$\beta(t) = \frac{U_1 N_s(t)}{A_{p1} E_p} - \frac{\gamma_p}{E_p} \tag{16.22}$$

Then, Eq. (16.20) can be rewritten as:

$$\frac{\partial^2 s}{\partial z^2} = \alpha^2(t)s + \beta(t) \tag{16.23}$$

The general solution of the homogeneous equation corresponding to the above differential equation can be expressed as:

$$S(z) = c_1 e^{\alpha(t)z} + c_2 e^{-\alpha(t)z} \tag{16.24}$$

A particular solution of the differential equation can be expressed as follows:

$$s^*(z) = -\beta(t) / \alpha^2(t) \tag{16.25}$$

Thus, the general solution to this differential equation can be expressed as:

$$s(z) = S(z) + s^*(z) = c_1 e^{\alpha(t)z} + c_2 e^{-\alpha(t)z} - \beta(t) / \alpha^2(t) \tag{16.26}$$

Substituting Eq. (16.26) into Eq. (16.12), we can write the axial force as:

$$P(z) = -\alpha(t) A_{p1} E_p \left(c_1 e^{\alpha(t)z} - c_2 e^{-\alpha(t)z} \right) \tag{16.27}$$

By applying the boundary conditions, the continuity conditions of force and displacement at the end of pile element 1 can be obtained:

$$\begin{cases} E_p A_{p1} \dfrac{\partial s}{\partial z} \bigg|_{z=0} = P_0 \\[2mm] s\big|_{z=0} = s_0 \end{cases} \tag{16.28}$$

Thus, we can obtain:

$$\begin{cases} c_1 = \dfrac{1}{2} \times \left[s_0 - \dfrac{P_0}{\alpha(t)E_p A_{p1}} + \dfrac{\beta(t)}{\alpha^2(t)} \right] \\[4mm] c_2 = \dfrac{1}{2} \times \left[s_0 + \dfrac{P_0}{\alpha(t)E_p A_{p1}} + \dfrac{\beta(t)}{\alpha^2(t)} \right] \end{cases} \tag{16.29}$$

Substituting c_1 and c_2 into Eq. (16.26) and Eq. (16.27), we obtain:

$$s(z) = \frac{1}{2} e^{\alpha(t)z} \left[s_0 - \frac{P_0}{\alpha(t)E_p A_{p1}} + \frac{\beta(t)}{\alpha^2(t)} \right] + \frac{1}{2} e^{-\alpha(t)z} \left[s_0 + \frac{P_0}{\alpha(t)E_p A_{p1}} + \frac{\beta(t)}{\alpha^2(t)} \right] - \beta(t) / \alpha^2(t) \tag{16.30}$$

$$P(z) = -\frac{1}{2}\alpha(t)A_{p1}E_p \left\{ e^{\alpha(t)z} \left[s_0 - \frac{P_0}{\alpha(t)E_pA_{p1}} + \frac{\beta(t)}{\alpha^2(t)} \right] - e^{-\alpha(t)z} \left[s_0 + \frac{P_0}{\alpha(t)E_pA_{p1}} + \frac{\beta(t)}{\alpha^2(t)} \right] \right\} (16.31)$$

Due to

$$\cosh[\alpha(t)z] = \frac{e^{\alpha(t)z} + e^{-\alpha(t)z}}{2} \tag{16.32}$$

$$\sinh[\alpha(t)z] = \frac{e^{\alpha(t)z} - e^{-\alpha(t)z}}{2} \tag{16.33}$$

The matrix form of Eqs. (16.30) and (16.31) can be expressed as:

$$
\left\{ \begin{array}{c} s(z) \\ P(z) \end{array} \right\} = \left[\begin{array}{cc} \cosh[\alpha(t)z] & \dfrac{-\sinh[\alpha(t)z]}{\alpha(t)A_{p1}E_p} \\ -\alpha(t)A_{p1}E_p \sinh[\alpha(t)z] & \cosh[\alpha(t)z] \end{array} \right] \left\{ \begin{array}{c} s_0 \\ P_0 \end{array} \right\}
$$

$$
+ \left\{ \begin{array}{c} \cosh[\alpha(t)z]\beta(t)/\alpha^2(t) - \beta(t)/\alpha^2(t) \\ -A_{p1}E_p \sinh[\alpha(t)z]\beta(t)/\alpha(t) \end{array} \right\}
$$

$$
= \mathbf{K}_1(z) \left\{ \begin{array}{c} s_0 \\ P_0 \end{array} \right\} + \mathbf{T}_1(z) \tag{16.34}
$$

When z is l_1, the load and displacement at the end of pile element 1 can be expressed as:

$$
\left\{ \begin{array}{c} s_1 \\ P_1 \end{array} \right\} = \mathbf{K}_1(l_1) \left\{ \begin{array}{c} s_0 \\ P_0 \end{array} \right\} + \mathbf{T}_1(l_1) \tag{16.35}
$$

After the load and displacement of pile element 1 were obtained, the load and displacement of pile element 2 were further solved:

$$
\left\{ \begin{array}{c} s_2 \\ P_2 \end{array} \right\} = \mathbf{K}_2(l_2)\mathbf{K}_1(l_1) \left\{ \begin{array}{c} s_0 \\ P_0 \end{array} \right\} + \mathbf{K}_2(l_2)\mathbf{T}_1(l_1) + \mathbf{T}_2(l_2) \tag{16.36}
$$

For the variable cross section of the LSTH pile, the pile element $j+1 (3 \le j+1 \le n+1)$ was taken as the object. As shown in Figure 16.11, when the soil at the variable section of the pile is in an elastic state, in accordance with the Boussinesq solution of the elastic theory of infinite elastic foundation under circular plate under uniformly distributed circular load, we can obtain:

$$
R_{bj} = \frac{M_{sb}(t)(d_j - d_{j+1})}{1 - v^2} s_j \tag{16.37}
$$

where $M_{sb}(t) = \dfrac{\eta_n}{\dfrac{\eta_n(1-D)}{E_k} \times \left(1 - \exp\left(-\dfrac{E_k}{\eta_k}t \right) \right) + Dt}$.

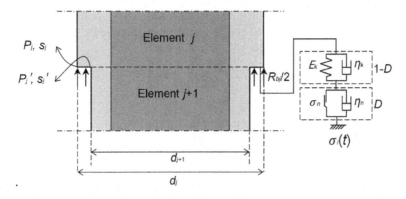

FIGURE 16.11   Calculation model at a variable cross section.

For any cross section at depth z of element $j+1$, the displacement satisfies the following differential equation and boundary conditions for element $j+1$ in the local coordinate system O_{j+1}:

$$\begin{cases} A_{p(j+1)}E_p \dfrac{\partial^2 s}{\partial z^2} - U_{j+1}\tau + \dfrac{\gamma_p}{E_p} = 0, 0 \le z \le l_{j+1} \\[4mm] P_j' = P_j - R_{bj} \\[2mm] s_j' = s_j \end{cases} \qquad (16.38)$$

According to the previous derivation, we can obtain:

$$\begin{Bmatrix} s(z) \\ P(z) \end{Bmatrix} = \begin{bmatrix} \cosh[\alpha(t)z] & \dfrac{-\sinh[\alpha(t)z]}{\alpha(t)A_{p(j+1)}E_p} \\[4mm] -\alpha(t)A_{p(j+1)}E_p \sinh[\alpha(t)z] & \cosh[\alpha(t)z] \end{bmatrix} \begin{Bmatrix} s_j \\ P_j - R_{bj} \end{Bmatrix}$$

$$+ \begin{Bmatrix} \cosh[\alpha(t)z]\beta(t)/\alpha^2(t) - \beta(t)/\alpha^2(t) \\ -A_{p(j+1)}E_p \sinh[\alpha(t)z]\beta(t)/\alpha(t) \end{Bmatrix} = \mathbf{K}_{j+1}(z) \begin{Bmatrix} s_j' \\ P_j' \end{Bmatrix} + \mathbf{T}_{j+1}(z) \qquad (16.39)$$

In this way, for the pile element i $(0 \le i \le n+2)$, the load and displacement at the end of pile element i can be expressed as:

$$\begin{Bmatrix} s_i \\ P_i \end{Bmatrix} = \mathbf{K}_i(l_i) \begin{Bmatrix} s_{i-1} \\ P_{i-1} \end{Bmatrix} + \mathbf{T}_i(l_i) \qquad (16.40)$$

As shown in Figure 16.10, the load and displacement at the tip of the pile can be expressed as:

$$\begin{Bmatrix} s_b \\ P_b \end{Bmatrix} = \mathbf{K}_{n+2}(l_{n+2})\mathbf{K}_{n+1}(l_{n+1}) \begin{Bmatrix} s_n' \\ P_n' \end{Bmatrix} + \mathbf{K}_{n+2}(l_{n+2})\mathbf{T}_{n+1}(l_{n+1}) + \mathbf{T}_{n+2}(l_{n+2}) \qquad (16.41)$$

When the soil at the pile tip is in the elastic state, the Boussinesq solution under a load of pile tip P_b based on the elastic theory of the infinite elastic foundation is as follows:

$$s_b = \frac{1-v^2}{M_{sb}(t)d_n}P_b \tag{16.42}$$

where $M_{sb}(t) = \dfrac{\eta_n}{\dfrac{\eta_n(1-D)}{E_k}\times\left(1-\exp\left(-\dfrac{E_k}{\eta_k}t\right)\right)+Dt}$.

Combining Eqs. (16.34)–(16.42), the relationship between s_b, P_b, s_0, and P_0 can be obtained by iterative calculation of each element, where s_b is the displacement of the pile tip, P_b is the axial force of the pile tip, s_0 is the displacement of the pile top, and P_0 is the axial force of the pile top. In order to calculate the long-term settlement of the LSTH pile, MATLAB can be used to compile the above calculation program.

In addition, the settlement calculation method of the LSTH piles in a layered foundation is alike to that of homogeneous soil. However, the characteristics of pile cross-sectional area and soil layer should be considered at the same time when subdividing the model, and load transfer functions with different parameters are introduced to describe the time-varying disturbance characteristics of pile–soil interaction between different layers.

16.5 VERIFICATION

The Φ 14 m and Φ 15 m LSTH piles installed to support the Shenzhen Bridge (Figure 16.2) were chosen to test the calculation method. As shown in Figure 16.12, FLAC^{3D} software is used to build two ¼-scale models of two piles. Allowing for the simplicity of modeling, the hollow parts of these piles are consistent with cylinders. The simulation method and related parameters are equivalent to the previous simulation results.

The calculation model shown in Figure 16.13 was used to calculate the long-term settlement of the two piles. Considering the cross-sectional area of the two piles, the piles were divided into five and seven parts. MATLAB™ was used to calculate the settlement of the two piles.

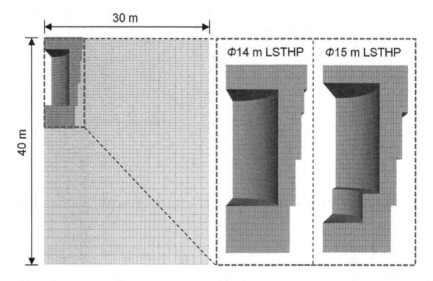

FIGURE 16.12 Numerical models of the large step-tapered hollow (LSTH) piles. (a) Φ14 m LSTH pile. (b) Φ15 m LSTH pile.

The settlement-time (s-t) curve of the $\Phi14\,\mathrm{m}$ LSTH pile under $40\,\mathrm{MN}$ load and that of the $\Phi15\,\mathrm{m}$ LSTH pile under $50\,\mathrm{MN}$ load in Figure 16.14 were processed by MATLAB™, yielding the parameters of the DSC time-dependent load transfer model. The parameters were written in Table 16.4.

The theoretical long-term settlement curves of the two piles comparing with the numerical simulation results were shown in Figure 16.12. Under the difference of the target load and the reference load, the difference between the numerical simulation curve and the calculated curve gradually increases with the accumulating fitting errors. The two results are pretty consistent: the average error percentages of these curves for $\Phi14\,\mathrm{m}$ LSTH pile are 3.65% (40MN), 1.64% (30MN), 3.83% (20MN), and 7.62% (10MN), and that for $\Phi15\,\mathrm{m}$ LSTH pile are 1.96% (50MN), 2.07% (40MN), 5.11% (30MN), and 9.25% (20MN). The results show that the theoretical method in this paper can be used to predict the long-term settlement of the LSTH pile within a certain period of time.

16.6 LIMITATIONS

Based on the DSC theory and load transfer method, the long-term load-settlement behavior of the LSTH pile in karst terrain is studied, but there are still some limitations:

i. In this work, the numerical simulation results of DSC load transfer model parameters and the time-dependent pile–soil interactions of LSTH piles are simplified as a function. However, considering the multistage deformation process of the surrounding soil and the uncertainty of pile–soil interaction, the applicability and application range of the proposed DSC load transfer model need more field data to verify.

ii. In this work, two numerical simulation examples are taken to clarify the application of the method. The proven results show that the theoretical method proposed can provide a reliable prediction for the long-term settlement of LSTH piles within time (within 1500 h in both cases).

iii. This work does not cover the bearing characteristics of LSTH piles under active traffic loads. The proposed model and method for evaluating the long-term settlement of LSTH piles was verified under static load; extending it to consider the effect of dynamic load will be a hot spot in future research.

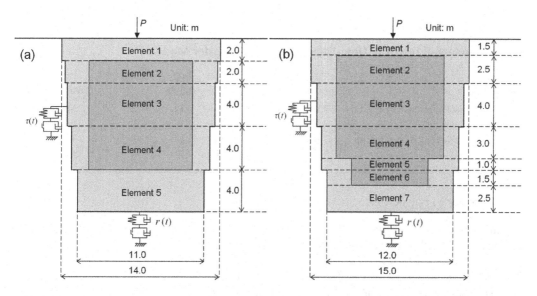

FIGURE 16.13 Settlement calculation models for the large step-tapered hollow (LSTH) piles. (a) $\Phi14\,\mathrm{m}$ LSTH pile. (b) $\Phi15\,\mathrm{m}$ LSTH pile.

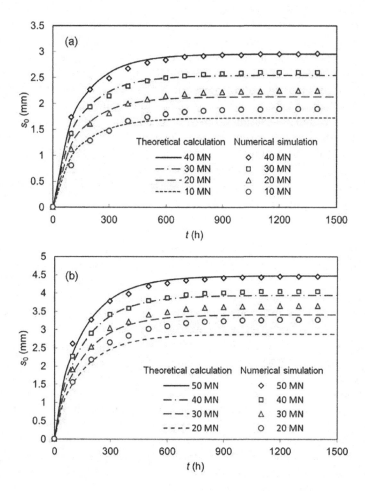

FIGURE 16.14 The s-t curves of the large step-tapered hollow (LSTH) piles. (a) Φ14 m LSTH pile. (b) Φ15 m LSTH pile.

TABLE 16.4

The Parameters of the DSC Time-Dependent Load Transfer Model

Pile (m)		A ($\times 10^{11}$)	B	E_k/G_k (MPa)	τ/σ (MPa)	τ_n/σ_n (MPa)	η_k (MPa·h)	η_n (MPa·h)
Φ 14	Variable cross section /Pile tip	7.188	0.388	425	1	0	27,230	18
	Pile side	7.188	0.388	163.462	1	0	27,230	18
Φ 15	Variable cross section/Pile tip	7.188	0.388	272	1	0	18,230	18
	Pile side	7.188	0.388	104.615	1	0	18,230	18

16.7 CONCLUSIONS

The time-dependent load transfer model and the theoretical solution for calculating the long-term settlement of the LSTH pile are presented. The conclusions are the following:

  i. A ¼-scale model of a Φ15 m LSTH pile was established using FLAC³ᴰ software. The results show that the stress state of the pile at different times corresponds to the disturbance process at different stages under vertical load, and the disturbance process of the LSTH pile has obvious time-dependent characteristics.

ii. The creep behavior of the RI state and FA state is described by the Kelvin model and complete viscoplastic model individually. Combining these two models with the disturbance function, a time-varying load transfer model by DSC is proposed to simulate the creep behavior of the LSTH pile–soil interface. This model can be used to accurately and reliably describe the time-disturbance characteristics of pile–soil interaction.

iii. Based on the proposed DSC load transfer model, the analytical solution of long-term settlement of the LSTH pile is also established. On the basis of a case study, the proposed theoretical method is cross-verified with numerical simulation.

16.8 APPENDIX

For a given set of soil mechanical parameters, the displacement s_{i1} corresponding to a given time t_i under a vertical load P can be obtained by the numerical simulation. An unknown function can be defined:

$$f(P, x_1, x_2, \ldots, x_n, t_i) = s_{i1} \tag{16.43}$$

where $f(P, x_1, x_2, \ldots, x_n, t_i)$ is an unknown function, P is a given vertical load, t_i is a given time, and s_{i1} is a displacement corresponding to a given time t_i. The total differential of the multivariate function can be written as:

$$\frac{\partial f}{\partial P} \cdot dP + \frac{\partial f}{\partial x_1} \cdot dx_1 + \frac{\partial f}{\partial x_2} \cdot dx_2 + \cdots + \frac{\partial f}{\partial x_n} \cdot dx_n + \frac{\partial f}{\partial t_i} \cdot dt_i = ds_i \tag{16.44}$$

where P and t_i are constants, $\dfrac{\partial f}{\partial P} = 0$ and $\dfrac{\partial f}{\partial t_i} = 0$. Δs_i is the difference value of displacement in the given points between the measured curve and the calculated curve, which can be expressed as (Figure 16.15):

$$\left[\frac{\partial f}{\partial x_1}\right]_{(P, t_1)} \cdot dx_1 + \left[\frac{\partial f}{\partial x_2}\right]_{(P, t_1)} \cdot dx_2 + \cdots + \left[\frac{\partial f}{\partial x_n}\right]_{(P, t_1)} \cdot dx_n = \Delta s_1$$

$$\left[\frac{\partial f}{\partial x_1}\right]_{(P, t_2)} \cdot dx_1 + \left[\frac{\partial f}{\partial x_2}\right]_{(P, t_2)} \cdot dx_2 + \cdots + \left[\frac{\partial f}{\partial x_n}\right]_{(P, t_2)} \cdot dx_n = \Delta s_2 \tag{16.45}$$

$$\vdots$$

$$\left[\frac{\partial f}{\partial x_1}\right]_{(P, t_n)} \cdot dx_1 + \left[\frac{\partial f}{\partial x_2}\right]_{(P, t_n)} \cdot dx_2 + \cdots + \left[\frac{\partial f}{\partial x_n}\right]_{(P, t_n)} \cdot dx_n = \Delta s_n$$

Eq. (16.45) can be rewritten in the matrix form:

$$[A] \cdot \{dx_j\} = [\Delta s_i] \tag{16.46}$$

where $\{dx_j\} = \{dx_1, dx_2, \ldots, dx_n\}^{\mathrm{T}}$, the element dx_j is the adjustment value for the parameter x_j, $\{\Delta s_i\} = \{\Delta s_1, \Delta s_2, \ldots, \Delta s_n\}^{\mathrm{T}}$, the matrix $[A]$ can be expressed as:

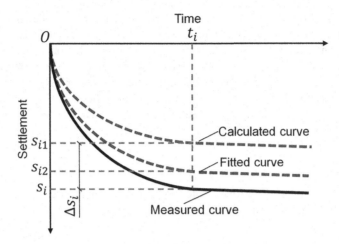

FIGURE 16.15   Relation between pile settlement s and time t under a vertical load.

$$[A] = \begin{bmatrix} A_{11} & A_{12} & \cdots & A_{1n} \\ A_{21} & A_{22} & \cdots & A_{2n} \\ \vdots & \vdots & \vdots & \vdots \\ A_{n1} & A_{n2} & \cdots & A_{nn} \end{bmatrix} \tag{16.47}$$

where $A_{ij} = \left[\dfrac{\partial f}{\partial x_j} \right]_{t_i}$, the element A_{ij} is the sensitivity of settlement to the parameter x_j under a vertical load P at a given time t_i, i is the subscript of time, and j is the subscript of the parameter.

The back-analysis process for the soil parameters is the following (Figure 16.16):

i. An initial set of parameters are input into the initial numerical model, and then the displacement s_i', corresponding to a time t_i under the vertical load P, is numerically calculated.

ii. Define

$$x_j' = x_j + \Delta x_j \tag{16.48}$$

where $\Delta x_j = 10\% x_j$, which is the increment of the parameter x_j. By substituting x_j' for x_j, with numerical simulation with these new parameters, a new displacement s_{ij}' (at the time t_i under a vertical load P) can be obtained.

iii. To obtain the matrix $[A]$, the corresponding element A_{ij} was calculated using Eq. (16.49):

$$A_{ij} = \frac{s_{ij}' - s_i'}{\Delta x_j} \tag{16.49}$$

iv. Solving the equations $[A] \cdot \{dx_j\} = [\Delta s_i]$, the correction value for the parameters $(\{dx_j\})$ can be obtained. Thus, we can calculate x_j':

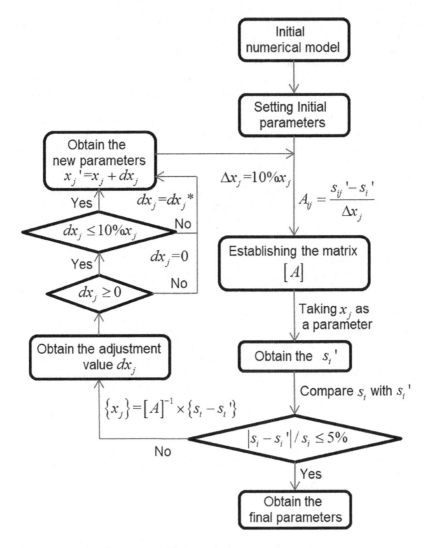

FIGURE 16.16 Process of soil parameters' back analysis.

$$\left\{ x_j' \right\} = \left\{ x_j \right\} + \left\{ dx_j \right\} \tag{16.50}$$

Calculating the modified parameters in the model, a new s-t curve closer to the measured curve can be obtained with the new parameters. The method above can be used to obtain the final parameters.

REFERENCES

Abbas, J.M., Chik, Z.H., Taha, M.R., and Shafiqu, Q.S.M. 2010."Time-dependent lateral response of pile embedded in elasto-plastic soil." *J. Cent. South Univ. Technol.*, 17(2), 372–380. https://doi.org/10.1007/s11771-010-0055-x.

Ai, Z.Y., and Yue, Z.Q. 2009. "Elastic analysis of axially loaded single pile in multilayered soils." *Int. J. Eng. Sci.*, 47(11), 1079–1088. https://doi.org/10.1016/j.ijengsci.2008.07.005.

Bažant, Z. P. 1994. "Nonlocal damage theory based on micromechanics of crack interactions." *J. Eng. Mech.*, 120(3), 593–617. https://doi.org/10.1061/(ASCE)0733-9399(1994)120:3(593).

Booker, J.R., and Poulos, H.G. 1976. "Analysis of creep settlement of pile foundations." *J. Geotech. Eng.*, 102: 1–14.

Bowman, E.T., and Soga, K. 2005. "Mechanisms of setup of displacement piles in sand: laboratory creep tests." *Can. Geotech. J.*, 42(5), 1391–1407. https://doi.org/10.1139/t05-063.

Chen, R.P., Zhou, W.H., and Chen, Y.M. 2009. "Influence of soil consolidation and pile load on the development of negative skin friction of a pile." *Comput. Geotech.*, 36(8), 1265–1271. https://doi.org/10.1016/j.compgeo.2009.05.011.

Cui, J., Li, J., and Zhao, G. 2019. "Long-term time-dependent load-settlement characteristics of a driven pile in clay." *Comput. Geotech.*, 112, 41–50. https://doi.org/10.1016/j.compgeo.2019.04.007.

Danno, K., and Kimura, M. 2009. "Evaluation of long-term displacements of pile foundation using coupled FEM and centrifuge model test." *Soils Found.*, 49(6), 941–958. https://doi.org/10.3208/sandf.49.941.

Deng, D.P., Li, L., Zhao, L.H., and Zou, J.F. 2012. "Back analysis and prediction of settlement-time curve of the pile foundation of beijing-shanghai high-speed railway." *J. Yangtze River Sci. Res. Inst.*, 29(04), 57–63 (in Chinese).

Desai, C.S., and Ma, Y. 1992. "Modelling of joints and interfaces using the disturbed state concept." *Int. J. Numer. Anal. Meth. Geomech.*, 16(9), 623–653. https://doi.org/10.1002/nag.1610160903.

Desai, C.S., and Toth, J. 1996. "Disturbed state constitutive modeling based on stress-strain and nondestructive behavior." *Int. J. Solids Struct.*, 33(11), 1619–1650. https://doi.org/10.1016/0020-7683(95)00115-8.

Desai, C.S. 2001. *Mechanics of Materials and Interfaces: The Disturbed State Concept.* Boca Raton, FL,: CRC Press.

Edil, T.B., and Mochtar, I.B. 1988. "Creep response of model pile in clay." *J. Geotech. Eng.*, 144(11), 1245–1260. https://doi.org/10.1061/(ASCE)0733-9410(1988)114:11(1245).

Fang, T., and Huang, M. 2019. "Deformation and load-bearing characteristics of step-tapered piles in clay under lateral load." *Int. J. Geomech.*, 19(6), 04019053. https://doi.org/10.1061/(ASCE)GM.1943-5622.0001422.

Feng, S.Y., Wei, L.M., He, C.Y., and He, Q. 2014. "A computational method for post-construction settlement of high-speed railway bridge pile foundation considering soil creep effect." *J. Cent. South Univ.*, 21(7), 2921–2927. https://doi.org/10.1007/s11771-014-2258-z.

Feng, S.Y., Li, X., Jiang, F., Lei, L., and Chen, Z. 2017. "A nonlinear approach for time-dependent settlement analysis of a single pile and pile groups." *Soil Mech. Found. Eng.*, 54(1), 7–16. https://doi.org/10.1007/s11204-017-9426-8.

Frantziskonis, G., and Desai, C.S. 1987. "Elasto-plastic model with damage for strain softening geomaterials". *Acta Mech.*, 68(3), 151–170. https://doi.org/10.1007/BF01190880.

Ghazavi, M., and Lavasan, A.A. 2006. "Bearing capacity of tapered and step-tapered piles subjected to axial compressive loading." *7th International Conference On Coasts, Ports & Marine Structures, ICOPMAS*, Tehran, Iran.

Guo, W.D. 2000. "Visco-elastic load transfer models for axially loaded piles." *Int. J. Numer. Anal. Method Geomech.*, 24(2), 135–163. https://doi.org/10.1002/(SICI)1096-9853(200002)24:2<135::AID-NAG56>3.0.CO;2-8.

Huang, M., Jiang, S., Xu, D.X., Deng, T., and Shangguan X. 2018. "Load transfer mechanism and theoretical model of step tapered hollow pile with huge diameter." *Chin. J. Rock Mech. Eng.*, 37(10), 2370–2383 (in Chinese). https://doi.org/10.13722/j.cnki.jrme.2018.0169.

Huang, M., Jiang, S., Xu, C.S., and Xu, D.X. 2020a. "A new theoretical settlement model for large step-tapered hollow piles based on disturbed state concept theory." *Comput. Geotech.*, 124, 103626. https://doi.org/10.1016/j.compgeo.2020.103626.

Huang, W., Liu, D.Y., Zhao, B.Y., and Feng, Y.B. 2014. "Study on the rheological properties and constitutive model of shenzhen mucky soft soil." *J. Eng. Sci. Tech. Rev.*, 7(3), 55–61. https://doi.org/10.3109/02652048.2011.642017.

Huang, M., Zhan, J.W., Xu, C.S., and Jiang, S. 2020b. "New creep constitutive model for soft rocks and its application in the prediction of time-dependent deformation in tunnels." *Int. J. Geomech.*, 20(7), 04020096.

ICGI (Itasca Consulting Group, Inc.). 2018. "Fast Language Analysis of Continua in 3 Dimensions. Version 6.0, Documentation." ICGI, Minneapolis, MN, USA.

Ismael, N.F. 2003. "Load Tests on Straight and Step Tapered Bored Piles in Weakly Cemented Sands." *6th International Symposium on on Field Measurements in Geomechanics (FMGM)*, Oslo, Norway. https://doi.org/10.1201/9781439833483.ch17.

Ismael, N.F. 2010. "Behavior of step tapered bored pile in sand under static lateral loading." *J. Geotech. Geoenviron. Eng.*, 136(5), 669–676. https://doi.org/10.1061/(ASCE)GT.1943-5606.0000265.

Jiang, S., Huang, M., Fang, T., Chen, W., and Shangguan, X. 2019. "A new large step-tapered hollow pile and its bearing capacity." *Proc. Inst. Civil Eng. Geotech. Eng.*, 173(3), 1–37. https://doi.org/10.1680/jgeen.19.00009.

Ju, J., and Francis, T. 2015. "Prediction of the settlement for the vertically loaded pile group using 3D finite element analyses." *Mar. Georesour. Geotechnol.*, 3(33), 264–271. https://doi.org/10.1080/10641 19X.2013.869285.

Li, Z., Wang, K.H., Li, S.H., and Wu, W.B. 2015. "A new approach for time effect analysis in the settlement of single pile in nonlinear viscoelastic soil deposits." *J. Zhejiang Univ. Sci. A (Appl Phys & Eng)*, 16(8), 630–643. https://doi.org/10.1631/jzus.A1400329.

Ling, X.C., and Cai, D.S. 2002. *Rock Mechanics*. Harbin, China: Hitp Harbin Institute of Technology Press. (in Chinese).

Ma, W.B., Rao, Q.H, Li, P., Guo, S.C., and Feng, K. 2014. "Shear creep parameters of simulative soil for deep-sea sediment." *J. Cent. South Univ.*, 21(12), 4682–4689. https://doi.org/10.1007/s11771-014-2477-3.

Michael, S.K. 2001. "A novel approach to predict current stress-strain response of cement based materials in infrastructure." The University of Arizona.

Mishra, A., and Patra, N.R. 2018. "Time-dependent settlement of pile foundations using five-parameter viscoelastic soil models." *Int. J. Geomech.*, 18(5), 04018020. https://doi.org/10.1061/(ASCE) GM.1943-5622.0001122.

Mylonakis, G. 2001. "Winkler modulus for axially loaded piles." *Géotechnique*, 51(5), 455–461. https://doi.org/10.1680/geot.51.5.455.39972.

Nanda, S., and Patra, N.R. 2014. "Theoretical load-transfer curves along piles considering soil nonlinearity." *J. Geotech. Geoenviron. Eng.*, 140(1), 91–101. https://doi.org/10.1061/(ASCE)GT.1943-5606.0000997.

Nian, T.K., Yu, P.C., Liu, C.N., Lu, M.J., and Diao, M.H. 2014. "Consolidation creep test and creep model of dredger fill silty sand." *J. Jilin Univ. Earth Sci. Ed.*, 44(3), 918–924 (in Chinese). https://doi.org/10.13278/j.cnki.jjuese.201403201.

Nie, G.X., and Chen, F. 2005. "Analytical solution for axial loading settlement curve of piles using extensive shear displacement method." *J. Cent. South Univ. (Sci. Technol.)*, 36(1), 163–168. (in Chinese). https://doi.org/10.3969/j.issn.1672-7207.2005.01.034.

Randolph, M.F., and Guo, W.D. 1999. "An efficient approach for settlement prediction of pile groups." *Géotechnique*, 49(2), 161–179. https://doi.org/10.1680/geot.1999.49.2.161.

Salgado, R., Seo, H., and Prezzi, M. 2013. "Variational elastic solution for axially loaded piles in multilayered soil." *Int. J. Numer. Anal. Methods Geomech.*, 37(4), 423–440. https://doi.org/10.1002/nag.1110.

Shibata, T. 1963. "On the volume changes of normally-consolidated clays." *Ann. Disaster Prev. Res. Inst. Kyoto Univ.*, 6(116), 264–266. https://doi.org/10.2472/jsms.12.264.

Small, J.C., and Liu, H.L.S. 2008. "Time-settlement behavior of piled raft foundations using infinite elements." *Comput. Geotech.*, 35(1), 187–195. https://doi.org/10.1016/j.compgeo.2007.04.004.

Taylor, D. W. 1942. "Research report on consolidation of clays." Dept. of Civil and Sanitary Engineering, Massachusetts Institute of Technology, Cambridge, MA, 82.

Tiago, G.S.D., and Adam, B. 2018. "Load-transfer method for piles under axial loading and unloading." *J. Geotech. Geoenviron. Eng.*, 144(1): 04017096. https://doi.org/10.1061/(ASCE)GT.1943-5606.0001808.

Wang, N., Le, Y., Hu, W.T., Fang, T., Zhu, B.T., Geng, D.X., and Xu, C.J. 2020. "New interaction model for the annular zone of stepped piles with respect to their vertical dynamic characteristics." *Comput. Geotech.*, 117, 103256. https://doi.org/10.1016/j.compgeo.2019.103256.

Wu, W.B., Wang, K.H., Zhang, Z.Q., and Chin J.L. 2012. "A new approach for time effect analysis of settlement for single pile based on virtual soil-pile model." *J. Cent. South Univ.*, 19(9), 2656–2662. https://doi.org/10.1007/s11771-012-1324-7.

Yang, Q., Leng W.M., Zhang, S., Nie, R.S., Wei L.M., Zhao, C.Y., and Liu, W.Z. 2014. "Long-term settlement prediction of high-speed railway bridge pile foundation." *J. Cent. South Univ.*, 21(6), 2415–2424. https://doi.org/10.1007/s11771-014-2195-x.

Yavari, N., Tang, A.M., Pereira, J.M., and Hassen, G. 2014. "A simple method for numerical modelling of mechanical behaviour of an energy pile." *Géotech. Lett.*, 4(April-June), 119–124. https://doi.org/10.1680/geolett.13.00053.

Zeng, Q.Y., Zhou, J., and Qu, J.T. 2005. "Method for long-term settlement prediction of pile-foundation in consideration of time effect of stress-strain relationship." *Rock Soil Mech.*, 26(8), 1283–1287 (in Chinese). https://doi.org/10.3969/j.issn.1000-7598.2005.08.019.

Zhang, Q.Q., and Zhang, Z.M. 2012. "Simplified calculation approach for settlement of single pile and pile groups." *J. Comput. Civil Eng.*, 26(6), 750–758. https://doi.org/10.1061/(ASCE)CP.1943-5487.0000167.

Zhang, C., Yang, J.S., Fu, J.Y., Ou, X.F., Xie, Y.P., Dai, Y., and Lei, J.S. 2019. "A new clay-cement composite grouting material for tunnelling in underwater karst area." *J. Cent. South Univ.*, 26, 1863–1873. https://doi.org/10.1007/s11771-019-4140-5.

Zhao, C.Y., Leng, W.M., and Zheng, G.Y. 2013. "Calculation and analysis for the time-dependency of the single-driven pile in double-layered soft clay." *Appl. Clay Sci.*, 79(Jul), 8–12.

Zhu, H., and Chang, M.F. 2002. "Load transfer curves along bored piles considering modulus degradation." *J. Geotech. Geoenviron. Eng.*, 128(9), 764–774. https://doi.org/10.1061/(ASCE)1090-0241(2002)128:9(764).

17 Fluid Flow
Seepage and Consolidation

Ahad Ouria
Dept. of Civil Engineering, University of Mohaghegh Ardabili

Vahab Toufigh
Dept. of Civil Engineering, Sharif University of Technology

17.1 INTRODUCTION

Soil mass as a granular material consists of three phases, including solid particles, pore fluid, and gas. The mechanical behavior of the soil is influenced by the interaction between these three phases. Sophisticated models formulate the mechanical behavior of the soil considering the interaction between these three phases in coupled form, which is complex for practical proposes. In order to avoid the complexity of unsaturated formulations, the fluid flow in unsaturated medium could be formulated based on a pressure-dependent permeability.

The mathematical models for transient seepage and consolidation problems are identical and share a similar equation as[1]:

$$\Delta q = \frac{\Delta V}{\Delta t} \tag{17.1}$$

where q, V, and t are the fluid flux passing through the soil element, volume of the soil element, and time, respectively. The governing partial differential equation for seepage and consolidation problems could be expressed in a general form as[1]:

$$\mathrm{div}\left(k_{(h)}\nabla h\right) = \frac{\partial V}{\partial t} \tag{17.2}$$

where h is the total head and k is the permeability matrix, which could be dependent on the total head.

The governing partial differential equation for a steady-state seepage problem is a special form of Eq. (17.2) with a neglected or zero volume change. Also, transient and steady-state unsaturated seepage problems could be treated using Eq. (17.2) with a permeability depending on the degree of saturation or the total head.[2,3] One of the most important applications of unsaturated flow problems is the determination of phreatic level or free surface within the earth dams or unconfined aquifers, which requires the solutions of Eq. (17.2) in a domain with unknown or unconfined boundaries.[2,3] Two classes of numerical schemes could be employed for determination of the free surface in unconfined domains including (i) linear numerical methods with a constant permeability but a deforming mesh and (ii) nonlinear numerical methods with a variable permeability but a fixed mesh.[2,3] The nonlinear methods are discussed here. In nonlinear methods, the domain is divided into three zones, including a saturated zone under the free surface with a constant permeability, a dry zone with a negligible permeability above the free surface, and a transition zone between the saturated and dry zones in which the permeability of the soil reduces from its value in the saturated zone to the negligible value in the dry zone as a function of the pressure head. Actually, the permeability of the soil in the unsaturated zone depends on the degree of saturation of the soil at

DOI: 10.1201/9781003362081-22

a particular position. In practice, in order to avoid the complexity of the unsaturated formulation, the permeability of the soil in the transition zone is regarded as the pressure head. Different forms of functions have been suggested for permeability of the soil in the transition zone. Desai and Li[2] suggested an idealized piecewise linear function. Figure 17.1 shows the assumptions required for solving the seepage problem in unconfined domains.

Using the assumption shown in Figure 17.1, the permeability of the soil in the transition zone would be[2]:

$$k_{us} = f(p)k_s \tag{17.3}$$

where k_{us} and k_s are the permeability of soil in unsaturated and saturated states, respectively.

The height of the transition zone depends on the type of the soil. A constant value or a pressure-dependent function could be employed for the permeability of the soil within the saturated zone.

In fully saturated soils, the deformation and failure behavior of soil is governed by the interaction of only two solid and fluid phases. One of the most important developments in the theoretical soil mechanics was the introduction of the effective stress concept by Terzaghi[4] in 1923 that opened a new field in geomechanics to study the behavior of saturated soils. According to Terzaghi's theory of consolidation, the volume change of the saturated soil in a time period is equal to the volume of the pore fluid leaving the soil. Assuming a constant coefficient of the compressibility of the solid phase, validity of Darcy's law with constant permeability, one-dimensional deformation, and instantons loading, Terzaghi[4] established a mathematical model and derived the governing partial differential equation (PDE) of the consolidation, which is another form of Eq. (17.1).

$$c_v \frac{\partial^2 u}{\partial z^2} = \frac{\partial u}{\partial t} \tag{17.4}$$

$$c_v = \frac{k}{m_v \gamma_w} \tag{17.5}$$

where
  u is the excess pore water pressure,
  c_v is the coefficient of consolidation,
  k is the permeability,
  m_v is the coefficient of volumetric compressibility, and
  γ_w is the unit weight of the water.

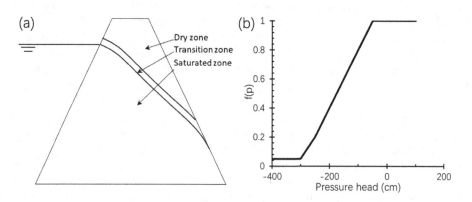

FIGURE 17.1 (a) Dividing the earth dam into three zones, namely, dry, transition, and saturated zones and (b) variation of the permeability function in transition zone[2] with a height of 300 cm.

Although Terzaghi's theory of consolidation is useful for a wide range of practical problems, there are some simplifications in his theory that makes its application restricted to a specific condition.[5] Based on the theory of Terzaghi, the compressibility and permeability of the soil have been assumed as constants, which is not the case in the real problems except for a very small step of loading that their changes could be neglected. Also, it could be inferred from the governing PDE and the coefficient of consolidation that in the Terzaghi's theory of consolidation, the compressibility and the permeability of the soil could have been considered variable but linearly dependent that their ratio remains constant as the void ratio of the soil changes, which is a rare case again.

Several studies have been conducted to implement sophisticated constitutive relationships in the theory of consolidation.[6–14] Gibson et al.,[7,8] derived the governing partial differential equation (PDE) for nonlinear consolidation using the Lagrangian method. They[7,8] took into account material nonlinearity due to variations in compressibility and permeability, as well as geometrical nonlinearity resulting from changes in the length of the drainage path caused by changes in the thickness of the soil layer as:

$$\frac{\partial}{\partial z}\left[\frac{k(e)}{\gamma_f(1+e)}\frac{d\sigma'}{de}\frac{e}{z}\right]+\frac{\partial e}{\partial t}-\left(\frac{\gamma_s}{\gamma_f}-1\right)\frac{d}{de}\left[\frac{k(e)}{(1+e)}\right]\frac{\partial e}{\partial z}=0 \tag{17.6}$$

where γ_f and γ_s are unit weights of fluid and soil solid particles, respectively, σ' is the effective stress, and

$$e = a - c_c \log\left(\frac{\sigma'}{\sigma'_0}\right) \tag{17.7}$$

$$e = b + \mathrm{Mlog}(k) \tag{17.8}$$

where a is the initial void ratio at the reference pressure σ'_0, c_c is the compression index of the soil, b is the void ratio at unit permeability, and M is constant parameters.

Substituting Eq. (17.7) and Eq. (17.8) in Eq. (17.6) results in a highly nonlinear PDE, and Eq. (17.4) could be a special case of that.

Pyrah[15] showed that nonlinear consolidation problems should be formulated using two independent parameters for compressibility and permeability rather than a parameter for the coefficient of consolidation. Abbasi et al.[9] showed that Eq. (17.6) could be expressed similar to Terzaghi's consolidation PDE with a stress-dependent coefficient of consolidations for homogeneous soils. Considering the effects of heterogeneity resulting from independent variations of permeability and compressibility of the soil in different depths of the soil layer, the governing nonlinear PDE for consolidation would be according to:

$$\frac{k_{(z,\sigma')}}{\gamma_w}\frac{\partial^2 u}{\partial z^2}=m_{v(z,\sigma')}\frac{\partial u}{\partial t} \tag{17.9}$$

where $m_{v(\sigma')}$ is the stress-dependent volumetric compression parameter as:

$$m_{v(\sigma')}=\frac{2.3(1+e_0)}{c_c\sigma'} \tag{17.10}$$

As discussed in this chapter, the nonlinearity of the consolidation problem originates from three sources of variations of permeability, compressibility, and length of the drainage path.[16–18] For normally consolidated soils, the variations of these three parameters are continuous. They have known values at the beginning and the end of the consolidation process. Also, their value at any particular time and depth are functions of the effective stress or the degree of the consolidation. Based on the disturbed state concept (DSC),[19] the response of the material at any particular state between two reference states of initially relative intact (RI), and fully adjusted (FA) final, could be represented by a coupling disturbance function. This chapter presents the application of the DSC for the solution of nonlinear flow problems including free surface flow and consolidation problems.

17.2   NUMERICAL SOLUTION FOR FREE SURFACE FLOW

Desai and Li[2] suggested a residual flow procedure for free surface problems in porous media in steady-state as well as transient conditions. Since the permeability of the soil was considered as a function of the pressure head, the resulted equations were nonlinear. Based on their proposed method, the unsaturation effect was implemented in the solution using a residual load vector.

$$[K_s]\{H\} = \{Q\} + \{R_r\} \tag{17.11}$$

where $[K_s]$ is the assembled saturated permeability matrix, $\{H\}$ is the assembled total head vector, $\{Q\}$ is the prescribed flow vector, formulated by the finite element or finite difference methods, and $\{R_r\}$ is the residual load vector that was defined as[2]:

$$\{R_r\} = [K_{us}]^{n-1}\{H\}^{n-1} \tag{17.12}$$

where $[K_{us}]$ is the assembled unsaturated permeability matrix calculated based on Eq. (17.3).
They proposed an iterative scheme to solve the resulted nonlinear equations as[2]:

$$[K_s]\{H\}^n = \{Q\} + [K_{us}]^{n-1}\{H\}^{n-1} \tag{17.13}$$

where n is the iteration number.
For the first iteration, the residual vector is neglected.

$$[K_s]\{H\}^0 = \{Q\} \tag{17.14}$$

17.3   NUMERICAL SOLUTION FOR NONLINEAR CONSOLIDATION

Consolidation of soft soils under a large load step is a nonlinear problem with complex PDEs. Finding a closed-form solution for nonlinear consolidation problem is almost impossible. Although there are several analytical and semi-analytical solutions, most of them have been achieved based on a series of simplified assumptions. Therefore, a numerical scheme is still the only comprehensive method for the solution of these problems. Using an explicate finite difference scheme, Eq. (17.9) could be discretized according to Figure 17.2, as[17]:

$$u_c^{t+\Delta t} = \alpha_B^t \beta_C^t u_B^t + \alpha_T^t \beta_C^t u_T^t +_C^t u_C^t \tag{17.15}$$

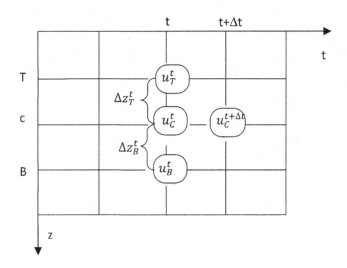

FIGURE 17.2   Finite difference discretization of the consolidation problem.

where

$$\alpha_B^t = \frac{4k_B^t \Delta t}{\gamma_w \left(mv_B^t + kmv_T^t\right)\Delta z_B^t \left(\Delta z_B^t + \Delta z_T^t\right)} \tag{17.16}$$

$$\alpha_T^t = \frac{4k_T^t \Delta t}{\gamma_w \left(mv_B^t + mv_T^t\right)\Delta z_T^t \left(\Delta z_B^t + \Delta z_T^t\right)} \tag{17.17}$$

$$\alpha_C^t = 1 - \alpha_B^t - \alpha_T^t \tag{17.18}$$

The effect of geometrical nonlinearity was implemented in the solution by updating Δz_T^t and Δz_B^t at any time step based on the effective stress. The convergence of this explicate scheme is conditional and subject to $\alpha_B^t \leqslant 0.5$ and $\alpha_T^t \leqslant 0.5$. A computer program was developed in a dimensionless form to work with the initial and final values of the parameters, and then was used for further studies. Geometrical and material nonlinearities have been studied independently to understand the nature of their behavior; finally, their coupled interaction was investigated using the DSC.

17.4   DISTURBED STATE CONCEPT (DSC)

The DSC was introduced by Desai[19] over four decades ago and since then has been developed for a variety of geotechnical and geological problems.[20–26] In the DSC framework, the observed response of material could be described based on the responses of its elements in certain reference states using an appropriate interpolating function called the disturbance function. Two reference states are generally initial and final states. DSC could be considered as a mechanical analogous to (but not exactly) squeeze theorem in mathematics. The DSC assumes that the response of the material to any excitation has resulted from the disturbance that occurred in the microstructure of the material. Therefore, it is supposed that the material in its initial state is RI without any disturbance in its microstructure and in its final state is FA with fully disturbed microstructure. Depending on the problem under consideration, different definitions and analogous could be adapted for the disturbance term. Based on the definitions of the DSC, RI, and FA states, the basic relationship of the DSC is[19]:

$$F_i = (1-D)F_{RI} + DF_{FA} \tag{17.19}$$

where F_i is the response of the material at any arbitrary state between RI and FA states, F_{RI} is the response of the material in a RI state, and F_{FA} is the response of the material in a FA state.

Equation (17.19) in its basic form could be adopted for establishing new solution methods for a wide range of problems, including nonlinear seepage and consolidation problems. In the case of nonlinear consolidation problems with known PDEs, the RI and FA states represent the solutions of the problem with boundary conditions or material properties related to two initial and final states.

17.5 DSC FOR UNCONFINED SEEPAGE AND FREE SURFACE FLOW IN POROUS MEDIA

As shown in Figure 17.1b, the permeability of the soil in the transition zone gradually changes from an initial value in the saturated zone to a final value in the dry zone. The fluid flow in the transitional zone is proportional to the permeability of the soil in this zone. Therefore, the DSC could be applied to this class of nonlinear flow problems.

Desai[19] considered the change of the permeability of the soil in the transition zone as a kind of disturbance as:

$$D_{sk} = \frac{k_s - k_{us}}{k_s - k_f} \tag{17.20}$$

where D_{sk} is the disturbance function related to the change of the permeability of the soil vs. degree of saturation, and k_f is the permeability of the soil in the dry zone. Therefore, the permeability of the unsaturated soil in the transition zone could be defined as:

$$k_{us} = D_{sk} k_f + (1 - D_{sk}) k_s \tag{17.21}$$

In Eq. (17.21), $D_{sk} = 0$ represents the relatively intact state in fully saturate condition under the free surface and $D_{sk} = 1$ represents the FA state in the dry zone.

Using Eq. (17.21) and formulating the PDE of steady-state seepage based on the finite element or the finite difference methods, the final form of matrix equation for soil elements would be:

$$\left[D_{sk} k_f + (1 - D_{sk}) k_s \right] \{H\} = \{Q\} \tag{17.22}$$

Rearranging Eq. (17.22) results:

$$[K_s]\{H\} = \{Q\} + \left[D_{sk} k_s - D_{sk} k_f \right]\{H\} \tag{17.23}$$

Comparing Eq. (17.23) with Eq. (17.11) and Eq. (17.12) reveals that the residual load vector is:

$$\{R_r\} = \left[D_{sk} k_s - D_{sk} k_f \right]\{H\} \tag{17.24}$$

Therefore, it can be seen that the residual flow procedure proposed by Desai and Li[2] is identical to the DSC introduced by Desai[19] for unsaturated flow. The iterative procedure shown in Eq. (17.13) and Eq. (17.14) could also be used for solving Eq. (17.22). The location of the free surface would be determined based on the fact that the pressure head at the free surface is zero or the total head is equal to the elevation head. Figure 17.3 shows the location of the free surface calculated for an isotropic and homogenous earth dam calculated by the DSC-based nonlinear finite element method and an adaptive mesh finite element method proposed by Ouria and Toufigh.[3] As can be seen, the locations of the free surfaces calculated by both methods are in the same range despite the fact that these two methods have employed different assumptions and procedures.

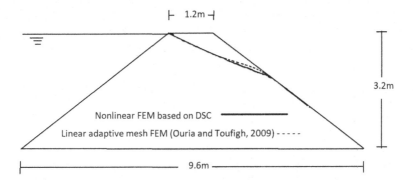

FIGURE 17.3 Comparing the location of the free surface within an earth dam calculated by the disturbed state concept (DSC)-based nonlinear and an adaptive mesh linear FEM method.[3]

17.6 DSC FOR NONLINEAR CONSOLIDATION DUE TO MONOTONIC LOADING

17.6.1 DSC AND LARGE STRAIN CONSOLIDATION DUE TO GEOMETRICAL NONLINEARITY

During the consolidation process, the effective stress increases in the depth of the soil layer as the degree of consolidation increases. Increasing the effective stress compresses the solid soil skeleton and reduces the thicknesses of the soil layer. If the total settlement of the soil layer is too large to be neglected, the effect of the change of the length of the drainage path should be considered in the solutions.[8] It was investigated by the following example. The consolidation of a single drained soil layer with unit height and unit coefficient of consolidation is considered in this example. It is supposed that the compressibility and the permeability of the soil layer remain constant during the consolidation process. The thickness of the soil layer reduces by 50% after the completion of the consolidation. Three analyses are conducted with three different conditions: (i) linear analysis based on Terzaghi's theory of consolidation with a constant thickness of the soil layer in its initial value of Hd=1 m; (ii) another linear analysis based on Terzaghi's theory of consolidation again with a constant thickness of the soil layer in its final value of Hd=0.5 m; and (iii) a nonlinear analysis with an initial thickness of 1m and the final thickness of 0.5 m. The results are shown in Figure 17.4. As can be seen in Figure 17.4, the result of the nonlinear analysis coincides with the result of the linear analysis with the initial thickness of the soil layer. As the degree of consolidation increases, the results of the nonlinear analysis approach the results of the linear analysis with the final thickness of the soil layer. These results resemble the DSC. The result of the nonlinear analysis at any particular time is between the results of two linear analyses with initial and final conditions. Therefore, the results of linear calculations with the initial thickness of the soil layer corresponded to the RI state, and the results of the linear calculations with the final thickness at any particular time corresponded to the FA state.

Finding a coupling mechanism to relate the results of the nonlinear behavior to the results of two linear behaviors in RI and FA states provides us with a simplified solution for large strain nonlinear consolidation problems based on Terzaghi's conventional theory of consolidation. The coupling mechanism is the disturbance function. Determination of the disturbance function is the most important step in the construction of a solution method for a problem using the DSC. Both mathematical procedure and experimental method could be employed for the determination of the form of the disturbance function.[14] Using the basic relationship of the DSC (Eq. 17.19), the disturbance function could be calculated at any particular time based on the results of the numerical solutions according to Eq. (17.25). The form of the calculated disturbance function is illustrated in Figure 17.5 for large strain consolidation with variable thickness of the soil layer.

$$U(t)_i = (1 - D_c)U(t)_{RI} + D_c U(t)_{FA} \qquad (17.24)$$

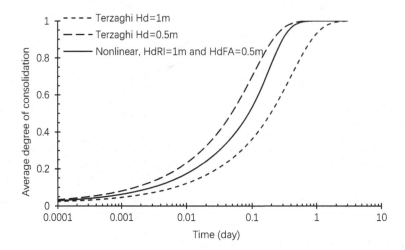

FIGURE 17.4 Average degrees of consolidation resulted from two linear solutions with initial and final thickness of the soil layer, and nonlinear solution with variable thickness of the soil varying from the initial thickness to the final thickness.

$$D_c = \frac{U(t)_i - U(t)_{RI}}{U(t)_{FA} - U(t)_{RI}} \qquad (17.25)$$

As shown in Figure 17.5, the plot of the disturbance function calculated for implementing the effect of geometrical nonlinearity resembles the shape of sigmoid functions. Sigmoid functions have great application in business, finance, machine learning, artificial intelligence, and neural networks.[27] Considering the form of the disturbance function, Eq. 17.26 is proposed for disturbance function for geometrical nonlinearity.

$$D_{gc} = \frac{1}{1 + e^{-\alpha_g(t - \beta_g)}} \qquad (17.26)$$

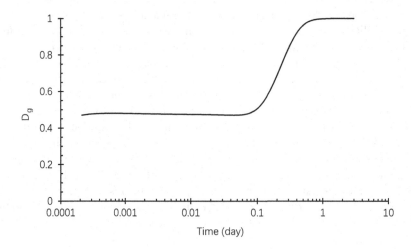

FIGURE 17.5 Form of the disturbance function calculated from the results of the numerical method.

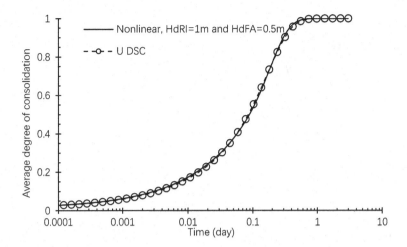

FIGURE 17.6 Results of nonlinear solution and the disturbed state concept (DSC)-based solution by Terzaghi's conventional theory considering the effect of nonlinearity due to the change of the layer thickness.

where D_{gc} is the consolidation disturbance function for geometrical nonlinearity, and α_g, and β_g are disturbance parameters.

Figure 17.6 shows the results of nonlinear calculations using the finite difference method and the results of calculations based on the DSC using Eqs. (17.24) and (17.26) with $\alpha_g = 6$ and $\beta_g = 0.02$. As can be seen, the results of the DSC-based solution are as accurate as the nonlinear finite difference method.

17.6.2 DSC and Nonlinear Consolidation Due to Material Nonlinearity

The effect of material nonlinearity on the consolidation process is investigated in a similar method with the investigation of geometrical nonlinearity. The effect of the changes of the compressibility and permeability on the consolidation process is shown in Figures 17.7 and 17.8, respectively. For the investigation of the effect of the change of compressibility, the permeability and the thickness of the soil were kept constant. For the investigation of the effect of permeability, the compressibility and the thickness of the soil were kept constant during the calculations. As can be seen in Figure 17.7a and b, the form of the consolidation disturbance function calculated for the effect of the change of compressibility and permeability are similar to the disturbance function calculated for the effect of the change of the thickness of the soil layer.

As can be seen in Figure 17.8b, the consolidation disturbance function related to permeability is a descending function that could be generated by adjusting the sign of the parameters. Therefore, disturbance functions for material nonlinearities that originated from the changes of compressibility and permeability are similar to disturbance function required for implementing the effects of the geometrical nonlinearity in the consolidation problems.

$$D_{mvc} = \frac{1}{1 + e^{-\alpha_{mv}(t - \beta_{mv})}}$$ (17.27)

$$D_{kc} = \frac{1}{1 + e^{-\alpha_k(t - \beta_k)}}$$ (17.28)

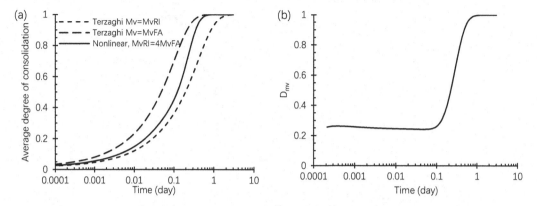

FIGURE 17.7 (a) Results of nonlinear consolidation analysis with variable compressibility, linear analysis with initial (RI), and final (FA) values of compressibility and (b) the calculated consolidation disturbance function for material nonlinearity that resulted from stress-dependent compressibility.

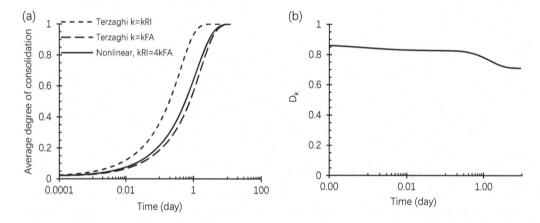

FIGURE 17.8 (a) Results of nonlinear consolidation analysis with variable permeability, linear analysis with initial (RI), and final (FA) values of permeability and (b) the calculated disturbance function for material nonlinearity that resulted from stress-dependent permeability.

Using $\alpha_{mv} = 7.86$ and $\beta_{mv} = 0.22$ in Eq. (17.27) and $\alpha_k = -0.2$ and $\beta_k = 8$ in Eq. (17.28), the average degrees of consolidation were calculated based on the DSC with deviations less than 0.01 for both variable compressibility (Figure 17.9a) and variable permeability (Figure 17.9b). These results show the potential of application of the DSC for the solution of nonlinear problems of seepage and consolidation.

17.6.3 DSC and Nonlinear Consolidation Due to Coupled Effect of Geometrical and Material Nonlinearities

As can be seen in Eqs. (17.26)–(17.28), the form of the disturbance functions for the changes of thickness of the soil layer (D_{cg}), reduction of the permeability (D_{ck}), and the compressibility (D_{cmv}) are similar but with different parameters. It must be noted that all these parameters are functions of the effective stress and the degree of consolidation; therefore, a similar disturbance function could be utilized for predicting their coupling effect on the nonlinearity of the consolidation problem.

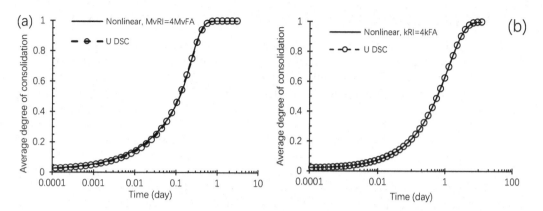

FIGURE 17.9 (a) Results of nonlinear consolidation analysis with variable permeability, linear analysis with initial relativity intact (RI), and final fully adjusted (FA) values of permeability and (b) the calculated disturbance function for material nonlinearity that resulted from stress-dependent permeability.

The following function is proposed for the solution of the nonlinear consolidation problems, including the material and geometrical nonlinearities.

$$D_c = D_c \left(D_{cg}, D_{cmv}, D_{ck} \right) = \frac{1}{1 + e^{-\alpha(t-\beta)}} \tag{17.29}$$

The results of a nonlinear consolidation analysis of a soft clay layer are shown in Figure 17.10. The initial values of layer thickness, coefficient of permeability, and coefficient of compressibility were of 1 m, 1 m/day, and 0.1 kPa^{-1}, respectively. The final values of these parameters were 0.5 m, 0.1 m/day, and 0.01 kPa^{-1}, respectively. The result of the DSC-based solution with $\alpha = 6$ and $\beta = 0.22$ coincided with the results of the nonlinear finite difference method with less than 1% deviation.

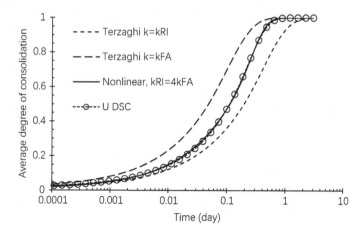

FIGURE 17.10 Comparing the results of a nonlinear consolidation problem with stress-dependent thickness, compressibility, and permeability obtained from the nonlinear finite difference method and the simplified solution based on the disturbed state concept (DSC) and Terzaghi's conventional theory of consolidation.

17.7   DSC FOR NONLINEAR CONSOLIDATION DUE TO CYCLIC LOADING

Nonlinear consolidation of normally consolidated soils under cyclic loading is a very complex phenomenon when the changes of the material properties during successive cycles of loading and unloading are considerable.[28–31] For highly overconsolidated soils subjected to cyclic loading with a small amplitude, changes of the void ratio and the permeability of the soil at any cycle of loading could be neglected, and also the stress–strain response of the material could be considered as elastic.[29] Therefore, numerical and analytical solutions could be achieved with less effort when compared to the normally consolidated soils. There are several approximate solutions for the consolidation of plastic soils under cyclic loading that have been achieved using several simplified assumptions.[28–30] Using the numerical methods for better approximation of the consolidation response of normally consolidated plastic soils under cyclic loading is inevitable.[31]

The consolidation response of a normally consolidated soil under a square wave cyclic loading is illustrated in Figure 17.11. The resulting degree of the consolidation should be a periodic function with limited amplitude at any cycle of loading, according to Figure 17.11b. Depending on the material properties and the period of cyclic loading, the change of the effective stress level at any cycle of loading would only be a fragment of the loading amplitude. At the first cycle of loading, the soil is in an overconsolidated state before the beginning of the unloading half cycle according to path 1–2 in Figure 17.11c. Neglecting the heterogeneity resulting from the change of the effective stress in the depth of the soil layer, after the first cycle, the state of the soil changes from the overconsolidated state at the beginning of any loading cycle into the normally consolidated state when the effective stress reaches the precompression pressure (points 2 and 4 in Figure 17.11 and point 4 in Figure 17.11d). The value of the precompression pressure changes in any loading cycle and depends on the maximum degree of consolidation at the end of the last loading cycle. There is considerable change in the compressibility coefficient of the soil when the state of the soil changes from the normally consolidated to the overconsolidated and vice versa, according to Figure 17.11d. Therefore, an appropriate adjustment is required to implement the effect of the state change of the soil in the solutions. Changes of the permeability of the soil under the cyclic loading are illustrated in Figure 17.11e.

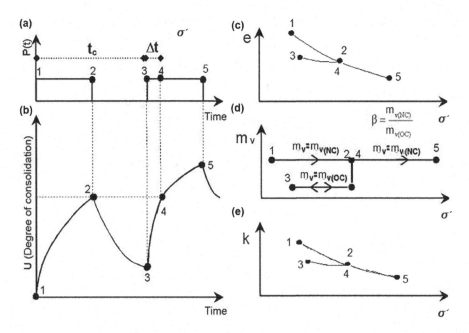

FIGURE 17.11   Schematic representation of the consolidation behavior of plastic soil under square wave cyclic loading: (a) square wave, (b) average degree of consolidation due to square wave loading, (c) change of the void ratio of the plastic soil under cyclic loading, (d) changes of the compressibility of the soil under cyclic loading, and (e) changes of the permeability of the soil under cyclic loading.

The permeability of the soil depends on the void ratio regardless of the state of the soil.[32] For cyclic loading with a relatively short period, the permeability of the soil decreases during the first half cycle and increases during the second half cycle of loading. The average permeability of the soil during any full cycle of loading could be assumed to be constant but decreases as the number of cycles increases.

Ouria,[17] Ouria and Toufigh,[33] and Toufigh and Ouria[34] assumed a constant consolidation coefficient by introducing the concept of virtual time and implemented the effect of the change of the coefficient of consolidation in the solution using a transformation in the time domain. Using the same argument and neglecting the remaining effective stress at the end of each full cycle of loading, each loading cycle could be considered independent from the previous cycles. The average degree of consolidation at the end of any half cycle of loading is equal to the maximum degree of consolidation at the end of the last half cycle of loading, thereby equating their time factors results:

$$\frac{\left(c_{v(NC)}\right)\left(\frac{tc}{2}\right)}{H_d^2} = \frac{\left(c_{v(OC)}\right)(\Delta t)}{H_d^2} \tag{17.30}$$

where H_d is the length of the drainage path in the clay layer and $c_{v(NC)} = \eta c_{v(OC)}$.

Subsuming the normally consolidated coefficient of consolidation with overconsolidated coefficient in Eq. (17.30) and adjusting the time results:

$$\frac{c_{v(NC)}t}{H_d^2} = \frac{\eta c_{v(OC)}t}{H_d^2} = \frac{c_{v(OC)}\eta t}{H_d^2} = \frac{c_{v(OC)}t'}{H_d^2} \tag{17.31}$$

where t' is the virtual or the adjusted time.

Using the normally consolidated value of c_v in the right side of Eq. (17.30) for any cycle of loading results:

$$\Delta t_N = \eta \frac{t_2}{2} \tag{17.32}$$

And, for any two following cycles will be as follows:

$$\Delta t_N = \eta t'_{N-2} \tag{17.33}$$

Therefore, virtual time for loading half-cycles using the normally consolidated value of c_v would be[34]:

$$t'_N = \frac{\Delta t'_N}{\eta} + \left(\frac{t_c}{2} - \Delta t'_N\right) \tag{17.34}$$

Extending Eq. (17.21) results[31]:

$$t'_N = \frac{\Delta t'_N}{\eta} + \left(\frac{t_c}{2} - \Delta t'_N\right) = \frac{\beta\left(\frac{\Delta t'_{N-2}}{\eta} + \left(\frac{t_c}{2} - \Delta t'_{N-2}\right)\right)}{\eta} + \left[\frac{t_c}{2} - \eta\left(\frac{\Delta t'_{N-2}}{\eta} + \left(\frac{t_c}{2} - \Delta t'_{N-2}\right)\right)\right] \tag{17.35a}$$

$$t'_N = \frac{t_c}{2\eta}\left[1 - (1-\eta)^{\frac{N+1}{2}}\right], \quad N = 3, 5, 7, \dots \tag{17.35b}$$

Recalling Eqs. (17.30) and (17.35b) could be used to calculate an equivalent value for the coefficient of consolidation for any cycle.

$$c_v' = \frac{c_v}{\beta}\left(1-(1-\eta)^{\frac{N+1}{2}}\right) \text{ or } t' = \frac{t_c}{2\eta}\left(1-(1-\eta)^{\frac{N+1}{2}}\right) \tag{17.36}$$

where c_v' is the equivalent coefficient of consolidation in which the effect of the change of the soil state from overconsolidated to normally consolidated has been implemented. The equivalent coefficient of consolidation for loading half-cycles and the coefficient of consolidation for unloading half-cycles for a soil sample with $c_{v(NC)} = 0.02 \text{ m}^2/\text{day}$ and $\eta = 0.1$ are shown in Figure 17.12.

As can be seen in Figure 17.12, the equivalent coefficient of consolidation has a constant value in unloading half-cycles. In loading half-cycles, its value starts from the value of the coefficient of consolidation in the normally consolidated state, gradually increases and approaches the value of the coefficient of consolidation in the overconsolidated state.

Figure 17.13 shows the results of the consolidation problem under cyclic loading solved by the finite difference method[21] with $H_d = 2.5 \text{ m}$ and $t_c = 10$ minutes in three different conditions: (i) linear condition with a constant coefficient of consolidation of $c_v = 0.02 \text{ m}^2/\text{minute}$, (ii) linear condition with a constant coefficient of consolidation of $c_v = 0.2 \text{ m}^2/\text{minute}$, and (iii) nonlinear condition with $c_{v(NC)} = 0.02 \text{ m}^2/\text{minute}$ and $c_{v(OC)} = 0.2 \text{ m}^2/\text{minute}$. For a simple comparison, only the upper bounds of solutions at any cycle of loading (the maximum degree of consolidation at each cycle) are shown. As can be seen, the results of nonlinear consolidation under cyclic loading with a variable coefficient of consolidation coincided with the results of linear consolidation under the same cyclic loading with a constant coefficient of consolidation in normally consolidated state at the beginning, and as time passed, its results coincided with the results of the linear problem with overconsolidated parameters.

The results shown in Figures 17.12 and 17.13 represent the basic definition of the DSC. For consolidation of normally consolidated soils under square wave cyclic loading, the coefficient of consolidation starts from an initial value (RI) and gradually approaches the finial value (FA). The response of the problem shown in Figure 17.11 is proportional to the equivalent coefficient of consolidation. In this case, the disturbance is comparable to the change of the equivalent coefficient of consolidation from the normally consolidated to the overconsolidated state. Ouria et al.[23] proposed the following disturbance function for the consolidation of plastic clays under cyclic loading:

$$D_{cc} = \left[1-(1-\eta)^{0.5T_c\xi t}\right] \tag{17.37}$$

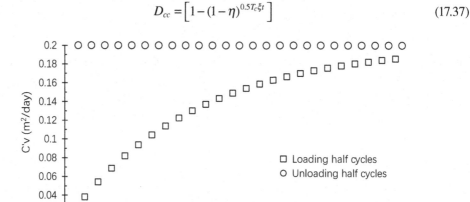

FIGURE 17.12 Graphical representation of the equivalent coefficient of consolidation.

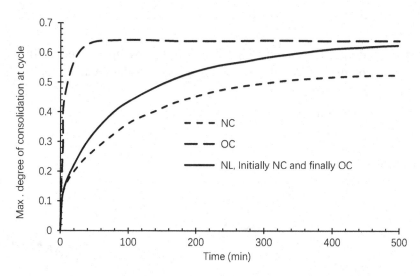

FIGURE 17.13 Consolidation under cyclic loading in linear conditions of initially NC, finally OC, and nonlinear conditions of initially NC, finally OC.

where T_c is the time factor for one cycle of loading, and ξ is a model parameter. Its value is $0 \leq \xi \leq 1$.

Using the basic relationship of the DSC, the average degree of consolidation of a plastic clay layer under square wave cyclic loading is calculated as[23]:

$$U = (1 - D_{cc}) U_{\text{RI(NC)}} + D_{cc} U_{\text{FA(OC)}} \tag{17.38}$$

Using Eq. (17.25), the nonlinear consolidation of plastic soils under square wave cyclic loading could be treated as a linear consolidation problem with constant material properties. Since the problem has been changed to the problem of a linear system, the superposition principle could be implemented for linear consolidation as[29]:

$$U_N = (-1)^N \sum_{n=1}^{N} (-1)^n U_{\left(t = n\frac{t_c}{2}\right)} \tag{17.39}$$

where N is the number of half-cycles.

For the sake of simplicity, the degree of consolidation is calculated at the end of each half cycle in Eq. (17.39). Substituting Eq. (17.39) into Eq. (17.38) for normally consolidated and overconsolidated states, the degree of consolidation in nonlinear conditions at the end of half-cycles would be[23]:

$$U_{\text{CN}} = (1 - D_{cc}) \left[(-1)^N \sum_{n=1}^{N} (-1)^n U_{\text{RI(NC)}} \left(t = \frac{nt_c}{2}\right) \right] + D_{cc} \left[(-1)^N \sum_{n=1}^{N} (-1)^n U_{\text{FA(OC)}} \left(t = \frac{nt_c}{2}\right) \right] \tag{17.40}$$

and the settlements at the end of half-cycles would be[23]:

$$\Delta S_N = \Delta U. m_{v(\text{NC})}.p.H, \quad N = 1, 3, 5, \ldots \tag{17.41}$$

$$\Delta S_N = \Delta U m_{v(\text{NC})} \alpha.p.H, \quad N = 2, 4, 6, \ldots \tag{17.42}$$

TABLE 17.1

Material Properties and Loading Conditions for Examples Used for the Verification of DSC-Based Solution for the Consolidation under Cyclic Loading with Nonlinear FDM

	$Cv(m^2/day)$	Tc (day)	β	$Hd(m)$	$mv(kPa^{-1})$	α	$P(kPa)$
Case1	0.02	30	0.1	5	0.0005	0.1	100
Case2	0.01	20	0.15	2	0.002	0.15	100

DSC, disturbed state concept.

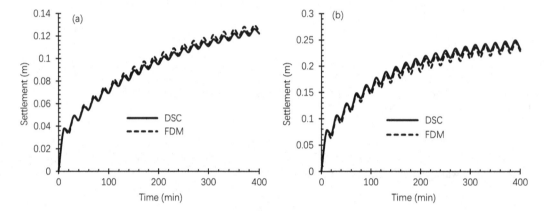

FIGURE 17.14 Comparing the results of nonlinear consolidation under cyclic loading calculated by the disturbed state concept (DSC)-based solution and nonlinear finite difference for (a) example case 1 and (b) example case 2.

The ability of the DSC for predicting the consolidation settlement of normally consolidated clays is verified by the results of the nonlinear finite difference[34] method for two cases with material properties and loading condition according to Table 17.1.

The results of calculations have been shown in Figures 17.14a and b.

As can be seen in Figure 17.14, the results of the method based on the DSC are in good correspondence with the results of the nonlinear finite difference method.

17.8 CONCLUSIONS

This chapter presented solution methods for nonlinear seepage and consolidation problems based on the DSC. Determination of the location of the free surface in unconfined seepage problems was conducted based on the DSC. Although there are several numerical methods for the solution of nonlinear seepage and consolidation problems, the application of nonlinear numerical methods requires complicated algorithms and sophisticated computer codes. In spite of numerical methods, the methods developed based on the DSC, are effortless and require only elementary mathematical operations. The methods introduced in this chapter involve the adoption of the DSC for residual flow procedure for the determination of the free surface within the earth dams.

Also, the DSC was used to adopt the solutions of Terzaghi's conventional theory of consolidation to achieve the solutions of the nonlinear consolidation problem with nonlinear material properties and geometry due to monotonic and cyclic loading. Unlike many other applications of the DSC, in this chapter, the DSC was employed to obtain solutions for certain forms of complicated nonlinear PDEs related to nonlinear seepage and consolidation problems. Therefore, it could be concluded that the application of the DSC could be developed to a wide range of problems in different fields of engineering and science.

REFERENCES

1. Terzaghi, K., Peck, R. B., and Mesri, G. *Soil Mechanics in Engineering Practice*, 3th ed. John Wiley, Hoboken, NJ, 1996.
2. Deasi, C.S., and Li, G.C., (1983) "A residual flow procedure and application for free surface flow in porous media." *Advances in Water Resources*, 6(1), 27–35.
3. Ouria, A., and Toufigh, M. M., (2009). "Application of Nelder-Mead simplex method for unconfined seepage problems." *Applied Mathematical Modelling*, 33(9), 3589–3598.
4. Terzaghi, K. *Erdbaumechanik Aufbodenphysikalischer Grundlage*, Franz Deuticke, Leipzig, p. 689 (1925)
5. Selvadurai, A. P. S. (2021). "Irreversibility of soil skeletal deformations: The Pedagogical Limitations of Terzaghi's celebrated model for soil consolidation." *Computers and Geotechnics*, Elsevier BV, 135, 104137.
6. Biot, M. A. (1941). "General theory of three-dimensional consolidation." *Journal of Applied Physics, American Institute of Physics*, 12(2), 155–164.
7. Gibson, R. E., England, G. L., and Hussey, M. J. L. (1967). "The theory of one-dimensional consolidation of saturated clays." *Géotechnique*, 17(3), 261–273.
8. Gibson, R. E., Schiffman, R. L., and Cargill, K. W. (1981). "The theory of one-dimensional consolidation of saturated clays. II. Finite nonlinear consolidation of thick homogeneous layers." *Canadian Geotechnical Journal, NRC Research Press*, 18(2), 280–293.
9. Abbasi, N., Rahimi, H., Javadi, A. A., and Fakher, A. (2007). "Finite difference approach for consolidation with variable compressibility and permeability." *Computers and Geotechnics*, 34(1), 41–52.
10. Abuel-Naga, H. M., and Pender, M. J. (2012). "Modified Terzaghi consolidation curves with effective stress-dependent coefficient of consolidation." *Géotechnique Letters*, 2(2), 43–48.
11. An-Feng, H., Chang-Qing, X., Jun, C., Chuan-Xun, L., and Kang-He, X. (2018). "Nonlinear consolidation analysis of natural structured clays under time-dependent loading." *International Journal of Geomechanics, American Society of Civil Engineers*, 18(2), 4017140.
12. Bekele, Y. W. (2021). "Physics-informed deep learning for one-dimensional consolidation." *Journal of Rock Mechanics and Geotechnical Engineering*, 13(2), 420–430.
13. Liu, Q., Deng, Y.-B., and Wang, T.-Y. (2018). "One-dimensional nonlinear consolidation theory for soft ground considering secondary consolidation and the thermal effect." *Computers and Geotechnics*, 104, 22–28.
14. Hall, K. M., and Fox, P. J. (2018). "Large strain consolidation model for land subsidence." *International Journal of Geomechanics, American Society of Civil Engineers*, 18(11), 6018028.
15. Pyrah, I. C. (1996). "One-dimensional consolidation of layered soils." *Géotechnique* 46(3), 555–560.
16. Ouria, A., (2004). "Numerical Modelling of Land Subsidence due to Periodic Changes of Groundwater Level." MSc thesis, University of Kerman, Kerman (in Persian).
17. Ouria, A. (2009). "Alternative Approaches to Nonlinear Consolidation and Seepage Problems." Ph.D thesis, University of Kerman, Kerman (in Persian).
18. Ouria, A., Toufigh, M.M., Fahmi, A. (2011). "Nonlinear analysis of transient seepage by the coupled finite element method." *International Journal of Mechanics*, 5(1), 35–39.
19. Desai, C. S. *Mechanics of Materials and Interfaces: The Disturbed State Concept*. Boca Raton, FL: CRC 2001.
20. Desai, C. S., and Wang, Z. (2003). "Disturbed state model for porous saturated materials." *International Journal of Geomechanics, American Society of Civil Engineers*, 3(2), 260–265.
21. Ouria, A. (2017). "Disturbed state concept-based constitutive model for structured soils." *International Journal of Geomechanics*, 17(7). doi:10.1061/(ASCE)GM.1943-5622.0000883.
22. Ouria, A., and Behboodi, T. (2017). "Compressibility of cement treated soft soils." *Journal of Civil and Environmental Engineering*, 47.1(86), 1–9.
23. Ouria, A., Desai, C. S., and Toufigh, V. (2015). "Disturbed state concept-based solution for consolidation of plastic clays under cyclic loading." *International Journal of Geomechanics*, 15(1). doi: 10.1061/ (ASCE)GM.1943-5622.0000336.
24. Ouria, A., Ranjbarnia, M., and Vaezipour, D. (2018). "A failure criterion for weak cemented soils." *Journal of Civil and Environmental Engineering*, 48.3(92), 13–21.
25. Farsijani, A., and Ouria, A. (2021). "Constitutive modeling the stress-strain and failure behavior of structured soils based on HISS model." *IQBQ*, 21(4), 231–250. https://mcej.modares.ac.ir/article-16-52042-en.html.
26. Farsijani, A., and Ouria, A. (2022). "Wetting-induced collapse behavior of unsaturated soils in disturbed state concept framework." *International Journal of Geomechanics*, 22, 4022014. doi: 10.1061/(ASCE) GM.1943-5622.0002327.

27. Anil, M., Kishan, M., Chilukuri, K. M., and Sanjay, R. (1996). "Characterization of a class of sigmoid functions with applications to neural networks." *Neural Networks* 9(5), 819–835.

28. Wilson, N. E., and Elgohary, M. M. (1974). "Consolidation of clays under cyclic loading." *Canadian Geotechnical Journal*, 11(3), 420–433.

29. Baligh, M. M., and Levadoux, J. (1978). "Consolidation theory for cyclic loading." *Journal of the Geotechnical Engineering Division, ASCE*, 104(4), 415–431.

30. Rahal, M.A, and Vueza, A.R. (1998). "Analysis of settlement and pore pressure induced by cyclic loading of a silo." *Journal of Geotechnical and Geoenvironmental Engineering,* 124, 1208–1210.

31. Ouria, A., and Toufigh, M. M., (2010). "Prediction of land subsidence under cyclic pumping based on laboratory and numerical simulations." *Journal of Geotechnical and Geological Engineering*, 28(2), 165–175.

32. Nagaraj, T.S., Pandian, N. S., and Narashimha Raju, P. S. R., (1993). "Stress state-permeability relationships for fine-grained soils." *Geotechnique*, 43(2), 333–336.

33. Ouria, A., and Toufigh, M. M. (2008). "An approximate solution for consolidation of inelastic clays under rectangular cyclic loading." *Journal of Applied Sciences*, 8(11), 2075–2082.

34. Toufigh, M. M., and Ouria, A. (2009). "Consolidation of inelastic clays under rectangular cyclic loading." *Soil Dynamics and Earthquake Engineering*, 29(2), 356–363.

18 A New Theoretical Settlement Model for Large Step-Tapered Hollow Piles Based on the Disturbed State Concept Theory

Ming Huang
College of Civil Engineering, Fuzhou University

Song Jiang
College of Civil Engineering, Fujian University of Technology

Chaoshui Xu
School of Civil, Environment and Mining Engineering,
University of Adelaide

Dexiang Xu
Urban Planning Design Institute of Ganzhou

18.1 INTRODUCTION

Conventional treatments when karst caves appear under bridge foundations include, for example, filling or grouting of the caves and wall protection by steel pipes. However, for bored piles, there are additional serious issues including mud loss, hole collapse and drill jamming, which can potentially endanger the safety of the project. Most previous studies on pile foundations in karst areas focus on the bearing capacity and failure characteristics of conventional piles under the influence of karst caves, leading especially to the calculation of a safe thickness required of the karst cave roof (Huang et al., 2017). The problem of pile foundations in karst areas has not been fundamentally solved due to the complexity and variability of caves.

The Shenzhen Bridge is an urban bridge with a length of 2,600 m, overpassing three railway lines in Ji'an, Jiangxi. Within 1.5 years (from 2012) of the commencement of engineering work, two complex problems were discovered, resulting in the halt of the bridge construction (Figure 18.1). Firstly, there were significant well-developed karst caves underneath the primal piers L1–L3. Further geological explorations revealed that there was no stable rock stratum that could be used as a stable holding layer for the bored pile even when the drilling depth reached 80 m, which made the construction of the bored pile extremely difficult. Secondly, the primal pier L1 was too close to Jing-Ji railway, with their shortest horizontal distance at only 1.6 m. In this situation, the stability of the railway would be compromised by the application of a percussive drill. In March 2013, two new large step-tapered hollow piles (LSTHPs) with diameters of 14 and 15 m, respectively, were used as the foundations for the Shenzhen Bridge. Pier L1 and L2 were relocated to sit on top of the 15 m diameter LSTHP (Figure 18.1). The percussion drill was replaced by a small excavator to reduce the

DOI: 10.1201/9781003362081-23

FIGURE 18.1   Cross-sectional view of the complex karst strata (a) and the bridge construction site (b).

ground vibration to ensure the stability of the railway. After adopting the LSTHP, the construction work continued and was completed quickly without any further disruption.

The LSTHP is a kind of bridge pile foundation used to tackle problems in complex karst strata. Compared with the normal uniform-section pile, the LSTHP with a diameter exceeding 14 m provides better side friction and vertical resistance, i.e., it can bear the much greater vertical load. It has a lower weight due to its hollow interior design, which can reduce the additional pressure at the pile tip. The cross section of the LSTHP is reduced stepwise with depth, which inherits all the advantages of a traditional step-tapered pile (Huang et al., 2018; Jiang et al., 2019). The large diameter of varied cross sections causes the transfer of load from the pile to the upper soil strata and therefore the stress created in front of the pile tip on the roof of karst caves is relatively small, so cave collapse can be avoided. The large diameter and relatively small slenderness ratio of the LSTHP make it look similar to the open caisson foundation, raft foundation or pier foundation. However, their basic load transfer mechanisms and construction methods are different. Because of the differences in the origin and bearing characteristics of the LSTHP, it is classified as a hand-dug pile (Jiang et al., 2019).

As a novel type of foundation, the LSTHP has been increasingly used as the foundation for bridge constructions in soft grounds in recent years. There are several published studies concerning the bearing characteristics of the pile under static or dynamic load (Ismael, 2003, 2010; Ghazavi and Lavasan, 2006; Wang et al., 2020; Fang and Huang, 2019). Jiang et al. (2019) introduced the construction method of the LSTHP and assessed its vertical bearing capacity using FLAC^{3D} software by considering the influences of width and number of varied cross sections as well as the soil characteristics around the pile. Huang et al. (2018) considered the enhancement effect of the shaft resistance in the lower part of varied cross sections of the pile, derived the theoretical solution of the enhancement coefficient and proposed a load transfer model based on the disturbed state concept (DSC) theory. However, there is a lack of research into the pile–soil interaction and load transfer mechanism, especially for a pile of such a large diameter with several varied cross sections. Pile settlement is an important factor for assessing the bearing capacity of piles and several calculation methods are available for settlement estimation of single conventional piles including, for example, the shear displacement method (Nie and Chen, 2005), load transfer method (Tiago and Adam, 2018; Zhang and Zhang, 2012), elasticity method (Ai and Yue, 2009; Salgado et al., 2011) and numerical method (Yavari et al., 2014; Ju, 2015). Among them, the load transfer method is the most widely used, which is based on the simple idea that a pile can be divided into a series of equal-length pile sections (elastic elements) and the pile–soil interaction at the side of these sections and the pile tip can be described by non-linear force-displacement relationships.

In this work, the load transfer mechanism of the LSTHP was analysed using the FLAC³ᴰ software. The DSC (Desai, 2001) theory was then introduced to establish a load transfer model to describe the non-linear characteristics of pile–soil interactions. An analytical solution for the settlement of the pile in homogeneous soil was derived using the proposed load transfer method. Based on a case study, the accuracy and applicability of the theoretical solution were cross-validated with numerical simulations. Finally, an extension of the proposed method for the evaluation of the settlement of the LSTHP in layered soils was discussed.

18.2   LOAD TRANSFER MECHANISM OF A LARGE STEP-TAPERED HOLLOW PILE (LSTHP)

To investigate the bearing characteristics and the mechanism of pile–soil interactions, a ¼-scale model of a $\Phi15$ m LSTHP used in the construction of the Shenzhen Bridge was constructed using the FLAC³ᴰ software package (Figure 18.2). The model size of 40 m × 40 m × 50 m is used to minimize the boundary effects. The stratum includes a silty clay layer from the surface to a depth of 45 m, followed by a limestone layer that contains a series of beaded karst caves (Jiang et al., 2019). The maximum height of the karst caves is more than 17 m at this construction site. The rocks within the zone of developed karst caves are broken and there is no stable bedrock that can be used as the holding layer within the depth of 80 m. As the LSTHP will be constructed within the silty clay layer, the numerical model will only use a single soil layer representing the silty clay. For boundary conditions, the top of the model ($z = 0$) is a free surface for x, y and z movement. The base of the model ($z = -50$ m) is restricted for movement in the z-direction but unrestricted for x and y movement. The central point at $x = 0$ m, $y = 0$ m and $z = -50$ m is restricted for any movement.

For mechanical behaviours, a linear elastic model is used for the pile while the commonly used Mohr–Coulomb criterion is applied to simulate the failure behaviour of the soil in the FLAC3D numerical model. The material parameters of corresponding elements are listed in Tables 18.1 and 18.2 (Jiang et al., 2019). Dynamic traffic load is not considered in this numerical study.

As shown in Figure 18.3a, under a vertical load of 50 MN, there is an abrupt change in the axial force distribution curve of the LSTHP at Sections B, C and D (Figure 18.2). A significant part of the load is transferred to the supporting soil due to the enlarged section profiles, which in turn decreases the load transferred downward through the pile (Section A). In addition, the friction resistance is also increased significantly due to the diameter of the pile. Therefore, the bearing capacity of the overlying soil can be better exploited for the benefit of the pile support. The increased diameter of the pile does result in increased overall weight but this is mitigated by using the hollow interior design (hence the weight is reduced by 40% compared with the solid pile). As shown in Figure 18.3b, the vertical stress inside the soil layer barely changes when the pile is constructed. After the application of the vertical load of 50 MN, the vertical stress in the soil layer increases slightly but the increment decreases gradually with depth until it becomes negligible at around 40 m. Therefore the effect of the pile on the stability of the karst caves becomes insignificant.

The load transfer process in the LSTHP has significant asynchronous characteristics. As shown in Figure 18.4, at the initial stage, only the top pile section (from 0 to 4 m) undergoes some relative displacement between the pile and soil. The side friction of the first pile section (Section ED) appears and gradually develops downward. As the load increases, the disturbed zone increases and Sections D, C and B will gradually take a significant portion of the pile load. It can be seen that the LSTHP retains the characteristics of load transfer of traditional piles, with the additional significantly increased resistance due to three varied cross sections (Section B, C and D) and enhanced friction because of the large pile diameter.

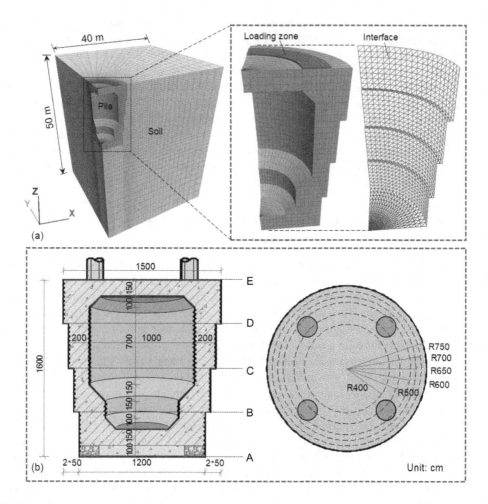

FIGURE 18.2   Model diagram of the Φ15 m large step-tapered hollow pile (LSTHP).

TABLE 18.1
Mechanical Parameters of the Pile and Soil

Element	Density (kg·m⁻3)	Poisson's Ratio	Elastic Modulus (MPa)	Bulk Modulus (MPa)	Shear Modulus (MPa)	Cohesion (kPa)	Internal Friction (°)
Pile	2,500	0.2	30,000	16,700	12,500	–	–
Clay	1,950	0.3	25	20.83	9.62	60	20

TABLE 18.2
Mechanical Parameters of the Pile–Soil Interface

Element	$K_n = K_s$ (MPa)	Cohesion (kPa)	Internal Friction Angle(°)
Interface	1,346	48	16

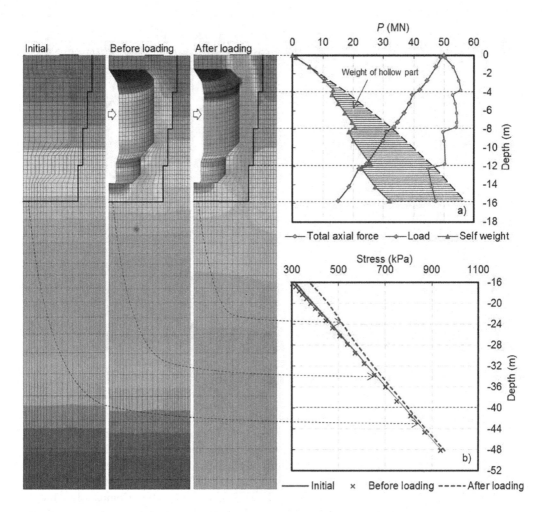

FIGURE 18.3 Bearing characteristics of large step-tapered hollow pile (LSTHP): (a) axial force and (b) vertical stress curves underneath the pile tip.

18.3 ESTABLISHMENT OF THE DSC LOAD TRANSFER MODEL

The DSC for geotechnical materials was first proposed by Frantziskonis and Desai (1987). Desai and Ma (1992) and Desai and Toth (1996) developed the idea further for studies of static and dynamic constitutive modelling of soils and soil–structure interfaces. In this conceptual framework, the soil under load is considered to be a material consisting of elements in two states: the relatively intact (RI) state and fully adjusted (FA) state, and the elements are randomly distributed within the material. Certain mechanical behaviour of the soil can then be expressed as a combined response of the elements in two states. For example, the basic equation of DSC theory (Desai, 2001) for stress can be expressed as:

$$\sigma = (1 - D)\sigma_i + D\sigma_c \qquad (18.1)$$

where σ is the total stress (tangential stress or normal stress), σ_i is the stress sustained by elements in the RI state, and σ_c represents the stress by elements in the FA state. D is a disturbance factor ($0 \leq D \leq 1$) reflecting load sharing between elements in two different states.

A load transfer model aims to establish a load transfer function (LTF) that can reflect the relationship between side friction and shear displacement of the pile–soil interface. However, the

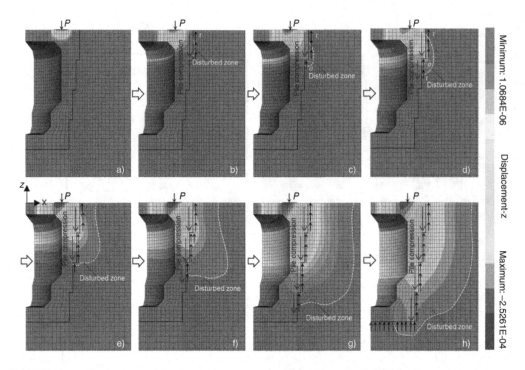

FIGURE 18.4   The load transfer process of large step-tapered hollow pile (LSTHP) (displacement in metre).

stress-displacement relationship at the pile–soil interface is in general non-linear, and it shows different characteristics under different geological conditions. For example, it can show hardening or softening behaviour under different confining pressures (Zhu and Shen, 1990). Therefore, in this case, it is difficult to establish a single LTF that can reflect different characteristics of the pile–soil interactions. To address this issue, several pile–soil load transfer models were proposed in the literature based on the DSC theory (Liu and Yang, 2006; Huang et al., 2018). Compared with other theoretical solutions, load transfer models based on DSC have clear physical meanings and offer a better analysis of the microscopic evolution of the pile–soil load transfer process. For the pile–soil load transfer mechanism of the LSTHP, Huang et al. (2018) proposed an LTF based on DSC, which can well describe the pile–soil interaction of the LSTHP. However, the function proposed is complex and it is an implicit function, which makes it difficult to be directly applied in the analysis of the settlement of the LSTHP. In this work, a relatively simple explicit LTF based on the DSC theory is established to describe the complex load transfer mechanism for the LSTHP.

The pile–soil interface model of the LSTHP incorporating DSC used in this work is shown in Figure 18.5. When the pile bears the load, part of the interface would be in the RI state and the other part in the failure state (FA state). The proportion of these two parts changes dynamically during loading and can be determined based on the deformation compatibility criterion. From Eq. (18.1), the pile–soil LTF for pile friction stress can be written as:

$$\tau = (1 - D)\tau_i + D\tau_c \tag{18.2}$$

where τ is the pile friction stress, τ_i is the friction stress borne by elements in RI state and τ_c is the stress by elements in FA state. D in this conceptual model can be defined as the ratio of the number of failure elements (n_p) to the total number of elements (n):

$$D = n_p/n \tag{18.3}$$

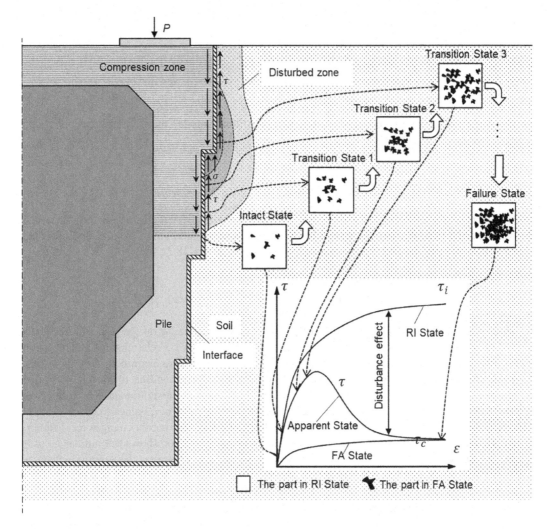

FIGURE 18.5 The pile–soil interface incorporating the disturbed state concept (DSC).

The damage to the pile–soil interface during loading is a continuous process. Based on the homogenization theory of composite materials, the strength of elements in the pile–soil interface can be described statistically using a Weibull distribution (Huang et al., 2018) with the probability density function:

$$f(x) = \frac{\eta}{\xi}\left(\frac{x}{\xi}\right)^{\eta-1} \exp\left[-\left(\frac{x}{\xi}\right)^{\eta}\right] \tag{18.4}$$

where ξ and η are distribution parameters. Variable x can be defined as one related to the loading process. In this work, this variable represents the pile settlement. When the settlement reaches s, the number of failed elements is:

$$n_p(s) = \int_0^s nf(x)dx \tag{18.5}$$

Therefore, the disturbance factor D can be obtained:

$$D = \frac{n_p(s)}{n} = \frac{n\left\{1 - \exp\left[-(s/\xi)^\eta\right]\right\}}{n} = 1 - \exp\left[-\left(\frac{s}{\xi}\right)^\eta\right] \qquad (18.6)$$

For the RI part of the interface, based on its linear constitutive model, the friction stress borne can be expressed as:

$$\tau_i = ks \qquad (18.7)$$

where k is a shear coefficient of the pile–soil interface, and s is pile settlement. For the FA part of the interface, the friction stress borne will be the residual friction strength of the pile–soil interface τ_c. Therefore, the explicit DSC LTF can be expressed as (from Eq. 18.2):

$$\tau(s) = ks \cdot \exp\left[-(s/\xi)^\eta\right] + \tau_c \cdot \left\{1 - \exp\left[-(s/\xi)^\eta\right]\right\} \qquad (18.8)$$

To better understand this LTF, the effects of parameters k, τ_c, η, and ξ on the load transfer characteristics were analysed using MATLAB™. The influences of k on τ-s relationships are shown in Figure 18.6a, where four cases of k values (4, 8, 16 and 32 kPa/mm) are examined. As k increases, the peak value of the τ-s curve increases, while the s value corresponding to the peak τ value decreases and the softening behaviour of the τ-s curve becomes more significant. The influences of ξ on τ-s relationships are shown in Figure 18.6b, where four cases of ξ values (0.5, 1.0, 1.5 and 2.0) are tested. As can be seen from the figure, the form of τ-s curves hardly changes for different ξ values except that the peak value of the curve increases. Figure 18.6c shows the influences of η values (0.2, 0.4, 0.6 and 0.8) on the τ-s relationships. Clearly, as η increases, the τ-s curve changes gradually from hardening to softening behaviour and the peak value decreases. Based on the influences of τ_c (10, 20, 30 and 40 kPa) on the τ-s relationships shown in Figure 18.6d, τ_c has no effect on the form of τ-s curves but as expected, the total friction stress increases with increasing τ_c. A preliminary conclusion from this exercise is that the macroscopic strength of the pile–soil interface can be described by parameters ξ and τ_c, and the form of τ-s curve can be described by parameters k and η. As demonstrated, this explicit DSC LTF can describe the non-linear, hardening or softening mechanical behaviour of the pile–soil interface (Huang et al., 2018).

18.4 CALCULATION OF THE SETTLEMENT OF AN LSTHP

18.4.1 CALCULATION MODEL AND BASIC ASSUMPTIONS

The idea is to divide the pile into several sections and each section is connected to the soil using the non-linear LTF described above to simulate the stress in the pile–soil interface. Figure 18.7 illustrates the configuration of such a model, which is a simplified conceptual representation of an LSTHP. The hollow section of the pile is divided into n segments with different cross sections of diameters from d_2 to d_{n+1}. The top cylindrical section of the pile is element 1 and the pile tip is element n+2. The elastic modulus of the pile is E_p and the total length of the pile is L_{n+2}. The perimeters and cross-sectional areas of the elements are $U_1, U_2, \ldots, U_{n+1}$, and $A_{p1}, A_{p2}, \ldots, A_{p(n+1)}, A_{p(n+2)}$, respectively.

The following basic assumptions are made:

1. The pile is divided into several elements along its axial direction and each pile element is a homogenous isotropic linear elastic body.

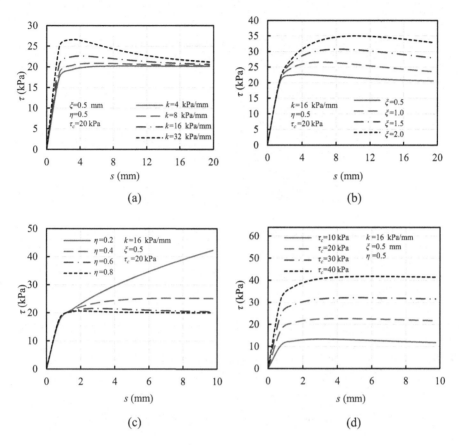

FIGURE 18.6 Influences of model parameters on the τ-s relationships: (a) parameter k; (b) parameter ξ; (c) parameter η; and (d) parameter τ_c.

2. The displacement at any point along the pile is only related to its side friction and the pile element–soil interaction can be simulated using an independent non-linear model.
3. The non-linear model mentioned in (2) can be expressed using a functional form.

18.4.2 Derivation

For the model shown in Figure 18.7, three different types of elements need to be considered in the calculation: (i) pile tip (element $n+2$); (ii) pile element with changing cross sections (element 2, …, $n+1$); and (iii) top pile element with a constant cross section (element 1). For the LSTHP, the weight of the pile is significant and must be incorporated in the settlement calculation.

i. Pile tip (element $n+2$)

This element has a solid bottom face and a constant cross section, as shown in Figure 18.8, where the free body diagram (FBD) of a thin section of the element is illustrated.

The equilibrium equation of the FBD can be written as:

$$\frac{dP(z)}{dz} = -U_{n+2}\tau(z) + \gamma_p A_{p(n+2)} \tag{18.9}$$

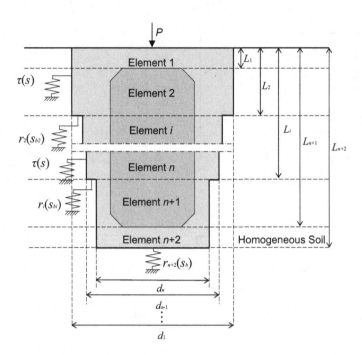

FIGURE 18.7   Settlement calculation model of a large step-tapered hollow pile (LSTHP).

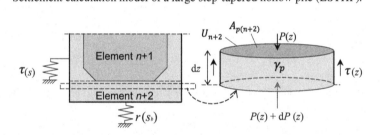

FIGURE 18.8   Calculation model for the cylindrical pile tip element.

where U_{n+2} is the perimeter of the pile element, $P(z)$ denotes axial force on the pile, $\tau(z)$ is the side friction, γ_p is the unit weight of the pile and $A_{p(n+2)}$ is the cross-sectional area of this pile element.

The elastic compression of this pile element is:

$$ds = -\frac{P(z)}{E_p A_{p(n+2)}} dz \tag{18.10}$$

where E_p is the elastic modulus of the pile. Combining Eqs. (18.9) and (18.10) leads to the following governing differential equation:

$$\frac{d^2 s}{dz^2} - \frac{U_{n+2}}{A_{p(n+2)} E_p} \tau(z) + \frac{\gamma_p}{E_p} = 0 \tag{18.11}$$

Adopting the new DSC load transfer model described in the previous section (Eq. 18.8), τ can be expressed as a function of the settlement s, and Eq. (18.11) can be written as:

$$\frac{d^2 s}{dz^2} - \frac{U_{n+2}}{A_{p(n+2)} E_p} \tau(s) + \frac{\gamma_p}{E_p} = 0 \tag{18.12}$$

By introducing incremental settlement ds, Eq. (18.12) can be rewritten as:

$$2\left(\frac{d^2s}{dz^2}\right)ds = d\left(\frac{ds}{dz}\right)^2 = 2\left(\frac{U_{n+2}}{A_{p(n+2)}E_p}\tau(s)-\frac{\gamma_p}{E_p}\right)ds \tag{18.13}$$

Integrating Eq. (18–13) against s, one can obtain:

$$\left(\frac{ds}{dz}\right)^2 = 2\int\left(\frac{U_{n+2}}{A_{p(n+2)}E_p}\tau(s)-\frac{\gamma_p}{E_p}\right)ds + C_1 \tag{18.14}$$

For compressed piles, strain is negative and therefore:

$$\frac{ds}{dz} = -\sqrt{\frac{2U_{n+2}}{A_{p(n+2)}E_p}T(s)-\frac{2\gamma_p}{E_p}s+C_1} \tag{18.15}$$

where C_1 is the constant after integration, and $T(s)$ is the integral primitive function of Eq. (18.8). The explicit form of $T(s)$ is difficult to obtain because Eq. (18.8) is a transcendental equation, and therefore, it is solved in this work using a numerical solution written in MATLAB codes.

From Eqs. (18.10) and (18.15), the axial force acting on the pile can then be expressed as:

$$P(z) = E_p A_p \sqrt{\frac{2U_{n+2}}{A_{p(n+2)}E_p}T(s)-\frac{2\gamma_p}{E_p}s+C_1} \tag{18.16}$$

Constant C_1 can be obtained as shown below. Assuming that the resistance r at the pile tip also follows the proposed load transfer model, but with different parameters, similar to Eq. (18.8), i.e.:

$$r = r(s_b) = k_b s_b \cdot \exp\left[-(s_b/\xi_b)^{\eta_b}\right]+\tau_{ch}\cdot\left\{1-\exp\left[-(s_b/\xi_b)^{\eta_b}\right]\right\} \tag{18.17}$$

For the pile tip element, the boundary conditions are:

$$\begin{cases} \left.\dfrac{ds}{dz}\right|_{s=s_b} = -\dfrac{r(s_b)}{E_p} \\[2ex] \left.s\right|_{z=L_{n+2}} = s_b \end{cases} \tag{18.18}$$

These conditions are incorporated into Eq. (18.15) to obtain C_1 as:

$$C_1 = \frac{r^2(s_b)}{E_p^2} - \frac{2U_{n+2}}{A_{p(n+2)}E_p}T(s_b)+\frac{2\gamma_p}{E_p}s_b \tag{18.19}$$

To evaluate the settlement of this element, Eq. (18.15) must be integrated. The equation can be rewritten as:

$$\frac{ds}{\sqrt{\dfrac{2U_{n+2}}{A_{p(n+2)}E_p}T(s)-\dfrac{2\gamma_p}{E_p}s+C_1}} = -dz \tag{18.20}$$

Integrate both sides of the equation with variable z from L_{n+2} to z, and variable s from s_b to s, we have:

$$\int_{s_b}^{s} \frac{ds}{\sqrt{\dfrac{2U_{n+2}}{A_{p(n+2)}E_p}T(s) - \dfrac{2\gamma_p}{E_p}s + C_1}} = L_{n+2} - z \qquad (18.21)$$

From this equation, the settlement s of the pile element at any depth z could be solved numerically. After the axial force $P(z)$ and settlement $s(z)$ of the pile tip element $(n+2)$ are obtained, the axial force and settlement of the next pile element $(n+1)$ can then be derived.

ii. Pile element with changing cross sections (element 2, ..., $n+1$)

The element i at the depth Li is used as an example to explain the calculation and its simplified conceptual model is shown in Figure 18.9. In this case, the axial force P_{i+1} and the settlement s_{i+1} of the next element $(i+1)$ are assumed to be known, as calculated in the previous iteration step. Therefore, we have:

$$\begin{cases} s_i = s_i' \\ P_i = P_i' + \dfrac{r_i(s_{bi}) \times (A_i - A_{i+1})}{A_i} = P_i' + \dfrac{r_i(s_{bi}) \times \pi \times (d_i^2 - d_{i+1}^2)/4}{\pi \times d_i^2/4} \end{cases} \qquad (18.22)$$

where $r_i(s_i)$ is the resistance created due to the increased cross section (Figure 18.9), A_i and A_{i+1} are the cross-sectional areas of the pile element i and $i+1$, and d_i and d_{i+1} are their corresponding diameters.

Using this relationship, plus the load transfer model for $r_i(s_{bi})$, the axial force and settlement of the pile element i can be calculated. The same calculation process can be iterated until element 2, the one below the top pile element.

iii. Top pile element (element 1)

This is a special case of the pile element with changing cross section as described above, where the cross section is constant and therefore the resistance r_1 created due to cross section change is zero. Eq. (18–22) can still be used so the axial force and pile settlement of this element are the same as the next element (element 2).

18.5   VERIFICATION

The $\Phi15$ m LSTHP used in the Shenzhen Bridge was used as a case study to verify the proposed calculation method. The simulation of the vertical static load test was carried out based on the ¼-scale pile–soil model shown in Figure 18.2. An initial vertical load of 16 MN is applied on top of the pile and the calculation is carried out until the system reaches its equilibrium state. Subsequently, 13 consecutive increments of 8 MN vertical loads are applied and the same

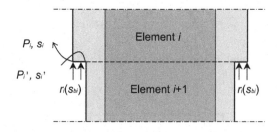

FIGURE 18.9   Calculation model at the pile element with variable cross sections.

calculation process is repeated. The results of load and settlement characteristics of the simulations are shown in Figure 18.10.

As can be seen from the figure, the side friction of the pile gradually increases with increasing load but the values fluctuate with depth. The overall average friction of the pile can be evaluated using a weighted average calculation:

$$\tau = \frac{\sum \tau_{li} \cdot l_i}{\sum l_i} \tag{18.23}$$

where l_i and τ_{li} are the length and side friction of element i. The relationship between this average value and pile settlement can be obtained based on the results at different load levels, as shown in Figure 18.10b. These values provide the basis to derive the parameters for the DSC load transfer model (Eq. 18.8) using the least square fitting technique. The final relationship thus obtained is:

$$\tau = 0.32223s \cdot \exp\left[-(s/75.34543)^{1.91034}\right] + 84.88255 \cdot \left\{1 - \exp\left[-(s/75.34543)^{1.91034}\right]\right\} \tag{18.24}$$

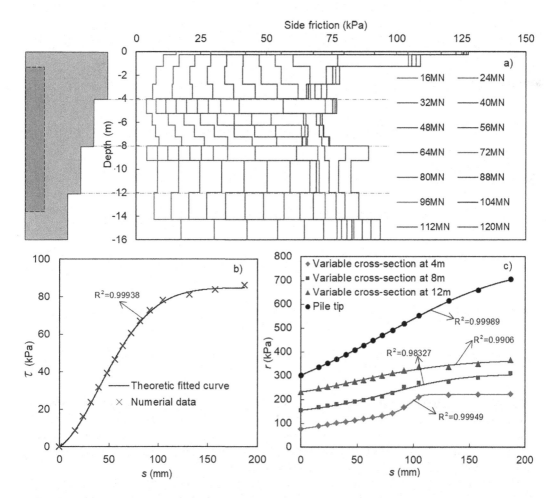

FIGURE 18.10 Simulation results of vertical static load test of a $\Phi15\,\mathrm{m}$ large step-tapered hollow pile (LSTHP): (a) side friction; (b) data points and fitted curve of the average side friction versus settlement; and (c) data points and fitted curve of the resistance at the pile tip and variable cross sections vs settlement.

On the other hand, the DSC LTF for the resistance at the pile tip and varied cross sections can be obtained from the numerical simulations by examining the normal stress in the soil underneath the pile tip or varied cross sections, as shown in Figure 18.10c. Note there are some initial resistances at the pile tip and varied cross sections before the application of the axial load, mainly due to the self-weight of the pile. Therefore, the complete DSC load transfer model for the pile tip and varied cross sections must also consider this element (r_{wi}):

$$r_i = r_{wi} + r(s_{bi}) = r_{wi} + k_{bi}s_{bi} \cdot \exp\left[-\left(s_{bi}/\xi_{bi}\right)^{\eta_{bi}}\right] + \tau_{cbi} \cdot \left\{1 - \exp\left[-\left(s_{bi}/\xi_{bi}\right)^{\eta_{bi}}\right]\right\} \quad (18.25)$$

Based on Figure 18.10c, the least square fitting of the data points using Eq. (18.25) for different depths gives the following relationships:

At the depth of 4 m:

$$r_3 = 76.83608 + 0.74111s_{b3} \cdot \exp\left[-\left(s_{b3}/100.07231\right)^{12.58329}\right]$$

$$+ 144.2906 \cdot \left\{1 - \exp\left[-\left(s_{b3}/100.07231\right)^{12.58329}\right]\right\} \quad (18.26)$$

At the depth of 8 m:

$$r_4 = 155.4455 + 0.57051s_{b4} \cdot \exp\left[-\left(s_{b4}/122.80539\right)^{2.71965}\right]$$

$$+ 149.48713 \cdot \left\{1 - \exp\left[-\left(s_{b4}/122.80539\right)^{2.71965}\right]\right\} \quad (18.27)$$

At the depth of 12 m:

$$r_7 = 232.4587 + 0.48221s_{b7} \cdot \exp\left[-\left(s_{b7}/121.21314\right)^{1.72881}\right]$$

$$+ 133.368 \cdot \left\{1 - \exp\left[-\left(s_{b7}/121.21314\right)^{1.72881}\right]\right\} \quad (18.28)$$

At the depth of 16 m (pile tip):

$$r_{10} = 302.1252 + 2.02505s_b \cdot \exp\left[-\left(s_b/204.89041\right)^{2.46426}\right]$$

$$+ 420.08226 \cdot \left\{1 - \exp\left[-\left(s_b/204.89041\right)^{2.46426}\right]\right\} \quad (18.29)$$

Based on the load transfer models obtained above, the pile settlement is calculated using the proposed method described in the previous section. The pile in this case is divided into ten elements based on the geometry of the pile, as shown in Figure 18.11. The DSC load transfer model was used to describe the non-linear characteristics of pile–soil interactions. The calculation procedure for the settlement of the LSTHP was programmed in MATLAB™.

The theoretical settlement of this pile calculated using the proposed method is compared with the numerical simulation result, as shown in Figure 18.12. It is encouraging to see that both results agree reasonably well and the theoretical model can adequately describe the non-linear behaviour of the settlement of the LSTHP.

18.6 DISCUSSION

The proposed method described above assumes only a homogeneous soil layer. However, the method can also be directly applied in cases of layered soils without any modification to the theoretical framework. The only differences are in the way the pile elements are constructed and the DSC LTFs used for different soil layers.

In a homogeneous soil, pile elements are constructed according to the geometry (cross-sectional areas) of the pile. However, in a layered soil, the pile elements must be formed considering both the pile cross-sectional areas and characteristics of the soil layer, see an example shown in Figure 18.13. For the DSC LTF, only one model is used in homogeneous soil. In a layered soil, the load transfer mechanisms in different soils are different and therefore different LTFs with different parameters must be used, see also Figure 18.13.

This work does not cover the bearing characteristics of the LSTHP under the action of active traffic load, which is an important aspect of bridge construction. Given the effectiveness of the proposed model and method to assess the settlement of the LSTHP under static load, their extension to account for the effects of dynamic loads should make an interesting topic for future research.

18.7 CONCLUSIONS

An LSTHP used in karst topography was introduced and its load transfer mechanism was analysed. A new load transfer model based on the DSC was formulated to describe the non-linear characteristics of pile–soil interactions. An analytical solution for the pile settlement was proposed using the load transfer model. Some works and conclusions from this research are summarized below:

- A ¼-scale numerical model of a Φ15m LSTHP was constructed in FLAC3D to investigate the mechanism of pile–soil interactions. The results indicate that the LSTHP can solve the problem of foundation construction in the case of complex karst topography. The load transfer process in the LSTHP has significant asynchronous characteristics which can be described by the proposed model.

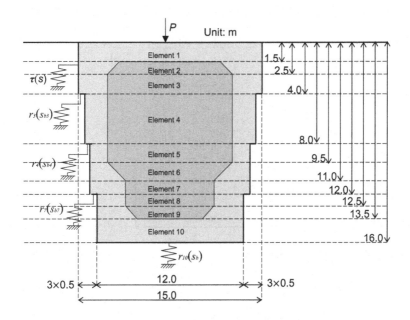

FIGURE 18.11 Settlement calculation model for a Φ15 m large step-tapered hollow pile (LSTHP).

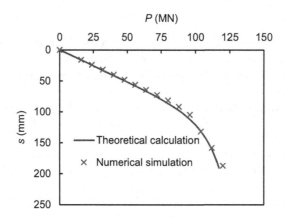

FIGURE 18.12 Comparison of the P-s curves between numerical simulation and theoretical calculation.

- A new load transfer model was developed based on the DSC theory. In this model, the RI state is defined by a linear elastic model, while the FA state is described by the Mohr–Coulomb criterion. The proposed DSC load transfer model is demonstrated to be capable of describing the load transfer mechanism of the pile–soil interface for the LSTHP. It can describe explicitly the non-linear, hardening or softening characteristics of the pile–soil interface.
- An analytical solution for the settlement of the LSTHP was also developed in this work based on the proposed DSC load transfer model. Based on a case study, the proposed theoretical method is cross-validated by comparing its outputs with results obtained from numerical simulations. A possible extension of the proposed method for the calculation of settlement of the LSTHP in layered soils is also discussed.

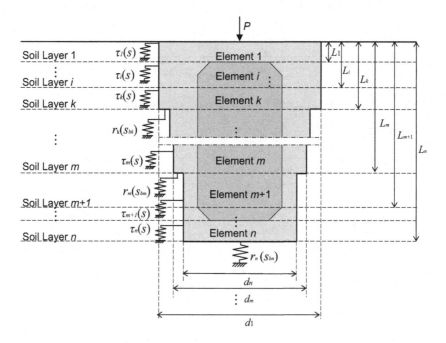

FIGURE 18.13 Settlement calculation model for a large step-tapered hollow pile (LSTHP) in layered soils.

REFERENCES

Ai Z., Yue Z. Elastic analysis of axially loaded single pile in multilayered soils. *International Journal of Engineering Science* 2009; 47(11): 1079–1088. https://doi.org/10.1016/j.ijengsci.2008.07.005.

Desai C.S. *Mechanics of Materials and Interfaces: The Disturbed State Concept.* CRC Press, Boca Raton, FL; 2001.

Desai C.S., Ma Y. Modelling of joints and interfaces using the disturbed state concept. *International Journal for Numerical and Analytical Methods in Geomechanics* 1992; 16: 623. https://doi.org/10.1002/nag.1610160903.

Desai C.S., Toth J. Disturbed state constitutive modeling based on stress-strain and nondestructive behavior. *International Journal of Solids and Structures* 1996; 33(11): 1619. https://doi:10.1016/0020-7683(95)00115-8.

Fang T., Huang M. Deformation and load-bearing characteristics of step-tapered piles in clay under lateral load. *International Journal of Geomechanics* 2019; 19(6): 04019053. https://doi:10.1061/(asce)gm.1943-5622.0001422.

Frantziskonis G., Desai C.S. Elastoplastic model with damage for strain softening geomaterials. *Acta Mechanica* 1987; 68: 151. https://doi:10.1007/bf01190880.

Ghazavi M., Lavasan A.A. Bearing capacity of tapered and step-tapered piles subjected to axial compressive loading. In *The 7th International Conference on Coasts. Ports & Marine Structures, ICOPMAS,* Tehran, Iran; 2006.

Huang M., Fu J.J., Chen F.Q., et al. Damage characteristics of karst cave roof and its safety thickness calculation under the coupling effect of pile-tip load and seismic wave. *Rock and Soil Mechanics* 2017; 38(11): 3154–3162. https://doi:10.16285/j.rsm.2017.11.010.

Huang M., Jiang S., Xu D.X., et al. Load transfer mechanism and theoretical model of step tapered hollow pile with huge diameter. *Chinese Journal of Rock Mechanics and Engineering* 2018; 37(10): 2370–2383.

Ismael N.F. Load tests on straight and step tapered bored piles in weakly cemented sands. *International Symposiumon on Field Measurements in Geomechanics*; 2003.

Ismael N.F. Behavior of step tapered bored piles in sand under static lateral loading. *Journal of Geotechnical & Geoenvironmental Engineering* 2010; 136(5): 669–676. https://doi:10.1061/(asce)gt.1943-5606.0000265.

Jiang S., Huang M., Fang T., et al. A new large step-tapered hollow pile and its bearing capacity. *Proceedings of the Institution of Civil Engineers-Geotechnical Engineering* 2019; 1–37. https://doi:10.1680/jgeen.19.00009.

Ju J. Prediction of the settlement for the vertically loaded pile group using 3D finite element analyses. *Mar Georesour Geotechnol* 2015; 33(3): 264–271. https://doi:10.1080/1064119x.2013.869285.

Liu Q.J., Yang L.D. New model of load transfer function for pile analysis based on disturbed state model. *Journal of Tong ji University (Natural Science)* 2006; 34(2): 165–169. https://doi:10.1007/s11709-007-0060-9.

Nie G.X., Chen F. Analytical solution for axial loading settlement curve of piles using extensive shear displacement method. *Journal of Central South University (Science and Technology)* 2005; 36(1): 163–168. https://doi:10.1016/j.cyto.2004.11.006.

Salgado R., Seo H., and Prezzi M. Variational elastic solution for axially loaded piles in multilayered soil. *International Journal for Numerical and Analytical Methods in Geomechanics* 2011; 37(4): 423–440. https://doi:10.1002/nag.1110.

Tiago G.S.D., Adam B. Load-transfer method for piles under axial loading and unloading. *Journal of Geotechnical and Geoenvironmental Engineering* 2018; 144(1): 04017096. https://doi:10.1061/(asce)gt.1943-5606.0001808.

Wang N., Le Y., Hu W.T., et al. New interaction model for the annular zone of stepped piles with respect to their vertical dynamic characteristics. *Computers and Geotechnics* 2020; 117: 103256.

Yavari N., Tang A.M., Pereira J.M., et al. A simple method for numerical modelling of mechanical behaviour of an energy pile. *Géotechnique Letters* 2014; 4: 119–124. https://doi:10.1680/geolett.13.00053.

Zhang Q.Q., Zhang Z.M. Simplified calculation approach for settlement of single pile and pile groups. *Journal of Computing in Civil Engineering* 2012; 26(6): 750–758. https://doi:10.1061/(asce)cp.1943-5487.0000167.

Zhu B.L., Shen Z.J. *Computational Soil Mechanics.* Shanghai Scientific and Technical Press, Shanghai; 1990.

19 Applications of DSC and HISS Soil Models in Nonlinear Finite Element Dynamic Analyses of Soil–Structure Interaction Problems

Bal Krishna Maheshwari and Mohd Firoj
Dept. of Earthquake Engineering, Indian Institute of Technology Roorkee

19.1 PREFACE

Numerous constitutive soil models have been developed in the past five decades to model the stress–strain behavior of soil. These models are implemented in the 3D nonlinear finite element modeling for soil–structure interaction (SSI) problems, e.g., for pile–soil interaction. The numerical modeling of the nonlinear soil behavior for geotechnical earthquake engineering problems is a challenging task. No analytical approach is readily available for the constitutive modeling of the geological materials. Therefore, it is essential to develop an efficient numerical algorithm for the modeling of these materials in the finite element formulation for practical geotechnical engineering problems.

In this chapter, a detailed finite element modeling of the hierarchical single surface (HISS) model for SSI problems is presented. The required parameters for the HISS model are evaluated using experimental data and presented. The applications of the HISS model in the finite element method (FEM), consistent infinitesimal finite element cell method (CIFECM), and coupled finite element method-scaled boundary element method (FEM-SBFEM) are presented. The capability of HISS to deal in the time domain is reviewed. Further, the disturbed state concept (DSC) parameters for the Solani sand, found using the results of cyclic triaxial tests, are presented.

This chapter presents a state-of-the-art compilation for the application of the HISS model in geotechnical earthquake engineering problems, which now has reached a reasonable stage of development. Implementation of the HISS model is reviewed for the rigorous CIFECM and FEM-SBFEM formulation schemes. Also, the HISS model is implemented for the substructure method for SSI problems. This has been incorporated in FORTRAN and MATLAB codes by the authors and co-workers.

19.2 INTRODUCTION

With the advancement of numerical methods, the FEM formulation is widely used in complex soil–structure interaction (SSI) and pile–soil–structure interaction (PSSI) problems. However, the formulation of analysis depends on the stress–strain relationship used for the geological materials. For the finite element formulation, this in turn depends on the constitutive model, which consists of the mathematical expression that represents the soil stiffness in a single finite element. The soil constitutive model in the FEM is implemented by integrating the constitutive equation with a known strain increment. The integration is carried out either using the implicit method or the explicit method. Various researchers [1–5] used the implicit and explicit methods for the integration of constitutive equations.

DOI: 10.1201/9781003362081-24

Significant development of the constitutive models occurred in the past five decades. Initially, the soil models were quite simple and with the progression of the research involvement, more advanced constitutive models have been developed for cyclic loading conditions and SSI problems. These models are developed based on experimental data and/or theoretical concepts of solid mechanics. Initially, a few parameter models (e, g., Von Mises, Tresca and Lade-Duncan) were introduced to quantify the soil plasticity in geotechnical engineering. These models are simple and easy to use and are appropriate to use for undrained saturated soil and cohesionless soil. Later two parameter models, viz., Mohr–Coulomb, Drucker–Prager (Extended Von Mises) and Lade's model were developed to overcome the limitations of one parameter models. Further, more advanced constitutive soil models (Cam-Clay, Modified Cam-Clay, Cap Plasticity, multiple nested yield surface, bounding surface, HISS, DSC models, etc.) were developed to capture the cyclic behavior of geological materials.

Drucker [6] proposed the cap model for the true yield surface of soil and using the same concept, various models were proposed for geological materials. The major limitation of these models is different yielding functions for yield and ultimate surfaces that cause the cutoff at the intersection of these surfaces. To overcome this limitation, Desai and co-authors [7–14] developed the HISS model for geological materials, which consists of a continuous single surface and has a unique normal direction at any point on the yielding surface. Further, Desai and co-authors [15,16] proposed the DSC for geological materials to consider the various factors such as creep deformations, volume change, and microcracking leading to fracture, hardening and softening, and mechanical and environmental forces.

From the past studies, it is observed that SSI affects the seismic response of the structure and soil. Many of the soil parameters affect the seismic response in SSI problems [17–21], i.e., of buildings, bridges, nuclear power plants (NPPs), etc. The selection of the constitutive soil model for the SSI problem is the most influencing geotechnical issue. Various researchers [22–24] used elastoplastic soil models for seismic soil–structure interaction problems. The nonlinear dynamic analysis of a pile group using the HISS soil model was carried out by Maheshwari et al. [25,26]. Maheshwari et al. [27] proposed the nonlinear substructure method for the dynamic analysis of a SSI system, and the nonlinearity of the soil was modeled using the HISS model. Maheshwari and Emani [28] implemented soil nonlinearity for the dynamic analysis of pile groups using the HISS model in the CIFECM. Maheshwari and Syed [29] verified the validity of the HISS model in the coupled FEM-SBFEM; further, Syed and Maheshwari [30] improved the algorithm for the time domain analysis of a single pile in nonlinear homogenous soil.

This chapter provides a state-of-the-art compilation of the application of the HISS model in finite element modeling of geotechnical engineering problems, which now has reached a reasonable stage of development. The HISS model implementation is reviewed for the FEM, rigorous coupled FEM-CIFECM, and FEM-SBFEM formulation schemes. Implementation of the HISS model in the substructure method of SSI problems is also described.

19.3   THE HISS MODEL

Desai and co-workers developed the HISS model for geological materials to incorporate plasticity behavior. A series of these models are available as reported by Desai et al. [31]. In the present chapter, application of δ_o^* version of the HISS model in SSI problems is reviewed. The δ_o version of the HISS model denotes the basic model for isotropic materials. Superscript * denotes the modified version of the model for cohesive soils. Both plasticity and hardening are considered in the HISS δ_o model. However, the HISS model does not account for softening behavior but this limitation can be overcome by incorporating the fully adjusted state of the DSC model. The DSC model is treated as an extended version of the HISS model in which both relatively intact (RI) and fully adjusted (FA) states are used. Here, only the RI state is considered for simplicity, which can be further enhanced by considering FA states for softening behavior of the soil.

In the HISS model, a material parameter (β) is used to define the yielding surface. For the δ_o version of the HISS model, setting the value of β to zero as was the case for Sabine clay, the dimensionless yielding surface (F) can be simplified as [27]:

$$F = \left(\frac{J_{2D}}{p_a^2}\right) + \alpha\left(\frac{J_1}{p_a}\right)^n - \gamma\left(\frac{J_1}{p_a}\right)^2 = 0 \tag{19.1}$$

where J_1 is the first invariant of the stress tensor, (σ_{ij}); J_{2D} is the second invariant of deviatoric stress tensor; p_a is atmospheric pressure; α is hardening function; and n and γ are the material parameters that define the shape of the yield surface. The hardening function, (α), in terms of the trajectory of volumetric plastic strain (ξ_v) can be defined as:

$$\alpha = h_1 / \xi_v^{h_2} \tag{19.2}$$

where h_1 and h_2 are the material parameters. The increment of trajectory of the volumetric plastic strain $(d\xi_v)$ is defined as:

$$d\xi_v = \left(\frac{1}{\sqrt{3}}\right) d\varepsilon_v^p, \text{ for } d\xi_v^p > 0 \tag{19.3a}$$

$$d\xi_v = 0 \text{ for } d\xi_v^p \leq 0 \tag{19.3b}$$

where $(d\xi_v^p)$ is the increment in volumetric plastic strain due to virgin loading.

19.3.1 FINITE ELEMENT FORMULATION OF THE HISS MODEL

In the finite element formulation of the constitutive modeling of the HISS model, the constitutive matrix of the material is updated in each element based on the strain in the respective element. The total strain increment is the sum of elastic and plastic strain increment as follows:

$$d\varepsilon = d\varepsilon_{ij}^e + d\varepsilon_{ij}^p \tag{19.4}$$

In the elastic region, the stress increment is defined using Hook's law as:

$$d\sigma_{ij}^e = C_{ij}^e d\varepsilon_{ij}^e \tag{19.5}$$

where C_{ij}^e is the elastic constitutive matrix of the material. In the plastic strain region, the stress increment using the associated flow rule is defined as:

$$d\sigma_{ij}^p = C_{ij}^p d\varepsilon_{ij}^p \tag{19.6}$$

where C_{ij}^p is the constitutive matrix of the material in the inelastic range and can be expressed in terms of the yielding surface (F) of the HISS model as follows:

$$C_{ij}^p = C_{ij}^e - \frac{C_{ij}^e\left(\frac{\partial F}{\partial \sigma_{ij}}\right)\left(\frac{\partial F}{\partial \sigma_{ij}}\right)^T . C_{ij}^e}{\frac{\partial F}{\partial \sigma_{ij}} . C_{ij}^e . \frac{\partial F}{\partial \sigma_{ij}} - \frac{\partial F}{\partial \xi_{ij}}\left[\left(\frac{\partial F}{\partial \sigma_{ij}}\right)^T \frac{\partial F}{\partial \sigma_{ij}}\right]^{1/2}} \tag{19.7}$$

Using the above elasto-plastic constitutive matrix of the HISS model, corresponding to the stress–strain in the element, the stiffness of the element is updated. Based on the new stress in the element, check for $F > 0$ or $F \le 0$. If, $F > 0$, the element is loading elasto-plastically, and if $F \le 0$, the element is undergoing elastic loading.

19.4 DSC PARAMETERS FOR SOLANI SAND

The DSC model is an extended version of the HISS model that uses two concepts, i.e., the RI state and the FA state using the critical state concept. DSC is used in characterizing the stress–strain behavior of soils including liquefiable saturated sand under monotonic and cyclic loadings. Here, DSC parameters for the Solani sand are evaluated using a cyclic triaxial test system.

In the DSC model, in addition to the parameters used for the elastic and HISS models, a parameter called the disturbance function D is used to define the deforming state with respect to RI and FA based on the laboratory test results. For RI, D is zero (means no disturbance). For FA, D is unity (means fully disturbed). The disturbance function, D, is defined as [32]

$$D = D_u \left[1 - e^{\left(-A \xi_D^Z \right)} \right] \tag{19.8}$$

where A, Z, and D_u are material parameters and ξ_D is the trajectory of the plastic deviatoric strains such that $\xi_D = \int \sqrt{d\varepsilon_{ij}^p d\varepsilon_{ij}^p}$ and $d\varepsilon_{ij}^p$ are the increments of plastic strain tensor. For calculating the DSC parameter using the cyclic triaxial test, first the phase change parameter, which is related to the point at which material starts compressing, is calculated. The value of n can be calculated using the following expression [33]:

$$\frac{J_{1a}}{J_{1m}} = \left(\frac{2}{n} \right)^{\frac{1}{n-2}} \tag{19.9}$$

where J_{1a} is found from the stress path of undrained test of saturated sand and J_{1m} depends on the confining pressure and is given as $J_{1m} = 3\sigma_o'$. The value of Young's modulus, E, can be determined using the tangent of the unloading stress–strain curve [33]. For confining pressure (σ_o') values of 100 and 160 kPa, the value of E is 320 and 400 MPa, respectively.

Based on the above procedure, the phase change parameter (n) for two different tests is calculated and listed in Table 19.1. The average of the two tests give the value of n as 100.5, which is much greater than the value of n obtained by Desai [13], i.e., $n = 3$, which may be due to a very low rise in excess pore pressure.

19.5 APPLICATION OF THE HISS MODEL IN FEM PROBLEMS

In this section, application of the HISS model is reviewed for SSI problems. The HISS model is used in the FEM, CIFECM, coupled FEM-SBFEM, and substructure method for SSI problems.

19.5.1 HISS Model in Single Pile and Piles Group Using FEM

Maheshwari et al. [25,26] used the FEM formulation for the modeling of pile and soil. The non-linearity of the soil was modeled using the HISS model while the pile element was assumed to be linear elastic. For modeling of the soil–pile system, a full 3D FEM model was used, as shown in Figure 19.1a. However, due to symmetrical geometry and loading conditions, only a quarter model is used by applying the axisymmetric and anti-symmetric boundary conditions at the inner nodes. To capture the shear deformation of the soil–pile system, eight-noded elements (Figure 19.1b) were

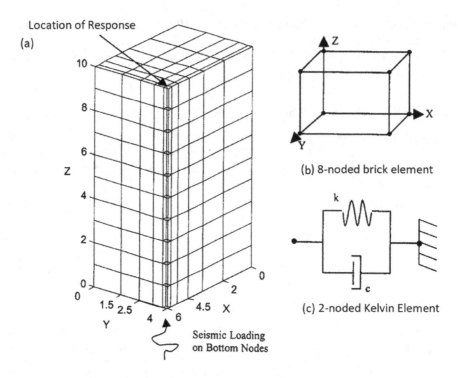

FIGURE 19.1 (a) 3D quarter model of soil–pile system. (b) Eight-noded brick element. (c) Two-noded Kelvin element. (After, Maheshwari et al. [26].)

used for soil–pile modeling [34]. For simplicity, it was assumed the soil and pile are rigidly connected with each other, thus no slip and no separation were considered.

In the dynamic analysis of an SSI system, it is desirable to reduce the number of elements or soil domain size. Therefore, forcing conditions at the boundaries' nodes are modeled in such a way that these nodes absorb the body and surface waves of all frequencies. Therefore, at the boundary nodes, Kelvin elements are attached as shown in Figure 19.1c. The coefficients of the spring and dashpot can be calculated using the suggestion of Novak and Mitwally [35], as follows:

$$K^* = \frac{G}{r_o}\left[S_1\left(a_r, v_s, D\right) + iS_2\left(a_r, v_s, D\right) \right] \tag{19.10}$$

where K^* is a complex stiffness matrix in which the real part corresponds to stiffness and the imaginary part corresponds to the damping coefficient; S_1 and S_2, which depend on the dimensionless frequency; and Poisson's ratio and damping of soil. These were obtained as closed form solutions of force displacement relationships of homogeneous infinite medium. G is shear modulus; r_o is the distance from the center of the pile foundation to the point where the spring is to be attached; v_s is the Poisson's ratio of soil; D is the damping ratio of the material, which is further used in calculating the Rayleigh damping parameter. α_r is the dimensionless frequency $\left(= \omega r_o / V_s\right)$, where ω is the angular frequency of excitation and V_s is the shear wave velocity of the soil.

The material properties of the soil as per Desai et al. [36] considered as follows: Shear modulus $G = 4147\,\text{kPa}$, Unit weight $\gamma_s = 17.5$ kN/m³, Poisson's ratio $v_s = 0.42$, and damping $D = 5\%$. The HISS model parameters for the soil are as follows: $\gamma = 0.047$, $n = 2.4$, $h_1 = 0.0034$, and $h_2 = 0.78$. The pile is assumed to behave elastically and has square cross section with dimension of each side as 0.5m. The length of the pile is 10 m. The mechanical properties of the pile are as follows: Modulus of elasticity $E_p = 25\,\text{GPa}$, Unit weight $\gamma_p = 24$ kN/m³, and Poisson's ratio $v_p = 0.25$.

TABLE 19.1

Phase Change Parameter (n) Using Cyclic Triaxial Tests

σ_o' (kPa)	J_{1a}(kPa)	J_{1m}(kPa)	n
100	290	300	120.5
160	458	480	80.5

For the seismic excitation, the 1940 El-Centro, Earthquake time history, with PGA = 0.32g and predominant frequency = 1.83 Hz, is used. To evaluate the complex dynamic stiffness of the soil–pile system, the harmonic excitation is applied at the pile cap. To determine the seismic response, earthquake time history is applied at the base nodes.

Considering the above data, Maheshwari et al. [25] demonstrated the predicted effect of nonlinearity using the HISS model on the dynamic impedance of a single pile and 2×2 pile group under harmonic excitation. It was observed that for a single pile, due to soil nonlinearity (HISS model), the stiffness and damping both reduced. The effect of nonlinearity is higher on the stiffness coefficient compared to the damping coefficient. Also, the damping coefficient varied linearly with frequency in the nonlinear case. For the 2×2 pile group also, a similar trend was observed as in the case of a single pile. However, for the pile group, a negative stiffness is observed due to the reduced stiffness (because of nonlinearity) and inertial forces due to mass on the top of the pile group.

Figure 19.2 shows the effect of the HISS model on the seismic response time history and corresponding Fourier spectrum at the pile head for a single pile. It is observed that the trends of pile head acceleration time histories are similar in the linear and nonlinear cases (Figure 19.2a). However, at some peaks, the HISS model predicts higher pile head response. The smoothed Fourier amplitude for the pile head is also shown in Figure 19.2b. It can be seen that the Fourier amplitude is higher for the nonlinear soil case compared to the elastic soil model.

19.5.2 HISS MODEL IN COUPLED FEM-CIFECM

In the CIFECM method of 3D modeling of the pile–soil system, the domain is divided into two substructure systems, namely, an unbounded far-field soil domain and the near field pile–soil system (Figure 19.3). The total dynamic stiffness of the system is obtained by coupling the near and far-field stiffness matrices. The size of the near field domain is fixed by the trial and error method. The near field is modeled using the FEM and the far field is modeled using the CIFECM so as to absorb the incoming incident wave. The nonlinearity of the near field is modeled using the HISS model while the far-field domain is in the elastic range. Detailed modeling of the coupled FEM and CIFECM for this pile–soil system can be found in Emani and Maheshwari [37]. The finite element modeling of the soil and pile system is identical as discussed in the previous section. The material properties of the soil and piles are given in Table 19.2. The diameter of the pile is 0.5 m. The HISS model parameters of the soil are as follows $\gamma = 0.047$, $n = 2.4$, $h_1 = 0.0034$, and $h_2 = 0.78$.

To evaluate the impedance function, the horizontal loading was applied harmonically at the pile cap. For evaluating the seismic response, the earthquake motion is applied at the interface node of the near field and far field. The 1995 Kobe Earthquake time history with PGA = 0.694g and predominant frequency = 2.43 Hz is used as input motion.

Figure 19.4 shows the horizontal impedances under the harmonic loading for a single pile and a 2×2 pile group. For the single pile, there is significant difference in the stiffness and damping values due to the plasticity of the soil (HISS model). The difference in stiffness of the soil–pile system is higher at the higher frequency (Figure 19.4a–b). For the 2×2 pile group, the effect of plasticity

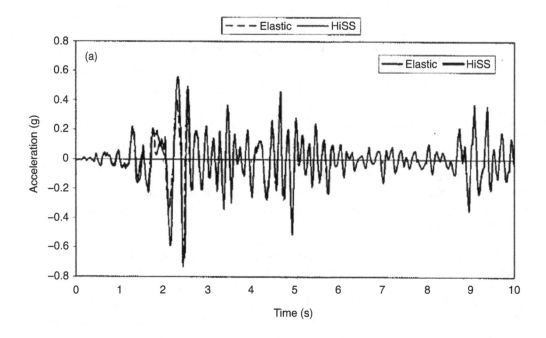

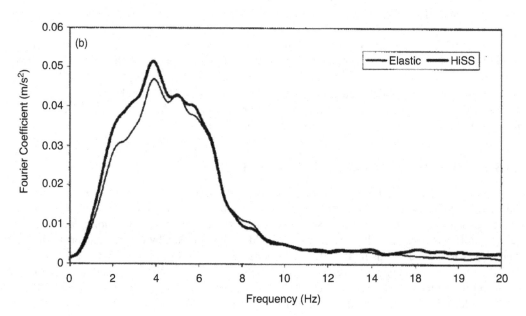

FIGURE 19.2 Linear and nonlinear seismic response of the pile head. (a) Acceleration time history. (b) Fourier Coefficient. (After, Maheshwari et al. [26].)

of the soil is similar to that of a single pile; however, there is negligible effect of plasticity on the damping of the pile–soil system.

Figure 19.5 shows the displacement time history at the pile head of 2 × 2 pile group under the real Kobe earthquake time history. It can be observed that there is considerable difference in the linear and HISS model response. In Figure 19.5, the peaks of the response at the same time indicate that the nonlinearity of soil was not sufficient to alter the total stiffness of the soil–pile system. This is attributed to the confining nature of the nonlinearity due to the HISS model.

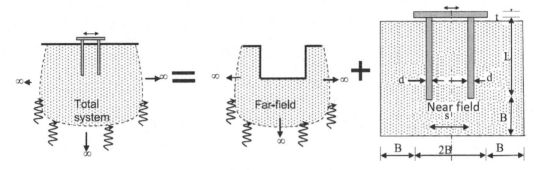

FIGURE 19.3 Sub-structuring of the full model into linear far field and nonlinear (hierarchical single surface, HISS model) for the near field. (After, Emani and Maheshwari [37].)

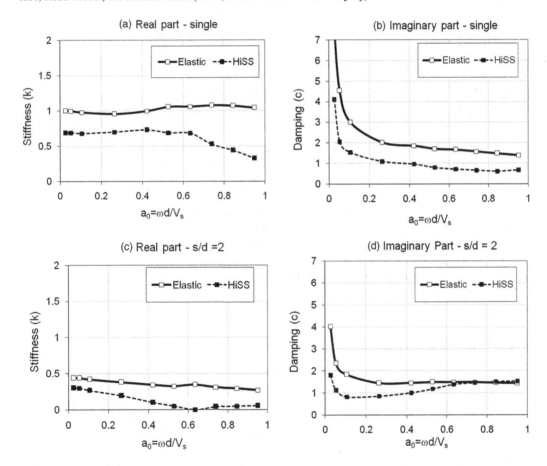

FIGURE 19.4 Horizontal Impedance for the (a) real part-single pile, (b) imaginary part-single pile, (c) real part-2 × 2 pile group, and (d) imaginary part-2 × 2 pile group with $s/d = 2$. (After, Maheshwari and Emani [28].)

19.5.3 HISS MODEL IN COUPLED FEM-SBFEM

In the coupled FEM-SBFEM modeling of the soil–structure system, the unbounded far field is modeled using the SBFEM by calculating the interaction force at the interface of the soil and structure system while the near field is modeled using the FEM approach. These interaction forces depend on the relative movement of the foundation with respect to the soil. Details of the coupled

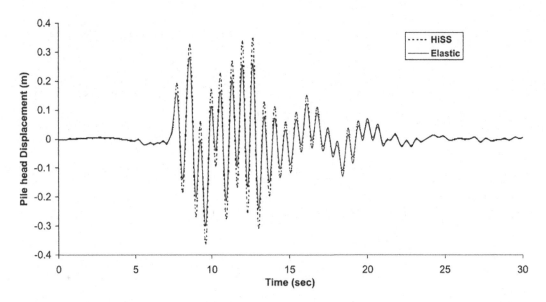

FIGURE 19.5 Pile head horizontal displacement time history for 2×2 pile group for $s/d = 5$. (After Maheshwari and Emani [28].)

FEM-SBFEM approach can be found in Syed and Maheshwari [38]. In order to evaluate the contact stress at the interface, in the HISS model, the Regula Falsi method is used. This is because in this method, the problem solution converges in relatively fewer iterations [5]. The near field soil nonlinearity is modeled using the HISS model. The geometrical dimensions of soil domain are shown in Figure 19.6. For the static loading condition, horizontal load is applied at the pile head level. The material properties of the soil–pile system are the same as shown in Table 19.2. The nonlinear HISS model properties for the different E_p/E_s values are listed in Table 19.3.

Figure 19.7 shows the predicted horizontal displacement at the pile head under the static loading for $E_p/E_s = 1000$. Beyond the 1500 kN load, the solution does not converge in the nonlinear case (HISS model). It can be observed that the pile head displacement is higher in the nonlinear case as compared to the linear elastic case at all the loading points.

19.5.4 HISS Model in Substructure Method

In the 3D seismic response of the PSSI problem, the full model is substructured into two parts: one is the superstructure and another is the pile–soil subsystem. The seismic response of the structure is calculated using the successive coupling of both the substructures at each time step. Details of the successive coupling incremental solution scheme can be found in Maheshwari et al. [27]. The nonlinearity of the soil is modeled using the HISS model in the FEM formulation. The selected problem of a 2×2 pile group is shown in Figure 19.8. The material properties used for the pile and soil are the same as discussed in Section 19.5.1. For nonlinear analysis, three kinds of conditions are

TABLE 19.2

Elastic Material and Geometrical Properties of Soil and Pile

	Young's Modulus (MPa)	Unit Weight (kN/m³)	Poisson's Ratio	Damping Ratio (%)	L/d
Soil	25	18	0.4	5	-
Pile	25,000	25	0.25	0	15

TABLE 19.3
The Nonlinear HISS Model Properties

	γ	n	h_1	h_2	E_s, MPa	Poisson's Ratio
$Ep/Es = 1000$	0.047	2.4	0.0034	0.78	25	0.4
$Ep/Es = 130$	0.123	2.45	0.845	0.0215	193	0.38

used consecutively to study the solution convergence, specifically the displacement condition [39], robustness of the HISS iterative solution [40], and the HISS model parameter λ [41]. The solution is assumed to converge when the absolute value of the dimensionless yield surface F is less than 10^{-10}.

Figure 19.9a and b shows the acceleration time history at the pile head and top of the structure, respectively. It is observed that due to the nonlinearity of the soil (HISS model), the response is increased by a significant amount. The peak acceleration of the pile head and structure is increased by 29.03% and 8.51%, respectively. Figure 19.9c shows the smoothed Fourier coefficient of the structure and pile head. It can be seen that the effect of the HISS model is significant at the low frequency and at the higher frequency (>7.5Hz); nonlinearity effect is negligible.

19.6 SUMMARY AND REMARKS

In this chapter, the HISS soil model is discussed for its implementation in the finite element formulation. Further, the applications of the HISS model for various geotechnical engineering problems such as a single pile, a pile group, and soil–structure interaction are illustrated. Furthermore,

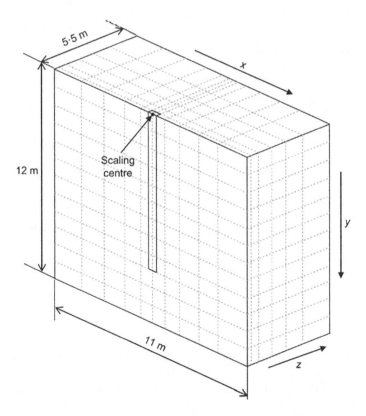

FIGURE 19.6 Half symmetric finite element model of single pile–soil system. (After, Syed and Maheshwari [30].)

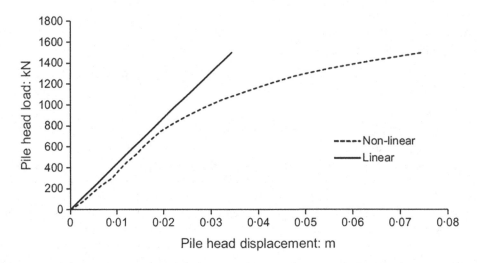

FIGURE 19.7 Horizontal load displacement at the pile head. (After, Syed and Maheshwari [30].)

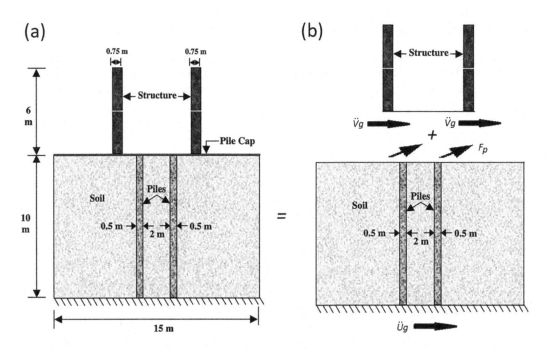

FIGURE 19.8 Pile–soil–structure system for 2 × 2 pile group. (a) Full model and (b) substructured model. (After, Maheshwari et al. [27].)

determination of DSC parameter for Solani sand is also discussed in detail using the cyclic triaxial test. Following conclusions can be drawn:

1. A 3D finite element model of a pile group in conjunction with the HISS model for the soil plasticity and hardening behavior can be used for harmonic and transient excitation loading. Soil nonlinearity reduced the dynamic impedance of the pile–soil system; however, its effect on the stiffness coefficient is larger than its effect on the damping coefficient. The nonlinear seismic response of the pile–soil system depends on the frequency of excitation; its effect is larger at low frequency as compared to higher frequency.

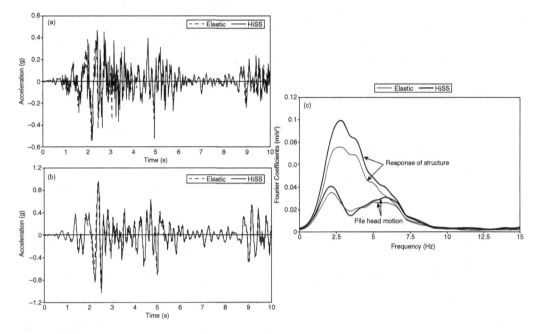

FIGURE 19.9   Effect of nonlinearity on (a) pile head acceleration time history. (b) Acceleration time history at the top of structure. (c) Fourier coefficient at the structure and pile head. (After, Maheshwari et al. [27].)

2. The coupled FEM-CIFECM method can be used for the nonlinear dynamic analysis of a PSSI system. In this modeling, the near field is modeled using the FEM with HISS model for soil nonlinearity and the far field is modeled using the CIFECM with elastic soil properties.
3. The coupled FEM-SBFEM approach can be used for nonlinear dynamic analysis of geotechnical engineering problems. In this modeling, the near field is modeled using the FEM with HISS model for soil nonlinearity and the far field is modeled using the SBEFM so as to absorb the incoming incident waves.
4. The HISS model can also be implemented in the substructure method using the successive increment technique, which has the advantage over the direct method of analysis, since in this technique two parallel programs can run simultaneously for two substructures.
5. The DSC parameter (phase change parameter, n) was calculated using the cyclic triaxial test for Solani sand.

ACKNOWLEDGMENTS

The leading author would like to acknowledge the contributions of his two former PhD students Dr. Pavan K. Emani and Dr. Madani Syed N. in the development of implementation of application of the HISS soil model for SSI problems. Without these contributions, the compilation of this book chapter would have not been possible. The authors extend their gratitude to both of them.

REFERENCES

1. Potts, D. M., and Gens, A. "A critical assessment of methods of correcting for drift from the yield surface in elasto-plastic finite element analysis." *International Journal for Numerical and Analytical Methods in Geomechanics*, 9, 2, 1985, 149–159.
2. Britto, A.M., and Gunn, M.J. Critical state soil mechanics via finite elements. Ellis Horwood series in civil engineering, Halsted Press, 1987A.

3. Borja, R. I., and Lee, S. R. "Cam-clay plasticity, part 1: implicit integration of elasto-plastic constitutive relations." *Computer Methods in Applied Mechanics and Engineering*, 78, 1, 1990, 49–72.

4. Borja, R.I. "Cam-Clay plasticity, Part II: Implicit integration of constitutive equation based on a nonlinear elastic stress predictor." *Computer Methods in Applied Mechanics and Engineering*, 88, 2, 1991, 225–240.

5. Sloan, S.W., Abbo, A.J., and Sheng, D. "Refined explicit integration of elastoplastic models with automatic error control." *Engineering Computations*, 18, 2, 2001, 121–194.

6. Drucker, D.C., "On uniqueness in the theory of plasticity." *Quarterly of Applied Mathematics*, 14, 1, 1956, 35–42.

7. Desai, C. S. "Finite element residual schemes for unconfined flow." *Int. J. Numer. Meth. Eng*, 10, 6, 1976, 1415–1418.

8. Desai, C. S. "A general basis for yield, failure and potential functions in plasticity." *Int. J. Numer. Anal. Methods Geomech.*, 4, 4, 1980, 361–375.

9. Desai, C. S., and Li, G. C. "A residual flow procedure and application for free surface flow in porous media." *Advances in Water Resources*, 6, 1, 1983, 27–35.

10. Desai, C. S., Somasundaram, S., and Frantziskonis, G. "A hierarchical approach for constitutive modelling of geologic materials." *Int. J. Numer. Anal. Methods Geomech.*, 10, 3, 1986, 225–257.

11. Desai, C. S., and Woo, L. "Damage model and implementation in nonlinear dynamic problems." *Computational Mechanics*, 11, 2, 1993, 189–206.

12. Desai, C. S., Basaran, C., and Zhang, W. "Numerical algorithms and mesh dependence in the disturbed state concept." *Int. J. Numer. Meth. Eng*, 40, 16, 1997, 3059–3083.

13. Desai, C. S. *Mechanics of Materials and Interfaces: The Disturbed State Concept*. Boca Raton, FL: CRC Press. 2001.

14. Xiao, Y., and Desai, C. S. "Constitutive modeling for overconsolidated clays based on disturbed state concept. I: Theory." *Int. J. Geomech.*, 19, 9, 2019, 04019101.

15. Desai, C. S., and Zhang, W. "Computational aspects of disturbed state constitutive models." *Comput Methods Appl Mech Eng.*, 151, 3, 1998, 361–376.

16. Desai, C.S., "Disturbed state concept as unified constitutive modeling approach." *Journal of Rock Mechanics and Geotechnical Engineering*, 8, 3, 2016, 277–293.

17. Lu, X.L., Chen, B., Li, P.Z. and Chen, Y.Q. "Numerical analysis of tall buildings considering dynamic soil-structure interaction." *Journal of Asian Architecture and Building Engineering*, 2, 1, 2003, 1–8.

18. Breysse, D., H. Niandou, S. Elachachi, and L. Houy. "A generic approach to soil–structure interaction considering the effects of soil heterogeneity." *Geotechnique*, 55, 2, 2005, 143–150.

19. Soneji, B. B., and R. S. Jangid. "Influence of soil–structure interaction on the response of seismically isolated cable-stayed bridge." *Soil Dynamics and Earthquake Engineering*, 28, 4, 2008, 245–257.

20. Çelebi, E., Göktepe, F., and Karahan, N. "Non-linear finite element analysis for prediction of seismic response of buildings considering soil-structure interaction." *Natural Hazards and Earth System Sciences*, 12, 11, 2012, 3495–3505.

21. Maheshwari, B. K., and M. Firoj. "Equivalent linear spring-dashpot model for embedded foundations of NPP." *In 17th World Conference on Earthquake Engineering, 17WCEE* Sendai, Japan, 4c-0014, 2020.

22. Sevim, B., Bayraktar, A., Altunişik, A.C., Atamtürktür, S., and Birinci, F. "Assessment of nonlinear seismic performance of a restored historical arch bridge using ambient vibrations." *Nonlinear Dynamics*, 4, 2011, 755–770.

23. Dhadse, G. D., Ramtekkar, G. D., and Govardhan, B. "Finite element modeling of soil structure interaction system with interface: A review." *Archives of Computational Methods in Engineering*, 2020, 1–18. doi:10.1007/s11831-020-09505-2.

24. Homaei, F., and Mahdi, Y. "The probabilistic seismic assessment of aged concrete arch bridges: The role of soil-structure interaction." *Structures*, 28, 2020, 894–904.

25. Maheshwari, B.K., Truman, K.Z., Naggar, M.H.E. and Gould, P.L. "Three-dimensional finite element nonlinear dynamic analysis of pile groups for lateral transient and seismic excitations." *Canadian Geotechnical Journal*, 41, 1, 2004, 118–133.

26. Maheshwari, B.K., Truman, K.Z., Gould, P.L. and El Naggar, M.H. "Three-dimensional nonlinear seismic analysis of single piles using finite element model: Effects of plasticity of soil." *International Journal of Geomechanics*, 5, 1, 2005, 35–44.

27. Maheshwari, B.K., Truman, K.Z., El Naggar, M.H. and Gould, P.L. "Three-dimensional nonlinear analysis for seismic soil–pile-structure interaction." *Soil Dynamics and Earthquake Engineering*, 24, 4, 2004, 343–356.

28. Maheshwari, B.K. and Emani, P.K. "Three-dimensional nonlinear seismic analysis of pile groups using FE-CIFECM coupling in a hybrid domain and HISS plasticity model." *International Journal of Geomechanics*, 15, 3, 2015, 04014055.

29. Maheshwari, B.K. and Syed, N.M. "Verification of implementation of HiSS soil model in the coupled FEM–SBFEM SSI analysis." *International Journal of Geomechanics,* 16, 1, 2016, 04015034.
30. Syed, N.M. and Maheshwari, B.K. "Non-linear SSI analysis in time domain using coupled FEM–SBFEM for a soil–pile system." *Géotechnique,* 67, 7, 2017, 572–580.
31. Desai, C.S., Sharma, K.G., Wathugala, G.W. and Rigby, D.B. "Implementation of hierarchical single surface δ_0 and δ_1 models in finite element procedure." *International Journal for Numerical and Analytical Methods in Geomechanics,* 15, 9, 1991, 649–680.
32. Armaleh, A. S. and Desai, C. S. *Model Include Testing of Cohesionless Soils Under Disturbed State Concept.* Report to the NSF, Dept. of Civil Eng. And Eng. Mechanics, Univ. of Arizona, Tucson, Arizona, USA, 1990.
33. Syed, N.M. and Maheshwari, B.K. "Evaluation of DSC parameters for solani sand." In *Proceedings of Indian Geotechnical Conference,* IIT Bombay, Mumbai, India, 2010.
34. Bentley, K. J., and El Naggar, M. H. "Numerical analysis of kinematic response of single piles." *Can. Geotech. J.,* 37, 2000, 1368–1382.
35. Novak, M., and Mitwally, H. "Transmitting boundary for axisymmetrical dilation problems." *J. Eng. Mech.,* 114, 1, 1988, 181–187.
36. Desai, C.S., Wathugala, G.W., and Matlock, H. "Constitutive model for cyclic behavior of clays. II: Applications." *Journal of Geotechnical Engineering,* 119, 4, 1993, 730–748.
37. Emani, P.K. and Maheshwari, B.K. "Dynamic impedances of pile groups with embedded caps in homogeneous elastic soils using CIFECM." *Soil Dynamics and Earthquake Engineering,* 29, 6, 2009, 963–973.
38. Syed, N.M. and Maheshwari, B.K. "Modeling using coupled FEM-SBFEM for three-dimensional seismic SSI in time domain." *International Journal of Geomechanics,* 14, 1, 2014, 118–129.
39. Bathe, K. J. *Finite Element Procedures in Engineering Analysis.* Englewood Cliffs, NJ: Prentice-Hall; 1982.
40. Wathugala, G.W. "Finite element dynamic analysis of nonlinear porous media with applications to piles in saturated clays." *PhD dissertation. Dept. of Civil Eng. and Eng. Mechanics,* Univ. of Arizona, Tucson, Arizona, 1990.
41. Chen, W.F., Baladi, G.Y. *Soil Plasticity: Theory and Implementation.* Amsterdam: Elsevier; 1985.

Section 6

Concrete, Pavement Materials, and Structural Engineering

20 Unified DSC/HISS Constitutive Model for Pavement Materials

Chandrakant S. Desai
Dept. of Civil and Architectural Engineering
and Mechanics, University of Arizona

20.1 INTRODUCTION

Appropriate constitutive models based on measured responses of materials in pavement systems are required for realistic prediction of the behavior of pavements. A unified model should allow for important factors such as elastic, plastic and creep deformations, and microcracking leading to softening, fracture and failure, and healing or strengthening under repetitive mechanical and thermal loadings, because the material experiences effects of these factors simultaneously, in a coupled manner.

Several models based on elasticity (e.g., resilient modulus (RM)), plasticity, and viscoelasticity have been developed in the past. However, most of these models cannot account for significant factors affecting pavement behavior in a unified manner. Their combination can result in ad hoc models, and increase the complexity and the number of parameters.

This paper presents a constitutive modeling approach, called the disturbed state concept (DSC) with hierarchical single surface (HISS) plasticity, which can account for the significant factors in a single unified framework for both pavement materials and interfaces [1–3]. Because of its unified and hierarchical nature, DSC/HISS presents a simplified approach with a lower number of parameters. One of the main advantages of the DSC/HISS is that most of its parameters have physical meanings and can be determined based on standard (laboratory) tests.

Furthermore, DSC can be used for most materials for which appropriate test data are available [1]. Hence, it is suitable for bound and unbound geologic materials that form the base of pavements. Moreover, the same framework for solid materials can be specialized to interfaces and joints [1].

The scope of this chapter included a brief review of existing models for materials and solution (computer) procedures, discussion of the limitations of existing procedures based mainly on empirical and/or empirical-mechanistic approaches, description of the unified modeling approach called the DSC/HISS that provides a mechanistic framework for pavement materials, a review of the use of the DSC/HISS model by others, descriptions of two- and three-dimensional nonlinear finite element (FE) computer procedure in which the DSC models are implemented, capabilities of the DSC model to handle major distresses: permanent deformation (rutting), microcracking and fracture, and reflection cracking under mechanical and thermal loading, and typical applications and validations.

20.2 APPROACHES FOR PAVEMENT ANALYSIS AND DESIGN

Figure 20.1 shows a schematic description of various approaches for analysis, design, and rehabilitation of pavements [2,3]. The empirical (*E*) approach is based on the experience and/or knowledge of certain index properties such as California bearing ratio (CBR), limiting shear failure, and limiting deflections [4]. Such index and empirical properties may not include effects of multidimensional geometry, loading, realistic material behavior and spatial distribution of displacements, stresses, and strains in the multilayered pavement systems. Hence, use of such empirical approaches can be considered to possess limited capabilities.

DOI: 10.1201/9781003362081-26

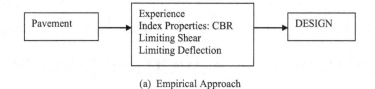

(a) Empirical Approach

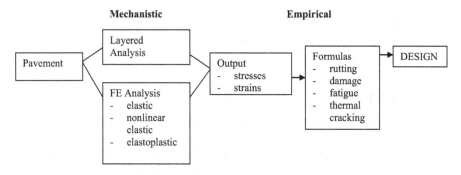

(b) Mechanistic-Empirical Approach

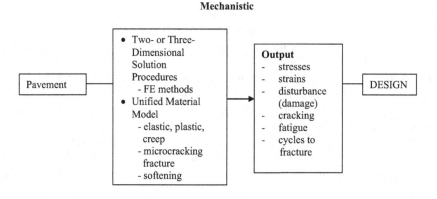

(c) Mechanistic Approach

FIGURE 20.1 Procedures for pavement analysis and design.

The mechanistic-empirical (M-E) approach is based on the limited use of the principles of mechanics such as elasticity, plasticity, and viscoelasticity. It involves two stages: In the first stage, the layered pavement system is analyzed by using a mechanistic model such as the layered elastic theory and FE procedure that includes elastic, nonlinear elastic (e.g., RM model), or elastoplastic models such as von Mises, Mohr–Coulomb, and hardening or continuous yielding [1,5,6]. The stresses and strains, usually under total or incremental application of the wheel load, computed from stage 1, are used in empirical formulas for calculation of rutting, damage, cracking under mechanical and thermal load, and cycles to failure, in the second stage. Very often, uniaxial quantities such as the tensile strain, ε_t, at the bottom of the asphalt layer, vertical compressive strain, ε_c, at the top of the subgrade layer, vertical stress, σ_y, under the wheel load, and tensile stress, σ_t, at the bottom of the asphalt layer, are used to compute various distresses using empirical formulas [4]. The M-E approach can lead to improved design compared to the empirical approach. Evaluation of distresses based on the uniaxial quantities in empirical formulas may not provide accurate predictions of the distresses as affected by multidimensional geometry, non-homogeneities, anisotropy, and nonlinear material response, which are dependent on stress, strain, time, environmental factors, and repetitive loading.

The full mechanistic (M) approach allows for geometry, nonhomogeneities, anisotropy, and non-linear material properties for all layers in a unified manner. As a result, the distresses are evaluated as part of the solution (e.g., FE) procedure, usually without the need of the empirical formulas.

The American Association of State Highway Officials (AASHTO) Design Guides [7] are commonly used for pavement design. Parts of the developments in the recently developed Design Guide include the M-E approach. The Strategic Highway Research Program (SHRP) [8] and the Superpave research attempts to develop general and unified material models. However, such unified models are very often based on ad hoc combinations of models for specific material properties such as linear elastic creep, viscoplastic creep, damage, and fracture [9–12]. Although these specialized models have been used commonly in the pavement engineering area, their ad hoc combination may not be suitable for the realistic behavior of materials in which the elastic, plastic, creep, damage, fracture, softening, and healing can occur simultaneously under the applied loading. They suffer from limitations such as the component models may not be consistently integrated, they can be relatively complex, and the material parameters involved can be large. Some of the parameters do not have physical meanings, they may not relate to the specific states during deformation, and hence, they need to be determined by using mainly curve fitting and least square procedures [8].

On the other hand, unified models that overcome many of the above limitations can be available in other engineering areas such as mechanics, geomechanics, and mechanical engineering. In the early 1980s, the author conducted a research project [13] where constitutive models, available at that time, were developed and implemented in mechanistic 2-D and 3-D FE procedures for track support systems. The models were calibrated by using comprehensive material tests, and computer codes were verified with respect to field observations. Indeed, there exists a need for advanced and unified mechanistic models. One such model based on DSC/HISS approach by Desai and coworkers [1,14–16] has been also used for pavement analysis by various investigators; some of the publications are reviewed below. This paper presents the DSC model, which includes the HISS model as a special case. It is unified and economical compared to many other models currently available.

The objective of this chapter is to present a full mechanistic approach with powerful and unified constitutive models based on the DSC that has been already developed and can be readily used for practical pavement applications. Before the presentation of the DSC/HISS models and use of associated two- and three-dimensional computer codes for the treatment of various distresses and applications, a brief review of some of the available models is first presented.

20.3 REVIEW OF THE RESILIENT MODULUS APPROACH

The RM approach has been used extensively in pavement engineering [4,17–19]. Although it can provide satisfactory predictions of elastic uniaxial displacements, it may not be capable of predicting the foregoing multidimensional effects such as rutting, microcracking and fracture. The RM approach relies on the test behavior of pavement materials in which it is observed that after a certain critical number of loading cycles, N_c, the material reaches the so-called *resilient state*. The behavior of the material in the resilient state is approximately elastic. Hence, it is possible to compute uniaxial (vertical) strains and displacements in the resilient state by using the M_R (RM). However, most materials can experience microcracking growth at cycles much before the resilient cycle, N_c. Microcracks would initiate, grow, and coalesce at the critical disturbance, D_c, which may occur before N_c, as described later. It may not be possible to predict the fracture behavior by using the RM approach.

The RM can be used to characterize the stress–strain behavior of a material in the resilient state by using mathematical functions such as hyperbola, parabola, and (combination of) exponential functions [1,4,8,18]. Then, it can be implemented as a nonlinear or piecewise linear elastic model in a solution (FE) procedure, which can yield stresses and strains under a given loading. They can be used in empirical formulas for various distresses, such as the use of the RM can represent the M-E approach, Figure 20.1b. A FE procedure with the piecewise nonlinear RM approach with interface and infinite elements is developed [20] for incorporation in the AASHTO 2002 Design Guide.

Functional Forms: One of the expressions for M_R that includes the combined effects of shear and mean stresses used commonly [18], and implemented in the FE code for 2002 Design Guide [20], is given by:

$$M_R = k_1 p_a \left(\frac{\theta - 3k_6}{p_a} \right)^{k_2} \left(\frac{\tau_{\text{oct}}}{p_a} \right)^{k_3} \tag{20.1}$$

where $k_1, k_2, k_3,$ and k_6 are material parameters, p_a = atmospheric pressure constant, $\theta = J_I = \mathrm{E}$ $\sigma_1 + \sigma_2 + \sigma_3$, which is proportional to the mean stress, τ_{oct} is the octahedral shear stress = $\left[\frac{1}{9} (\sigma_1 - \sigma_2)^2 + (\sigma_2 - \sigma_3)^3 + (\sigma_3 - \sigma_1)^2 \right]^{1/2}$, and $\sigma_1, \sigma_2,$ and σ_3 are the principal stresses.

Poisson's Ratio: When M_R is used in the context of nonlinear elasticity, it can take the place of the traditional tangent elastic modulus, E_t. Then, for isotropic materials, Poisson's ratio, v, can be assumed to be constant or can also be expressed as a function of stress [8]. In the nonlinear elastic formulation, the behavior of the material is still treated as elastic during each increment of loading. Hence, in the context of the theory of elasticity, the value of the Poisson's ratio needs to be less than 0.5; otherwise, the formulation will collapse due to the singularity in the stress–strain matrix.

Efforts have been made to express a strain ratio of the lateral strain, ε_3, to the axial strain, ε_1, whose value may be greater than 0.50. Indeed, such a ratio can be termed as the Poisson's ratio only up to the behavior when $v < 0.50$. Such formulations in the context of linear elastic theory may not be realistic. Theories such as plasticity can be used to accommodate (dilative) behavior beyond the point at which the compactive response transits to the dilative state [1].

20.3.1 Other Models

In addition to the RM approach, several other empirical approaches are used in pavement analysis for the evaluation of important distresses such as rutting, damage, and fracture. These approaches are usually based on the computed stresses and strains at selected locations in the pavement, often obtained by using layered elastic analysis or nonlinear elastic FE procedures. In conjunction with empirical factors based on (field) observations, such approaches can sometimes provide reasonable predictions. However, they may not be considered mechanistic because they do not involve calculation of distresses based on evolving stresses and strains in the multidimensional pavements as affected by nonlinear material response.

In addition to the linear and nonlinear elastic models, plasticity models have often been used for pavement materials. Plasticity models can include classical (e.g., von Mises, Mohr–Coulomb, and Drucker–Prager) and enhanced (e.g., continuous hardening or yielding critical state and Cap, Vermeer, and hierarchical single surface – HISS) models. Although these models can provide improvements, particularly regarding prediction of permanent deformations, they are not capable of handling other important factors such as microcracking, fracture, and softening or degradation.

The SHRP project [8] employed the viscoelastic models used commonly for pavement materials such as asphalt [10,11] and other models for plasticity, fracture, and damage. Each of these models was essentially an independent one, and the overall model for the combined (elastic, plastic, creep, damage, and fracture) responses may be considered to represent an ad hoc combination of various models. Therefore, the overall model can be complex and involve a great number of parameters, many of which do not have physical meanings. In other words, they were not related to specific states of material behavior, and their determination could involve mainly curve fitting procedures. In short, the overall model used may be unrealistic for rational and realistic modeling of the unified behavior of bound and unbound (geologic) materials. It was suggested that the RM model can be used for the unbound materials. As implied before, this is unrealistic because previous research for geologic materials showed that nonlinear elastic (RM) models cannot represent the actual behavior to allow for factors such as plastic strains, volume change, stress paths and repetitive loading, microcracking, and fracture [1].

It is believed that the approaches based on separate viscoelasticity with plasticity, damage and fracture mechanics models may also lead to an ad hoc combination and may not lead to unified and economical models for pavement materials [21].

Hence, although continuing improvements have occurred for pavement distress analysis, not many unified mechanistic models have yet been reported for analysis, design, maintenance, and rehabilitation. A unified model should be able to characterize all significant material responses in a *single framework*. This chapter presents an integrated methodology based on the unified constitutive model called the DSC/HISS for the characterization of the behavior of pavement materials, interfaces, and joints. It is believed that the DSC/HISS with two- and three-dimensional computer (FE) procedures can provide a fully mechanistic approach considered to be desirable in pavement engineering.

20.3.2 FACTORS IN THE MECHANISTIC MODEL

The basic issue is the prediction of the performance of the pavement under repetitive mechanical and environmental (thermal, fluid, etc.) loadings. The mechanical loading is due mainly to the repeated application of traffic-wheel load. The thermal loading arises due to daily and seasonal variation of temperature. The fluid in pavement materials can be due to the ingress of water, which may lead to full or partial saturation of materials.

In some conventional procedures, the materials in the pavement are assumed to be linearly elastic and isotropic; then, the models such as elastic layered theory are used to predict displacements, stresses, and strains [4]. However, both the bound and unbound materials in the pavement exhibit nonlinear behavior, which is affected by factors such as the state of stress and strain; initial or in situ conditions like stress, pore water pressure and inhomogeneities; irreversible (plastic) deformations; viscous or creep response; stress path; volume change; anisotropy; temperature; and fluid and type of loading. Hence, although the assumption of elastic behavior may sometimes yield satisfactory results, their validity is highly limited. For a full mechanistic characterization, it is necessary to use constitutive or material models that allow for the foregoing factors, particularly those that significantly affect the pavement response.

20.3.3 THE DSC/HISS APPROACH

The DSC is based on the idea that the behavior of a deforming material (element) can be expressed in terms of the behavior of the relative intact (RI) or continuum part and the microcracked (or healed) part called the fully adjusted (FA). During the deformation, the (initial) RI material transforms continuously into the FA part and at the limiting condition (load), the entire material element approaches the FA state. Details of the DSC and HISS plasticity equations are presented in Chapter 1. Some relevant and important topics are presented below.

20.4 CREEP BEHAVIOR

The DSC model allows incorporation of the creep behavior using the hierarchical multicomponent DSC (MDSC) system. The rheological model that includes both elastic and plastic creep is depicted in Figure 20.2. This model allows the choice of elastic, Maxwell, viscoelastic (ve), elastoviscoplastic (evp) [22], and viscoelastic-viscoplastic (vevp) models depending upon the user's need. Often, the evp (Perzyna) model that essentially allows for the secondary creep has been used. However, for asphalt concrete, the vevp model that allows for both elastic and plastic creep is considered more suitable. Details of this model are given by References [1,23]. The major advantages of the MDSC [1] and overlay [24] models are that they are consistent with the FE analysis; their material

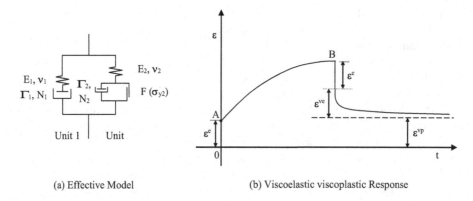

(a) Effective Model   (b) Viscoelastic viscoplastic Response

FIGURE 20.2 Multicomponent (vevp) disturbed state concept (DSC) model: Viscoelastic and viscoplastic models are special cases.

parameters are similar as those for elastic, plastic, and viscoplastic models; and they can be determined from standard laboratory tests. These are important advantages compared to the ad hoc combination of models used in the previous pavement research.

For the viscoplastic (Perzyna) version of the MDSC, there are two parameters, Γ and N, given in the following equations:

$$d\dot{\underline{\varepsilon}}^{up} = \Gamma \langle \varphi \rangle \frac{\partial F}{\partial \underline{\sigma}}$$

(20.2a)

$$\phi = \left(\frac{F}{F_o} \right)^N$$

(20.2b)

where $\dot{\underline{\varepsilon}}^{up}$ is the incremental vector of the viscoplastic strain, and F_o is the reference value of F.

20.4.1 VISCOELASTIC-VISCOPLASTIC (VEVP) MODEL

The vevp model in the MDSC, Figure 20.2, allows for both viscoelastic and viscoplastic creep. It is based on the two units in the MDSC system. Unit 1 has parameters $E_1, v_1, \Gamma_1, N_1, \sigma_{y1}(=0)$; and Unit 2 has $E_2, v_2, \Gamma_2, N_2, F(=\sigma_{y2})$. Here, E and v = elastic modulus and Poisson's ratio, Γ and N = parameters in the viscoplastic model (Eq. 20.2), σ_y = yield stress, and F = yield function. The procedure for finding these parameters and the mechanism in this model can be described as follows [1]:

At time $t = 0$, the dashpots are not operational and only the springs deform. The instantaneous elastic strain, ε^e, Figure 20.2b, is given by:

$$\varepsilon^e = \frac{\sigma_0}{E_1 t_1 + E_2 t_2}$$

(20.3)

where σ_0 = applied (constant) stress and t_1 and t_2 = thicknesses of (two) units. When $t>0$, the dashpot in unit 1 will operate first because the yield stress in the slider in unit 2 is nonzero; this leads to viscoelastic creep. Later ($t>0$), when the yield condition in slider 2 is reached, the dashpot in unit 2 will become operational. When a continuous yield model is used to model the slider, both viscoelastic and viscoplastic creep can take place simultaneously during A–B, Figure 20.2b. Upon unloading, ε^e will be recovered first, then the viscoelastic strain, ε^{ve}, will be recovered, and finally the viscoplastic strain will remain, Figure 20.2b.

Although the evp (Perzyna) model has been used for many materials, particularly when the viscoelastic creep is insignificant, for materials like asphalt the above vevp model is more appropriate.

20.4.2 THERMAL EFFECTS

The thermal effects involve responses due to the temperature change (ΔT) and the dependence of material parameters on the temperature. The former is obtained by expressing the incremental strain vector as [1,25]:

$$d\underset{\sim}{\varepsilon}^t(T) = d\underset{\sim}{\varepsilon}^e(T) + d\underset{\sim}{\varepsilon}^p(T) + d\underset{\sim}{\varepsilon}(T) \tag{20.4}$$

where $\underset{\sim}{\varepsilon}$ is the strain vector; t, e, and p denote total, elastic, and plastic strains; T denotes temperature dependence; and $d\underset{\sim}{\varepsilon}(T)$ is the strain vector due to the temperature change, ΔT.

The parameters in the DSC/HISS model (Chapter 1) are expressed as the function of T. The temperature dependence is expressed by using a single function [1,26]:

$$P(T) = p(T_r)\left(\frac{T}{T_r}\right)^{\lambda} \tag{20.5}$$

where P is *any* parameter (elastic, plastic, creep, or ultimate), T_r is the reference temperature, e.g., room temperature (= 27°C or 300 K), and λ is a parameter. The values of $P(T_r)$ and λ are found from laboratory test data at different temperatures.

20.4.3 RATE EFFECTS

Details of the rate effects in constitutive DSC/HISS modeling are given in Reference [27].

20.4.4 MICROCRACKING AND FRACTURE

In the classical fracture mechanics approach, it is usually necessary to introduce, in advance, a crack(s) of arbitrary dimensions at a selected location in the pavement, say, at the interface between the asphalt pavement and base. Then, the equations from fracture mechanics are used to evaluate the initiation and propagation of the fracture. This approach is limited because the selection of the initial crack and its location are arbitrary because the crack(s) may initiate at other locations in the pavement depending on the geometry, material properties, loading, and existence of initial cracks. Also, the fracture theories may not allow adequately for the nonlinear (elastoplastic and creep) behavior of the materials.

On the other hand, the DSC/HISS allows evaluation of the initiation and location of microcracks and their growth depending on the geometry, nonlinear properties, and loading conditions; Figure 20.3a shows a schematic stress–strain curve with the representation of initiation and growth of cracks. Here, the initiation of (micro) cracks is identified by critical disturbance, D_{cm}, and the final fracture by the critical disturbance, D_c. The values of D_{cm} and D_c are determined from laboratory tests under quasistatic or cyclic loading that exhibit softening or healing response, Figure 20.4.

In the computer analysis, the initiation of microcracking in the elements is identified based on the given value of D_{cm}. The microcracks coalesce and grow, and at the given value of D_c, the final fracture can occur. Values of disturbance greater than D_c indicate growth of fractures leading to failure at $D = D_f$, Figure 20.3b. The analysis provides a complete picture of the growth of microcracks and fracture in the elements with (cyclic) loading. Based on the values of D_c from test data, the DSC has provided successful results for computation of microcracking to fracture, failure, and liquefaction; for example, for failure analysis of computer chip-substrate problems in electronic packaging [1,26,28] and for liquefaction in saturated soils [1,29]. Thus, the DSC provides a natural and holistic progression of microcracking and fracture without the need for arbitrary selection of the location and geometry of cracks, as it is required in the fracture mechanic's approach.

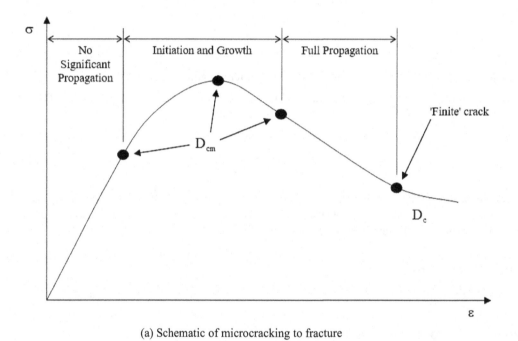

(a) Schematic of microcracking to fracture

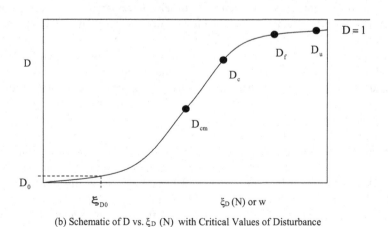

(b) Schematic of D vs. ξ_D (N) with Critical Values of Disturbance

FIGURE 20.3   Initiation of microcracking to fracture and disturbance states.

20.4.5 COMPRESSION AND TENSION RESPONSES

The behaviors of pavement materials, such as asphalt, concrete, and some soils, also exhibit different softening or degradation in compression and tension; hence, the disturbance and HISS functions can be different for both. Hence, it may be necessary to derive and use different yield functions, depending on compressive or tensile stress–strain behavior.

In the previous use of the HISS model for compressive yield, the stress transfer approach has often been used [1,30] if the computed stress indicates tension. Scarpas et al. [31] have used the HISS for compressive yield and the Hoffman model to characterize the tensile behavior.

However, different HISS surfaces for compression and tension can be derived and implemented. Figure 20.5 shows schematic of such yield surfaces in compression and tension [3]. In Figure 20.5, positive J_1 denotes compression and negative J_1 denotes tension. The sign of J_1 would depend on the magnitudes and nature (positive or negative) of the principal stresses.

When the initial state of stress is compressive (J_1- positive) and it becomes tensile (J_1- negative) during loading, the dashed part of the compressive yield surfaces is not applicable, and the tensile yield surfaces become operational. If the initial stress state is tensile (J_1-negative), and during loading, it becomes compressive (J_1- positive), the dashed portion of the tensile yield surfaces is not applicable, and the compressive yield surfaces become operational.

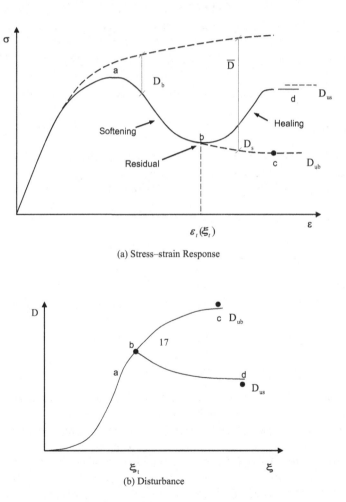

(a) Stress–strain Response

(b) Disturbance

FIGURE 20.4 Schematic of softening and healing (stiffening) response in the disturbed state concept (DSC).

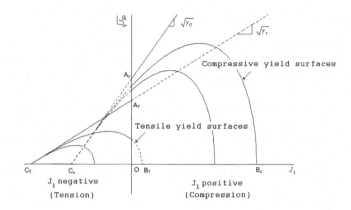

FIGURE 20.5   Schematic of compressive and tensile yield surfaces.

In a computational (FE) procedure, it would be necessary to provide material parameters for both compressive and tensile yield surfaces. Such parameters need to be determined from compression and tension tests. Then, depending upon the sign of the computed J_1, appropriate compressive or tensile yield surfaces need to be used to derive the constitutive matrix. It may be necessary to introduce required modifications in the code. For instance, points A_c and A_T in Figure 20.5 represent singularity; they can be identified only by extending test points when J_1 approaches zero.

It may be noted that unloading due to reduction in J_1 (without shear) would involve inward plastic strain increment to the yield surface for associative plasticity. Thus, the current model will predict only elastic response.

20.4.6   HEALING OR STIFFENING

Both softening and healing have been incorporated in the DSC/HISS by modifying the disturbance function, D (Chapter 1) to include the effect of healing or stiffening depicted after the residual state in Figure 20.4b. The approach has been successfully applied for temperature-induced stiffening in dislocated silicon with impurities [32]. It can be easily applied for healing in pavement when the test results to characterize the healing are available.

20.5   INTERFACES AND JOINTS

A significant attribute of the DSC/HISS model is that the above framework can be applied to model the behavior of interfaces and joints [1,14,33]. Details of the DSC/HISS model for interfaces and joints are given in Chapter 1. Here, the main attention is given to "solid" materials like asphalt, concrete, and soils.

20.6   MATERIAL PARAMETERS

The material parameters for the DSC/HISS model for the RI, FA, and disturbance function can be derived from standard triaxial, multiaxial, hydrostatic, and/or shear tests. Details of the procedures for their determination are presented in Chapter 1. Details for determination of material parameters for the DSC/HISS model are given in Chapter 1. It may be noted that for the general and significant capabilities provided by the DSC/HISS model, the number of parameters is not large [1]. For similar capabilities, other models often entail a greater number of parameters, e.g., in SHRP/SUPERPAVE approach [8]. *Also, the DSC/HISS parameters have physical meanings; in other words, most are related to the specific states during the material behavior.*

20.7 COMPUTER IMPLEMENTATION

The DSC model has been implemented in two- and three-dimensional computer (FE) procedures [1,20,28,30,34,35]. The computer codes allow for the nonlinear material behavior, in situ or initial stresses, static, repetitive, and dynamic loading, thermal and fluid effects, and sequential construction. They include computation of displacements, strain (elastic, plastic, creep), stresses, pore water pressures, and disturbance during the incremental and transient loading. Specifications of critical values of disturbance, D, permit identification of the initiation of microcracking leading to fracture and softening, and cycles to fatigue failure. Plots of the growth of disturbance, i.e., microcracking to fracture, are obtained as part of the computation [28,29,36]. Accumulated plastic strains lead to the evaluation of the growth of permanent deformations and rutting.

20.7.1 LOADING

The codes allow for quasistatic and dynamic loading for dry and saturated materials. The repetitive loading on pavement can involve a great number of cycles; an approximate procedure is described below.

20.7.2 REPETITIVE LOADING – ACCELERATED PROCEDURE

Computer analysis for 3- and 2-D idealizations can be time-consuming and expensive, especially when significantly greater number of cycles of loading need to be considered. Therefore, approximate and accelerated analysis procedures have been developed from a wide range of problems in civil (pavements) [4,8], mechanical engineering, and electronic packaging [26,28]. Here, the full computer (FE) analysis is performed for only a selected (reference) initial cycle [10,20], and then the growth of plastic strains is evaluated based on empirical relation between plastic strains and number of cycles obtained from laboratory test data. A general procedure with some new factors has been developed [28]. This procedure is modified for pavement analysis and is described below.

From experimental cyclic tests on various engineering materials, the relation between plastic strain (in the case of DSC, the deviatoric plastic strain trajectory, ξ_D, Eq. (1.3), Chapter 1), and the number of loading cycles can be expressed as:

$$\xi_D(N) = \xi_D(N_r) \left(\frac{N}{N_r} \right)^b \tag{20.6}$$

where N_r = reference cycle, and b is a parameter, depicted in Figure 20.6. The disturbance equation can be written as:

$$D = D_u \left[1 - \exp\left(-A\left\{ \xi_D(N)^Z \right\} \right) \right] \tag{20.7}$$

Substitution of $\xi_D(N)$ from Eq. (20.6) in Eq. (20.7) leads to:

$$N = N_r \left[\frac{1}{\xi_D(N_r)} \left\{ \frac{1}{A} \ell n \left(\frac{D_u}{D_u - D} \right) \right\}^{1/Z} \right]^{1/b} \tag{20.8}$$

Eq (20.8) can be used to find the cycles to failure, N_f, for the chosen critical value of disturbance $= D_c$ (say, 0.50, 0.75, 0.80).

The accelerated approximate procedure for repetitive load assumes that during the repeated load applications, there is no inertia due to "dynamic" loading. The inertia and time dependence

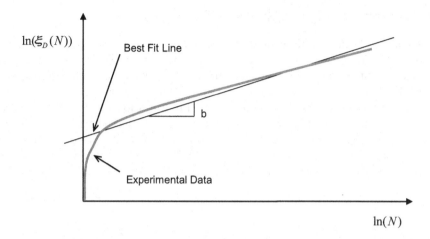

FIGURE 20.6 Accumulated plastic strain vs. number of cycles for approximate accelerated procedure.

can be analyzed by using the 3-D and 2-D procedures; however, for millions of cycles, it can be highly time-consuming. Applications of repeated load in the approximate procedure involve the following steps:

1. Perform full 2-D or 3-D FE analysis for cycles up to N_r and evaluate the values of $\xi_D(N_r)$ in all elements (or at Gauss points).
2. Using Eq. (20.6), compute $\xi_D(N)$ at selected cycles in all elements.
3. Compute disturbance in all elements using Eq. (20.7).
4. Compute cycles to failure N_f by using Eq. (20.8) for the chosen value of D_c.

The above value of disturbance allows plotting of contours of D in the FE mesh, and based on the adopted value of D_c, it is possible to evaluate extent of fracture at N_f.

Loading-Unloading-Reloading. Special procedures are integrated in the codes to allow for loading, unloading, and reloading during the repetitive loads; details are given in [1,37].

20.8 VALIDATIONS AND APPLICATIONS

The DSC model and its specialized versions have been used and validated with respect to laboratory test data for a wide range of materials, e.g., soils, rocks, concrete, asphalt, metal (alloys), silicon, and interfaces and joints. The validations include test data used to determine the parameters, independent tests not used in the determination, and field and simulated problems in the laboratory by using the computer (FE) procedures. A review and details of applications and validation of the DSC/HISS model related to pavement materials are provided below. Here, reviews of the use of the DSC/HISS model by other investigators are followed by details of a number of applications by Desi and coworkers.

20.8.1 Review of the Use of DSC/HISS by Various Researchers

Scarpas et al. [31] have used the DSC/HISS modeling by using the yield function in the HISS model. They observed that, "…. After some extensive literature survey, it became evident that the (HISS yield) proposed by Desai and coworkers fulfills this requirement and hence consists of ideal candidate for implementation (for pavement analysis)." They presented comprehensive validation for the uniaxial behavior of asphalt concrete. Figure 20.7 shows typical comparison between predicted and observed stress–strain curve for $T = 25°C$ and strain rate = 5 mm/sec [30,31,34]. The modeling

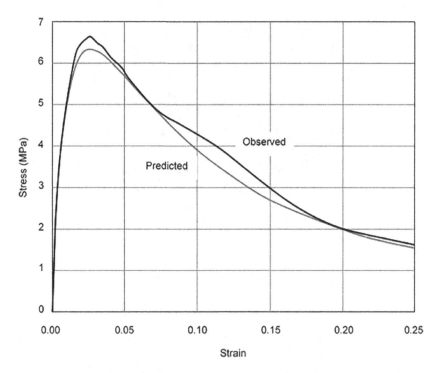

FIGURE 20.7 Comparison between predicted and observed test data for $T = 25°C$, $\dot{e} = 5$ mm/sec.

approach proposed in Reference [31] was used by Dunhill [38] for quasistatic characterization of the behavior of asphalt mixtures.

Bonaquist [39] and Bonaquist and Witczak [40] have presented the use of the DSC/HISS model as a comprehensive constitutive model for granular materials in flexible pavement systems. Uzan [41] has compared a constitutive model by Vermeer [42] and the HISS plasticity approach. It has been reported that the HISS model fits the test results better than that by the Vermeer model.

Subgrade soils from four field sites were modeled by using the HISS plasticity model with the extension proposed in Reference [39] for permanent deformations under repeated loading by Schwartz and Yau [43]. This study analyzed correlations between the HISS model parameters and other more conventional soil models. Wei et al. [44] employed the DSC/HISS model to characterize the behavior of large stone asphalt mixtures (LASMs) for fatigue response under cyclic loading.

The HIS viscoplastic model was used to characterize the behavior of asphalt mixtures with the effect of temperature and loading rate by Huang et al. [45]. Comprehensive validations were reported between the model predictions and test data. Aggarwal et al. [46] used the HISS plasticity model for defining the tests behavior of triaxial specimens of unreinforced and reinforced (with geogrids) pavement composite materials.

The DSC/HISS model was used to predict RM (M_r) for crushed rock base (CRB) as a road material by Khobklang et al. [47]. A series of compression static triaxial and RLT (repeated load triaxial) tests were used; typical test results are shown in Figure 20.8. The RLT is a laboratory tests procedure that can measure the performance of granular pavement materials in terms of permanent deformation (rut resistance) and RM (stiffness).

The equation for the RM was derived as:

$$M_r^a = C_1 \frac{\sigma_1}{\sigma_2} + C_2 \tag{20.9}$$

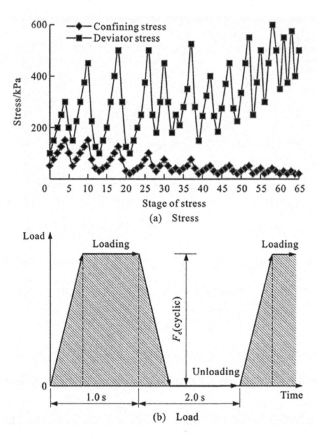

FIGURE 20.8 (a) Stress vs. stage of stress and (b) applied stress distribution according to the Austroads APRG 00/33–2000 standard.

where $C_1 = 0.22\sigma_3 + 3.30$ and $C_2 = 1.1\sigma_3 + 3.30$, and disturbance D is given by:

$$D = 1 - \frac{1.1\sigma_3 + 33.0}{1.32\sigma_3 + 36.3} \tag{20.10}$$

The equation for predicting the RM is expressed as:

$$M_r^a = (1 - D)M_r^i + DM_r^c \tag{20.11}$$

where $M_r^i = M_r^c \dfrac{\sigma_1}{\sigma_3}$ and $M_r^c + 1.32\sigma_3 + 36.3$.

Figure 20.9 shows the predictions of the stress–strain responses and RM for the CRB obtained by using the DSC equation. They show excellent agreement with tests' data.

20.8.2 DESCRIPTIONS OF THE USE OF DSC/HISS BY DESAI AND COWORKERS

The 2- and 3-dimensional procedures have been used to predict laboratory and/or field behavior of a wide range of engineering problems, e.g., static, and dynamic soil–structure interaction, dams and

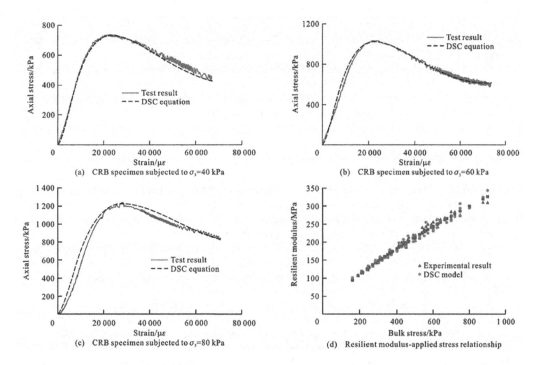

FIGURE 20.9 Back predictions for stress–strain curves and resilient modulus.

embankments, reinforced earth, tunnels, composites in chip-substrate systems in electronic packaging [1,30,34], and microstructure instability or liquefaction [28,29]. In the area of multicomponent systems like rail tracks and pavements, its specialized versions have been used to predict the field behavior [13,16,48].

The DSC/HISS models have been used to illustrate capabilities of 2-D and 3-D analyses of rigid and flexible pavements subjected to monotonic and repetitive loading including permanent deformations, fracture, and reflection cracking [2]. Only typical examples for asphalt pavements are included below using the DSC/HISS models. The DSC multicomponent model can include the viscoelastic and viscoplastic creep [1]; however, they are not included here, so also the thermal effects.

Table 20.1 shows the DSC parameters for pavement and unbound materials for the following 2-D and 3-D analyses. The parameters for the asphalt concrete were determined from the comprehensive triaxial tests reported by Reference [49]. The quasistatic and creep tests were conducted under various confining pressures and temperatures. Details of the determination of the parameters are given in References [50,51]. The unbound materials were characterized as elastic-plastic by using the HISS model; their parameters were determined from the triaxial tests reported by Reference [39]. Scarpas et al. [31] have reported uniaxial tests for asphalt concrete in which the prepeak and postpeak (softening) behaviors were observed; these tests were used to evaluate the parameters for disturbance, D.

20.8.3 Validation of the DSC Model

Validations are presented for stress–strain data and practical problems.

Example 1: Stress–Strain Data

Figure 20.10 shows comparisons between the observed and predicted stress–strain curves for strain rate = 1 in/min at three typical temperatures (*T*) and confining pressures (σ_3): (a) *T* = 40°F,

σ_3 = 43.8 psi 9 (300 kPa), (b) T = 77°F, σ_3 = 0.0 psi (0.0 kPa), and (c) T = 140°F, σ_3 = 250 psi (1.72 MPa)(49–51). A typical comparison between predicted and observed typical stress–strain curve for T = 25°C and strain rate = 5 mm/sec [31] is presented in Figure 20.7.

The above comparisons show that the DSC model can provide very good simulation of the behavior of asphalt concrete. Validations for a wide range of other materials: concrete, geologic, and metal alloys are given by Reference [1]. Validations for use of the HISS model for unbound materials (Table 20.1) have been presented by Bonaquist [39,40].

Example 2: Multilayered Pavement: Asphalt

Linear and nonlinear elastic models such as the RM have limited capabilities, and they cannot account for plastic deformations or rutting and fracture. Hence, it is necessary to use models that can allow for plastic deformations. The example below is intended to illustrate the difference between the results from the elasticity and plasticity models.

Figure 20.11 shows the FE mesh for a four-layer system with asphalt (flexible) layer [30,34]. The elastic, plastic (HISS model), and DSC properties are shown in Table 20.1. The wheel load equal to 200 psi (1.4 MPa) is applied near the center with axisymmetric idealization. The load is applied in increments and the incremental iterative procedure is used [1].

Figure 20.12 shows the displacements of the surface at the final load of 200 psi. The deformations within all layers are elastic and are much lower than those within all layers with the elastic-plastic (HISS-δ_0)model. Since rutting and microcracking leading to fracture are dependent on plastic deformations, it is essential to use models that allow for plastic or irreversible deformations. The load of 200 psi may be much lower than the ultimate load and may not cause significant disturbance. Hence, the results from the DSC are not significantly different from those from the HISS plasticity model. As will be seen later, a similar load for the repetitive case can cause higher plastic strains and disturbance at higher cycles.

Example 3: Two- and Three-Dimensional Analyses

Generally, the pavement problem and wheel loading would require three-dimensional analysis, particularly to predict microcracking and fracture response. However, for an economic analysis, a two-dimensional procedure may provide satisfactory but approximate solutions for certain applications.

Figure 20.13 shows a problem that was idealized as two- and three-dimensional; the former involved plane strain assumption with unit thickness in the y-direction, Figure 20.13c. The asphalt concrete layer (as in Figure 20.11) was simulated by using the DSC (HISS) model and the unbound layers were simulated by using the plasticity (HISS-δ_0 model) [30,34]. The total wheel load, which is applied incrementally, is 200 psi. The material properties of the four layers are shown in Table 20.1.

Figure 20.14 shows load vs. displacement curves at the central node, and stress (σ_z) vs load in the element near the top center for the 2-D and 3-D analyses. The computed stresses from both analyses do not show significant difference, while the displacements show a difference of the order of 20%. However, for some practical problems, such a difference may be acceptable, particularly because the 3-D analysis consumes more time and effort.

Example 4: 3-D Analysis of Flexible Pavement

The 3-D FE mesh is shown in Figure 20.13. Table 20.1 gives the materials' properties for the four layers.

LOADING

Two analyses were performed: (i) monotonic load up to 200 psi (1.4 MPa) in which the load was applied in 50 increments and (ii) repetitive loading, Figure 20.15, with different amplitudes, P.

The cyclic (repetitive) load (loading, unloading, and reloading) was applied sequentially; however, time dependence was not included. In the accelerated procedure described before, full FE analysis was performed for each load amplitude up to $N_r = 20$ cycles. Then, the deviatoric plastic strain trajectory (ξ_D) at the given cycle (N) was computed for the subsequent cycles using Eq. (20.6). The disturbance, D, was computed at the given cycles using Eq. (20.7). This allowed for the analysis of disturbance with cycles and computation of the cycles to failure, N_f, depending upon the chosen criteria for critical disturbance, D_c, Eq. (20.8).

Figure 20.16 shows surface permanent deformation for load = 100 and 200 psi, respectively. Figure 20.17 shows the contours of disturbance at load = 200 psi. Under the monotonic load of 200 psi, the maximum disturbance (D) is of the order to 0.024. In other words, no microcracking and fracture occur. However, as shown below, for repetitive load with similar amplitude, fracture would occur at higher cycles.

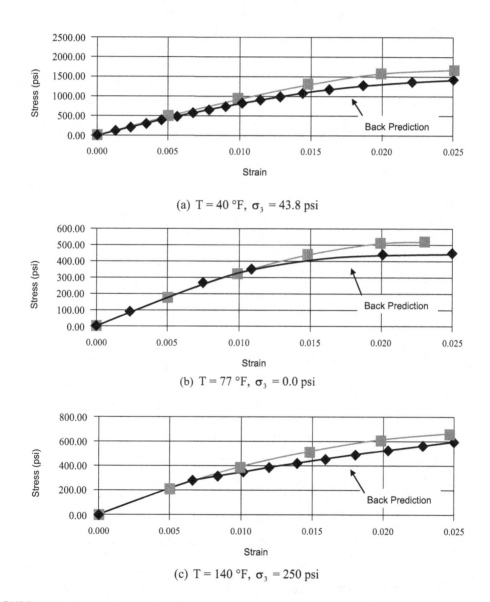

(a) T = 40 °F, σ_3 = 43.8 psi

(b) T = 77 °F, σ_3 = 0.0 psi

(c) T = 140 °F, σ_3 = 250 psi

FIGURE 20.10 Comparison between the disturbed state concept (DSC) predictions and test data for various temperatures and confining stresses (1 psi = 6.89 kPa).

TABLE 20.1

Material Parameters for Pavement Materials for 2- and 3-D Analyses of Asphalt Pavements

Parameter	Asphalt Concrete	Base	Subbase	Subgrade
E	500,000 psi	56,533 psi	24,798 psi	10,013 psi
ν	0.3	0.33	0.24	0.24
Γ	0.1294	0.0633	0.0383	0.0296
β	0.0	0.7	0.7	0.7
n	2.4	5.24	4.63	5.26
$3R$	121 psi	7.40 psi	21.05 psi	29.00 psi
a_1	1.23E-6	2.0E-8	3.6E-6	1.2E-6
η_1	1.944	1.231	0.532	0.778
D_u	1			
A	5.176			
Z	0.9397			

1 psi = 6.89 kPa.

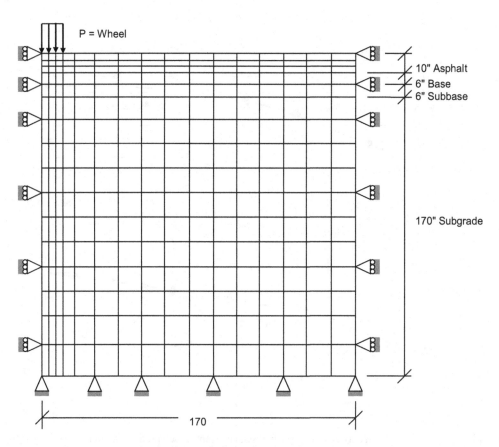

FIGURE 20.11   Finite element mesh for plasticity analysis (1 inch = 2.54 cm).

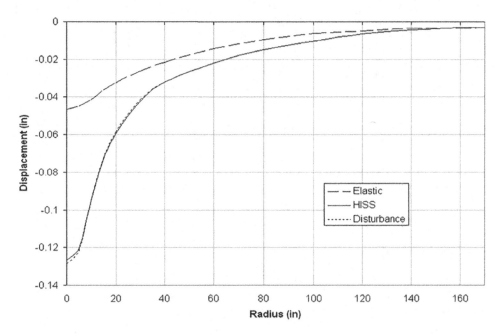

FIGURE 20.12   Computed surface displacements from elasticity, plasticity, and DSC models (1 inch = 2.54 cm).

Figure 20.18 shows contours plots of disturbance after 10, 1000, and 20,000 cycles, respectively, with load amplitude = 70 psi (480 kPa) and $b = 1.0$. After about 20,000 cycles, the disturbance $Dc = 0.8$ and greater develops, and a portion of pavement experiences cracking and fracture.

REFLECTION CRACKING

It is often appropriate to place an overlay on the existing pavements [4,52–54]. Often, under static and repetitive loading, cracks develop into the overlay directly above the cracks in the existing pavement. The DSC model is capable of predicting such "reflection" cracking.

Figure 20.19 shows a four-layered flexible pavement system with the introduction of (three) cracks at different locations in the asphalt. The concrete overlay and asphalt were modeled by the DSC while the unbound layers were modeled by the HISS-δ_0 plasticity model [30,34]. The DSC properties for concrete are given in Desai [2]. The material parameters for asphalt are given in Table 20.1. For the repetitive load behavior, the value of b, Eq. (20.6), was adopted as 0.80 for concrete and 0.30 for asphalt.

The pavement with the overlay was subjected to a repetitive load amplitude = 5.0 MPa. Figure 20.20 shows the contours of disturbance around the existing cracks for $N = 100$, 1000, and 10^6 cycles. At lower number of cycles ($N = 100$), the pavement does not experience significant cracking as $D \leq 0.80$. However, the maximum disturbance, which is greater than about 0.6 around the cracks, attracts relatively higher disturbance above the cracks. As the cycles increase, the trend continues and around $N = 1000$ (and 10^6) cycles, the disturbance around the cracks and above them show fracture ($D_c \geq 0.8$). Indeed, the value of the amplitude of the repetitive load used here is high; for lower amplitude (say, 0.69 MPa), the number of cycles will be higher.

UNIFIED METHODOLOGY

Although (other) efforts have been and are being made to develop unified models for materials in pavement engineering, it is believed that they are not yet successful in characterizing major significant responses in a single framework. However, based on the mechanistic considerations,

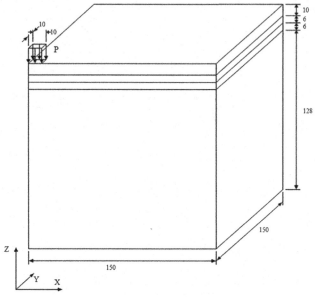

(a) Layered System (dimensions in inches): 1 inch = 2.54 cm.

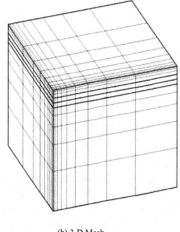

(b) 3-D Mesh

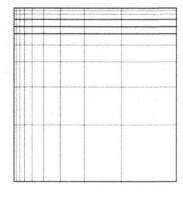

(c) 2-D Mesh

FIGURE 20.13   Four-layered pavement: Two- and three-dimensional analyses.

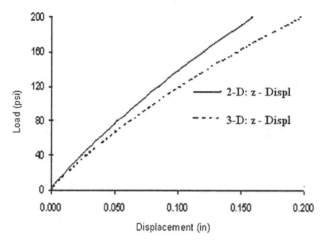

(a) Load-Displacement Curve at Center Node

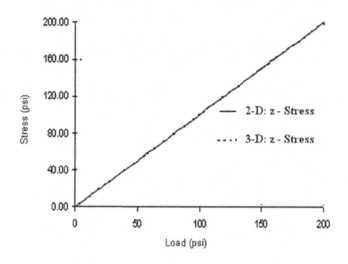

(b) Stress σ_z vs. Load in Element at Center

FIGURE 20.14   Load displacement and vertical stress from 2-D and 3-D computations (1 psi = 6.89 kPa; 1 inch = 2.54 cm).

and successful development of the DSC/HISS models and computer programs (e.g., DSC-SST2D, DSC-DYN2D, DSC-SST3D) for pavement and other engineering disciplines, it is now possible to employ the DSC/HISS unified models for a number of significant distresses. This paper presents various applications for distress predictions. In view of the length limitations, applications for other factors such as thermal cracking, moisture effect, and healing are not included. However, they can be included in the framework of the DSC/HISS capabilities [1]. Indeed, additional research developments including comparisons with analytical and field or simulated pavement performance are desirable for analysis, design, and rehabilitation applications,

20.9   CONCLUSIONS

A unified constitutive model called the DSC/HISS has been developed and used to characterize the elastic, plastic, creep, fracture, softening, and healing behavior of materials in both rigid and flexible pavements. The model has been implemented in 2-D and 3-D nonlinear FE procedures and used for

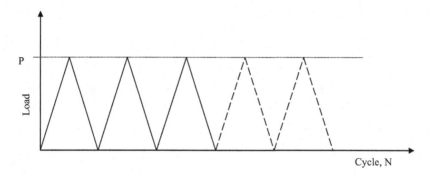

FIGURE 20.15   Schematic of repetitive loading.

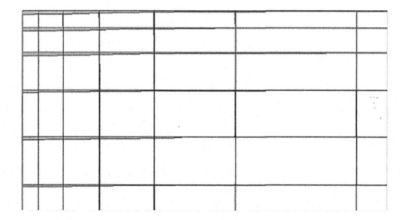

(a) Permanent Displacement (x10) after Steps = 25 and Load = 100 psi

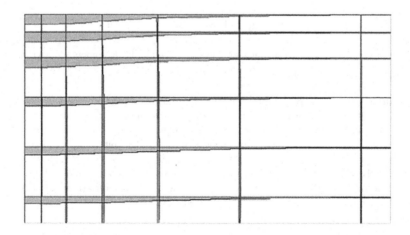

(b) Permanent Displacement (x10) after Steps = 50 and Load = 200 psi

FIGURE 20.16   Permanent displacement for 3-D computations under monotonic loading (1 psi = 6.89 kPa; 1 inch = 2.54 cm).

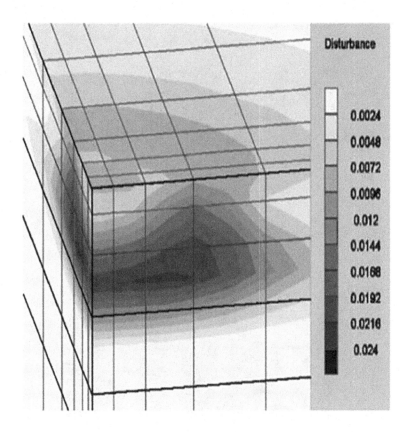

FIGURE 20.17 Contours of disturbance after 50 steps of monotonic loading (= 120 psi).

predicting the behavior of a wide range of problems in civil, mechanical, and electrical engineering. It is applied here to perform 2-D and 3-D analyses for the important analysis and design aspects for pavements such as rutting (permanent deformations), fracture, and reflection cracking.

It is believed that the DSC/HISS approach possesses significant advantages over other available models. Its concept is simple, it involves lesser number of parameters compared to any other model with similar capabilities, its parameters have physical meanings, and it is suitable for application for analysis, design, and maintenance.

The models and programs have been validated with respect to field and laboratory tests for a wide range of problems in civil and mechanical engineering. They have also been verified for several problems in pavement engineering.

ACKNOWLEDGMENTS

The development of the constitutive models and applications has been supported by grants from various government and private agencies such as NSF and DOT. The computer results reported herein were obtained with assistance from Dr. R. Whitenack, Ms. A. Bozorgzadeh, Mr. D. Cohen, and Mr. B. Simon.; their help is gratefully acknowledged. The assistance of Dr. H. B. Li and Dr. S. S. Sane for the 2- and 3-D analyses are acknowledged.

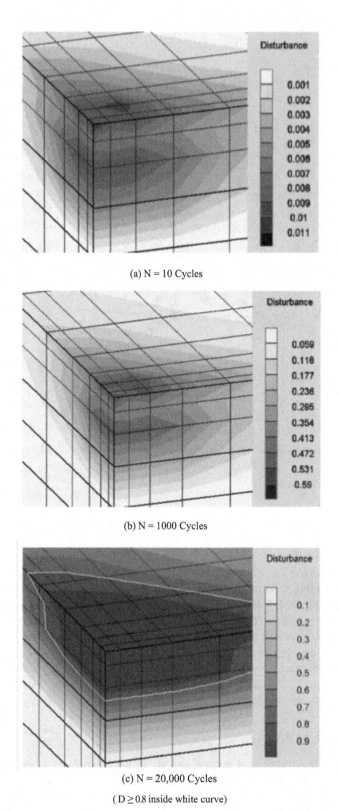

(a) N = 10 Cycles

(b) N = 1000 Cycles

(c) N = 20,000 Cycles

(D ≥ 0.8 inside white curve)

FIGURE 20.18   Contours of disturbance at various cycles: Load amplitude = 70 psi, b = 1.0. 1psi = 6.89 kPa.

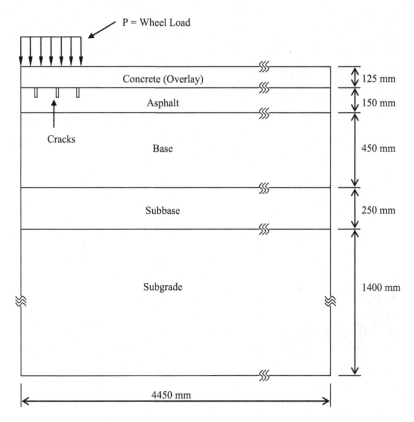

FIGURE 20.19   Reflection cracking: Layered system with three existing cracks in asphalt.

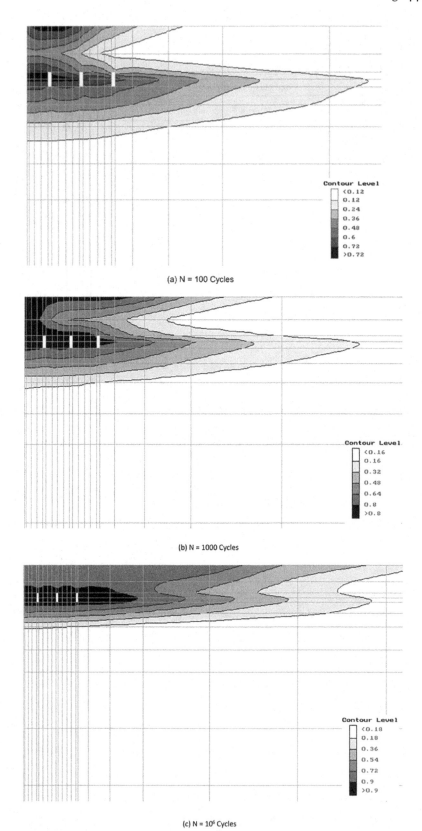

FIGURE 20.20   Reflection cracking: Contours of disturbance at various cycles.

REFERENCES

1. Desai, C.S. *Mechanics of Materials and Interfaces: The Disturbed State Concept.* CRC Press, Boca Raton, FL, 2001.
2. Desai, C.S. "Mechanistic pavement analysis and design using unified material and computer model." Keynote Paper, *Proceedings of Third International Symposium on 3-D Finite Element for Pavement Analysis, Design and Research,* Amsterdam, The Netherlands, 2002.
3. Desai, C.S. "Unified DSC model for pavement materials with numerical implementation." *Int. J. Geomech., ASCE,* 7, 2, 2007, 83–101.
4. Huang, Y.H. *Pavement Analysis and Design.* Prentice Hall, Englewood Cliffs, NJ, 1993.
5. Schofield, A.N. and Wroth, C.P. *Critical State Soil Mechanics.* McGraw-Hill, London, UK, 1968.
6. Vermeer, P.A. "A five-constant model unifying well-established concepts." *Proceedings of International Workshop on Constitutive Relations for Soils,* Grenoble, France, 1982, 175–197.
7. American Association of State Highway Officials (AASHTO). *Guide for Design of Pavement Structures.* Washington, DC, USA, 1986; 1993.
8. Lytton, R.L. et al. "Asphalt concrete pavement distress prediction: Laboratory testing, analysis, calibration and validation." *Report No. A357, Project SHRP RF. 7157-2,* Texas A&M Univ., College Station, TX, 1993.
9. Kim, Y.R., Lee, H.J., Kim, Y. and Little, D.N. "Mechanistic evaluation of fatigue damage growth and healing of asphalt concrete: Laboratory and field experiments." *Proceedings of 8th International Conference on Asphalt Pavements,* University of Washington, Seattle, Washington, USA, 1997, 1089–1107.
10. Schapery, R.A. "A theory of mechanical behavior of elastic media with growing damages and other changes in structure." *J. Mech. Psys. Solids,* 28, 1990, 215–253.
11. Schapery, R.A. "Nonlinear viscoelastic and viscoplastic constitutive equations with growing damage." *Int. J. Fracture,* 97, 1999, 33–66.
12. Secor, K.E. and Monismith, C.L. "Viscoelastic properties of asphalt concrete." *Proceedings of 41st Annual Meeting,* Highway Res. Board, Washington, DC, USA, 1962.
13. Desai, C.S., Siriwardane, H.J. and Janardhanam, R. "Interaction and load transfer through track support systems, Parts 1 and 2." *Final Report, DOT/RSPA/DMA-50/83/12,* Office of University Research, Department of Transportation, Washington, DC, USA, 1983.
14. Desai, C.S. and Ma, Y. "Modelling of joints and interfaces using the disturbed state concept." *Int. J. Num. Anal. Methods Geomech.,* 16, 1992, 623–653.
15. Desai, C.S., Somasundaram, S., and Frantziskonis, G. "A hierarchical approach for constitutive modeling of geologic materials." *Int. J. Num. Anal. Methods Geomech.,* 10, 3, 1986, 225–257.
16. Desai, C.S., Rigby, D.B. and Samavedam, G. "Unified constitutive model for materials and interfaces in airport pavements." *Proceedings of ASCE Specialty Conference on Airport Pavement Innovations – Theory to Practice,* Vicksburg, Mississippi, USA, 1993.
17. Uzan, J. "Characterization of granular materials." *NRC TRB 1022 Transp. Res. Board,* Washington, DC, 1985, 52–59.
18. Witczak, M.W. and Uzan, J. "The universal airport pavement design system. Granular material characterization." *Reports I to IV,* Univ. of Maryland, College Park, Maryland, USA, 1988.
19. Barksdale, R.D., Rix, G.J., Itani, S., Khosla, P.N., Kim, R., Lamb, P.C. and Rahman, M.S. "Laboratory determination of resilient modulus for flexible pavement design." *NCHRP Report 1–28,* Georgia Inst. of Technology, Atlanta, GA, 1990.
20. Desai, C.S. "Finite element code (DSC-2D) for 2002 design guide." *Report submitted to AASHTO 2002 Design Guide,* Arizona State Univ., Tempe, Arizona. USA, 2000.
21. Desai, C.S. "Application of unified constitutive model for pavement materials based on hierarchical disturbed state concept." *Report submitted to SUPERPAVE:* Univ. of Maryland, College Park, Maryland, USA, 1998.
22. Perzyna, P. "Fundamental problems in viscoplasticity." *Adv. Appl. Mech.,* 9, 1966, 243–277.
23. Desai, C.S., Samtani, N.C. and Vulliet, L. "Constitutive modelling and analysis of creeping slopes." *J. Geotech. Eng., ASCE,* 121, 1995, 43–56.
24. Pande, G.N., Owen, D.R.J. and Zienkiewicz, O.C. "Overlay models in time dependent nonlinear material analysis". *Comput. Struct.,* 7, 1977, 435–443.
25. William, G.W. and Shoukry, J.N. "3D finite element analysis of temperature-induced stresses in dowel jointed concrete pavements." *Int. J. Geomech.,* 3, 3, 2001, 291–307.
26. Desai, C.S., Chia, J., Kundu, T. and Prince, J. "Thermomechanical response of materials and interfaces in electronic packaging: Parts I and II." *J. Elect. Packag., ASME,* 119, 4, 1997, 294–300, 301–309.
27. Desai, C.S., Sane, S., and Jenson, J. "Constitutive modeling including creep and rate dependent behavior and testing of glacial till for prediction of motion of glaciers." *Int. J. Geomech.,* 11, 6, 2011, 465–476.
28. Desai, C.S. and Whitenack, R. "Review of models and the disturbed state concept for thermomechanical analysis in electronic packaging." *J. Elect. Packag., ASME,* 123, 2001, 1–15.

29. Desai, C.S. "Evaluation of liquefaction using disturbed state and energy approaches." *J. Geotech. Environ. Eng., ASCE,* 126, 7, 2000, 618–631.

30. Desai, C.S. "Applications of DSC-SST2D code for two-dimensional static, repetitive and dynamic analysis with user's manual I to III." Report, Tucson, Arizona, USA, 1998.

31. Scarpas, A., Al-Khoury, R., Van Gurp, C.A.P.M. and Erkens, S.M.J.G. "Finite element simulation of damage development in asphalt concrete pavements." *Proceedings of 8th International Conference on Asphalt Pavements*, University of Washington, Seattle, Washington, USA, 1997, 673–692.

32. Desai, C.S., Dishongh, T. and Deneke, P. "Disturbed state constitutive model for thermomechanical behavior of dislocated silicon with impurities." *J. Appl. Phys.*, 84, 11, 1998, 5977–5984.

33. Desai, C.S., Zaman, M.M., Lightner, J.G. and Siriwardane, H.J. "Thin-layer element for interfaces and joints." *Int. J. Num. Anal. Methods Geomech.*, 8, 1, 1984, 19–43.

34. Desai, C.S. "Applications of DSC-SST3D code for three-dimensional coupled static, repetitive and dynamic analysis with user's manual I to III." *Report, Tucson*, Arizona, USA, 2000.

35. Desai, C.S., Sharma, K.G., Wathugala, G.W. and Rigby, D.B. "Implementation of hierarchical single surface δ_0 and δ_1 models in finite element procedure." *Int. J. Num. Anal. Methods Geomech.*, 15, 1991, 649–680.

36. Pradhan, S.K. and Desai, C.S. "DSC model for soil and interface including liquefaction and prediction of centrifuge test." *J. Geotech. Geoenviron. Eng., ASCE*, 132, 2, 2006, 214–222.

37. Shao, C., and Desai, C.S. "Implementation of DSC models and application for analysis of field pile tests under cyclic loading." *Int. J. Num. Anal. Methods Geomech.*, 24, 6, 2000, 601–624.

38. Dunhill, S.T. "Quasi-static characterization of asphalt mixtures." *Ph. D. Thesis*, Univ. of Nottingham, U. K., 2002.

39. Bonaquist, R.J. "Development and application of a comprehensive constitutive model for granular materials in flexible pavement structures." *Doctoral Dissertation*, University of Maryland, College Park, Maryland, USA., 1996.

40. Bonaquist, R.F. and Witczak, M.W. "A comprehensive constitutive model for granular materials in flexible pavement structures." *Proceedings of 8th International Conference on Asphalt Pavements*, Seattle, Washington, USA, 1997, 783–802.

41. Uzan, J. "Permanent deformation of granular base material." *Transp. Res. Rec.*, 1673, 1, 1999, 89–94.

42. Vermeer, P.A. "Non-associated plasticity for soils, concrete and rock." In *Physics of Dry Granular Materials*, Herrmann, H.J., et al. (Editors), NATO ASI Series (Series E: Applied Sciences), Springer, Dordrecht, 350, 1998, 163–196.

43. Schwartz, C.W., and Yau, A.Y.Y. "Cyclic plasticity characterization of unbound pavement materials." *Proceedings of Computer Methods in Geomechanics*, Desai, C.S. et al. (Editors), Balkema, Rotterdam, 2001.

44. Wei, J., Fu, T., Meng, Y., and Xiao, C. "Investigating the fatigue characteristics of large sone asphalt mixture based on the disturbed state concept." *Adv. Mater. Sci. Eng.*, 2002, 2002, 10. Article ID: 3873174.

45. Huang, B., Mohammad, L.N., and Wathugala, G.W. "Application of a temperature dependent viscoplastic hierarchical single surface model for asphalt mixtures." *J. Mater. Civil Eng.*, 16, 2004, 2.

46. Aggarwal, P., Sharma, K.G., and Gupta, K.K. "Modeling of unreinforced and reinforced pavement composite material using the HISS model." *Res. Note, IJE Trans. Appl.*, 20, 1, 2007, 13–22.

47. Khobklang, P., Vimonsatit, V., Jitsnagian, P., and Nikraz, H. "DSC modeling for predicting resilient modulus of crushed rock base as road base material for western australia roads." *J. Traffic Transp. Eng.*, 13, 2, 2013, 1–7.

48. Desai, C.S. and Siriwardane, H.J. "Numerical models for track support structures." *J. Geotech. Eng. Div., ASCE*, 108, GT3, 1982, 461–480.

49. Monismith, C.L. and Secor, K.E. "Viscoelastic behavior of asphalt concrete pavements." *Proceedings of International Conference on Structural Design of Asphalt Pavements*, Ann Arbor, MI, 203, 1, 1962, 728–760.

50. Desai, C.S. and Cohen, D. "Determination of DSC Parameters for Asphalt Concrete." *Report*, Tucson, Arizona, USA, 2000.

51. Simon, B. "Analysis of distresses in flexible pavements using the disturbed state concept." *Report*, Department of Civil Engineering and Engineering Mechanics, The University of Arizona, Tucson, Arizona, USA, 2001.

52. Molenaar, A.A.A. "Structural performance and design of flexible road constructions and asphalt concrete overlays." *Ph.D. Dissertation*, Delft University of Technology, Netherlands, 1983.

53. FHWA "Crack and seat performance. review report, demonstration projects and pavement divisions." *Federal Highway Administration*, Washington, DC, USA, 1987.

54. Kilareski, W.P. and Bionda, R.A. "Structural overlays strategies for jointed concrete pavements, Vol. 1, sawing and sealing of joints in A-C overlay of concrete pavements." *Report No. FHWA-RD-89-142*, Federal Highway Administration, Washington, 1990.

21 Application of DSC/HISS Models for Behavior and Prediction of Rutting in Asphalt Pavements

Shivani Rani
City of Lawton Engineering Division

Musharraf Zaman
Civil Engineering and Petroleum Engineering
University of Oklahoma Norman

21.1 INTRODUCTION

More than 90% of the roads in the United States are paved using asphalt mixes (USDOT, 2017). An asphalt mix is a composite material consisting of mineral aggregates and asphalt binder. While aggregates provide a load-bearing structure for an asphalt mix, the asphalt binder holds the aggregates together and provides tensile strength to the asphalt mix. The common distresses in an asphalt pavement are rutting, fatigue cracking, reflective cracking, low-temperature cracking, and moisture-induced damage (Lu and Harvey, 2006; Abed and Al-Azzawi, 2012; Dugan et al., 2020; Rani, 2019; Ghabchi et al., 2021; Rani et al., 2021). Under high temperature and heavy vehicular loading, however, rutting or permanent deformation is one of the primary distresses in asphalt pavements.

Rutting in asphalt pavements can be caused by permanent deformations in multiple layers (surface, base, subbase, and subgrade), although the surface layer is a major contributor. Also, repeated traffic loading and high temperature are important contributors. Rutting deformation can occur either through plastic flow in the hot-mix asphalt (HMA) layer or shearing in one or more of the underlying layers (Brown et al., 2009). According to Krugler et al. (1985), three different mechanisms are responsible for rutting: (i) consolidation due to traffic; (ii) plastic deformation due to instability of the mix; and (iii) instability due to stripping of the binder underneath the surface course. A pavement usually experiences rutting in three different stages (Figure 20.1): (i) primary rutting or wear rutting due to environmental conditions and traffic loading; (ii) secondary rutting or structural rutting due to permanent deformation of structural layers; and (iii) tertiary rutting or instability rutting due to lateral movement of materials within an asphalt pavement (Dawley et al., 1990; Brown et al., 2009; Rani et al., 2018). A significant amount of rutting can cause structural failure, loss of control while driving, and hydroplaning caused by the accumulation of water (Huang, 2004; Hoffman and Sargand, 2011). With increased traffic volume and tire pressure, heavy overloading, and severe weather conditions, rutting has become a major problem across the globe (Qing-lin, 2001). In view of widespread impacts, rutting is one of the criteria for the design of asphalt pavements in the United States (AASHTO, 1993). Other Countries such as India, China, and Europe also incorporate rutting as a design criterion (Zu-Kang, 2003).

Several asphalt binder modifiers, namely, polymers (elastomers or rubbers and plastomers or plastics), fibers, and oxidants are used to enhance the resistance of asphalt mixes to rutting. For example, Jamshidi et al. (2012) and Arshad et al. (2013) found that the addition of Sasobit® wax to

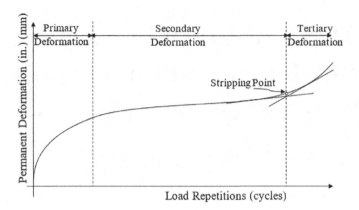

FIGURE 21.1   Schematic of typical permanent deformation curve for asphalt mixes.

a binder enhances its resistance to rutting by increasing stiffness and decreasing viscosity. Gandhi and Amirkhanian (2007) also reported that adding Sasobit® and Asphamin® to a binder leads to an increase in stiffness. However, the extent of stiffness improvement depends on the amount of WMA (warm-mix asphalt) additives as well as their types. Zhang and Yu (2010) reported an increase in stiffness when modifying a binder using Styrene–Butadiene Rubber (SBR). Several studies have been conducted on the use of Polyphosphoric Acid (PPA) to improve the mechanical properties of binders (e.g., Baldino et al., 2013; Ali et al., 2018, 2019; Ghabchi et al., 2021; Rani et al., 2019a, 2021). The addition of PPA to an asphalt binder is found to increase its stiffness at high temperatures (Baldino et al., 2013; Rani et al., 2019b). Also, the addition of PPA decreases the sensitivity of binders to temperature variation and increases their resistance to rutting.

21.1.1   Factors Influencing Rutting

For prediction of rutting, it is important to understand the influencing factors. Rutting is affected by several factors such as air voids, mineral filler content, dynamic modulus of mix, complex modulus of binder, binder grade, temperature, axle load, tire pressure, traffic level, and compaction level. These factors have been studied by many researchers (e.g., Tarefder and Zaman, 2002; Tutumluer et al., 2005; Lee et al., 2007; Hossain, 2017; Rani et al., 2020; Rani, 2019). For instance, the Federal Highway Administration (FHWA, 1988) reported that high binder content is one of the main factors for increased rutting. Tarefder et al. (2003) ranked the binder grade at the top of the list of parameters contributing to rutting, followed by temperature, aggregate gradation, moisture, and binder content. However, Hussan et al. (2017) ranked temperature as the most influential factor and binder content as the third influential factor. These researchers also observed that the flakiness index of aggregates affects the rutting potential; an increase in the flakiness index increases rutting. Zou et al. (2017) reported mix type as the most influential factor when considering rutting performance of asphalt pavements. Within an asphalt mix, binder grade is more influential than aggregate gradation (Zou et al., 2017). In contrast, Stakston and Bahia (2003) reported that the resistance to rutting primarily depends on the aggregate gradation. A mix with strong aggregates but inferior gradation would likely fail in the field due to rutting. According to Golalipour et al. (2012), rutting resistance increases when aggregate gradation is near the upper limit curve of the job mix formula. According to Kim and Souza (2009) and Leon and Charles (2015), an increase in angularity reduces rutting due to better interlocking between aggregates. Also, Ramli et al. (2013) studied the effects of fine aggregate angularity on rutting and noted that mixes with more angular aggregates are less susceptible to rutting. In a recent study by Hossain (2017), it was observed that rutting is more sensitive to change in traffic intensity instead of variation in material or geometric properties. A summary of factors influencing rutting is presented in Table 21.1. These influencing factors show the complexities of predicting rutting mechanistically.

TABLE 21.1
Factors Influencing Rutting

Factor	Change in Factor	Effect on Rutting
Asphalt Binder		
Performance grade	Increase	Decrease
Viscosity	Increase	Decrease
Complex modulus	Increase	Decrease
Aggregate		
Gradation	Gap-graded to continuous	Increase
Flakiness index	Increase	Increase
Surface texture	Smooth to rough	Decrease
Size	Increase in NMS	Increase
Shape	Angular to regular	Decrease
Fine aggregate angularity	Higher	Decrease
Asphalt Mixture		
Binder content	Low to optimum	Decrease
	Optimum to high	Increase
Fine content	Increase	Decrease
Dynamic modulus	Softer to stiffer	Decrease
Air void content	Optimum to low/high	Increase
Voids in mineral aggregate	Increase	Decrease
Test Conditions		
Temperature	Increase	Increase
Moisture	Dry to wet	Increase
Load repetitions	Increase	Increase
Tire pressure	Increase	Increase
Tire inclination	Increase (a certain limit)	Increase
Compaction level	Increase	Decrease

21.1.2 Available Techniques for the Determination of Rutting

21.1.2.1 Laboratory Testing

Several laboratory tests are available for measuring contributions of both binders and mixes to rutting. For example, dynamic shear rheometer (DSR) tests can be used to determine the rutting potential of asphalt binders at different temperatures and under different aging conditions (AASHTO T 315-12). The DSR measures the complex modulus (G^*) and phase angle (δ) of an asphalt binder at test temperatures. These properties are used to determine the rutting factor ($G^*/\sin \delta$) – an indicator of asphalt binder's contributions to rutting. According to Bahia and Anderson (1995), mechanistically rutting is directly proportional to the work dissipated in each traffic loading cycle, which is inversely proportional to the factor ($G^*/\sin\delta$). A higher $G^*/\sin\delta$ value, which may be achieved by either increasing G^* or decreasing δ, suggests a lower potential of work dissipation in a loading cycle, a higher recovery of the displacement due to vehicular traffic, and a lower rutting.

Several previous studies have shown that Hamburg Wheel Tracking (HWT) tests, Asphalt Pavement Analyzer (APA) tests, and Flow Number (FN) tests that allow measurements of deformation under repetitive loading of desired magnitude and cycle can be used to determine the rutting potential of asphalt mixes at the design stage (NCHRP 508, 2003; Lu and Harvey, 2006; Ghabchi et al., 2021). Ideally, rutting performance of constructed pavements can be determined using a Full-Scale Accelerated Pavement Testing (APT) facility, where a heavy vehicle simulator runs over a full-scale pavement to induce rutting (Metcalf, 1996; Khan et al., 2013).

21.1.2.2 Field Measurements

Rutting can be measured in the field using a straight edge-rut gauge combination. It usually measures the difference in elevation between the center of the wheel path and a line connecting two points located at a distance of 2 feet (60.96 cm) from the center of the wheel path (ASTM E 1703–10). A better way of measuring rutting manually is to use a Face Dipstick®. A Face Dipstick® provides a more precise rut profile than the straight edge-rut gauge combination (FHWA, 2013; Hossain, 2017). In recent years, several automated techniques (digital image-based) have been developed for measurement of rutting more accurately. For example, the Texas Department of Transportation (TxDOT) developed a five-point acoustic sensor system for rut data collection using Pavement Management Information System (PMIS) (AASHTO PP 38-05). In 2009, TxDOT designed a system, named VRUT, to measure rutting from continuous transverse profiles at highway speed. The system utilizes a 'high-power infrared laser line projector and a high-speed 3D digital camera with built-in laser line image processing capability' (Huang et al., 2013). Similarly, a Laser Rut Measurement System (LRMS) such as Pave 3D-8K characterizes rutting along the transverse direction using laser profilers (Hoffman and Sargand, 2011; Serigos et al., 2012).

21.1.2.3 Empirical Models

Several empirical models are available for the prediction of rutting in asphalt pavements. These models basically consider the number of load repetitions and limit displacement in the top asphalt mix layer or the accumulation of displacement in each layer. However, these models do not include the fundamental properties of materials and effect of mixed traffic levels. Some of those empirical models are listed in Table 21.2. Table 21.2 also presents the rut prediction model used in the Mechanistic Empirical Pavement Design Guide (MEPDG) software, also called AASHTOW are Pavement ME, for the design of asphalt pavements. The MEPDG rut model can be calibrated for local conditions to accommodate the effect of local traffic, material properties, and environmental conditions. For example, Hossain (2017) determined the local calibration parameters for the prediction of rut depth in asphalt pavements based on the traffic and environmental conditions in Oklahoma. The material properties were calculated using laboratory testing.

21.1.2.4 Numerical Modeling

Permanent deformation or rutting can be predicted using two-dimensional (2-D) or three-dimensional (3-D) numerical models of pavements, as reported by several researchers (Bakheet et al., 2001; Desai, 2007; Abed and Al-Azzawi, 2012; Nahi et al., 2014; Hu et al., 2017). For instance, Wu et al. (2011) developed a Finite Element (FE) model using commercially available software, ABAQUS, to investigate the permanent deformation of flexible pavements including a cementitiously stabilized base or subbase. Temperature dependency of the asphalt material was considered by adjusting the loading modulus at every 25,000 load repetitions. Abed and Al-Azzawi (2012) developed a 2-D plane strain FE model using ANSYS to estimate the permanent deformation of asphalt pavements. The estimated stress parameters from the FE model were used in the local empirical models to estimate rutting. Using a 3-D FE model, Hu et al. (2017) found that inclined tires and decelerating vehicles generate the maximum shear stress, resulting in an increased vertical strain in a pavement. Nahi et al. (2014) reported that creep models can be used to determine the rutting potential of mixes based on the FE simulation of a dynamic creep test. Zhou and Scullion (2002) proposed a rut prediction model based on the FE analyses of 49 different pavement sections in Texas capturing the three-stage permanent deformation of mixes.

To develop numerical models, selection of proper constitutive model is important. The selected constitutive model(s) shall describe the behavior of the associated materials such as elasticity, plasticity, creep, hardening, softening, macro- or micro-cracking, and failure. In actual pavements, some of these behaviors can occur simultaneously due to heavy vehicular traffic loading, oxidation due to aging or temperature variation, and moisture-susceptibility. The simplest constitutive models are Hooke's law, Darcy's law, or Ohm's law for linear behavior (Hooke, 1678; John and Mayo, 1913;

TABLE 21.2
Available Empirical Rut Models

No.	Rut Model		Reference
1	$\varepsilon_p = a_1 + b_1 \log N$ $\varepsilon_p = \dfrac{b_1}{N} (N > 1)$	a_1, b_1 = regression coefficients	Barksdale (1972)
2	$\varepsilon_p = aN^b$	a, b = regression coefficients	Monismith et al. (1975)
3	$\varepsilon_p^l = a\left(\dfrac{N}{1000}\right)^b$	a = permanent strain after 1,000 cycles b = permanent strain rate	Huurman (1977)
4	$RD = \alpha \cdot N^\beta \cdot T^\theta$	α, β, θ = coefficients used in equation	SHRP (1993)
5	$\log \dfrac{\varepsilon_P}{\varepsilon_r} = \log C + 0.4262 \log N$ $C = \dfrac{T^{2.02755}}{5615.391}$ (AASHTO 2002 Model)	ε_r >= resilient strain	Witczak (2001)
6	$\dfrac{\varepsilon_P(N)}{\Delta \varepsilon_r} = a_1 \cdot T^{a_2} \cdot N^{a_3}$	$\Delta \varepsilon_r$ = resilient strain a_1, a_2, a_3 = material parameters	ARA (2004), Salama et al. (2007), Hu et al. (2011)
7	$RD = \displaystyle\sum_{i=1}^{n} 10^{-5.72} \cdot T_i^{2.512} \cdot \left\{ \dfrac{0.58}{V} \cdot N \right\}^{0.743 \left(\frac{\tau_i^{0.472}}{\tau_{0i}} \right)}$	τ = shear stress V = traffic speed	Su et al. (2008)
8	$RD = C \displaystyle\sum_{i=1}^{n} N_i^a \cdot L_i^b \cdot T_i^c \cdot t_i^d$	C = adjustment factor for traffic wander effect L = load ratio t = loading duration a, b, c, d = coefficients used in equation	Fwa et al. (2004)
9	$RD = \displaystyle\sum_{i=1}^{n} \varepsilon_i^p h_i$ (total rut depth) $\dfrac{\varepsilon_p}{\varepsilon_r} = K_z \beta_{r1} 10^{K_{r1}} T^{\beta_{r2} k_{r2}} N^{\beta_{r3} k_{r3}}$ (MEPDG Model)	h = thickness of sublayer ε_r = resilient strain $\beta_{r1}, \beta_{r2}, \beta_{r3}$ = local calibration coefficients k_{r1}, k_{r2}, k_{r3} = national coefficients K_z = depth coefficient factor	MEPDG (2004)

$RD \,(\varepsilon_p) = total\ rut\ depth\ (permanent\ strain),$
$N = number\ of\ load\ repetitions,$
$T = temperature,$
$n = number\ of\ layers$

Whitaker, 1986). When using Hooke's law, the material behavior is simulated using elastic modulus and Poisson's ratio. However, most materials behave nonlinearly exhibiting creep effects, temperature variation effects, microcracking, fracture, moisture-induced damage effects, and other effects. Classical plasticity models such as von Mises, Mohr–Coulomb (M-C), and Drager–Prager (D-P) can simulate the nonlinear failure behavior of a material. However, these material models do not consider the hardening effect, which occurs due to the changing physical state of the material due to internal microstructural formations. Models such as the critical-state model or Cam Clay model

(DiMaggio and Sandler, 1971) can be used to describe the continuous yielding of materials as a function of volumetric plastic strain. These models, however, do not incorporate the effect of deviatoric plastic strain, which is an important component of the total accumulated plastic strain. The effect of both volumetric and deviatoric plastic strains have been included in the hierarchical single surface (HISS) model developed by Desai and coworkers (Desai et al., 1986, 1990; Desai, 2015a,b). The HISS model also incorporates other material models as special cases, providing a better way to define the nonlinear behavior of a material. The HISS model assumes that various behavioral characteristics contribute different magnitudes of deviation from normality of the plastic strain to the yield surface (Desai et al., 1986). That deviation can also be introduced by using the disturbed state concept (DSC) proposed by Desai (2001). Consequently, higher order nonassociative HISS models are not needed because the DSC allows for a part of nonassociative response similar to the nonassociative plasticity material models. The HISS or DSC/HISS model has been implemented successfully for clays, sands, rocks, glacial till, and unbound granular materials in asphalt pavements (Katti and Desai, 1995; Bonaquist and Witczak, 1996; Uzan, 1999; Desai, 2001; Wei et al., 2020). The HISS model can be included in the DSC, which allows for softening or degradation and healing.

21.1.2.5   FE Model Used in the Present Study

In this chapter, a 2-D FE model was developed for a typical pavement section for predicting the vertical displacement as a measure of rutting in asphalt pavements. The FE model was subjected to both static and repetitive loadings considering a single axle and single tire exerting a contact pressure of 200 psi (1,379 kPa) on the pavement surface. The DSC approach was used to define the transition of material behavior from the relatively intact (or undisturbed) state to the fully adjusted (or disturbed) state. The relatively intact state of pavement materials was defined using both linear elastic and the HISS models.

21.2   DISTURBED STATE CONCEPT

The DSC is a unified approach, developed by Desai and his coworkers, to model the behavior of complex materials such as asphalt, soils, rocks, concrete, joints, and interfaces (Desai et al., 1986; Desai, 2001). The DSC can be used for introducing the disturbance, micro-cracking, and thermal cracking into the material due to static or repetitive loadings. The major advantages of using the DSC are its capability of describing elasticity, plasticity, creep, microcracking, degradation, softening, and healing or stiffening behavior of different materials (Desai, 2001, 2007). As noted previously, the DSC/HISS incorporates other plasticity models such as von Mises model, D-P model and M-C model as special cases.

The DSC considers the equilibrium of a material in relative intact (RI) and fully adjusted (FA) states. The RI state represents the undisturbed or solid reference of a material, and the FA state represents a cracked or fully disintegrated material. The RI state transits continuously into the FA state while deforming due to microstructural changes caused by rotation, translation of particles, and other phenomena. At the limiting conditions, the FA state is reached by all the material elements. Therefore, intact (continuous) and broken (discontinuous) states of material and their interaction are coupled in the DSC. The interaction between the RI state and the FA state is defined by the disturbance function, D, where $D=0$ for the RI state and $D \approx 1$ for the FA state (Figure 21.2).

The progression of the disturbance function due to transition from the RI state to the FA state is schematically represented in Figure 21.2 and defined as given in Eq. (21.1).

$$D = D_u \left(1 - e^{-A\xi_D^Z} \right) \tag{21.1}$$

where D_u, A, and Z = disturbance parameters. The disturbance parameters can be estimated from the stress–strain curve of a material. For example, a typical stress–strain curve is shown in Figure 2.

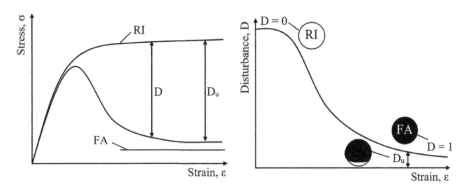

FIGURE 21.2   A schematic representation of the disturbed state concept.

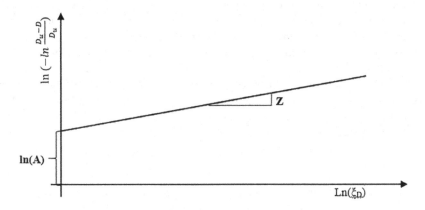

FIGURE 21.3   Determination of disturbance parameters A and Z (Desai, 2001).

The disturbance at different strains, i.e., D_{ci} (disturbance at the initiation of crack), D_{cf} (disturbance at finite crack), D_f (disturbance at failure), and D_u (ultimate disturbance) can be calculated using Eq. (21.2). The calculated disturbance at strains can be plotted on a double logarithmic scale, namely, $\ln[-\ln(D_u - D)/D_u]$ vs. $\ln \xi_D$, as shown in Figure 21.3. The intercept of the line on ordinate axis and its slope provides the parameter A and parameter Z, respectively.

$$D = (\sigma_i - \sigma)/(\sigma_i - \sigma_f) \text{ at a strain,} \qquad (21.2)$$

where σ_i = stress tensor for the RI state, σ = actual stress tensor, and σ_f = stress tensor for the FA state.

21.2.1   DETERMINATION OF DISTURBANCE PARAMETERS FOR ASPHALT MIXES

To determine the disturbance parameters, i.e., A, Z, and D_u, for asphalt mixes, the stress–strain results presented by Pellinen et al. (2004) were used. Pellinen et al. (2004) conducted triaxial tests on dense-graded asphalt mix samples at varying air voids of 0%, 4%, 8%, and 12%. Since the target % air voids for an asphalt mix is considered as 4.0% by many departments of transportation (DOTs), only the stress–strain results for 4.0% air voids were considered. A constant test temperature (55°C) and a constant loading rate (50 mm/min) were used in this testing. The asphalt mix samples were loaded up to an axial strain of 30% to obtain post-peak behavior.

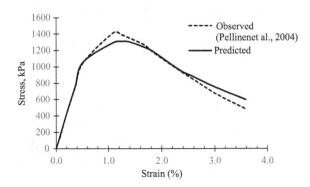

FIGURE 21.4 Predicted and observed stress–strain curves for 4.0% air void asphalt mix.

A linear elastic model was used to define the RI state of the asphalt mix. The elastic modulus was determined using the initial tangent of the stress–strain curve. The FA state was assumed to occur when the cracking level reached 95% or the sample failed, i.e., $D_u = 0.95$. The disturbance at different stress–strain levels was calculated using Eq. (21.2). As mentioned above, the experimental data were plotted using a logarithmic scale, as shown in Figure 21.3. Based on Pellinen et al.'s data, the disturbance parameters Z and A were calculated as 0.883 and 0.765, respectively. These parameters were used to back-calculate the stress at different strain levels (Figure 21.4). Overall, the agreement between the predicted and observed stress–strain responses of the asphalt mix was considered highly satisfactory.

21.2.2 The RI State

The RI state defines the undisturbed or continuum state of a material (Desai, 2001). One can characterize the RI state of material as elastic, nonlinear elastic, elastoplastic, viscoplastic, thermoviscoplastic, and hardening or softening (Desai, 2001). The RI response is relative depending upon which material model is selected for simulation. In this study, a linear elastic model was used to define the elastic behavior of materials. The HISS model was also used to simulate the elastoplastic hardening of materials and to include the irreversible displacement, coupled shear, and volumetric responses (Desai, 2001).

The HISS model is a single yield surface model developed by Desai and coworkers (Desai et al., 1986, 1990; Desai, 2001). It provides a general formulation for the characterization of elastoplastic behavior of materials. The HISS model is capable of simulating both associated and nonassociated behavior, and isotropic and anisotropic hardening of materials. There are different versions of HISS model such as δ_0- and δ_1- models (Desai et al., 1986). The δ_0-model allows for isotropic hardening and associative response, whereas δ_1-model allows for isotropic hardening and nonassociative response. In the HISS model, various behavioral characteristics contribute to different levels of deviation (subscript to δ) from normality of the plastic strain to the yield surface (Desai et al., 1986). The formulation of the yield surface F in the HISS model can be found in Desai (2001).

21.2.2.1 Determination of Parameters for the HISS-δ_0 Model

The HISS-δ_0 model requires seven to eight parameters to formulate the yield surface in different stress spaces, and disturbance function, D, requires three parameters (Desai, 2001). A list of these parameters and their descriptions is presented in Table 21.3. These parameters can be determined from the stress–strain response of materials from laboratory tests such as uniaxial, shear, hydrostatic, and cylindrical triaxial or other advanced laboratory tests such as multiaxial cubical tests (Desai, 2001, 2007). Details of the procedure for the determination of material constants for these models can be found in other chapters of this book as well as in Desai and Wathugala (1987), Desai (1990), or Desai (2001).

TABLE 21.3
List of Parameters for the HISS/DSC Model

Parameter	Description
	Elastic Behavior Parameters
E	Elastic modulus
ν	Poisson's ratio
	Plastic Behavior Parameters
γ	Relates to ultimate strength of materials
B	Relates to the shape of yield surface in principal stress space
N	Relates to the state of stress at which material transits from nondilative behavior to dilative behavior
$3R$	Defines the shift of yield surface to the negative J_1- axis in a stress space
	Hardening Behavior Parameters
α_1	Defines evolution of yield function
η_1	
	Disturbance Parameters
D	Defines transition from the RI state to the FA state
A	
Z	

HISS = hierarchical single surface, DSC = disturbed state concept, RI = relative intact, FA = fully adjusted.

21.2.3 FULLY ADJUSTED STATE

Asphalt mixes when subjected to heavy traffic loading and oxidation due to aging or temperature variation could result in a disintegrated system in which loose materials tend to create their specific volume. These loose materials can hardly sustain any load unless they are confined. The FA state represents a cracked or fully disintegrated state of materials and can be defined by various material models. For example, it can be assumed to possess no strength like classical damage material model (Kachanov, 1986). It can also be assumed to possess only hydrostatic strength, or it can be assumed to be at the critical state (Schofield and Wroth, 1968; Desai, 2001). For simplicity, the material in the FA state can be assumed to possess no strength or only hydrostatic strength. Therefore, the same material model used for the RI response can be modified to simulate the FA response, as recommended in Desai (2001).

21.3 DSC/HISS MODELS FOR ASPHALT PAVEMENTS

Several constitutive models have been proposed to predict damages or distresses in asphalt pavements (Salama and Chatti, 2011; Luo et al., 2013; Zhang et al., 2019). However, these models are not fully capable of considering the damage evolution in material structures (Wei et al., 2020). The DSC provides a unified framework to model the disturbance (softening or healing) evolution in material structures considering various material responses such as elasticity, plasticity, creep, microcracking, and fracturing (Desai, 2001, 2007; Wei et al., 2020).

Desai (2001, 2007) introduced the DSC/HISS model for asphalt pavements in 2-D and 3-D FE frameworks. The 2-D model was subjected to an incremental loading of 200 psi (1,379 kPa). The behavior of the RI state was simulated using the HISS model. The material parameters for asphalt mix and aggregate base layers were determined from the triaxial test results presented by Monismith and Secor (1962) and Bonaquist (1996), respectively. To determine the disturbance parameters, uniaxial test data reported by Scarpas et al. (1997) were used. The corresponding material parameters are listed in Table 21.4.

TABLE 21.4

Material Parameters for the HISS Model (Desai, 2007)

Parameter	Asphalt Mix	Aggregate Base	Stabilized Subgrade	Natural Subgrade
E	500,000 psi	56,533 psi	24,798 psi	10,013 psi
ν	0.3	0.33	0.24	0.24
γ	0.12940	0.0633	0.0383	0.0296
β	0.00	0.7	0.7	0.7
n	2.4	5.24	4.63	5.26
3R	121.0 psi	7.4 psi	21.5 psi	29 psi
a_1	1.23E-06	2.00E-08	3.60E-06	1.20E-06
η_1	1.944	1.231	0.532	0.778

1 psi = 6.89 kPa.

HISS = *hierarchical single surface.*

Previous studies have reported that the DSC/HISS model can simulate the stress–strain behavior of asphalt mixes under different temperatures, confining pressures, or strain rates. Also, the HISS model can predict the plastic displacements or rutting. Such deformations can be higher than elastic displacements due to plastic flow in pavement materials. It was reported that inclusion of disturbance did not make a significant difference in the predicted displacement when subjected to a static load of 200 psi (1,379 kPa) (Desai, 2007). The predicted displacement using the combined DSC/HISS model was close to the predicted displacement using the HISS model. When subjected to a repetitive loading of 200 psi (1,379 kPa), the evolution of fracture and microcracking was observed in the pavement layers. However, the analysis was conducted for only 20 cycles. The plastic strain and disturbance up to 20,000 loading cycles were computed using an accelerated approximate procedure, developed by Desai and Whitenack (2001) and modified by Desai (2007) for analysis of pavement.

Desai (2008) used the DSC/HISS and vevp models to simulate the creep and rate dependence behavior of subgrade materials. The model parameters were determined using triaxial tests results presented by Sane and Desai (2007) and Sane et al. (2008) and are summarized in Table 21.5. The FE model was developed using the DSC-SST2D code written by Desai and his coworkers (Desai, 1999). Overall, a good correlation was observed between the predicted and observed stress–strain results at confining pressures of 150 and 400 kPa. It was also reported that the vevp (viscoelastic viscoplastic) model predicted strains closer to laboratory test data as compared to evp (elastic viscoplastic) model because asphalt mixes exhibit both viscoelastic and viscoplastic behavior using a vevp model, which may be a preferred option if model parameters are available.

Vallejo (2012) predicted the failure behavior of an asphalt pavement overlaid by a HMA layer with and without pre-existing reflection cracks (Figure 21.5). A damage material model (Figure 21.6) was used based on the DSC-type approach accounting for volume change, stress state, displacement stress path, cracking, and fracturing that allows the prediction of failure behavior using both classical damage and fracture failure principles. The extended FE model (Belytschko and Black, 1999) was developed using the ABAQUS platform. The model was validated by comparing the predicted results with the Indirect Tensile Strength (IDT) results from laboratory tests on asphalt samples. The RI state of the associated pavement materials was modeled using a linear elastic model. The elastic limit was defined based on the yield stress of the material. The FA state was defined by complete material loss with no more strength. The material failure was defined in two stages: damage initiation and damage evolution. The damage evolution was defined using Eq. (21.3).

$$\sigma = (1 - D_e)\sigma_e \qquad (21.3)$$

TABLE 21.5
DSC/HISS and vevp Material Model Parameters (Desai, 2008)

DSC/HISS Parameters for Sky Pilot Till			Parameters for vevp Model			
Category	Symbol	Sky-Pilot Till	vevp Creep Parameters	Confining Pressure ($\sigma_2 = \sigma_3$)		
				100 kPa	200 kPa	400 kPa
Elastic	E	50,470				
	ν	0.45	E_1 (kPa)	6,666	24,244	30,000
Plastic-HISS	γ	0.0092	E_2 (kPa)	42,500	83,333	63,333
	β	0.52	ν_1	0.45	0.45	0.45
	n	6.85	ν_2	0.45	0.45	0.45
	$\bar{c}$	16.67	Γ_1 (kPa^{-1}min^{-1})	0.00145	0.00184	4.24E-6
Plastic-hardening	α_1	1.35E-07	Γ_2 (kPa^{-1}min^{-1})	0.006	0.006	0.006
	η_l	0.14	N_1	2.28	4	1.4
FA (critical) state	$\bar{m}$	0.09	N_2	1.63	2.2	2
	λ	0.16				
	e_0	0.52				
Disturbance	A	5.5				
	Z	1				

HISS = hierarchical single surface, DSC = disturbed state concept

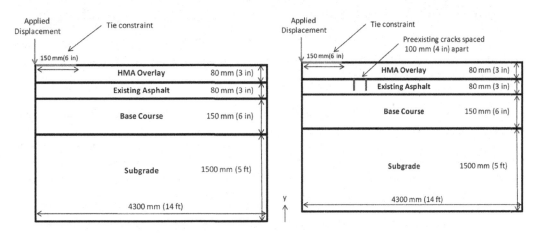

FIGURE 21.5 Asphalt pavement sections (Vallejo, 2012).

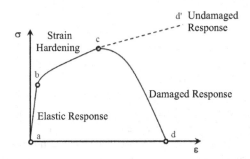

FIGURE 21.6 Typical stress–strain curve for the damage material model (Vallejo, 2012).

TABLE 21.6

Elastic and Damage Properties of Asphalt Mixes (Vallejo, 2012)

	Elastic Material Properties			
	Undamaged Asphalt	Damaged Asphalt	Base Course	Subgrade
Elastic Modulus, E (ksi)	1,015	1,015	56.5	10
Poisson's Ratio, ν	0.35	0.35	0.33	0.24
	Damage Material Properties			
Peak Traction Stress, t_0^{max} (psi)	242	135		
Cohesive Fracture Energy, G_c (lb-in/in²)	3.71	2.57		

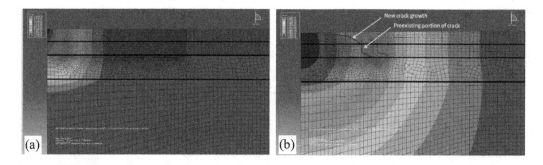

FIGURE 21.7 Vertical displacement in asphalt pavement section without pre-existing cracks (a) and with pre-existing cracks (b) (Vallejo, 2012).

where σ=stress during loading, De=damage evolution variable, and σ_e=stress within undamaged material.

The damage initiation properties were calculated based on the IDT data from laboratory tests on asphalt mixes. The samples were compacted at 6.0% air voids with a 4.0-inch diameter and a 1.8-inch thickness. The load was applied at a constant displacement rate of 2-inch/minute (50 mm/minute). The IDT test was modeled in the ABAQUS software subjected to a constant displacement at the top of the model. The fracture damage study conducted by Aragão et al. (2011) was used to calculate the asphalt evolution material properties. The corresponding elastic and damage properties are listed in Table 21.6. The crack initiation and propagation showed good agreements with the laboratory observations.

The FE model indicated stress concentrations at the bottom of wheel load and asphalt layer in the pavement section without pre-existing cracks (Figure 21.7) and had a vertical displacement of 0.006-inch (0.15-mm). However, for the pavement section with pre-existing cracks, the stresses were mainly concentrated at a short distance from the point of applied loading. The cracks were found to start at the bottom layer at the location of the existing crack and then propagated toward the top layer (Figure 21.7). The total displacement for the pavement section with pre-existing cracks was found to be 0.5-inch (12.7 mm), which was significantly greater than the total displacement for the pavement section without pre-existing cracks.

Khobklang et al. (2013) used the DSC approach to define the mechanical behavior of hydrated cement–treated crushed rock base used for asphalt pavement sections in Australia. The internal angle of friction of 43° and cohesion of 150 kPa were measured based on drained triaxial compression tests under static loading following the Austroads APRG 00/33–2000 method. Three different confining pressures of 50, 100, and 150 kPa were used. The maximum dry density was reported as 2.27 T/m³. The corresponding optimum moisture content was 5.5%. The samples were also subjected to 10,000

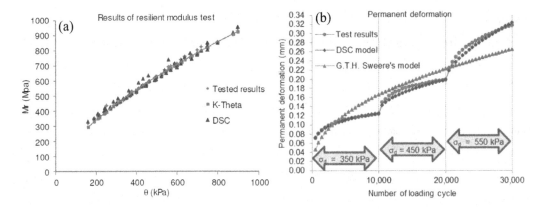

FIGURE 21.8 Predicted and observed resilient modulus (a) and permanent displacement (b) for hydrated cement–treated crushed rock base used for asphalt pavement sections in Australia (Khobklang et al., 2013).

repetitive loading cycles at three different deviatoric stresses of 350, 450, and 550 kPa. The linear elastic response was considered for the RI state. Permanent displacement and resilient modulus (M_r) of samples when subject to repetitive loadings were back-predicted. The DSC exhibited a good match with the measured resilient modulus (Figure 21.8). The model developed by Samris (2004) was also used to predict the permanent displacement. It was found that the DSC predicted responses were closer to the experimental results than the Samris (2004) model.

In a recent study conducted by Wei et al. (2020), the DSC approach was used to define the disturbance evolution and to investigate the fatigue performance of large stone asphalt mixes when subjected to repetitive loading. The material parameters were calculated from the Overlay tester fatigue test results conducted at varying levels of frequency, material aging, and loading. The test temperature was kept at 25°C. The peak load was measured with loading that decreased gradually until fracture. The elastic plastic model was used for the RI state. The FA state was defined based on the approximation of the ultimate asymptotic response of the material. The disturbance was defined based on the external work done during the first loading cycle, external work done per cycle, and external work done in the last cycle. The disturbance evolution was expressed using Eq. (21.4).

$$D = 1 - \left[1 - \left(\frac{N}{N_f} \right)^r \right]^a \tag{21.4}$$

where N=number of loading cycles, N_f=fatigue life, and r and a=fitting parameters.

The measured parameters under different test conditions are listed in Table 21.7. The results indicated that disturbance developed rapidly in early stages of loading and attenuated in the latter stage of loading. The magnitude of the disturbance parameter D increased with decreasing frequency and increasing loading and aging.

21.4   DETAILS OF THE FE MODEL

21.4.1   Geometric and Material Parameters

A typical four-layered pavement section was considered in this study, Figure 21.9. As shown, the section consisted of a 10-inch (25.4 cm) thick asphalt mix layer, a 6-inch (15 cm) thick aggregate base layer, a 6-inch (15 cm) thick subbase layer, and a semi-infinite subgrade layer. The linear elastic model was used to represent the elastic behavior of materials. The HISS model was used to simulate the elastoplastic hardening behavior of materials. The disturbance function was used to simulate cracking,

TABLE 21.7

Calculated Fitting Parameters at Different Test Conditions (Wei et al., 2020)

Test Conditions		r	a	A	Z
Standard		0.4	1.93	0.159	−0.422
Group 1	1 Hz	0.053	0.82	0.154	−0.285
	5 Hz	0.25	1.8	0.125	−0.379
Group 2	0.1 mm	0.28	0.85	0.598	−0.173
	0.2 mm	0.17	1.17	0.204	−0.318
Group 3	3.60%	0.22	1.93	0.131	−0.321
	3.90%	0.25	1.34	0.194	−0.253
Group 4	Short aging	0.26	1.96	0.113	−0.375
	Long aging	0.24	1.18	0.151	−0.435

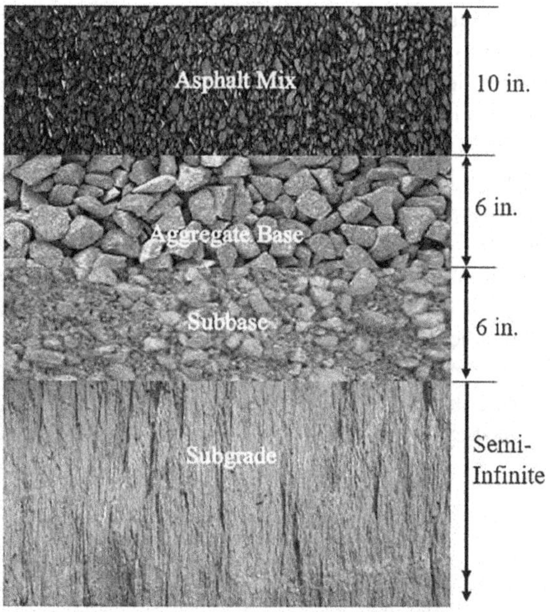

FIGURE 21.9   Geometry of asphalt pavement section (1-in. = 2.54 cm).

softening, and hardening response of asphalt mixes. Since no triaxial tests were conducted in this study, the elastic and the HISS model parameters for the pavement materials were taken from the literature (Desai, 2007), as listed in Table 21.4.

21.4.2 Model Type

The developed FE model used to predict the rutting of the pavement section is shown in Figure 21.10. As noted in the preceding sections, both 2-D and 3-D FE models have been used for modeling a pavement structure depending upon the complexity of the problem and the available resources (Lu and Wright, 1998; Bakheet et al., 2001; Wu et al., 2011). Although a 3-D model may be preferred for a more accurate representation of rutting and micro-cracking in a pavement under vehicular traffic

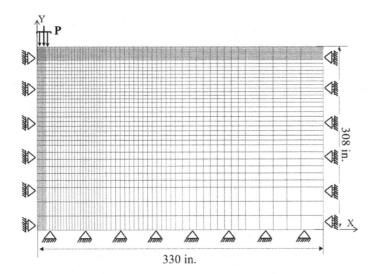

FIGURE 21.10 Two-dimensional finite element model for pavement section (1-in=2.54 cm). (The authors would like to express their gratitude to Dr. Desai for providing the DSC-SSD2D code including the input preparation code and the user manual, without which the analyses reported in this chapter would not have been possible.)

loading (Desai, 2007), 2-D analyses are used frequently for convenience with acceptable results. For example, Hua (2000) reported less than 2% difference in rut depths estimated by using a 2-D plane-strain model and a 3-D model subjected to 5,000-wheel passes. Also, 2-D analyses require significantly less computing time and effort compared to 3-D analyses (Desai, 2007). In this study, a 2-D FE model was developed, as shown in Figure 21.10. A MATLAB software vR2018a and DSC-SSD2D code* (Desai, 1999) was used to construct the FE model and perform analyses.

21.4.3 BOUNDARY CONDITIONS

Following the boundary conditions used in previous studies (Al-Khateeb et al., 2011), the base of the FE model was fixed in both horizontal and vertical directions, as shown in Figure 21.10. The sides of the model, however, were restrained from moving in the horizontal direction but were free to move in the vertical direction. Sliding and separation at the interface between the layers were not considered, which means a perfect bond was assumed at the connecting nodes of the consecutive layers.

21.4.4 LOADING PATTERNS

The loading pattern used in the FE model should be representative of the field conditions to predict rutting. Because an asphalt pavement is subjected to repetitive vehicular loading, the FE model used repetitive loading (Helwany et al., 1998; Wu et al., 2011; Abed and Al-Azzawi, 2012). Static loading was also used for validation of the FE model and for comparison purposes. The Federal Highway Administration (FHWA) classifies vehicles into 13 different classes, based on vehicle type and axle configurations such as single axle, tandem axle, tridem axle, and quad axle (FHWA, 2014). In this study, a single axle with a single tire configuration carrying 25,000 lb. (110 kN) load was selected for simplicity, as shown in Figure 21.11.

Considering symmetry about the vertical axis, simulations were performed for a single tire carrying 12,500 lb. (55 kN) load with 200 psi (1,379 kPa) tire pressure. The effect of tire wall stiffness on the contact pressure was neglected resulting in the contact pressure being equal to the tire pressure (Huang, 2004; NCHRP, 2002). The loading width corresponding to 200 psi (1,379 kPa) contact

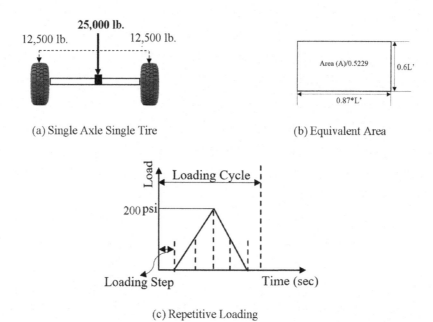

(a) Single Axle Single Tire   (b) Equivalent Area

(c) Repetitive Loading

FIGURE 21.11   Loading condition for finite element analyses (1 lb=0.004448 kN and 1 psi=6.89 kPa).

pressure was calculated as 10 inches (25.4 cm) using Eq. (21.5) (Huang, 2004). The repetitive load-ing was applied in six steps, following a triangular pattern, keeping the loading pressure at 200 psi (1,379 kPa), as shown in Figure 21.11c. The analyses were performed up to 200 loading cycles, and the results were compared with the static analyses.

$$\text{Equivalent Length } (L') = \sqrt{\frac{\text{Area}(A)}{0.5229}} \tag{21.5a}$$

$$\text{Area}(A) = \frac{\text{Load}}{\text{Contact Pressure}} \tag{21.5b}$$

$$\text{Length}(L) = 0.87 * L' \tag{21.5c}$$

$$\text{Width}(w) = 0.6 * L' \tag{21.5d}$$

21.4.5 Domain and Mesh Size Convergence Analysis

Selection of appropriate domain, mesh size, and boundary conditions is important for FE analysis (Helwany et al., 1998; Wu et al., 2011). For this purpose, a sensitivity analysis was performed con-sidering various domain and mesh sizes. The combined thickness (T) of the asphalt mix, aggregate base, and subbase layers were kept fixed at 22-inch (56 cm), while the thickness of the subgrade layer was varied in multiples of T, i.e., 9T, 11T, 13T, 15T, and 17T. Similarly, four different domain widths, namely, 9T, 11T, 13T, and 15T, were selected for these analyses. The predicted vertical dis-placement at the top of the surface layer is presented in Table 21.8. It is evident from Table 21.8 that the vertical displacement increases with an increase in the subgrade thickness. However, there is a

TABLE 21.8
Domain Size Convergence Results

	Depth, in.	Width, in.	Vertical Displacement, in.	% Change
Depth variation	9T	11T	−0.1370	
	11T	11T	−0.1398	2.04
	13T	11T	−0.1415	1.20
	15T	11T	−0.1430	1.10
	17T	11T	−0.1445	1.03
Width variation	15T	11T	−0.1430	
	15T	13T	−0.1436	0.38
	15T	15T	−0.1426	−0.65
	15T	17T	−0.1427	0.04

(1-in=2.54 cm)

small change in the displacement value due to increasing the thickness of the subgrade layer from 11T to 13T or 13T to 15T. For instance, from Table 21.8, the change in the vertical displacement was 1.20% when increasing the subgrade thickness from 11T to 13T. Similarly, the vertical displacement increased only by 0.04% due to increasing the domain width from 13T to 15T. Therefore, a domain size of 14T×15T (308-inch×330-inch) (782.32×838.2 cm) was selected for further mesh size convergence analysis. In addition, X-stress, Y-stress, and shear stress at the selected domain size were mostly confined in the top 100-inch (254 cm) of the pavement section and were not impacted by the domain boundaries.

For the mesh size convergence analysis, the element size was varied from 10 inches (25.4 cm) to one inch (2.54 cm). Only 2% variation was observed in the predicted vertical displacement for various mesh sizes. Therefore, the element size of 1-inch×2.5-inch (2.54 cm×6.45 cm) was used under the applied load and increased in both horizontal and vertical directions with distance from the location of the applied load, see Figure 21.10.

21.5   RESPONSE UNDER STATIC LOADING

As noted earlier, the RI state can be simulated using either an elastic or inelastic model. Figure 21.12 presents the vertical displacement profile of the pavement section in both horizontal and vertical directions when using elastic and the HISS or the DSC/HISS models and subjected to a constant pressure of 200 psi (1,379 kPa). As expected, the vertical displacement was maximum at the point of load application and decreased in both horizontal and vertical directions. The vertical displacement for the elastic model was approximately the same as the vertical displacement computed using the KENLAYER software, which is used widely to predict pavement performance (Huang, 2004). A 5.5% difference was observed between the predicted vertical displacement using the elastic model and the KENLAYER results, as shown in Figure 21.12. The KENLAYER software uses elastic multilayer asphalt pavements with full bonding between layers and is subjected to a circular wheel load (Huang, 2004). The software can also be used for multilayer systems subjected to single, dual, and dual-tandem wheel loadings, where the material behavior of each layer can be defined as linearly elastic, inelastic, or viscoelastic (Huang, 2004). The software computes stresses and displacements based on a fourth-order differential equation satisfying the boundary conditions and continuity conditions (Huang, 2004).

A significant difference between the predicted vertical displacement between elastic and inelastic simulation was observed. For instance, the maximum vertical displacement was 0.073-inch (0.185 cm) for the elastic simulation and 0.142-inch (0.36 cm) for the HISS or the HISS/DSC

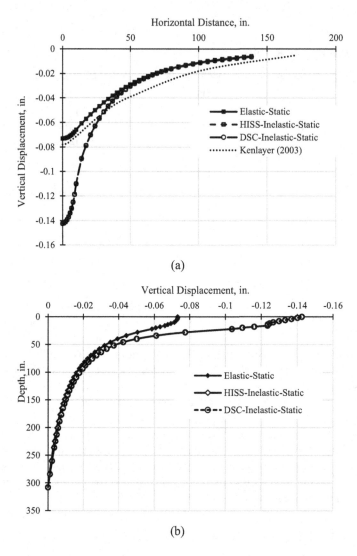

FIGURE 21.12 Vertical displacement in horizontal (a) and vertical (b) directions for elastic and inelastic (hierarchical single surface (HISS) and HISS/disturbed state concept (DSC)) models under static loading (1-in=2.54 cm).

simulation, indicating that the associated material was reaching the inelastic zone under the applied loading that could not be captured using the linear elastic model. No significant difference was observed in the vertical displacement when using the HISS model and the HISS/DSC model. As shown in Figure 21.12, the maximum vertical displacement for the HISS and HISS/DSC models were equal (0.142-inch or 0.36 cm), and displacement profiles overlapped in both horizontal and vertical directions. It was evident that, in this case, the inclusion of disturbance did not impact the predicted vertical displacement of the pavement. Thus, either the HISS or the DSC/HISS model can be used to simulate the material behavior under static loading.

21.6   RESPONSE UNDER REPETITIVE LOADING

Figure 21.13 presents the predicted vertical displacement of the considered pavement section when subjected to static and repetitive loadings. The repetitive loading analyses were performed by

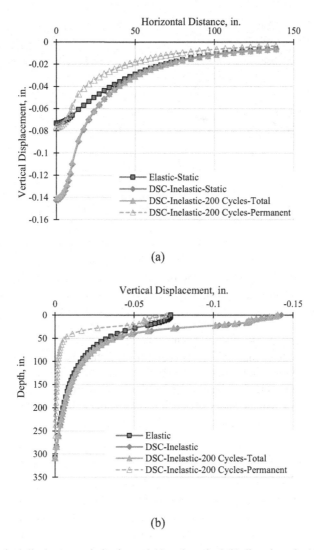

(a)

(b)

FIGURE 21.13 Vertical displacement in horizontal (a) and vertical (b) directions for hierarchical single surface (HISS)/disturbed state concept (DSC) model under static and repetitive loadings (1-in = 2.54 cm).

applying the same loading pressure of 200 psi (1,379 kPa) in a six-step triangular loading cycle, as shown in Figure 21.11c. The analyses were performed for 200 loading cycles. The DSC/HISS model was considered for the asphalt layer. The RI state for the other layers was simulated using the HISS model. As shown in Figure 21.13, the vertical displacement was maximum underneath the applied load and decreased in both horizontal and vertical directions, as expected. Most displacement/deformation was concentrated in the top 100-inch (254 cm) of the pavement. From Figure 21.13, the maximum vertical displacement under the applied repetitive loading was equal to 0.14-inch (0.356 cm), which is very close to the predicted displacement under static loading. However, the 0.064-inch (0.163 cm) of the total displacement was recovered after unloading of the applied 200 loading cycles. As a result, the plastic (or permanent) vertical displacement was approximately 0.076-inch (0.193 cm) measured after loading and unloading of the applied 200 loading cycles. The permanent vertical displacement diminished to zero at approximately 50 inches (127 cm) below the top surface of the pavement section. This behavior, as expected, could not be seen in the static loading cases.

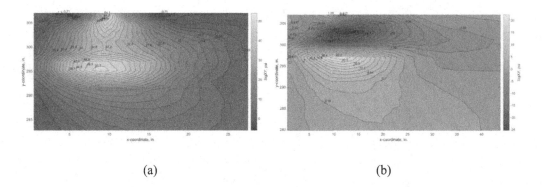

(a)   (b)

FIGURE 21.14   Shear stress contour maps in pavement section at peak (a) and at end (b) of the 200th loading cycle (1-in = 2.54 cm).

Furthermore, the shear stress was found to be concentrated at the edge of the applied load and at the interface between the asphalt layer and the aggregate base layer. As shown in Figure 21.14, the maximum shear stress was measured as 54.1 psi (373 kPa) at the edge of the applied load followed by 51.7 psi (356.5 kPa) at the interface between the asphalt layer and the aggregate base layer. After unloading, the shear stress reduced to approximately 22–24 psi (151.7–165.5 kPa) and concentrated mostly in the middle of the asphalt mix layer and at the interface between the asphalt layer and the aggregate base layer. Directional flows of the shear stress at the interface and within the asphalt mix layer were in the opposite directions suggesting the chances of slippage or debonding at the interface. These results indicate the need for an interface model in this region to account for possible slippage or debonding.

21.6.1   Effect of Phase Change Parameter under Repetitive Loading

As noted earlier, six different plastic parameters are required to define the plastic behavior of materials using the DSC/HISS model. Among them, the effect of phase change parameter 'n' was studied here as it defines the state of zero volume change and relates to the state of stress at which the material transits from nondilative behavior to dilative behavior. Hence, the parameter 'n' relates to the % air voids present in an asphalt mix (Desai, 2001), which is one of the parameters governing the asphalt mix design for pavements besides the other parameters such as temperature and aging (Brown et al., 2009; ODOT, 2009).

According to Desai (2001), the n-value should be greater than 2.0 for a convex yield surface. Therefore, the analysis was conducted at four different values of 'n', i.e., $n = 2.4$, 3.0, 3.5, and $n = 4.0$ for the asphalt mix layer. The asphalt mix having $n = 4$ would have more % air voids than the one having $n = 2.4$. The other properties of the asphalt mix layer and properties of the other layers were kept constant, as listed in Table 21.4. The predicted vertical displacements for varying n-values are presented in Figure 21.15. The total, elastic, and plastic vertical displacements at the top of the pavement layers are given in Table 21.9.

It is evident from Figure 21.15 that the vertical displacement at the top of the pavement surface increased significantly with an increase in the n-value. For instance, the maximum vertical displacement increased by 10% (from 0.14-inch to 0.154-inch) (from 0.356 cm to 0.391 cm) after increasing the n-value from 0.24 to 3.0. Further increase in the n-value from 3.0 to 3.5 increased the vertical displacement by 18.8% (from 0.154-inch to 0.183-inch) (from 0.391 cm to 0.465 cm). When using $n = 4.0$, the vertical displacement is found to be 0.25-inch (0.635 cm), which was 78.5% higher than the vertical displacement at $n = 2.0$. In addition, it was observed that the contribution of the underlying aggregate base layers to the total permanent displacement of the pavement section increased

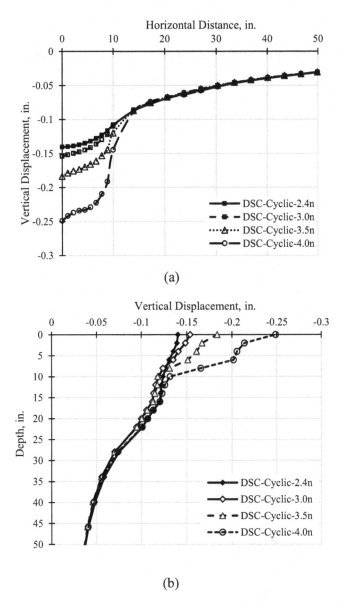

FIGURE 21.15   Vertical displacement in horizontal (a) and vertical (b) directions at *n*-values (1-in=2.54 cm).

with an increase in the n-value. From Table 21.9, the permanent displacement of the aggregate base layer increased by 14% after increasing the n-value from 2.4 to 4.0. These results suggest that the asphalt mixes with high % air voids are expected to experience significantly higher rutting along with increased rutting potential from aggregate base or subgrade layers.

21.6.2   THE EFFECT OF CONTACT TIRE PRESSURE

A pavement is often subjected to various vehicle types with different configurations during its life-time exerting varying tire pressures on the pavement surface. Passenger cars exert a contact pres-sure as small as 35 psi (241.3 kPa) (FHWA, 2014). Trucks, however, can develop a contact pressure as high as 200 psi (1,379 kPa) (FHWA, 2014). In this study, the effect of contact tire pressure on

TABLE 21.9
Vertical Displacement (in inch) at the Top of Pavement Layer at *n*-Values

Analysis	Asphalt Mix	Aggregate Base	Subbase	Subgrade
		Total Displacement – Repetitive		
DSC-2.4n	−0.140	−0.124	−0.121	−0.101
DSC-3.0n	−0.154	−0.119	−0.112	−0.095
DSC-3.5n	−0.184	−0.118	−0.113	−0.095
DSC-4.0n	−0.249	−0.131	−0.120	−0.101
		Permanent Displacement – Repetitive		
DSC-2.4n	−0.071	−0.056	−0.058	−0.046
DSC-3.0n	−0.084	−0.052	−0.050	−0.039
DSC-3.5n	−0.114	−0.050	−0.050	−0.040
DSC-4.0n	−0.180	−0.064	−0.058	−0.046
		Elastic Displacement – Repetitive		
DSC-2.4n	−0.070	−0.068	−0.063	−0.056
DSC-3.0n	−0.070	−0.068	−0.063	−0.056
DSC-3.5n	−0.069	−0.068	−0.063	−0.056
DSC-4.0n	−0.069	−0.067	−0.062	−0.055

(1-in = 2.54 cm)

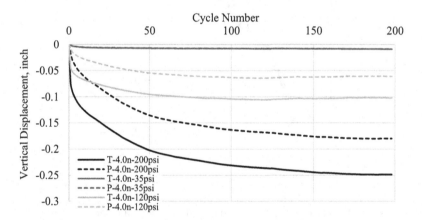

FIGURE 21.16 Total (T) and permanent (P) vertical displacement at different contact tire pressures (1-inch = 2.54 cm and 1-psi = 6.89 kPa).

the vertical displacement of the pavement section was investigated by varying the applied loading pressure from 35 psi (241.3 kPa) to 200 psi (1,379 kPa). The results are presented in Figure 21.16. As shown in Figure 21.16, an increase in the applied pressure increased the vertical displacement at the top of pavement surface, as expected. For instance, the maximum vertical deflection was just 0.008-inch (0.02 cm) when subjected to 35 psi (241.3 kPa) loading pressure. Increasing the loading pressure from 35 psi (241.3 kPa) to 120 psi (1,379 kPa) increased the displacement to 0.102-inch (0.259 cm). At high loading pressure of 200 psi (1,379 kPa), the maximum vertical displacement was 0.249-inch (0.632 cm), which is 31 times higher than the vertical displacement at 35 psi (241.3 kPa) pressure. These results indicate that increased contact pressure can cause significantly larger vertical displacement or rutting in a pavement. This is primarily attributed to the plastic flow of asphalt mixes at considerably high tire pressures.

It is noted that a pavement is generally subjected to millions of loading cycles during its lifetime. Hossain (2017) reported that a pavement section at the southbound lane of Interstate-35 (I-35) in Oklahoma is subjected to 8,200 trucks per day. The present study was conducted only for 200 loading cycles due to limitation of the DSC-SSD2D code in handling large data. To estimate the permanent displacement and damage at higher loading cycles, a procedure suggested by Desai (2007) can be used. Accordingly, the analysis can be conducted for fewer cycles (e.g., 100 cycles) and then results can be translated to a higher number of cycles using the deviatoric displacement at a reference cycle and the ratio of targeted and reference number of cycles, as illustrated in Eq. (21.6). However, the verification of results would be imperative.

$$\epsilon_D(N) = \epsilon_D(N_r) \left(\frac{N}{N_r} \right)^b \tag{21.6}$$

where N_r=reference cycle, N=deign number of cycles, and b=constant parameter. It defines the displacement rate in an asphalt pavement. A value of 0.3 was recommended by Desai (2007) for a cracked pavement section. Accordingly, the permanent deformations were measured at 1,000 cycles and 10,000 cycles as 0.123-inch (0.312 cm) and 0.245-inch (0.622 cm), respectively. However, verification of these results would be imperative using the laboratory test data such as HWT test or field measurements.

21.7 SUMMARY

The DSC/HISS model with the HISS model for the RI was used in this chapter to predict the rutting potential of asphalt pavements. Rutting is the permanent displacement in the asphalt pavements caused by repetitive vehicular traffic loading. Although the asphalt layer is a primary contributor, other layers (base, subbase, and subgrade) can also contribute to rutting. A detailed review of previous studies pertaining to rutting is included. The rutting deformation can occur either through plastic flow in the asphalt layer or shear in one of the underlying layers. The DSC/HISS is a unified approach introduced by Desai and co-workers (Desai et al., 1986; Desai and Wathugala, 1987; Desai, 2001; Wei et al., 2020) to model the behavior of complex materials such as asphalt, soils, rock, and concrete. The DSC/HISS can effectively model disturbance, microcracking, and fracture of materials due to static and repetitive loadings. Thermal and time-dependent loadings may be incorporated as well.

A 2-D FE model was developed by the authors for a four-layered pavement section and analyzed using the DSC-SSD2D code to predict its rutting performance under both static and repetitive loadings. The RI state or undamaged state of the pavement materials was defined using both linear elastic and the HISS models. The DSC was used in combination with the HISS model for the asphalt mix layer to introduce disturbance into the material behavior. The material parameters for both linear elastic and the HISS or the HISS/DSC models were taken from the previous studies. The analyses were performed for both static and repetitive loading cycles (200 cycles) assuming a single axle and single tire that carries a 25,000 lb. (111.2 kN) load with a contact tire pressure of 200 psi (1,379 kPa) at the pavement surface. The interface between the asphalt layer and the aggregate layer was assumed to be fully bonded. The results indicated that under the applied static loading, the total vertical displacement predicted using the inelastic material model was significantly higher compared to the total vertical displacement predicted using the elastic model. However, no significant differences were observed in the predicted vertical displacements when disturbance was introduced to the HISS model through the DSC approach.

Although, the total vertical displacement after 200 loading cycles was close to the maximum vertical displacement under static loading, for repetitive loading, a recovery of the total vertical displacement was observed after unloading of the applied load. The computed permanent displacement after 200 loading cycles was less than the maximum vertical displacement predicted under static loading. The shear stress under repetitive loading was found to be concentrated at the interface between the asphalt layer and the aggregate base layer, suggesting a lateral movement potential at the interface that could lead to slippage and potential failure at the interface.

The effect of phase change parameter, n, was studied as it relates to the % air voids present in the asphalt mix, which is one of the parameters governing the asphalt mix design for asphalt pavements. It was observed that the asphalt mix with a larger n-value would exhibit a higher rutting potential. Also, the contribution of underlying aggregate base layers to the permanent displacement of the pavement section was found to increase with an increase in the n-value.

This study was limited to 200 loading cycles due to limitation of the DSC-SSD2D code in handling larger data. Incorporation of the HISS/DSC models in commercial FE software such as ABAQUS would allow larger loading cycles and inclusion of thermal effects. The rut values were extrapolated for higher loading cycles using a procedure suggested by Desai (2007).

ACKNOWLEDGEMENT

The authors would like to acknowledge the help and support provided by Dr. Chandrakant S. Desai, Regents Professor Emeritus of Civil and Architectural Engineering and Mechanics at the University of Arizona and Dr. Rouzbeh Ghabchi, Assistant Professor at the South Dakota State University for their assistance.

REFERENCES

AASHTO (1993). "AASHTO guide for design of pavement structures, 1993 (Vol. 1)." *American Association of State Highway and Transportation Officials*, Washington, DC.

Abed, A. H. and Al-Azzawi, A. (2012). "Evaluation of rutting depth in flexible pavements by using finite element analysis and local empirical model." *American Journal of Engineering and Applied Sciences*, 5(2), 163–169.

Ali, S. A., Ghabchi, R., Rani, S., and Zaman, M. (2019). "Characterization of effect of aging on polymer-and polyphosphoric acid-modified asphalt binders using X-ray diffraction (XRD)." *Journal of Testing and Evaluation*, 48(6), 4190–4203.

Ali, S. A., Ghabchi, R., Zaman, M., Steger, R., Rani, S., and Rahman, M. A. (2018). "Mechanistic evaluation of effect of PPA on moisture-induced damage using SFE and XRF." In *International Conference on Transportation and Development 2018: Airfield and Highway Pavements*, Reston, VA: American Society of Civil Engineers, 411–421.

Al-Khateeb, L. A., Saoud, A., and Al-Msouti, M. F. (2011). "Rutting prediction of flexible pavements using finite element modeling." *Jordan Journal of Civil Engineering*, 5(2), 173–190.

ARA (2004). "Guide for mechanistic-empirical design of new and rehabilitated pavement structures: final report." *National Cooperative Highway Research Program, Transportation Research Board*, Washington, DC.

Aragão, F. T. S., Kim, Y. R., Lee, J., and Allen, D. H. (2011). "Micromechanical model for heterogeneous asphalt concrete mixtures subjected to fracture failure." *Journal of Materials in Civil Engineering*, 23(1), 30–38.

Arshad, A. K., Kridan, F. A. M, and Rahman, M. Y. A. (2013). "The effects of Sasobit modifier on binder at high and intermediate temperatures." *International Journal of Engineering and Advanced Technology*, 2(3), 81–84.

Bakheet, W., Hassan, Y., and El-Halim, A. O. Abd (2001). "Modelling in situ shear strength testing of asphalt concrete pavements using the finite element method." *Canadian Journal of Civil Engineering*, 28, 541–544.

Baldino, N., Gabriele, D., Lupi, F. R., Rossi, C. O., Caputo, P., and Falvo, T. (2013). "Rheological effects on bitumen of polyphosphoric acid (PPA) addition." *Construction and Building Materials*, 40, 397–404.

Barksdale, R. D. (1972). "Laboratory evaluation of rutting in basecourse materials." *Proceedings of the 3rd International Conference on Structure Design of Asphalt Pavements*, University of Michigan, 161–174.

Belytschko, T., and Black, T. (1999). "Elastic crack growth in finite elements with minimal remeshing." *International Journal for Numerical Methods in Engineering*, 45(5), 601–620.

Bonaquist, R. and Witczak, M. (1996). "Plasticity modeling applied to the permanent deformation response of granular materials in flexible pavement systems." *Transportation Research Record: Journal of the Transportation Research Board*, 1540, 7–14.

Bonaquist, R. J. (1996). "Development and application of a comprehensive constitutive model for granular materials in flexible pavement structures." Doctoral dissertation, Univ. of Maryland, College Park, MD.

Brown, E. R., Kandhal, P. S., Roberts, F. L., Kim, Y. R., Lee, D. Y., and Kennedy, T. W. (2009). "Hot mix asphalt materials, mixture design and construction: Third edition." NAPA Research and Education Foundation Lanham, Maryland.

Dawley, C. B., Hogewiede, B. L., and Anderson, K. O. (1990). "Mitigation of instability rutting of asphalt concrete pavements in Lethbridge, Alberta, Canada". *Journal of the Association of Asphalt Paving Technologists*, 59, 481–508.

Desai, C. S. (2001). *Mechanics of Materials and Interfaces: The Disturbed State Concept*. CRC Press, Boca Raton, FL.

Desai, C. S. (2008). "Unified constitutive modeling for pavement materials with emphasis on creep, rate and interface behavior." *Proceedings of the International conference on Advances in Transportation Geotechnics*, CRC, Nottingham, UK, 15–25.

Desai, C. S. and Whitenack, R. (2001). "Review of models and the disturbed state concept for thermomechanical analysis in electronic packaging." *J. Electron. Packag.*, 123, 1–15.

Desai, C. S., Somasundaram, S., and Frantziskonis, G. (1986). "A hierarchical approach for constitutive modelling of geologic materials." *International Journal for Numerical and Analytical Methods in Geomechanics*, 10(3), 225–257.

Desai, C. S., Wathugala, G. W., Sharma, K. G., and Woo, L. (1990). "Factors affecting reliability of computer solutions with hierarchical single surface constitutive models." *Computer Methods in Applied Mechanics and Engineering*, 82(1), 115–137.

Desai, C. S. (1990). "Modelling and testing: Implementation of numerical models and their application in practice" in *Numerical Methods and Constitutive Modelling in Geomechanics*, C.S. Desai and G. Gioda (eds.), Springer-Verlag, Vienna.

Desai, C. S. (1999). "DSC-SST2D code for two-dimensional static repetitive and dynamic analysis." User's Manual I to III, Tucson, AZ, USA.

Desai, C. S. (2007). "Unified DSC constitutive model for pavement materials with numerical implementation." *International Journal of Geomechanics, ASCE*, 7(2), 83–101.

Desai, C. S. (2015a). "Constitutive modelling of materials and contracts using the disturbed state concept: Part 1 – Background and analysis." *Computers and Structures*, 146, 214–233.

Desai, C. S. (2015b). "Constitutive modelling of materials and contracts using the disturbed state concept: Part 2 – Validations at specimen and boundary value problem levels." *Computers and Structures*, 146, 234–251.

Desai, C. S. and Wathugala, G. W. (1987). "Hierarchical and unified models for solids and discontinuities (joints/interfaces)," *Notes for Short Course*, Tucson, Arizona.

Dugan, C. R., Sumter, C. R., Rani, S., Ali, S. A., O'Rear, E. A., and Zaman, M. (2020). "Rheology of virgin asphalt binder combined with high percentages of RAP binder rejuvenated with waste vegetable oil." *ACS Omega*, 5(26), 15791–15798.

DiMaggio, F. L. and Sandler, I. S. (1971). "Material model for granular soils." *Journal of the Engineering Mechanics Division*, 97(3), 935–950.

FHWA (1988). "Performance of coarse graded mixes at WesTrack – premature rutting." *Report No. FHWA-RD-99-134*, Federal Highway Administration, Washington, D.C.

FHWA (2013). "LTPP manual for profile measurements and processing: profile measurements using the Face Dipstick®." *Report No. FHWA-HRT-08-056*, Federal Highway Administration, Washington, DC.

FHWA (2014). "Traffic Monitoring Guide, Appendix C: Vehicle types." Federal Highway Administration, Washington, DC.

Fwa, T. F., Tan, S. A., and Zhu, L. Y. (2004). "Rutting prediction of asphalt pavement layer using C-ϕ model." *Journal of Transportation Engineering*, 130(5), 675–683.

Gandhi, T. S. and Amirkhanian, S. N. (2007). "Laboratory investigation of warm asphalt binder properties–a preliminary analysis." In *The Fifth International Conference on Maintenance and Rehabilitation of Pavements and Technological Control (MAIREPAV5)*, Park City, Utah, 475–480.

Ghabchi, R., Rani, S., Zaman, M., and Ali, S. A. (2021). "Effect of WMA additive on properties of PPA-modified asphalt binders containing anti-stripping agent." *International Journal of Pavement Engineering*, 22(4), 418–431.

Golalipour, A., Jamshidi, E., Niazi, Y., Afsharikia, Z., and Khadem, M. (2012). "Effect of aggregate gradation on rutting of asphalt pavements." *Procedia-Social and Behavioral Sciences*, 53, 440–449.

Helwany, S., Dyer, J., and Leidy, J. (1998). "Finite-element analyses of flexible pavements." *Journal of Transportation Engineering*, 124(5), 491–499.

Hoffman, B. R. and Sargand, S. M. (2011). "Verification of rut depth collected with the INO laser rut measurement system (LRMS)." *Report No. FHWA/OH-2011/18*, Ohio Research Institute for Transportation and the Environment, Ohio University, Athens, OH.

Hooke, R. (1678). *De Potentia Restitutiva, or of Spring. Explaining the Power of Springing Bodies*. London U.K: John Martyn.

Hossain, N. (2017). "Mechanistic input parameters and model calibration for design and performance evaluation of flexible pavements in Oklahoma." *Ph.D. Dissertation*, The University of Oklahoma, Norman, OK.

Hu, S., Zhou, F., and Scullion, T. (2011). "Development, calibration, and validation of a new ME rutting model for HMA overlay design and analysis." *Journal of Materials in Civil Engineering*, ASCE, 23(2), 89–99.

Hu, X., Faruk, A. N., Zhang, J., Souliman, M. I., and Walubita, L. F. (2017). "Effects of tire inclination (turning traffic) and dynamic loading on the pavement stress–strain responses using 3-D finite element modeling." *International Journal of Pavement Research and Technology*, 10(4), 304–314.

Hua, J. (2000). "Finite element modeling and analysis of accelerated pavement testing devices and rutting phenomenon." *Ph.D. Thesis*, Purdue University, West Lafayette.

Huang, Y., Copenhaver, T., and Hempel, P. (2013). "Texas Department of Transportation 3D transverse profiling system for high-speed rut measurement." *Journal of Infrastructure Systems, ASCE*, 19(2), 221–230.

Huang, Y. H. (2004). *Pavement Analysis and Design: 2nd Edition*. Hoboken, NJ: Prentice-Hall.

Hussan, S., Kamal, M. A., Hafeez, I., Farooq, D., Ahmad, N., and Khanzada, S. (2017). "Statistical evaluation of factors affecting the laboratory rutting susceptibility of asphalt mixtures." *International Journal of Pavement Engineering*, 20, 1–15.

Huurman, M. (1997). "Permanent deformation in concrete block pavements." *Ph.D. Dissertation*, Delft University of Technology, Netherlands.

Jamshidi, A., Hamzah, M. O. and Aman, M. Y. (2012). "Effects of Sasobit content on the rheological characteristics of unaged and aged asphalt binders at high and intermediate temperaturs." *Materials Research*, 15(4), 628–638.

John C. S. and Mayo D. H. (1913). "The History of Ohm's Law." Popular Science, Bonnier Corporation ISSN 0161–7370, gives the history of Ohm's investigations, prior work, Ohm's false equation in the first paper, illustration of Ohm's experimental apparatus, 599–614.

Kachanov, L. M. (1986). *Introduction to Continuum Damage Mechanics*. Martinus Nijhoft, Dordrecht.

Katti, D. R. and Desai, C. S. (1995). "Modeling and testing of cohesive soil using disturbed-state concept." *Journal of Engineering Mechanics*, 121(5), 648–658.

Khan, S., Nagabhushana, M. N., Tiwari, D., and Jain, P. K. (2013). "Rutting in flexible pavement: An approach of evaluation with accelerated pavement testing facility." *Procedia-Social and Behavioral Sciences*, 104, 149–157.

Khobklang, P., Vimonsatit, V., Jitsangiam, P., and Nikraz, H. (2013). "A preliminary study on characterization of mechanical behavior of hydrated cement treated crushed rock base using the disturbed state concept." *Scientific Research and Essays*, 8(10), 404–413.

Kim, Y. R. and Souza, L. T. (2009). "Effects of aggregate angularity on mix design characteristics and pavement performance." *Report No. MPM-10*, University of Nebraska-Lincoln, Lincoln, NE.

Krugler, P., Mounce, J., and Bentenson, W. (1985). "Asphalt pavement rutting in the western states: Two Texas lectures and the WASHTO report." *Special Study*, 26, State Department of Highways and Public Transportation, Austin, TX.

Lee, S. J., Amirkhanian, S. N., Putman, B. J., and Kim, K. W. (2007). "Laboratory study of the effects of compaction on the volumetric and rutting properties of CRM asphalt mixtures." *Journal of Materials in Civil Engineering, ASCE*, 19(12), 1079–1089.

Leon, L. and Charles, R. (2015). "Impact of coarse aggregate type and angularity on permanent deformation of asphalt concrete." *Proceedings of the Computational Methods and Experimental Measurements XVII*, 59, 303–313.

Lu, Q. and Harvey, J. (2006). "Evaluation of Hamburg wheel-tracking device test with laboratory and field performance data." *Transportation Research Record: Journal of the Transportation Research Board* 1970, 25–44.

Lu, Y. and Wright, P. J. (1998). "Numerical approach of visco-elastoplastic analysis for asphalt mixtures." *Computers and Structures*, 69, 139–147.

Luo, X., Luo, R., and Lytton, R. L. (2013). "Characterization of asphalt mixtures using controlled-strain repeated direct tension test." *Journal of Materials in Civil Engineering*, 25(2), 194–207.

MEPDG (2004). "National Research Council, guide for mechanistic-empirical design (MEPDG)." National Cooperative Highway Research Program (NCHRP), Washington, DC.

Metcalf, J. B. (1996). "Application of full-scale accelerated pavement testing." *NCHRP Synthesis of Highway Practice 235*, Transportation Research Board, National Research Council, Washington DC.

Monismith, C. L. and Secor, K. E. (1962). "Viscoelastic behavior of asphalt concrete pavements." In *International Conference on the Structural Design of Asphalt Pavements*, University of Michigan, Ann Arbor, 203(1).

Monismith, C. L., Ogawa, N., and Freeme, C. (1975). "Permanent deformation characterization of subgrade soils due to repeated loading." *Transportation Research Record* 537, 1–17.

Nahi, M. H., Kamaruddin, I., Ismail, A., and Al-Mansob, R. A. E. (2014). "Finite element model for rutting prediction in asphalt mixes in various air void contents." *Journal of Applied Sciences*, 14(21), 2730–2737.

NCHRP Report 468 (2002). "Contributions of pavement structural layers to rutting of hot mix asphalt pavements." *Transportation Research Board- National Research Council*, Washington, DC.

NCHRP Report 508 (2003). "Accelerated laboratory rutting tests: evaluation of the asphalt pavement analyzer." *Transportation Research Board- National Research Council*, Washington, DC.

ODOT (2009). "2009 Standard Specifications for Highway Construction." Oklahoma Department of Transportation, USA.

Pellinen, T. K., Song, J., and Xiao, S. (2004). "Characterization of hot mix asphalt with varying air voids content using triaxial shear strength test." *In Proceedings of the 8th Conference on Asphalt Pavements for Southern Africa (CAPSA'04)*, 12, 16 pages.

Qing-lin, S. H. A. (2001). *Premature Damage Phenomena and Preventive Techniques in Expressway Asphalt Pavement*. Beijing: People's Transportation Press.

Ramli, I., Yaacob, H., Hassan, N. A., Ismail, C. R., and Hainin, M. R. (2013). "Fine aggregate angularity effects on rutting resistance of asphalt mixture." *Jurnal Teknologi*, 65(3), 105–109.

Rani, S. (2019). "Characterization of rutting in asphalt pavements using laboratory testing." Ph.D. Dissertation, University of Oklahoma Norman.

Rani, S., Ghabchi, R., Ali, S. A., and Zaman, M. (2018). "Effect of anti-stripping agents on asphalt mix performance using a mechanical approach." In Zhang, K., Xu, R., Chen, S.H. (eds.) *Civil Infrastructures Confronting Severe Weathers and Climate Changes Conference*, Springer, Cham, 21–31.

Rani, S., Ghabchi, R., Ali, S. A., and Zaman, M. (2019a). "Laboratory characterization of asphalt binders containing a chemical-based warm mix asphalt additive." *Journal of Testing and Evaluation*, 48(2), 1334–1349.

Rani, S., Ghabchi, R., Ali, S. A., Zaman, M., and O'Rear, E. A. (2021). "Moisture-induced damage potential of asphalt mixes containing polyphosphoric acid and antistripping agent." *Road Materials and Pavement Design*, 23, 1–21.

Rani, S., Ghabchi, R., Zaman, M., and Ali, S. A. (2019b). "Effect of polyphosphoric acid on stress sensitivity of polymer-modified and unmodified asphalt binders." In *Airfield and Highway Pavements 2019: Testing and Characterization of Pavement Materials*, Reston, VA, 238–247.

Rani, S., Ghabchi, R., Zaman, M., and Ali, S. A. (2020). "Rutting performance of PPA-modified binders using multiple stress creep and recovery (MSCR) test." In Prashant, A., Sachan, A., Desai, C. (eds.) *Advances in Computer Methods and Geomechanics*, Springer, Singapore, 607–616.

Salama H. K. and K. Chatti, (2011). "Evaluation of fatigue and rut damage prediction methods for asphalt concrete pavements subjected to multiple axle loads." *International Journal of Pavement Engineering*, 12(1), 25–36.

Salama, H. K., Haider, S. W., and Chatti, K. (2007). "Evaluation of new mechanistic-empirical pavement design guide rutting models for multiple-axle loads." *Transportation Research Record: Journal of the Transportation Research Board*, 2005, 112–123.

Samris (2004). "Selection and evaluation of models for prediction of permanent deformations of unbound granular materials in road pavement." Sustainable and Advanced Materials for Road Infrastructure.

Sane, S. M. and Desai, C. S. (2007). "Disturbed state concept based constitutive modeling for reliability analysis of leadfree solders in electronic packaging and for prediction of glacial motion." *Report to NSF*, Department of Civil Engineering and Engineering Mechanics, University of Arizona, Tucson, AZ, USA.

Sane, S. M., Desai, C. S., Jenson, J. W., Contractor, D. N., Carlson, A. E., and Clark, P. U. (2008). "Disturbed state constitutive modeling of two Pleistocene tills." *Quaternary Science Reviews*, 27(3–4), 267–283.

Scarpas, A., Al-Khoury, R., Van Gurp, C. A. P. M., and Erkens, S. M. J. G. (1997). "Finite element simulation of damage development in asphalt concrete pavements." *Proceedings of 8th International Conference on Asphalt Pavements*, University of Washington, Seattle, 673–692.

Schofield, A. N. and Wroth, C. P. (1968). *Critical State Soil Mechanics*. McGraw-Hill, London.

Serigos, P. A., Prozzi, J. A., Nam, B. H., and Murphy, M. R. (2012). "Field evaluation of automated rutting measuring equipment." *Report No. FHWA/TX-12/0–6663-1*, Center for Transportation Research, Austin, TX.

SHRP (1993). "Summary report on permanent deformation on asphalt concrete." *Report No. SHRP-A-318*, Strategic Highway Research Program, National Research Council, Washington DC.

Stakston, A. D. and Bahia, H. (2003). "The effect of fine aggregate angularity, asphalt content and performance graded asphalts on hot mix asphalt performance." *Highway Research Study 0092-45-98*, University of Wisconsin, Madison.

Su, K., Sun, L. J., and Hachiya, Y. (2008). "Rut prediction for semi-rigid asphalt pavements." *Transportation and Development Innovative Best Practices* 2008, 486–491.

Tarefder, R. A. and Zaman, M. (2002). "Evaluation of rutting potential of hot mix asphalt using the asphalt pavement analyzer." *Report No. ORA 125–6660*, University of Oklahoma, Norman, OK.

Tarefder, R. A., Zaman, M., and Hobson, K. (2003). "A laboratory and statistical evaluation of factors affecting rutting." *International Journal of Pavement Engineering*, 4(1), 59–68.

Tutumluer, E., Pan, T., and Carpenter, S. H. (2005). "Investigation of aggregate shape effects on hot mix performance using an image analysis approach." *A final study report on the Transportation Pooled Fund Study TPF-5 (023)*, University of Illinois at Urbana-Champaign Urbana, IL.

USDOT (2017). "The Superpave system: new tools for designing and building more durable asphalt pavements." *U.S. Department of Transportation, Federal Highway Administration*, Washington, DC. *Online access: http://www.library.unt.edu/gpo/OTA/featproj/fp_super.html (Last accessed: February 21, 2017)*.

Uzan, J. (1999). "Permanent deformation of a granular base material." *Transportation Research Record: Journal of the Transportation Research Board*, 1673, 89–94.

Vallejo, M. (2012). "Predicting failure behavior of polymeric and asphalt composites using damage models." *M.S. Thesis*, University of New Mexico, USA.

Wei, J., Fu, T., Meng, Y., and Xiao, C. (2020). "Investigating the fatigue characteristics of large stone asphalt mixtures based on the disturbed state concept." *Advances in Materials Science and Engineering*, 2020, 10. https://doi.org/10.1155/2020/3873174.

Whitaker, S. (1986). "Flow in porous media I: A theoretical derivation of Darcy's law". Transport in Porous Media, 3–25. doi:10.1007/BF01036523. S2CID 121904058.

Witczak, M. W. (2001). "Development of the 2002 guide for the design of new and rehabilitated pavements-flexible pavements overview." *Hot topics*.

Wu, Z., Chen, X., Yang, X., and Zhang, Z. (2011). "Finite element model for rutting prediction of flexible pavement with cementitiously stabilized base–subbase." *Transportation Research Record*, 2226(1), 104–110.

Zhang, F. and Yu, J. (2010). "The research for high-performance SBR compound modified asphalt." *Construction and Building Materials*, 24(3), 410–418.

Zhang, J., Wang, Y. D., and Su, Y. (2019). "Fatigue damage evolution model of asphalt mixture considering influence of loading frequency." *Construction and Building Materials*, 218, 712–720.

Zhou, F. and Scullion, T. (2002). "Three stages of permanent deformation curve and rutting model." *International Journal of Pavement Engineering*, 3(4), 251–260.

Zou, G., Xu, J., and Wu, C. (2017). "Evaluation of factors that affect rutting resistance of asphalt mixes by orthogonal experiment design." *International Journal of Pavement Research and Technology*, 10(3), 282–288.

Zu-kang, Y. A. O. (2003). "A review on design criteria of asphalt pavements." *Highway*, 2, 43–49.

22 Concrete

Vahab Toufigh
Dept. of Civil Engineering, Sharif University of Technology

Massoud Hosseinali
Dept. of Civil and Environmental Engineering, University of Utah

Ahad Ouria
Dept. of Civil Engineering, University of Mohaghegh Ardabili

22.1 INTRODUCTION

Concrete is one of the most utilized materials for the construction of infrastructure, footings, retaining structures, pavements, bridges, and many other types of structures. According to Mehta,[1] concrete consumption is about seven billion tons a year worldwide, and it is expected to be increasing until 2050. Therefore, improving our understanding of the behavior of concrete is essential. In this respect, many models have been proposed to predict the mechanical properties of concrete.

Several models have been proposed to predict the complete stress–strain curve of concrete. However, they are often restricted to a specific type of concrete or loading condition. On the other hand, most of those models are mathematical models and are only able to predict the behavior of concrete under uniaxial loading. More than 110 stress–strain models have been developed for predicting the axial compressive behavior of concrete in unconfined or actively confined stress states.[2] This multiplicity of models may confuse engineers and designers in choosing an appropriate model. In addition to the mathematical models, there are elastic and elasto-plastic models, which can be used to model two- or three-dimensional problems. However, most of them are not capable of modeling the softening behavior of concrete and need experimental studies to determine the required parameters. For a more accurate and cost-effective design, it is necessary to consider other types of models. One of the models capable of capturing the hardening/softening behavior of materials is the disturbed state concept (DSC) proposed by Desai.[3] A distinguishing feature of the DSC is that it takes into account the interactions between the parts of a material as affected by the microstructural self-adjustment of particles. However, its formulation does not require particle or micro-level characterization. As a result, the need of particle-level constitutive laws, which are required in the micromechanical and microcrack interaction models, is eliminated.[4] The only advantage of the mathematical models, in comparison with the DSC, is that they do not require experimental tests to determine parameters.

In this chapter, a simple relation is given for the relatively intact (RI) state of the DSC. Then, the distribution of disturbance in the DSC is simplified to eliminate the need of experimental tests to determine disturbance parameters. The proposed disturbance model is simply based on basic parameters of concrete, such as ultimate strength, density, and confinement conditions and does not require further experimental tests.

22.2 THE DISTURBED STATE CONCEPT

The DSC was first introduced by Chandrakant S. Desai[3] and since then has been developed for a variety of problems in different deciplines.[5–12] According to the DSC, the observed behavior of a system could be described based on its responses in two reference states using an appropriate state

DOI: 10.1201/9781003362081-28

function called the disturbance function. Two reference states are the initial RI state and the final fully adjusted (FA) state. Based on the DSC, the response of the material to any excitation has resulted from the disturbance that occurred in the microstructure of the material. In the initial state, the material is relative intact (RI) without any disturbance in its microstructure and is FA with fully disturbed microstructure in the final state. Depending on the nature of the problem, different definitions could be adapted for the disturbance term. The basic relationship of the DSC is[3]:

$$F_i = (1-D)F_{RI} + DF_{FA} \tag{22.1}$$

where F_i is the response of the material at any arbitrary state between RI and FA states, F_{RI} is the response of the material in a RI state, and F_{FA} is the response of the material in a FA state.

Equation (22.1) has been adopted for various problems such as consolidation of plastic clays,[5] constitutive modeling of structured soils,[6,7] unsaturated soils,[8] cementitious soils,[9,10] interfaces,[11] and polymer concrete.[12]

22.3 CONSTITUTIVE MODELING FOR CONCRETE USING THE DSC

The DSC model is based on the idea of decomposition of a material into its RI and FA components.[3] The decomposition of typical concrete and its relation to the cracks of concrete are depicted in Figure 22.1. As can be seen, before the peak, there is almost no significant disturbance in concrete (compared to that of post peak), that is, material shows RI behavior. Although there are some cracks and the response is not entirely elastic, it is insignificant and can be considered as elastic response altogether. After the peak, as the cracks in concrete grow (propagation of disturbance), concrete loses its strength until it reaches the FA state, which has different definitions corresponding to the confinement condition; this will be discussed in the following paragraphs.

22.3.1 DISTURBANCE

In this study, disturbance expression is similar to the one used conventionally,[13] and the definition of disturbance is based on the concept proposed by Desai[14] as an extension to the uniaxial model proposed by Kachanov[13] and Rabotnov.[15] Thus, the disturbance parameter, D, is[16,17]:

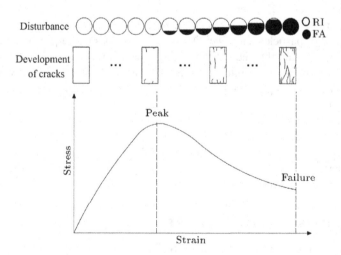

FIGURE 22.1 Stress–strain curve of concrete including the disturbed state concept (DSC) decomposition and cracks' distribution.

$$D = D_u\left(1 - e^{-A\varsigma_D^Z}\right)$$ (22.2)

where D_u is the ultimate value of D; A and Z are material parameters; and D is the trajectory of deviatoric plastic strain:

$$\varsigma_D = \int\left(de_{ij}^p de_{ij}^p\right)^{0.5}$$ (22.3)

where e_{ij} is the deviatoric strain tensor of total strain tensor ε_{ij}.

As proposed by Desai[13] and Ma and Desai,[14] the disturbance based on the stress–strain response can be expressed generally as follows:

$$D = \frac{\sigma^{RI} - \sigma^{obs}}{\sigma^{RI} - \sigma^{FA}}$$ (22.4)

where σ^{RI}, σ^{obs}, and σ^{FA} are RI, observed, and FA stresses, respectively.

Equating Eq. (22.2) with Eq. (22.4) for two arbitrary points of the experimental stress–strain curve yields A and Z values. After calculating material parameters, the stress–strain curve can be obtained through the following incremental equation:

$$d\sigma_{ij} = (1 - D)d\sigma_{ij}^{RI} + Dd\sigma_{ij}^{FA} + dD\left(d\sigma_{ij}^{RI} - d\sigma_{ij}^{FA}\right)$$ (22.5)

At the pre-failure stage ($D=dD=0$), Eq. (22.4) yields the RI stress–strain response.

22.4 THE HIERARCHICAL SINGLE-SURFACE (HISS) MODEL FOR CONCRETE

A failure surface is the boundary of elastic behavior and is usually written in terms of stresses. Common simple failure criteria that have been used in the literature are Tressa, Von-Mises, Mohr–Coulomb, and Drucker–Prager. Most of the plasticity models provided for concrete consist of two surfaces resulting in singularity. The HISS plasticity model is an advanced model without any singularity, which provides a general formulation for the elasto-plastic behavior of materials. This model can be adopted for materials with both isotropic and anisotropic hardening behavior as well as associated or nonassociated plasticity characterizations to predict the material response based on the continuum plasticity theory. The yield function of HISS δ_0 plasticity model is[3]:

$$F = \bar{J}_{2D} - \left(-\alpha\bar{J}_1^n + \gamma\bar{J}_1^2\right)\left(1 - \beta S_r\right)^m = 0$$ (22.6)

in which $\bar{J}_{2D} = \bar{J}_{2D}/p_a$, p_a is the atmospheric pressure constant; $\bar{J}_1 = (J_1 + 3R)/p_a$, R is the bonding stress; n is the phase change parameter where the volume change transits from compaction to dilation or vanishes; $m=-0.5$ is often used; γ is the parameter related to the (ultimate) yield surface; β corresponds to the shape of yield function in the $\sigma_1 - \sigma_2 - \sigma_3$ space; Sr is the stress ratio and is $\left(\frac{\sqrt{27}}{2}\right)\left(J_{3D}/J_{2D}^{3/2}\right)$; α is the hardening parameter that can be expressed in terms of deviatoric plastic strain trajectory; J_1 is the first invariant of the stress tensor; and J_{2D} is the second invariant of deviatoric stress tensor.

Using the elasto-plastic constitutive relations, the incremental form of stress–strain could be expressed as:

$$\Delta\{\sigma\}_i = C_{ep}\Delta\{\varepsilon\}_i \tag{22.7}$$

where C_{ep} is the elasto-plastic stiffness matrix and is:

$$C_{ep} = C_e - \frac{C_e \mathrm{nn}^\mathrm{T} C_e}{H + \mathrm{n}^\mathrm{T} C_e \mathrm{n}} \tag{22.8}$$

where C_e is the elastic stiffness matrix.

22.5   THE DSC-BASED SIMPLIFIED MODEL FOR CONCRETE

22.5.1   THE RI RESPONSE

The constitutive relationships based on the HISS model described in the previous section are a strong tool for predicting the stress–strain and failure behavior of concrete. For practical purposes, the RI and FA states of the material could be represented by simplified failure criteria such as the Mohr–Coulomb model. In this section, the RI response was assumed to be a nonlinear elastic-perfectly plastic response to one-dimensional behavior, which can be achieved easily in two- and three-dimensional problems using Mohr–Coulomb yield function with only a slight difference (elastic-perfectly plastic). Although sophisticated models can also be used for RI behavior, the goal of this method is to propose a model with the least computational costs.[18]

For one-dimensional analysis, the RI state should satisfy the following conditions:

1. The value of 28-day compressive strength of concrete (stress at peak for normal-weight concrete, NWC) can be obtained from the following empirical expression[2]:

$$f_c' = \left(\frac{21}{w/c} + 32\sqrt{sf/c}\right)\left(\frac{\rho}{2400}\right)^{1.6} \tag{22.9}$$

   where w/c is the water–cement ratio, ρ is the density of concrete, and sf/c is the silica fume-cement ratio;

2. Initial slope of the stress–strain curve is equal to the initial modulus of elasticity of concrete, which can be determined based on the empirical relation derived from Ozbakkaloglu and Lim[2]:

$$E_i = 4400\sqrt{f_c'}\left(\frac{\rho}{2400}\right)^{1.4} \tag{22.10}$$

   where f_c' is the 28-day compressive strength of concrete from Eq. (22.10), and ρ is the density of concrete;

3. Slope of the curve at peak is equal to zero due to the perfectly plastic nature of the post-peak response;

4. The initial value of stress is equal to zero, where the strain is equal to zero.

Having these four conditions for one-dimensional behavior, a third-order polynomial can perfectly satisfy the pre-peak response as follows:

$$f = +\gamma\varepsilon + \beta\varepsilon^2 + \alpha\varepsilon^3 \tag{22.11}$$

Substituting the conditions mentioned above into Eq. (21.11), the coefficients of the RI response are:

$$\lambda = 0 \tag{22.12a}$$

$$\gamma = E_i \tag{22.12b}$$

$$\beta = \frac{3f_{\text{peak}} - 2\gamma\varepsilon_{\text{peak}}}{\varepsilon_{\text{peak}}^2} \tag{22.12c}$$

$$\alpha = \frac{-2f_{\text{peak}} + \gamma\varepsilon_{\text{peak}}}{\varepsilon_{\text{peak}}^3} \tag{22.12d}$$

For the post-peak response to the RI behavior, it is assumed that concrete can no longer undertake additional stresses; therefore, stress is constant due to the lack of disturbance in the RI behavior (perfectly plastic).

22.5.2 THE FA RESPONSE

The FA state is directly related to the confinement condition.[18] With regard to the behavior of concrete, if the tests are unconfined, zero stress condition represents the FA state, i.e., the concrete specimen can no longer bear any stress at the final stage. On the other hand, if the tests are confined, then concrete has residual stress at the steady-state response, which is the FA state.

22.5.3 DISTURBANCE SIMPLIFICATION

In order to simplify the disturbance distribution, it is assumed that concrete fails when it completely reaches the FA state, which means that $D_u = 1$. Moreover, based on empirical observations, taking $Z = 1{:}622$ (constant number) allows us to obtain different types of distribution by changing parameter A; therefore, the only remaining variable is A. Disturbance distribution with different values of A is shown in Figure 22.2.[18]

For evaluating the variations of disturbance distribution with changes of parameter A, more than 50 stress–strain curves of the behavior of light-weight concrete (LWC) and NWC were modeled using Du = 1:0, Z = 1:622, and the best possible value of A. The results are shown in Figure 22.3. The relation between parameter A and peak and residual stresses is given in Eq. (22.12) based on the trend line in Figure 22.3 for NWC. For the case of LWC, the relation can be obtained by simply using an additional constant term, while it was observed that the second term in Eq. (22.12) yields more accurate results:

$$\log(A) = 1.1652\left(f_{\text{peak}} - f_{\text{residual}}\right)^{0.244} - \left(f_{\text{residual}}\right)^{0.08} + \left(1 - \frac{\rho}{2400}\right)^{0.65} \tag{22.13}$$

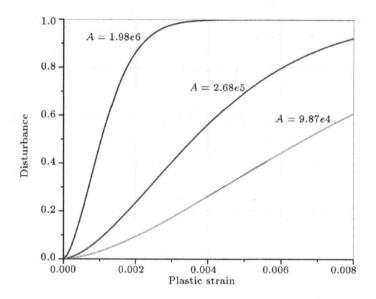

FIGURE 22.2   Disturbance function for different values of A.

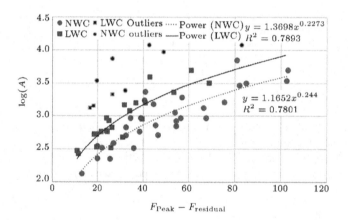

FIGURE 22.3   Log(A) versus $\left(f_{\text{peak}} - f_{\text{residual}} \right)$.

where the value of the stress at peak (f_{peak}) for NWC can be obtained using Eq. (22.9), while this value for LWC is as follows[2]:

$$f_{\text{peak}} = f_c' + 5.2 f_c'^{0.91} \left(\frac{f_l^*}{f_c'} \right)^a \tag{22.14}$$

where $\dfrac{f_l^*}{f_c'}$ is the residual stress ratio, $a = f_c'^{-0.06}$, and residual stress (f_{residual}) can be obtained using the equation below[3]:

$$f_{\text{residual}} = 1.6 f_{\text{peak}} \left(\frac{f_l^{*0.24}}{f_c'^{0.32}} \right) \le f_{\text{peak}} - 1.5 f_c' \tag{22.15}$$

22.6 RESULTS

22.6.1 PREDICTING THE STRESS–STRAIN RESPONSE OF CONCRETE BY THE HISS MODEL

The HISS model was used to predict the behavior of two types of concrete,[19] namely, high-performance concrete, Noori et al.,[20] and high-strength concrete, Farnam et al.[21] The parameters of the HISS model and the disturbance parameters for these two types of concrete were according to Table 22.1. Calculations have been done using the nonlinear finite element model in 2-D cylindrical coordinate system with 900 elements. The results are illustrated in Figures 22.4 and 22.5 for high-strength concrete and high-performance concrete, respectively.

TABLE 22.1
Parameters of HISS Model Used in Simulation of HSC and HPC.[19]

Elastic Prams.		Plastic Params.			Ph. Change	Growth		Disturbance Params.		
E (mPa)	ν	γ	β	m	n	a1	$\eta 1$	Du	A	Z
13,000	0.285	1.1e-12	0.75	−0.5	5.032	4.64e-11	0.8267	0.875	247	1.625

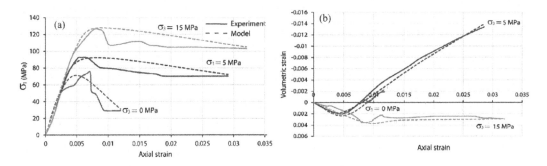

FIGURE 22.4 Axial stress and volumetric strain versus axial strain for high-strength concrete under 0, 5, and 15 MPa confining pressures.[21]

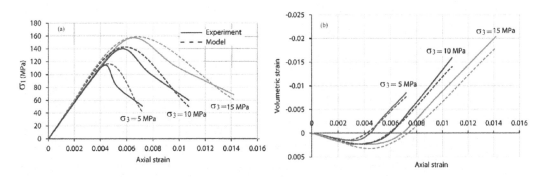

FIGURE 22.5 Axial stress and volumetric strain versus axial strain for high-performance concrete under 5, 10, and 15 MPa confining pressures.[20]

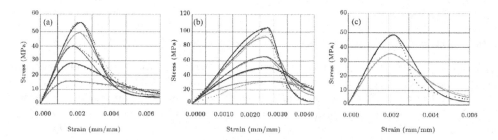

FIGURE 22.6 Comparison of the predicted curves (continuous line) with experimental unconfined results (dotted line) of different normal-weight concretes (NWCs) from: (a) Wischers,[22] (b) Dahl,[23] and (c) Taerwe.[24]

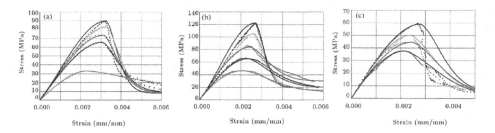

FIGURE 22.7 Comparison of the predicted curves (continuous line) with experimental unconfined results (dotted line) of different normal-weight concretes (NWCs) from: (a) Hsu and Hsu,[25] (b) Wee et al.,[26] and (c) Desnerck et al.[27]

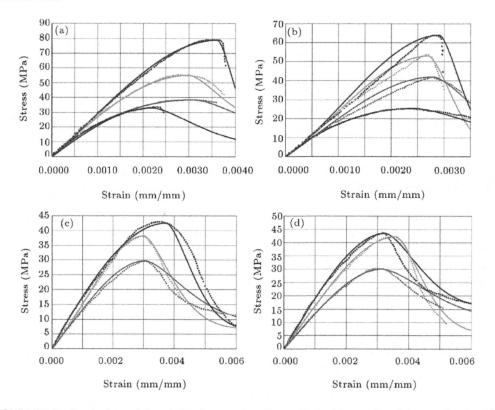

FIGURE 22.8 Comparison of the predicted curves (continuous line) with experimental unconfined results (dotted line) of different light-weight concretes (LWCs) from: (a and b) Kaar et al.[28] and (c and d) Shah et al.[29]

22.6.2 PREDICTING THE STRESS–STRAIN RESPONSE OF CONCRETE BY THE SIMPLIFIED MODEL

Experimental data of more than 50 tests in the literature[21–36] were predicted using Eqs. (22.2), (22.5), (22.10), and (22.15). The results of NWC are shown in Figures 22.6 and 22.7; the results of LWC are shown in Figures 22.8 and 22.9; and the results of actively confined concrete are shown in Figures 22.10 and 22.11.

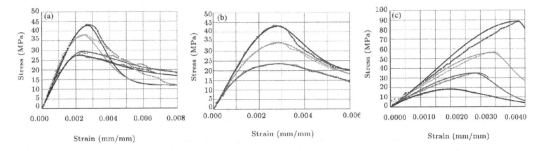

FIGURE 22.9 Comparison of the predicted curves (continuous line) with experimental unconfined results (dotted line) of different light-weight concretes (LWCs) from: (a and b) Shannag[30] and (c) Zhang and Gjorv.[31]

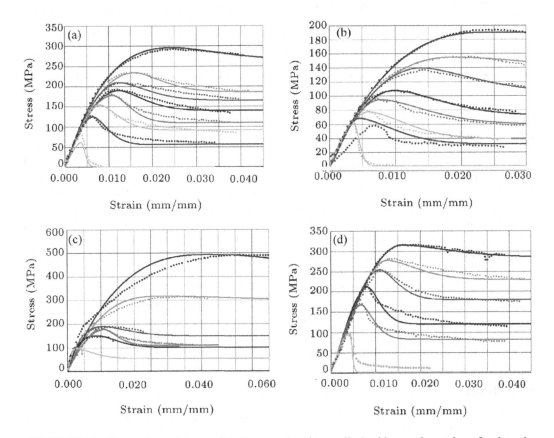

FIGURE 22.10 Comparison of the predicted curves (continuous line) with experimental confined results (dotted line) of different types of concrete from: (a) Newman[32] and (b–d) Xie et al.[33]

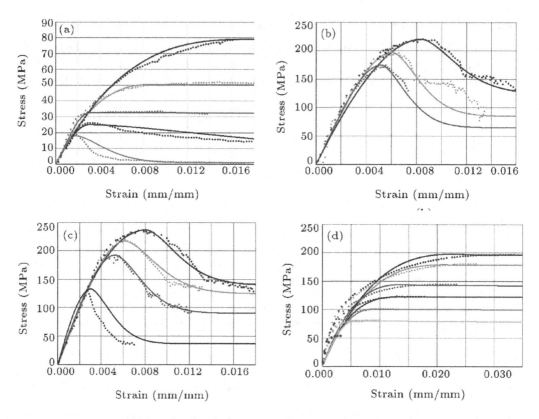

FIGURE 22.11 Comparison of predicted curves (continuous line) with experimental confined results (dotted line) of different types of concrete from: (a) Hurlbut,[34] (b and c) Attard and Setunge,[35] and (d) Bellotti and Rossi.[36]

22.7 CONCLUSIONS

In this chapter, the DSC was employed to describe the stress–strain and failure behavior of different types of concrete. The HISS model was evaluated and a simplified model with least computational cost was presented to predict the behaviors of various types of concrete. In this respect, a simple RI state was presented for the DSC. Then, the number of parameters used in the disturbance was reduced from three to one to simplify the parameter determination procedure. An empirical equation was then given to calculate parameter A based on different parameters of concrete such as ultimate strength, density, and confinement conditions. Accordingly, more than 50 experimental tests of different types of concrete and confinement conditions from other researchers were predicted using the proposed model. In most cases, predictions show good agreement with experimental results. The proposed model has the least computational cost and does not require experimental tests. Moreover, since there is a physical justification for the DSC, it can also be used for two- and three-dimensional problems. The results of these studies indicate that the DSC and HISS plasticity provide a useful tool for understanding and predicting the stress–strain response of different types of concrete.

REFERENCES

1. Mehta, P. K. Greening of the concrete industry for sustainable development, *Concrete International*, 24(7), pp. 23–28 (2002).
2. Ozbakkaloglu, T., and Lim, J. Stress-strain model for normal- and light-weight concretes under uniaxial and triaxial compression, *Construction and Building Materials*, 71, pp. 492–509 (2014).

3. Desai, C.S., *Mechanics of Materials and Interfaces: The Disturbed State Concept*, CRC Press, Boca Raton, FL (2010).

4. Bazant, Z.P. Nonlocal disturbance theory based on micromechanics of crack interactions, *Journal of Engineering Mechanics, ASCE*, 120(3), pp. 593–617 (1994).

5. Ouria, A., Desai, C.S., and Toufigh, V. Disturbed state concept-based solution for consolidation of plastic clays under cyclic loading, *International Journal of Geomechanics*, 15(1), pp. 208–214 (2014).

6. Ouria, A. Disturbed state concept-based constitutive model for structured soils, *International Journal of Geomechanics*, 17(7) (2017). doi:10.1061/(ASCE)GM.1943–5622.0000883.

7. Farsijani, A., Ouria A. Constitutive modeling the stress-strain and failure behavior of structured soils based on HISS model. *IQBQ*, 21(4): 231–250 (2021). http://mcej.modares.ac.ir/article-16-52042-en.html.

8. Farsijani, A., and Ouria, A. Wetting-induced collapse behavior of unsaturated soils in disturbed state concept framework, *International Journal of Geomechanics* 22(4), 04022014 (2022).

9. Ouria, A., and Behboodi, T. Compressibility of cement treated soft soils. *Journal of Civil and Environmental Engineering*, 47.1(86), 1–9 (2017).

10. Ouria, A., Ranjbarnia, M., and Vaezipour, D. A failure criterion for weak cemented soils. *Journal of Civil and Environmental Engineering*, 48.3(92), 13–21 (2018).

11. Toufigh, V., Shirkhorshidi, S.M., and Hosseinali, M. Experimental investigation and constitutive modeling of polymer concrete and sand interface, *International Journal of Geomechanics, ASCE* (2016). doi:10.1061/(ASCE)GM.1943-5622.0000695.

12. Toufigh, V., Hosseinali, M., and Shirkhorshidi, S.M. Experimental study and constitutive modeling of polymer concrete's behavior in compression, *Construction and Building Materials*, 112(1), pp. 183–190 (2016).

13. Kachanov, L.M., *The Theory of Creep*, (English Translation edited by A.J. Kennedy), Chapters IX and X. National Lending Library, Boston (1958).

14. Desai, C.S. A consistent finite element technique for work-softening behavior, *Proceeding International Conference on Computational Methods in Nonlinear Mechanics*, J.T. Oden et al., Eds., Austin, TX, USA (1974).

15. Rabotnov, Y.N., *Creep Problems in Structural Members*, North-Holland, Amsterdam (1969).

16. Desai, C.S., Somasundaram, S., and Frantziskonis, G. A hierarchical approach for constitutive modelling of geologic materials, *International Journal for Numerical and Analytical Methods in Geomechanics*, 10(3), pp. 225–257 (1986).

17. Frantziskonis, G. and Desai, C.S. Constitutive model with strain softening, *International Journal of Solids and Structures*, 23(6), pp. 733–750 (1987).

18. Hosseinali, M. and Toufigh, V. A simple model for various types of concretes and confinement conditions based on disturbed state concept. *Scientia Iranica, Transactions on Civil Engineering (A)*, 25(2), 557–564 (2018).

19. Toufigh, V., Jafarian Abyaneh, M., and Jafari, K. Study of behavior of concrete under axial and triaxial compression, *ACI Materials Journal*, 114(4) (2017). doi: 10.14359/51689716.

20. Noori, A., Shekarchi, M., Moradian, M., and Moosavi, M. Behavior of steel fiber-reinforced cementitious mortar and high-performance concrete in triaxial loading, *ACI Materials Journal*, 112(1), pp. 95–104 (2015).

21. Farnam, Y., Moosavi, M., Shekarchi, M., Babanajad, S. K., and Bagherzadeh, A. Behaviour of slurry infiltrated fibre concrete (SIFCON) under triaxial compression, *Cement and Concrete Research*, 40(11), pp. 1571–1581 (2010).

22. Wischers, G. Application and effects of compressive loads on concrete, *Betontechnische Berichte*, 2, 31–58 (1979).

23. Dahl, K.K.B. A constitutive model for normal and high strength concrete", Project 5, Report 5.7, American Concrete Institute, Detroit (1992).

24. Taerwe, L.R. Influence of steel bars on strain softening of high-strength concrete, *ACI Materials Journal*, 89(1), 54–60 (1992).

25. Hsu, L.S. and Hsu, C.T. Complete stress-strain behavior of high-strength concrete under compression, *Magazine of Concrete Research*, 46(169), 301–312 (1994).

26. Wee, T.H., Chin, M.S., and Mansur, M.A. Stress strain relationship of high-strength concrete in compression, *Journal of Materials in Civil Engineering*, 8(2), pp. 70–76 (1996).

27. Desnerck, P., Schutter, G.D., and Taerwe, L. Stressstrain behavior of self-compacting concrete containing limestone llers, *Structural Concrete*, 13(2), pp. 95–101 (2012).

28. Kaar, P.H., Hanson, N.W., and Capell, H.T. Stress strain characteristic of high strength concrete", Research and Development Bulletin RD051-01D, Portland Cement Association, Skokie, Illinois (1977).

29. Shah, S.P., Naaman, A.E., and Moreno, J. Effect of confinement on the ductility of lightweight concrete, *International Journal of Cement Composites and Lightweight Concrete*, 5(1), pp. 15–25 (1983).

30. Shannag, M.J. Characteristics of lightweight concrete containing mineral admixtures, *Construction and Building Materials*, 25(2), pp. 658–662 (2011).

31. Zhang, M.H. and Gjorv, O.E. Mechanical properties of high-strength lightweight concrete, *ACI Materials Journal*, 88(3), pp. 240–247 (1991).

32. Newman, J.B., Concrete Under Complex Stress, London, UK, Department of Civil Engineering, Imperial College of Science and Technology, London, UK (1979).

33. Xie, J., Elwi, A.E. and Macgregor, J.G. Mechanical properties of high-strength concretes containing silica fume, *ACI Materials Journal*, 92(2), 135–45 (1995).

34. Hurlbut, B. Experimental and computational investigation of strain-softening in concrete", PhD Dissertation, University of Colorado (1985).

35. Attard, M.M. and Setunge, S. Stress-strain relationship of conned and unconned concrete, *ACI Materials Journal*, 93(5), pp. 432–442 (1996).

36. Bellotti, R. and Rossi, P. Cylinder tests: Experimental technique and results, *Materials and Structures*, 24(1), pp. 45–51 (1991).

Section 7

Interfaces and Joints

23 Interfaces and Joints

Chandrakant S. Desai
Dept. of Civil and Architectural Engineering and
Mechanics, University of Arizona

23.1 INTRODUCTION

Interfaces and joints that can occur between two materials composing an engineering structure are special cases of contact problems in mechanics. They represent a discontinuity that may pre-exist or is induced during loading. Existence of discontinuities in a material body may render the system to be discontinuous. Hence, the theories based on continuum mechanics may not be applicable to govern its response. Interfaces can occur in metals, ceramics, concrete, rocks and soils, composites, and structured-geologic media (soil or rock). In civil engineering and geomechanics, such contact interfaces can occur between structures and geologic foundations, concrete and reinforcement, and joints in rock masses. The schematic of a typical interface is shown in Figure 23.1.

Although several models have been proposed for the contact problem, contact behavior has not yet been fully understood when several significant factors are considered that influence the behavior. Details including reviews of contact mechanics, interfaces, and joints have been presented in Reference [1]. In the following, we provide mainly the use of the DSC/HISS model for simulating the behavior of interfaces and joints.

23.2 REVIEW

Models based on strength, limit equilibrium, elastic and classical elastoplastic theories have been presented in various publications for areas of friction, contact mechanisms, and structure–media (soil or rock) problems. A detailed review is provided in Desai [1].

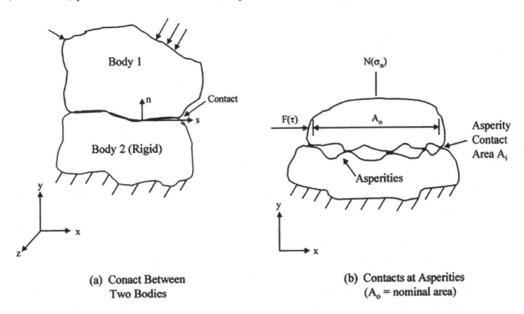

(a) Conact Between Two Bodies

(b) Contacts at Asperities (A_o = nominal area)

FIGURE 23.1 Schematic of interface.

23.3 COMMENTS

Several significant factors influence the behavior at the interface (Figure 23.1). They include nonlocal behavior; effects of elastic, plastic, and creep strains; microcracking; degradation and softening; stiffening or healing; existence of filler materials (oxides, gouge, etc.); and environmental factors and type of loading (static, cyclic, environmental, etc.). Although the available conventional models [1] allow for strength, elastic, and limited plasticity response, they usually do not allow for factors such as continuous yielding or hardening, microcracking, and softening leading to post-peak degradation, stiffening, and viscous (time) effects.

The objective of this chapter is to present the unified DSC/HISS model that allows for the above factors, in addition to those included in conventional models. It is also noted that many of the previous models can be derived as special cases of the DSC/HISS approach. It is believed that the interface behavior may be treated as a problem in constitutive modeling in which the foregoing factors are integrated. This contrasts with some of the previous models, which have treated contact or interface behavior by introducing constraints (kinematic and/or force) to allow for the effects of special characteristics such as relative motions (sliding, debonding, etc.).

Another important issue is the implementation of the interface models in solution (computer) procedures taking into consideration factors such as robustness, accuracy, and stability of the numerical predictions. This aspect is handled by using the disturbance and the concept of the thin-layer element (TLE), in which the interface is simulated as a thin zone of finite thickness, t [2]. In the finite-element procedure, the interface zone is treated as a regular element whose constitutive behavior (with the DSC/HISS) is defined based on appropriate laboratory tests, e.g., special shear testing devices.

Models based on the DSC/HISS that include elastic, plastic, and creep deformations; microcracking; (asperity) degradation response; and liquefaction have been presented in References [3–13]. The DSC/HISS approach allows for the nonlocal effects and characteristic dimension and hence leads to computations that are unaffected by spurious mesh dependence [14,15].

The following descriptions include the DSC/HISS models for interfaces and joints, laboratory testing for the calibration of material parameters, and implementation of the models in numerical procedures using the TLE. The latter is described first, as the DSC/HISS is often formulated with respect to the thin-layer simulation of interfaces and joints.

23.4 THIN-LAYER INTERFACE ELEMENT

An interface can involve several configurations. The interface between two metallic bodies can be considered clean because there is no third material between them. Then, the smeared interface zone can entail different levels of roughness defined by the surface (microlevel) asperities (Figures 23.1b and 23.2a). The contact can be smooth to medium rough to very rough, depending on the characteristics (height, length) of irregular asperities. It may be noted that even a "smooth" contact involves asperities at different levels (macro, micro, etc.); hence, an ideally smooth surface may not exist.

It may happen that when two bodies made of different materials are in contact, like steel (pile) and soft clay (Figure 23.2b), there exists a finite "smeared" zone between the two that behaves as an interface (Figure 23.2e). In the case of rock joints, the contact may be filled with a third material (e.g., gouge) that acts as a bulk interface (Figure 23.2c). Similar conditions can occur in the case of electronic chip-substrate system joined together, say, by solders; here, the filled zone can be treated as the bulk interface (Figure 23.2d), [16,17]. The TLE has also been used to simulate interfaces in assembled metallic structures with bolted connections, lap joints, and interfaces between sleepers (ties) and ballast in railroad track support systems [18–22].

The purpose of a TLE is to present a model for bulk interface or joint that can provide a realistic simulation of relative motions between the two bodies. The interface zone is referred to as the TLE [2,23] with thickness, t, that can be treated as an equivalent or smeared zone between two materials

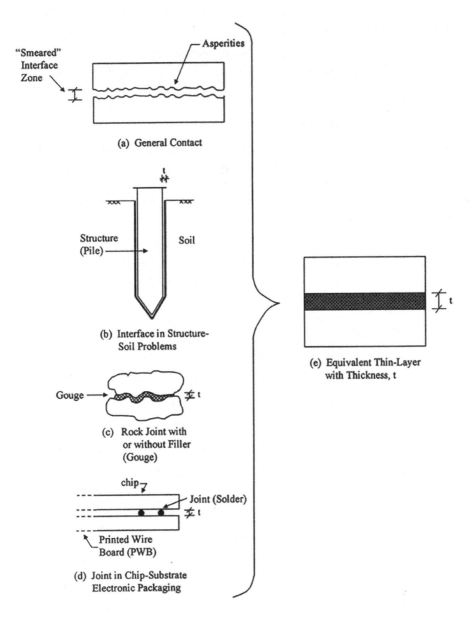

FIGURE 23.2   Schematics of various interfaces and joints.

(Figure 23.2e). It can provide, in a weighted sense, a realistic approach to model the behavior of an interface. Even in the case of clean contact (Figure 23.1a), it can be possible to develop a dimension (thickness) of the equivalent "smeared" zone, as affected by the asperities, to represent the interface behavior. Thus, in all cases in Figure 23.2, it is assumed that the interface zone can be represented by an equivalent dimension with thickness, t. Then, a question arises as to how to determine the thickness, t. This can be difficult. One of the direct ways would be to perform tests in which the deformation behavior of the interface is measured. Then, the thickness (t) is determined by performing numerical (finite element) predictions and comparing them with observations using parametric variations of the thickness [2]. In the case of nondestructive tests such as X-ray computerized tomography and acoustic methods, it can be possible to measure the dimensions of the influence zones around the asperities, which can be used to estimate the thickness.

23.5   DISTURBED STATE CONCEPT

The interface (Figure 23.3) can be considered as a deforming material that is composed of the relative intact (RI) and fully adjusted (FA) parts (Figure 23.3c). Now, it is necessary to define the RI and FA behaviors.

23.5.1   RELATIVE INTACT BEHAVIOR

The RI behavior can be represented by using theories such as linear (or nonlinear) elasticity, elastoplasticity (HISS δ_o -model), thermo-elastoplasticity, and thermo-viscoplasticity [1]. If the (linear) theory of elasticity is used, the RI behavior can be simulated by using two moduli, shear stiffness, k_s, and normal stiffness, k_n (Figure 23.4). The values of these moduli can depend on such factors as the (initial) normal stress (σ_{no}) and roughness (R_o)

$$k_s = k_s(\sigma_{no}, R_o) \tag{23.1a}$$

$$k_n = k_n(\sigma_{no}, R_o) \tag{23.1b}$$

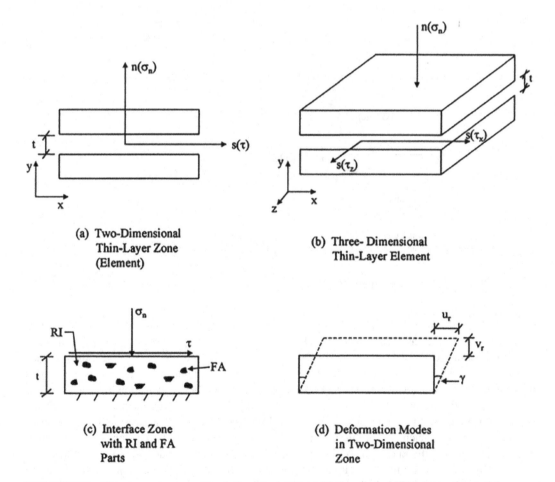

(a) Two-Dimensional
    Thin-Layer Zone
    (Element)

(b) Three- Dimensional
    Thin-Layer Element

(c) Interface Zone
    with RI and FA
    Parts

(d) Deformation Modes
    in Two-Dimensional
    Zone

FIGURE 23.3   Thin-layer element with relative intact (RI) and fully adjusted (FA) states of material.

23.5.2   Nonlinear Elastic as RI Response

For a two-dimensional idealization (Figure 23.3a), the incremental equations for the RI behavior with nonlinear piecewise elastic response are expressed as:

$$
\left\{ \begin{array}{c} d\tau^i \\ \sigma_n^i \end{array} \right\} = \left[\begin{array}{cc} k_{ss}^i & k_{sn}^i \\ k_{ns}^i & k_{nn}^i \end{array} \right] \left\{ \begin{array}{c} du_r^i \\ dv_r^i \end{array} \right\}
\tag{23.2a}
$$

where τ and σ_n are the shear and normal stresses, respectively, k_s and k_n denote tangent shear and normal stiffnesses, respectively, u_r and v_r are the relative shear and normal displacements,

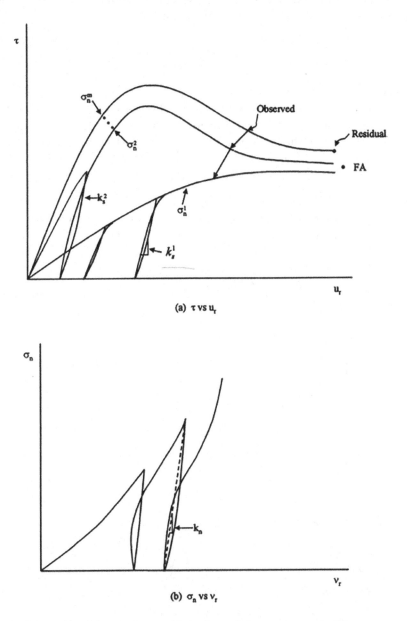

(a) τ vs u_r

(b) σ_n vs v_r

FIGURE 23.4   Schematics of shear and normal behavior and shear and normal stiffnesses.

respectively, i denotes RI or continuum response, and d denotes an increment. If it is assumed that the elastic shear and normal responses are uncoupled, the cross terms in Eq. (23.5a) will be zero. In matrix notation, Eq. (23.5a) can be expressed as:

$$\{d\sigma\}^i = \left[C_j\right]^i \{du\}^i \tag{23.2b}$$

where $\left[C_j\right]^i$ is the RI constitutive.

23.5.3 THE HISS MODEL FOR RI BEHAVIOR

When the HISS plasticity model is adopted for the RI part, the yield function for a two-dimensional interface is given by References [1,3] (Chapter 1) (Figure 23.5):

$$F = \tau^2 + \alpha\sigma_n^n - \gamma\sigma_n^q = 0 \tag{23.3a}$$

where σ_n can be modified as $\sigma_n + R$, R is the intercept along the σ_n axis (Figure 23.5), γ is the slope of the ultimate response, n (subscript) is the phase change parameter, which designates the transition from compressive to dilative response, q governs the curve of the ultimate envelope (if the ultimate envelope is linear, $q=2$), and α is the growth or yield function given by:

$$\alpha = h_1 / \xi^{h_2} \tag{23.3b}$$

where h_1 and h_2 are hardening parameters, and ξ is the trajectory of (accumulated) plastic relative shear and normal displacements (u_r and v_r), respectively, given by:

$$\xi = \int \sqrt{\left(du_r^p \cdot du_r^p + dv_r^p \cdot dv_r^p\right)} = \xi_D + \xi_v \tag{23.3c}$$

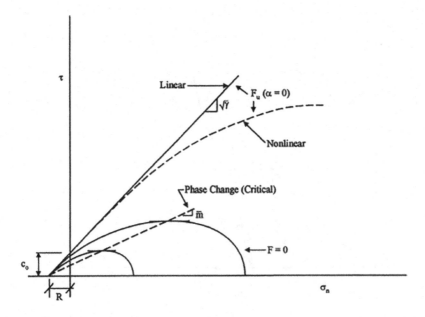

FIGURE 23.5　Yield and ultimate surfaces in the HISS model.

where ξ_D and ξ_v denote accumulated plastic shear and normal displacements, respectively, and p denotes plastic.

Details regarding the definitions of RI and FA behaviors of interfaces, disturbance functions, and parameter determination are given in Chapter 1 and in Reference [1].

23.5.4 FULLY ADJUSTED BEHAVIOR

The critical state concept can be used to simulate the FA behavior using the following equations:

For shear and normal stress at the critical state (Archard [24]):

$$\tau^c = C_o.\sigma_n^m \tag{23.4a}$$

and dilation at critical state (Schneider [25]):

$$v^c = v^o = \exp\left(-k^{\sigma_n}\right) \tag{23.4b}$$

where c denotes critical state, C_o, m and k are material constants, and v^o is the ultimate dilation when $\sigma_n = 0$.

23.5.5 DISTURBANCE FUNCTION

The disturbance function for interfaces can be expressed like that for solids (Chapter 1) as:

$$D = D_u\left[1 - \exp\left(-A\xi^Z\right)\right] \tag{23.5}$$

where A, Z, and D_u are parameters and ξ is the sum of the accumulated shear and normal relative plastic displacements, ξ_D and ξ_v, respectively.

23.5.6 DEFINITIONS

The definitions of stress, relative displacement, and stiffnesses used with the TLE and the DSC/HISS model are described below.

The net contact area involves contacts at the asperities between the two bodies (Figure 23.1b) and is usually smaller than the total area of the contact. To account for the nonlocal effects, it is necessary to introduce weighting functions. In the DSC, such weighting is introduced through the disturbance function, D. However, based on the nominal area, A_o, the shear, and normal stress are first defined as:

$$\tau = \frac{F}{A_0} \tag{23.6a}$$

$$\sigma_n = \frac{N}{A_0} \tag{23.6b}$$

with the TLE, the stiffness moduli can be expressed approximately as:

$$k_s = \frac{G}{t} \tag{23.7a}$$

$$k_n = \frac{E}{t} \tag{23.7b}$$

where G and E are the equivalent shear and elastic moduli for the interface, respectively.

Now, the relative shear displacements are expressed as Figure 23.3d

$$u_r \approx \gamma t \tag{23.8a}$$

$$v_r \approx \varepsilon_{nt} \tag{23.8b}$$

23.5.7 DETERMINATION OF PARAMETERS

The DSC model with the elastoplastic δ_o -version for the RI response for interfaces involves the constants identified in Table 23.1. The procedures for finding these parameters are essentially like that those for solid materials [1] and Chapter 1.

23.6 VALIDATION FOR INTERFACE AND JOINT MODELS

The DSC/HISS models have been used including validations by several investigators. For example, interfaces for dynamic analysis, three-dimensional model, and unsaturated soils are given in Chapters 24, 25, respectively. In the following, typical examples are presented from the work of the author and co-workers.

Example 1: Joints in Concrete

Fishman and Desai [3,26] performed a comprehensive series of shear tests on joints in concrete with a weight ratio of 1: 3.26: 2.9: 0.75 (cement: sand: coarse aggregate: water) with unconfined compressive strength of about 17.26 MPa (2,500 psi). The roughness was identified by the angle (0°, 5°, 7°, 9°) of the saw-shaped teeth (Figure 23.6). The cyclic multi-degree of freedom (CYMDOF-P) (Figure 23.7) device was used with sample size: upper block 30×30×14 cm and lower block 40×40×14 cm. The tests were conducted under normal stresses of 0.345, 0.139, 0.069, and 0.35 MPa.

TABLE 23.1
Parameters for Model

Model	Constants	Comment
Elastic	k_s, k_n	
Plasticity (RI)	a, b, n, and γ	Straight ultimate envelope
	a, b, n, γ, and q	Curved ultimate envelope
Critical state (FA)	C_0, m, v_0, and k	
Disturbance	D_u, A and Z	If $D = D_\tau = D_n$

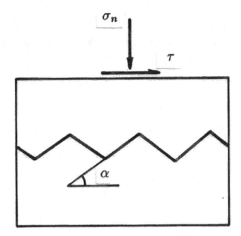

FIGURE 23.6 Saw tooth–shaped concrete joint [24].

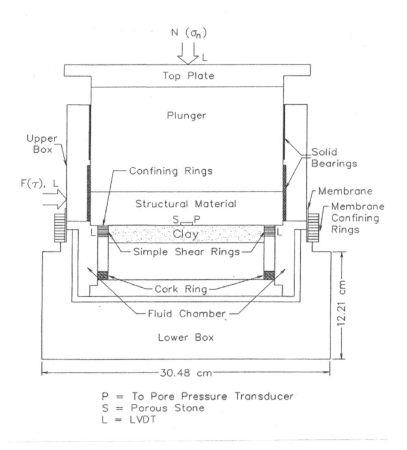

P = To Pore Pressure Transducer
S = Porous Stone
L = LVDT

FIGURE 23.7 Cross section and components of cyclic multi-degree of freedom (CYMDOF-P) device.

Ma and Desai [27,28] used the DSC/HISS model to predict the measured behavior under the above tooth angles and various normal stresses. Typical comparisons between predictions and measurements for the asperity angle equal to 5 degrees are presented in Figures 23.8 and 23.9: shear responses (Figure 23.8) and typical dilatant response, v vs u, for $\sigma_n = 0.138$ MPa (Figure 23.9).

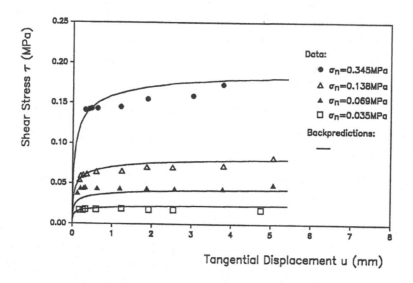

FIGURE 23.8   Shear stress vs. tangential displacement for tooth angle 5°.

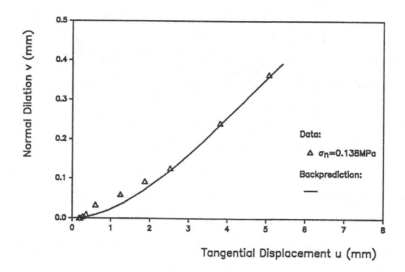

FIGURE 23.9   Dilatant v vs. u response for tooth angle 5°, $\sigma_n = 0.138$ MPa.

Example 2: Smooth to Rough Rock Joints

Several different normal stresses were applied to three types of joints, namely, smooth, medium rough, and rough, in tests conducted by Schneider [25]. Here, typical comparisons between predictions and tests data for typical Type B (rough in sandstone) are presented for normal stresses equal to 1.29, 0.93, and 0.32 MPa. Figure 23.10 shows comparisons for the shear responses for the three normal stresses. In Figure 23.11 are shown typical comparisons for the dilatant responses, for $\sigma_n = 0.93$ MPa.

Comparisons for an independent test for $\sigma_n = 0.61$ MPa are shown in Figures 23.12 and 23.13; here, the predictions were obtained based on the parameters found from three other tests.

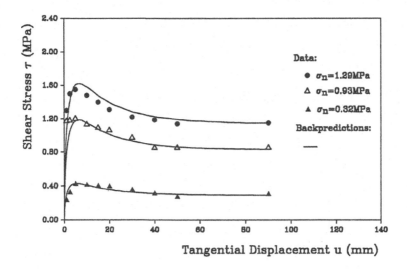

FIGURE 23.10 Shear stress vs. tangential displacements for Type B joint for various normal stresses.

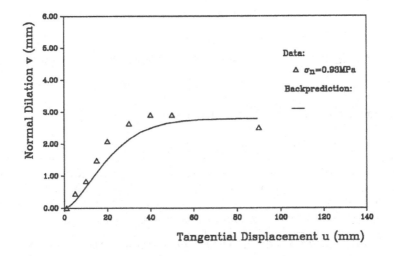

FIGURE 23.11 Normal dilation vs. tangential displacement for Type B joints, normal stress=0.93 MPa.

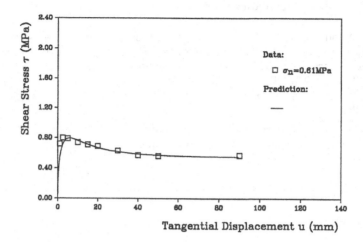

FIGURE 23.12   Shear stress vs. tangential displacement for Type B joint, normal stress=0.61MPa; independent test.

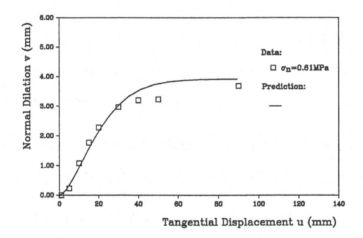

FIGURE 23.13   Normal dilation vs. tangential displacement for Type B Joint, normal stress=0.61 MPa; independent test.

Example 3: Joints for Scale Effects and Different Roughnesses

Bandis et al. [29] reported a comprehensive series of shear tests for studying scale effects with surfaces ranging from rough undulating to almost smooth under various normal stresses. The joints' specimens were made of a mixture of plaster of Paris, silver sand, calcined alumina, and water.

Four types of joints were used and identified using JRC (Joint Roughness Coefficient) with values of 16.6, 10.6, 7.5, and 6.5. The shear tests were conducted under normal stresses equal to 90, 30, and 10 kPa.

Back-predictions for the four JRCs were obtained using the DSC/HISS model. Typical comparisons between back-predictions and measurement are presented in Figure 23.14 for shear stress vs. tangential (relative) displacement, and in Figure 23.15 for dilatant response for JRC=6.5 and $\sigma_n = 30.0$ kPa.

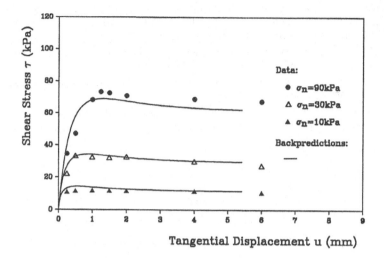

FIGURE 23.14   Shear stress vs. tangential displacement for JRC=6.5 under various normal stresses.

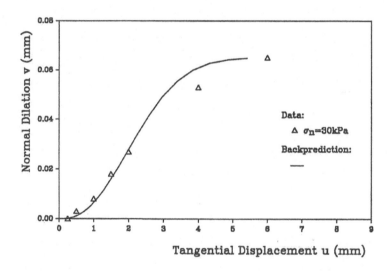

FIGURE 23.15   Normal dilation vs. tangential displacement for JRC=6.5 and normal stress=30 kPa.

Example 4: Interface between Sand and Steel

A series of laboratory shear tests for interfaces between sand and steel using the CYMDOF device have been reported by Alanazy and Desai [10]. The top specimen consists of a smooth steel block, 20.3 cm (8.0 in) in diameter with thickness of 7.76.cm (3.05 in). The roughness, R_{max}, which represents the relative height between the highest peak and the lowest trough over a length equal to the mean diameter of sand particles (D_{50}) was 7.4 μm, for which the normalized roughness R_n, which was obtained by dividing R_{max} by D_{50}, was 20×10^{-3}.

Uniform fine Ottawa sand with rounded to subrounded grains constituted the bottom specimen. The initial density of the specimens was 1.66 gm/cc (103.60 lb/ft³) corresponding to the relative density of 60%.

One-way static and two-way cyclic loading testing were performed, the latter under normal stress σ_n equal to 69 (10), 138 (20), and 207 (30) kPa (psi). Only typical predictions using the DSC/HISS model for one-way and two-way tests are presented here.

Figure 23.16 shows the comparisons between predictions and test data for shear stress vs. relative tangential displacement for one-way loading under $\sigma_n = 138$ kPa. Comparisons between relative normal displacements vs. relative shear displacement are presented in Figure 23.17 for the same one-way loading. Comparisons between shear stress and relative shear displacement are displayed in Figure 23.18 for various cycles for two-way cyclic loading under $\sigma_n = 69$ kPa.

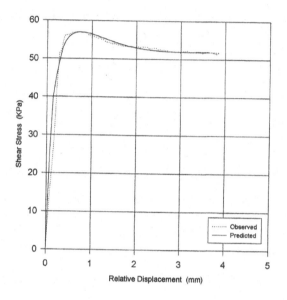

FIGURE 23.16   Predicted and observed shear stress vs. relative displacement: one-way monotonic loading, $\sigma_n = 138$ kPa.

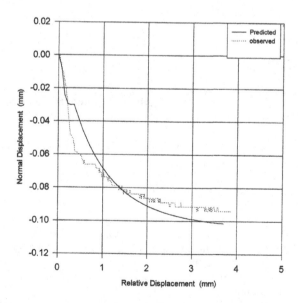

FIGURE 23.17   Predicted and observed normal displacement vs. relative (shear) displacement: one-way monotonic loading, $\sigma_n = 138$ kPa.

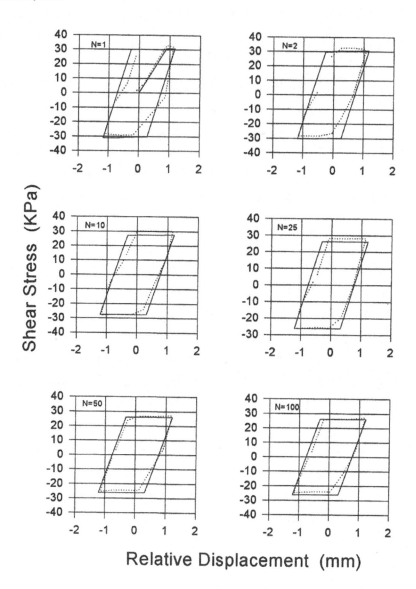

FIGURE 23.18   Predicted and observed shear stress vs. relative shear displacement; two-way cyclic loading, $\sigma_n = 69$ kPa.

Example 5: Clay–Steel Interface Tests

Undrained shear tests for saturated clay–steel interfaces were conducted by Rigby and Desai [9,30] by using the CYMDOF-P device. Three types of tests included consolidation, simple shear, and direct shear tests.

Figure 23.19 shows the schematic of the clay-interfaces tests. The clay was obtained from pile load tests site at Sabine Pass, Texas conducted by Earth Technology Corporation [31]. Here, the simple shear tests are considered for prediction by using the DSC/HISS model [32].

The specimens were consolidated at normal stresses of 69, 138, and 207 kPa [9,30]. Then, shear displacement was applied at the rate of 1.5 mm/min to the steel part of the specimen with the normal stress kept constant during the test. The water valves were closed during the test implying undrained condition.

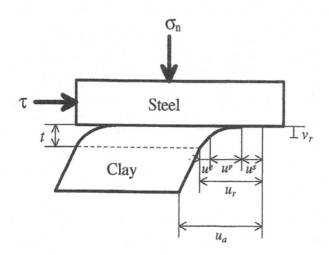

FIGURE 23.19 Schematic of interface between clay and steel.

The DSC/HISS model was used to predict the test behavior for the three normal stresses. Here, typical comparisons between predictions and tests data are presented for the normal stress = 138 kPa. The parameters for the clay–steel interface based on the tests data reported in References [9,30,32] are given below:

Parameters for Clay–Steel Interface

Elastic: $E = 4300\,kPa$; $\vartheta = 0.42$

Relative Intact state:
$n = 2.6$, $\gamma = 0.077$, $\beta = 0.0$, $R = 0.0$,
$h_1(*) = 0.000408$, $h_2 = 2.95$, $h_3 = 0.0203$, $h_4 = 0.0767$,
Fully Adjusted (Critical state):
$\lambda = 0.298$, $e^{oc} = 1.359$, $\bar{m} = 0.123$,

Disturbance function:
$D_u = 1.0$
$A = 0.816$,
$Z = 0.418$,
Unloading and Reloading (Figure 23.20)
$G^{BUL} = 4300\,kPa$,
$G^{EUL} = 400\,kPa$,
$\varepsilon_1^p = 0.0305$.
(*) The modified hardening function for the clay–steel interface was adopted as:
$$\alpha = \frac{h_1}{P} \quad (23.9)$$
where $P = \left(\xi_v + h_3\, \xi_D^{h_4} \right)^{h_2}$.

The back-predictions were obtained by simulating the test behavior using a two-dimensional two-phase (solid/fluid) dynamic finite element procedure in which the DSC/HISS model was implemented [33]. The finite element mesh for interface is shown in Figure 23.21, in which the incremental displacement ($u_a = 1.5\,mm$) was applied. Typical comparisons between predictions and test data are presented here.

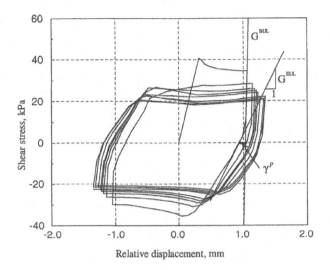

FIGURE 23.20 Determination of unloading and reloading parameters, $\sigma_n = 138$ kPa.

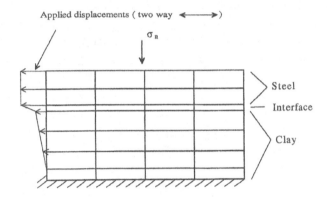

FIGURE 23.21 Finite element mesh for clay–steel interface.

Figure 23.22 shows comparisons between predictions and cyclic test data for shear stress vs. relative (tangential) displacements for $\sigma_n = 138$ kPa and $u_a = 1.5$ mm. Comparisons between pore water pressures from test data and FEM (finite element method) calculations are depicted in Figure 23.23.

Example 6: Interface Model for Viscoplastic Behavior

Figure 23.24 shows the setup for the interface between the soil (in moving mass) and bed rock in the case of creeping landslides and slopes [7,8]. A brief description of creep behavior is given below.

The incremental elastoplastic equation for associative $\left(\delta_o\right)$ and nonassociative $\left(\delta_1\right)$ is given in Eq. (23.3). For nonassociative behavior, the plastic potential function, Q, is expressed as:

$$Q = \tau^2 + \alpha_Q\, \sigma_n^n - \gamma_n^2 \tag{23.10}$$

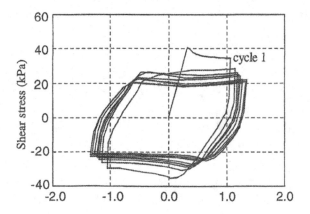

(a). Shear Stresses from Test

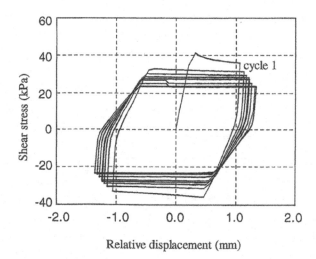

(b). Shear Stresses from FEM

FIGURE 23.22   Comparisons for cyclic shear stress vs. relative displacements from tests and finite element predictions, $\sigma_n = 138$ kPa, $u_a = 1.5$ mm.

and the hardening function, α_Q, as:

$$\alpha_Q = \alpha + \mathbb{k}\left(\alpha_0 - \alpha\right)\left(1 - r_v\right) \tag{23.11}$$

For relatively smooth interfaces, Eq. (23.9) can be expressed as:

$$\alpha_Q = \alpha + \alpha_{ph}\left[1 - \frac{\alpha}{\alpha_o}\right]^{\mathbb{k}} \tag{23.12}$$

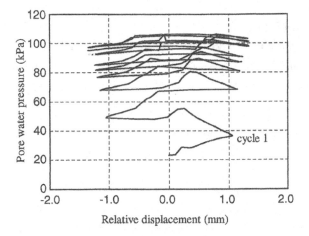

(a). Pore water pressure from test

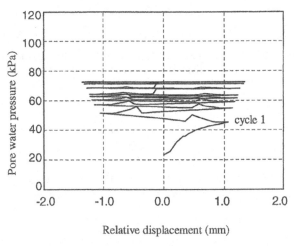

(b). Pore water pressure from FEM

FIGURE 23.23 Comparisons between pore water pressure from test and FEM predictions, $\sigma_n = 138$ kPa $u_a = 1.5$ mm.

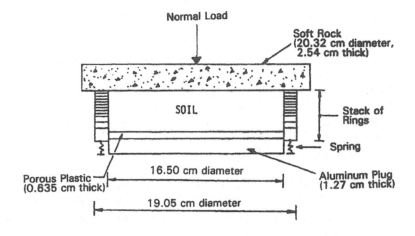

FIGURE 23.24 Test setup cross section for viscoplastic interface.

TABLE 23.2
Parameters for Viscoplastic Interface Model

Model	Parameters	Symbol	Value
Elastic	Normal stiffness	k_n	8×10^6 kPa/cm
	Shear stiffness	k_s	2800 kPa/cm
Plastic	Ultimate	γ	0.24
	Phase transition	n	2.04
	Hardening	a	143.00
		b	10.00
	Nonassociative	$\bar{\kappa}$	0.57
Viscous	Fluidity	Γ	0.057/min
	Exponent	N	3.15

where $\alpha_{ph} = \left(\dfrac{2}{n}\right)\gamma\, \sigma_n^{2-n}$ and $\alpha_o = \gamma\, \sigma_n^{2-n}$ are the values of α at the end of the transition point in dilative response and the start of the shear loading, that is, at the end of the normal loading, respectively, and $\bar{\kappa}$ is the nonassociative parameter. The hardening function is now expressed as:

$$= \alpha = \frac{\gamma e^{-a\xi_v^{vp}}}{\left(1 + \xi_D^{vp}\right)^b}$$

(23.13)

where ξ_v^{vp} and ξ_D^{vp} are the accumulated volumetric and deviatoric viscoplastic relative displacements, respectively, and a and b are hardening parameters. According to the Perzyna's model, the viscoplastic strain rate is given by:

$$\dot{\varepsilon}^{vp} = \Gamma\phi\left\langle\left(\frac{F}{F_o}\right)\right\rangle\frac{\partial Q}{\partial \sigma}$$

(23.14)

and

$$\phi\left\langle\left(\frac{F}{F_o}\right)\right\rangle = \left(\frac{F}{F_o}\right)N$$

(23.15)

where Γ and N are the fluidity and viscoplastic exponent parameters, respectively.

A comprehensive series of laboratory tests were performed for the creep response using triaxial, CYMDOF (Cyclic Multi-Degree of Freedom), and direct shear devices [7,8]. Drained tests using the CYMDOF device (Figure 23.7) were conducted under different normal stresses, $\sigma_n = 103$, 207, and 345 kPa with different amplitudes of shear displacements, $u_r^a = 0.19$, 0.64, and 1.27 cm. The parameters for the viscoplastic model were determined by following the procedure presented in Desai [1] (DSC) and are given in Table 23.2.

Comparisons between predictions and test data were found to be excellent. Typical comparisons are shown in Figures 23.25 and 23.26: the former shows comparisons for shear stress and relative shear displacements for $\sigma_n = 103$ kPa, and the latter shows comparisons between viscoplastic shear strain vs. time for two pre-consolidation stresses.

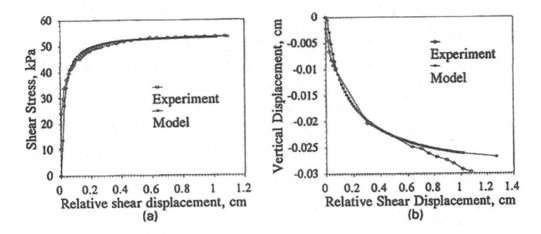

FIGURE 23.25 Comparisons between predictions and test data for interface test, $\sigma_n = 103$ kPa.

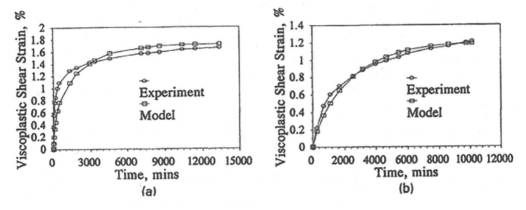

FIGURE 23.26 Comparisons between prediction and test data for creep stress ratio=0.6: reconsolidation stress=(a) 200 kPa and (b) 400 kPa.

REFERENCES

1. Desai, C.S., *Mechanics of Materials and Interfaces: The Disturbed State Concept.* CRC Press, Boca Raton, FL, 2001.
2. Desai, C.S., Zaman, M.M., Lightner, J.G., and Siriwardane, H.J., Thin-layer element for interfaces and joints. *Int. J. Num. Anal. Meth. Geomech.*, 8, 1, 1984, 19–43.
3. Desai, C.S. and Fishman, K.L., Plasticity based constitutive model with as- sociated testing for joints. *Int. J. Rock Mech. Min. Sci.*, 4, 28, 1991, 15–26.
4. Navayograjah, N., Desai, C.S., and Kiousis, P.D., Hierarchical single surface model for static and cyclic behavior of interfaces. *J. Eng. Mech.*, ASCE, 118, 5, 1992, 990–1011.
5. Desai, C.S., "Constitutive modelling using the disturbed state as microstruc- ture self-adjustment." Chap. 8 in *Continuum Models for Materials with Micro-Structures*, H.B. Mühlhaus (Editor), John Wiley, Chichester, UK, 1995.
6. Desai, C.S., "Behavior of interfaces between structural and geologic media." A state-of-the-art paper, *Proc., Int. Conf. on Recent Advances in Geotech. Earthquake Eng. and Soil Dynamics*, St. Louis, MO, 1981.
7. Desai, C.S., Samtani, N.C., and Vulliet, L., Constitutive modelling and analysis of creeping slopes. *J. Geotech. Eng.*, ASCE, 121, 1, 1995, 43–56.

8. Samtani, N.C., Desai, C.S., and Vulliet, L., An interface model to describe viscoplastic behavior. *Int. J. Num. Anal. Meth. Geomech.*, 20, 1996, 231–252.

9. Desai, C.S. and Rigby, D.B., Cyclic interface and joint shear device including pore pressure effects. *J. Geotech. Geoenviron. Eng., ASCE*, 123, 6, 1997, 568–579.

10. Alanazy, A. and Desai, C.S., "Testing and modelling of sand-steel interfaces under static and cyclic loading," Report, Dept. of Civil Eng. and Eng. Mechs., The Univ. of Arizona, Tucson, AZ, 1996.

11. Pal, S. and Wathugala, G.W., Disturbed state model for sand-geosynthetic interfaces and application to pull-out tests. *Int. J. Num. Anal. Meth. Geomech.*, 23, 15, 1999, 1873–1892.

12. Desai, C.S. and Toth, J., Disturbed state constitutive modeling based on stress-strain and nondestructive behavior. *Int. J. Solids Struct.*, 33, 11, 1996, 1619–1650.

13. Desai, C.S. Evaluation of liquefaction using disturbance state and energy approaches. *J. Geotech. Geoenviron. Eng., ASCE*, 126, 7, 2000, 618–631.

14. Desai, C.S., Basaran, C. and Zhang, W., Numerical algorithms and mesh dependence in the disturbed state concept. *Int. J. Num. Meth. Eng.*, 40, 1997, 3059–3083.

15. Desai, C.S. and Zhang, W., Computational aspects of disturbed state constitutive models. *Int. J. Comp. Meth. Appl. Mech. Eng.*, 151, 1988, 361–376.

16. Desai, C.S., Chia, J., Kundu, T., and Prince, J. Thermomechanical response of materials and interfaces in electronic packaging: Part I and Part II. *J. Electron. Packag., ASME*, 119, 4, 1997, 294–300, 301–309.

17. Desai, C.S., and Whitenack, R. Review of models and the disturbed state concept for thermomechanical analysis in electronic packaging. *J. Electron. Packag., ASME*, 123, 2001, 1–15.

18. Ahmadian, H., Jalali, H., Morttershead, J.E., and Friswell, M.I. "Dynamic modeling of spot welds using thin layer interface theory." *Proc., 10th Int. Congress on Sound and Vibration*, Stockholm, Sweden, 2003, 3439–3446.

19. Bograd, S., Schmit, A., and Lothar, G. "Joint damping prediction by thin layer elements." In: Allen M., Mayes R., Rixen D. (eds.), *Dynamics of Coupled Structures*, Vol. 1, International Publishers, Basel, 2014, 239–244.

20. Yao, X., Wang, J., and Zhai, X. "Research and application of improved thin layer element method for aero-engine bolted joints." *Proc., Institution of Mechanical Engineers, Part G: J. of Aerospace Eng.*, May 9, 2016.

21. Jamia N., Jalali H., Friswell M.I., Khodaparast H.H., Taghipour J. "Modelling the effect of preload in a lap-joint by altering thin-layer material properties." In: Kerschen G., Brake M.R., Renson L. (eds.), *Nonlinear Structures & Systems*, Volume 1, 2022, 211–217. Conference Proceedings of the Society for Experimental Mechanics Series. Springer, Cham. https://doi.org/10.1007/978-3-030-77135-5_25.

22. Pour, P.M., Yeat, W.F., Thanern, W.A.M., Noorzaie, J., and Jaafari, M.S. Numerical modeling of railway track support system using finite element and thin-layer elements. *Int. J. Eng.*, 22, 2, 2009, 131–144.

23. Sharma, K.G. and Desai, C.S., An analysis and implementation of thin-layer element for interfaces and joints. *J. Eng. Mech., ASCE*, 118, 1992, 545–569.

24. Archard, J.I., Elastic deformation and the laws of friction. *Proc., Roy. Soc., Lon., Series A*, 243, 1959, 190–205.

25. Schneider, H.J., The friction and deformation behavior of rock joint. *Rock Mech.*, 8, 1976, 169–184.

26. Fishman, K.L., and Desai, C.S. Constitutive modeling of idealized rock joints under static and cyclic loading. Report to NSF, CEEM Department, Univ. of Arizona, Tucson, AZ, USA, 1988.

27. Ma, Y., and Desai, C.S. "Constitutive modeling of joints and interfaces by using disturbed state concept." Report to NSF, CEEM Department, Univ. of Arizona, Tucson, AZ, USA, 1990.

28. Desai, C.S. and Ma, Y., Modelling of joints and interfaces using the disturbed state concept. *Int. J. Num. Analyt. Meth. Geomech.*, 16, 1992, 623–653.

29. Bandis, S., Lumsden, A.C., and Barton, N.R., Experimental studies of scale effects on the shear behavior of rock joints. *Int. J. Rock Mech. Min. Sci.*, 18, 1981, 1–21.

30. Rigby, D.B., and Desai, C.S. "Testing and Constitutive Modelling of Saturated Clay- Steel Interfaces." Report to NSF, CEEM Department, Univ. of Arizona, Tucson, AZ, USA, 1996.

31. The Earth Technology Corporation. Pile Segment Tests- Sabine Pass. RTC Report No. 85–007, December 1986.

32. Shao, C., and Desai, C.S. Implementation of the DSC model for dynamic analysis of soil-structure interaction problems. Report to NSF, CEEM Department, Univ. of Arizona, Tucson, AZ, USA, 1998.

33. Desai, C.S. DSC-SST2D-Computer Code for Static and Coupled Dynamic Analysis: Solid-Fluid Soil-Structure Problems. Reports and Manuals, C. Desai, Tucson, AZ, 1999.

24 Interfaces and Joints
Laboratory Tests and Applications

Innjoon Park
School of Civil Engineering and Environmental
Science, Hanseo University

Changwon Kwak
Dept. of Civil and Environment Engineering, Inha Technical College

24.1 INTRODUCTION

The shear strength of the interface between geosynthetics and soils is one of the most important elements to investigate the performance and stability of civil structures, such as waste landfills. The interface between geosynthetics and soils can contain weak points, such as sliding surfaces. Therefore, various studies on the behavior of this interface have been performed previously (Seo et al. 2004; Triplett and Fox 2001; De and Zimmie 1997; Stark et al. 1996). The lining and cover system, which typically consists of a combination of natural and geosynthetic materials, is one of the most critical parts of waste landfill designs. The shear strength at the interface between the various materials determines both the overall stability of such systems and the integrity of the geosynthetics. There are many examples of landfill slope failures caused by interfaces with poor shear strength across the world (Jones and Dixon 1998). A few factors that can affect the shear strength and durability of geosynthetics, such as the biodegradation of organic matter, temperature, and air pressure, were studied by Zhou and Rowe (2003). It is reported that the acidity and basicity in the leachate cause the chemical degradation of geosynthetics over time (Rowe and Sangam 2002; Cassidy et al. 1992). This chemical degradation involves a bond scission in the backbone of macromolecules, intermolecular cross-linking, and chemical reactions in the sidechains (Schnabel 1981) that will eventually lead to a decrease in the mechanical properties and failure. Therefore, chemical degradation is the most important destructive mechanism and requires attention.

One of the most trustworthy and evident methods to understand the behavior of geosynthetic-soil interfaces is a laboratory test. The use of a large shear box (300×300 mm) is commonly accepted for the interface testing of a soil-geotextile system (e.g., Grag and Saran 1990; Koerner 1994). Some researchers (Lee and Manjunath 2000; Shallenberger and Filz 1996) conducted large-size direct shear tests on soil-geotextile interfaces in woven and nonwoven cases, and the strength and frictional characteristics were compared. Seo et al. (2007) studied the shear behavior of the interfaces between geosynthetics based on large direct shear tests. In this chapter, the shear behavior was clearly influenced by the presence of water at the interface, the type of the interface, and the material involved.

Apart from conventional static direct shear tests, the dynamic approach to considering the interface behavior under traffic and seismic conditions was tried. Yegian et al. (1995) researched the seismic response of geosynthetics and the slipping of geosynthetic systems. The dynamic interface characteristics of geosynthetics were also analyzed and laboratory tests were performed in previous reports (Castelli et al. 2001; Yegian et al. 1995). Zhi-ling and Chao (2011) performed a series of direct shear tests under cyclic loads and studied the effects of the frequency and the amplitude on the geogrid–soil interface shear strength.

DOI: 10.1201/9781003362081-31

In addition to laboratory tests, the formulation of constitutive laws for the interface was also studied by Kaliakin and Li (1995). Park et al. (2000) employed the disturbed state concept (DSC), based on the HiSS model, to predict the soil–structure interaction.

24.2 DISTURBED STATE CONCEPT (DSC)

It is important to utilize the appropriate constitutive model in order to obtain more precise representations of geosynthetic-soil interface behavior. In the geotechnical engineering field, many advanced constitutive models are introduced such as the Mohr–Coulomb model, the Ramberg–Osgood model, the HiSS model, among others. However, these conventional models are not enough to predict or simulate a realistic damage progress at the geosynthetic-soil interface (Park 1997). The DSC and disturbance function were used to investigate the effect of damage on the interface, and disturbance function parameters were determined as well. The disturbance factor is a key factor of the DSC for defining dynamic shear stress degradation quantitatively. The convenience and reliability of the DSC were successfully verified in prior studies (Armaleh and Desai 1990; Ma 1990; Rigby and Desai 1995; Park et al. 2000), which also showed the valuable accomplishment of the numerical formulation of dynamic interface behavior (Seo et al. 2007).

The early theory of DSC was originated by Desai (1974) to characterize the softening behavior of an over consolidated soil, using its behavior in a normally consolidated state as reference. This initial concept was later developed and unified as the DSC to model the behavior of a wide range of materials (Desai and Ma 1992; Desai 1995). There are two reference states, namely relatively intact (RI) and fully adjusted (FA), which express the initial and final conditions of a material in the DSC. The initial condition is defined as the RI state, and the material continuously approaches the FA state as a result of the external force. The observed state can be expressed in terms of RI and FA states. Self-adjustment occurred during this procedure, and the disturbance increased in the material's microstructure. Dynamic and static loads caused different types of shear stress–strain responses; however, the overall trends were identical. Shear stress degradation due to the accumulation of internal damage can be observed in both loading conditions. Typical shear stress–strain behavior under static and dynamic load conditions is described through some figures by Desai (2001).

The disturbance, D, represents the deviation of the current state with respect to the initial and final states of the material, and it can be expressed by the disturbance function. Therefore, the disturbance function is a convenient tool to express the degree of damage quantitatively. The direct way to estimate the disturbance is to measure the total volume of failed regions and calculate the ratio to the total volume of intact regions. However, it is not practically possible to define D based on such physical measurements because there is no direct way to measure the internal volume of FA material directly. In lieu of direct measurements, a few equations were proposed to define disturbance in terms of variables such as mechanical energy, accumulated plastic strains, normal or shear stresses, and so on (Desai, 2001). Those equations were based on the idea that the decay in a material is like the decay trend in natural systems and as a result, mathematical functions that express decay can be adopted to describe material damage.

The disturbance, D, is known to be the function of certain internal factors that affect the constitutive behavior in a broad sense. In this study, the internal microstructural damage was assumed to be mainly affected by the deviatoric plastic strain trajectory, ξ_D, according to the following equation:

$$D = D\left[\xi_D, (t, \alpha_i)\right] \tag{24.1}$$

where t is the time or the number of loading cycles and α_i represents environmental factors. The functional form is based on a modification of Weibull (1951):

$$D = D_u\left[1 - e^{\left(-A\xi_D^Z\right)}\right] \tag{24.2}$$

where A and Z are intrinsic material parameters and D_u asymptotically approaches 1.0 but can never reach 1.0. This equation is commonly accepted for growth and decay processes (Armaleh and Desai, 1990). The disturbance, D, has a value between 0.0 and 1.0 at each strain but many materials fail before D reaches 1.0 (Park and Desai, 2000).

The disturbance, D, can be considered as the deviation of the current shear state with respect to the initial and final states of the material. Accordingly, in the cyclic shear test, D value can be defined based on the shear stresses at the RI, FA, and the observed cycle as:

$$D = \frac{\tau^i - \tau^a}{\tau^i - \tau^c} \tag{24.3}$$

where i, a, and, c denote RI, observed cycle, and FA states, respectively, and the equation evaluates the normalized shear stress degradation, D, from each hysteretic loop. In this study, the curve at the 50th cycle was regarded as the FA state since shear stress–strain curves after 50 cycles showed almost identical shape and very little variation to each other. The deviatoric plastic strain trajectory, ξ_D, means the accumulated plastic strain at each stress–strain loop. Deviatoric plastic strain trajectory, ξ_D, is the accumulated length of the strain traveling inside the hysteretic loops. Typical shape of D function curve and the mathematical sensitivity of A and Z parameters are demonstrated in the study by Kwak et al. (2013). According to that study, when A increases, the curve moves to the left. This indicated that the material experienced more damage at the same strain level. However, if D increases, the slope of curve also increases, indicating a rapid increase in the material damage. Accordingly, the shear strength characteristics of the geosynthetic-soil interface can be defined using the disturbance function visually and quantitatively.

24.3 LABORATORY TEST APPARATUS

It is reported that different types of degradation mechanisms can act on geomembranes based on the exposure condition and/or polymer properties (Koerner, 2012). The ultraviolet, chemical, biological, and thermal degradation are a few examples. Generally, there is more than one degradation mechanism acting at a given time, and the synergistic effects can accelerate the degradation (Rowe et al., 2008). Chemical degradation of geomembranes is one of the major degradation mechanisms in waste landfill sites. Leachates generated through chemical and biological processes contain aggressive chemical substances, which affect the shear behavior of the interface by leaching or by combining with other substances in the interface material. Such an effect has yet to be studied intensively. In addition, the effect of dynamic loading, which is mainly caused by an earthquake, on the geosynthetic-soil interface can be highly significant, since dynamic loading can induce various modes of motion, such as slip, a loss of contact or de-bonding, and re-contact or re-bonding at the interface (Navayogarajah et al., 1992).

The properties of polymers are known to be sensitive to temperature. Biodegradation or hydration, in addition to the decomposition of organic components, are the primary factors contributing to the generation of heat in waste landfills (Klein et al., 2001; Yesiller et al., 2005). Wide variations of the temperatures in waste landfill sites are reported in the literature. Temperatures between 30°C and 50°C or higher were reported around the bases of landfills (Rowe, 1998), temperatures between 20°C and 30°C or higher were reported for liner systems (Rowe, 1998), and maximum temperatures between 70°C and 85°C were reported by Dach and Jager (1995). Hanson et al. (2015) investigated the effects of temperature and moisture on the shear strength of a double nonwoven geosynthetic clay liner (GCL) and a high-density polyethylene (HDPE)-textured geomembrane (T-GM) interface for municipal solid waste (MSW) landfills. Based on the experiment study, the measured interface shear strengths varied by up to 54% with temperature and up to 43% with moisture content due to the geosynthetic damage including surface texture of the geomembranes and bentonite extrusion. Stark et al. (2010) described input parameters and stability analyses for a MSW landfill experiencing elevated temperatures due to an aluminum waste reaction, and thermal degradation could consume, burn, and remove the reinforcing material from MSW. Jafari et al. (2014) presented the elevated temperature that can reduce service life or effectiveness of HDPE geomembranes by accelerating antioxidant depletion of geomembranes and polymer degradation.

The variation in temperature generally results in sensitive and direct changes of the engineering properties of the soil and geosynthetics. Most degradation processes, such as chemical and occidental processes, develop in a slow and gradual manner. However, the thermal condition affects the behavior of geosynthetics instantly. A few experimental studies were conducted to determine how a rise in temperature can affect geosynthetics in waste landfills (Rowe and Islam, 2009; Rowe et al., 2009, 2008; Koerner and Koerner, 2006; Lefebvre et al., 2000). However, the focus was rather on the geosynthetic material. Very few studies were performed to assess the effect of temperature on the interface behavior. Akpinar and Benson (2005) conducted laboratory shear tests using a double-interface shear device enclosed in a constant-temperature chamber to investigate the effect of temperature on the shear strength of a geomembrane–geotextile interface. An increase in the peak and post-peak interface friction angle was observed between 0°C and 33°C, a relatively low temperature condition.

Numerous efforts have been made to develop an acceptable apparatus, as laboratory testing to enhance our understanding of the behavior of these interfaces. Various types of shear devices and test methods were proposed, and numerous experimental approaches for investigating the shear behavior at the geosynthetic-soil interfaces were suggested. They are, however, restricted by the type of loading and the application of a temperature rise. In this study, major structures, and parts of the multi-purpose interface apparatus (M-PIA) were modified to consider cyclic loading and elevated temperatures.

The initial version and a minor improvement of the M-PIA were introduced in previous work (Kwak et al., 2013).[b] The present version was focused on the convenient preparation of specimens, avoiding bending moment on the specimen by shear force, improvements of the data acquisition system, and adding a supplemental heating system to consider thermal condition. Details of the specifications, schematic, and the appearance of the modified M-PIA system are described in the previous study (Kwak et al., 2016).

The circular-shaped shear device consisted of two components: a shear box and shear rings. The geosynthetic and soil were set in the circular shear box. The geosynthetic sample was placed on top of the soil and a rubber membrane encapsulated the geosynthetic and soil, creating a complete specimen with a volume of $314 \, cm^3$ ($D = 100$ mm, $H = 40$ mm). Though a minimum dimension of 300 mm is required according to ASTM D5321 recommendations, containers that are smaller than 300 mm can also be used if they can show that data obtained by the smaller devices are free of bias (ASTM D5321–08). Twelve independent shear rings with a thickness of 3 mm were coated with TFE-fluorocarbon to avoid any frictional resistance between the rings, which may cause inaccuracies in the test results.

In this research, a static test was used to verify the displacement capacity to reach peak strength of geosynthetic-soil interface and the results are as shown in the previous study (Kwak et al., 2016). It is important to note that the height-to-diameter ratio has no significant influence on the shear stress behavior in simple shear tests (Vucetic & Lacasse, 1982).

Normal load and shear displacement were applied through two separate modules: the upper module controlled the normal load applied at the upper plate, which was in contact with the top surface of the geosynthetic specimen, and the lower module controlled the shear displacement with a maximum frequency of 0.5 Hz. After reaching the target normal force, 5% of shear strain was exerted by a linear motion slider. The linear motion slider was directly connected to a horizontal servo motor to apply shear displacement with a desired constant rate or frequency based on a closed-loop computer control system. Railways were placed at the bottom of the linear motion slider to guide the linear movement. Two separate load and displacement application systems allowed for independent measurements of the variation of the normal force, the shear force, and the shear displacement. The data acquisition system contained a software package, and the minimum data logging interval was 0.1 second. All measured forces and displacements were automatically displayed on a computer monitor and stored by the embedded software.

Despite the important accomplishments of the previous versions of M-PIA (Kwak et al., 2013; Park et al., 2010), some performance limitations such as unintended moment in the specimen and rough data acquisition due to open loop system, etc., were discovered. A major improvement to overcome these limitations was subsequently presented. The first improvement was the redesign of the basic structure in the load and displacement application system. The load and displacement application system were separated into the upper and lower modules. Though the separation of the loading system was not an epochal modification, the unintended moment in the specimen was observed in the previous versions of the apparatus and it was confirmed that the moment became negligible by dividing the loading unit into independent modules. The second improvement was that the raw material of the shear ring was TFE-fluorocarbon instead of steel to minimize the frictional resistance between each ring, which may cause some inaccuracy in the shear response readouts. The coefficient of friction for TFE-fluorocarbon was 0.04, which was considerably small compared with steel, which showed a coefficient of friction of 0.57 (Serway and Jewett, 2004). There were 12 shear rings, and the thickness was 3.0 mm each, which was enough to maintain the stiffness of the shear ring itself and prevented the membrane from tearing while shearing. As TFE-fluorocarbon also showed good dimensional stability and chemical resistance (Furukawa et al., 1952), it was considered an appropriate material for the shear ring. All soil and geosynthetic specimens were set to be encircled by 12 independent shear rings, with a sill thickness of 2 mm at the upper base plate to avoid any leakage of sand or liquid; therefore, the total height of the geosynthetic-soil specimen was 40 mm, as the summation of 36 mm of shear rings and 4 mm of sill thickness. The inner diameter of the shear ring was 100 mm, and the outer diameter was 140 mm.

The third modification was the invention of an outer mold for a convenient setup of the specimen. The shear rings and the specimen were placed inside the mold and moved to the top of the lower module. The outer mold was then split into two halves after which they were removed for minimum disturbance of the geosynthetic-soil specimen. The outer mold enabled an easy and rapid set up of the specimen without the inceptive displacement of the shear rings. Accordingly, the reliability of the test increased. The shape of shear box and the test-ready state of the specimen are described in the previous study (Kwak et al., 2016).

Another major modification included the adoption of a closed-loop controlling algorithm in the load control system. During the test, the closed-loop controlling algorism took feedback from the output of the system into account. It continuously monitored the output of the processes and then compensated for the errors of the system automatically. A proportional-integral-derivative (PID) controller was employed as a closed-loop controlling device to manifest the sinusoidal shear displacement in the apparatus. The sinusoidal responses obtained in the closed-loop and shear displacement under closed-loop system in the modified apparatus are demonstrated in the previous study (Kwak et al., 2016).

The M-PIA was designed to operate between 20°C and 80°C to quantify the shear response at temperatures that may be reasonably encountered in a landfill. The heating system consisted of heating water tubes, a water cistern, temperature sensors, and a control system. The water cistern was located on the right side of the apparatus, and water was heated by electrical wires in the cistern. Water tubes transferred heated water to the inside of the base and top plates contacted to the shear box. Bakelite plates were placed at the bottom and top of the shear box for insulation. All parts of the shear box were made of aluminum to enhance the heat transfer capabilities. The circulation of the heated water maintained the temperature of the specimen. The temperature inside the specimen was measured by a T-type temperature sensor, which was inserted into the specimen. Details of the heating system of the shear box around the specimen and the schematic of the overall heating system are described in the previous study (Kwak et al., 2016).

24.4 LABORATORY TEST CONDITIONS

A series of cyclic simple shear tests were conducted in accordance with the chemical conditions, the submerging periods, and the normal stress applied to the specimen. In this study, all laboratory tests were performed at room temperature of 23±2°C for the geosynthetic-soil interface. Each test condition was presented and the specifications of the geosynthetic material and soil were also described.

The materials in the test of the geosynthetic-soil interface were the geosynthetics and granular soil. Jumunjin sand (Korea Standard Sand) was used as the type of granular soil in the present study. The particle size of Jumunjin sand was uniform with the grain-size distribution presented and the physical characteristics were summarized in the previous study by Kwak et al. (2014, 2016). It should be noted that this study was limited to the Jumunjin sand in terms of the test material, and the behavior of the in situ sand (SM) may differ. However, the particle-size distribution curve of the poorly graded sand (SP) spread on the geosynthetics based on the waste management regulations showed similar trend to the curve of Jumunjin sand. A composite type of geosynthetics, i.e., geo-composite, was most applied to the waste landfill site in Korea, because it exhibited vigorous chemical resistance and diverse combination of geosynthetic materials enabled the waste landfill design subjected to in situ condition.

The specifications of the geo-composite are also described (Kwak et al., 2016). Nonwoven fabric was laminated on both sides of a geonet by thermal bonding. The main ingredient of the geonet was HDPE and the filament of the nonwoven fabric was made of polyethylene and polypropylene.

Both geosynthetic and soil specimens were immersed in basic, neutral, and acid solutions for 30 and 850 days, respectively, to study the short- and long-term responses of the interface. The predominant frequency range of the earthquake load was known to be between 0.1 and 15 Hz, based on previous studies (Shibuya et al., 1995; Araei et al., 2009; Silva et al., 1988). Soil behavior is often assumed to be independent of the frequency of seismic loading between 0.1 and 10 Hz. In addition, ASTM D3999 recommends the frequency of 0.5 or 1.0 Hz in case of applying uniform sinusoidal loading. Based on those research and recommendation, 0.5 Hz of sinusoidal loading was chosen to perform the cyclic simple shear test.

A series of strain-controlled tests were required to investigate the tendency of cyclic shear behavior even close to the failure continuously. In this study, some preliminary tests were performed to determine the appropriate amplitude of shear strain. The amplitude was more than 5%, indicating that the specimen failed very quickly, making it impossible to acquire a significant quantity of data to study the shear stress–strain behavior. The shear stress–strain curves were too steep and congested to investigate when shear strain was less than 1.5%. Since many cycles were required to reach failure, and the specimen did not even show the failure, 3.0% of shear strain was then applied in this study. Table 24.1 describes the test schedule.

Apart from this test schedule, additional prototype tests considering the elevated temperature were performed and the results were also demonstrated. Temperatures of 20°C and 60°C, which represented normal and high temperatures, respectively, of the waste landfills were applied to verify the effect of thermal degradation, which was combined with chemical conditions.

TABLE 24.1
Test Schedule

Submerging Period (days)	Normal Stress (MPa)	Chemical Condition
30	0.3	Acid
		Neutral
		Basic
	0.6	Acid
		Neutral
		Basic
850	0.3	Acid
		Neutral
		Basic
	0.6	Acid
		Neutral
		Basic

24.5 LABORATORY TEST RESULTS

Obvious shear stress degradation with an increasing number of cycles was observed in all cases. The total number of cycles was 200 and 100 for the specimen submerged for 30 and 850 days, respectively, because significantly rapid shear degradation occurred in the case of 850 days of the submerging period. The rate of shear degradation tended to decrease with increasing number of cycles, but it was not possible to estimate the rate directly, yet. In all cases, shear stresses' convergence appeared after 50 cycles, and this was considered as a threshold of the FA state. The maximum shear strain at each cycle showed almost identical value (3%); therefore, the performance of the apparatus seemed to be acceptable.

As a result, the typical shear stress degradation curves according to vertical stress and normal stress are shown in Figure 24.1. In Figure 24.1, the maximum shear stresses at each cycle increase in case 0.6 MPa of vertical stress is applied.

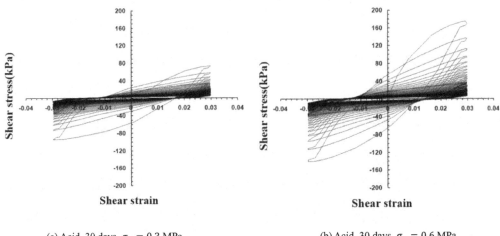

(a) Acid, 30 days, $\sigma_n = 0.3$ MPa   (b) Acid, 30 days, $\sigma_n = 0.6$ MPa

FIGURE 24.1   Typical shear stress–strain relationships.

The shape of the stress path can be found in the previous study (Kwak, 2014). It demonstrated a gradual decrease in both shear stress and normal stress with the load cycles. At failure, the stress path touched the failure envelope and subsequently stabilized with asymptotic normal stress toward the residual strength. The secant angle of peak envelope decreased by around 1.6% and 4.4%, respectively, for specimens submerged for 850 and 30 days, as normal stress increased. The friction angle of granular soil under constant volume condition decreased as normal stress increased, and this result was consistent with the previous theories on friction angle. However, it is important to note that the variation of friction angle may show some uncertainty and inconsistent tendency under the dynamic loading condition. For normal stresses of 0.3 and 0.6 MPa, each secant angle of peak envelope increased by 4.3% and 7.3%, respectively, according to the increase of the submerging period.

24.6 EVALUATION OF DISTURBANCE FUNCTION AND PARAMETERS

Based on the experimental results, shear stress–strain curves were obtained by the cyclic simple shear test. The disturbance, D, was calculated by using shear stress–strain curves and the deviatoric plastic strain trajectory, ξ_D, was estimated. The relationships between D and ξ_D, i.e., the disturbance function curves, were also evaluated. The shapes of the typical disturbance function curves (Figure 24.2) were shown to be parabolic in all conditions.

The degree of damage at the geosynthetic-soil interface was represented by the increase in disturbance, D. In the experiment results, the shape of the disturbance function varied in accordance with the chemical conditions. In most cases, the largest interface disturbance appeared under the basic condition.

The characteristics of the interface shear behavior can be expressed by the disturbance function curves as described. Conversely, the disturbance function curve can be reproduced by a mathematical combination with the functional form of the curve and disturbance parameters, A and Z. The parameters represented the intrinsic material characteristics and determined linear regression of the curves from the laboratory test results.

The A and Z parameters calculated via the linear regression were suggested and the average values including the standard deviations (SD) were also estimated. The lower limit (LL) and upper limit (UP), which were the boundary values far from the average value by 1.0 SD, were also

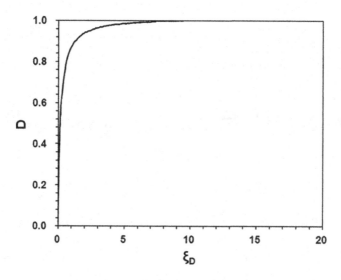

FIGURE 24.2 Typical disturbance function.

suggested. The variable A increased with the increase of the submerging period under the same chemical and normal stress conditions, indicating that the shape of the disturbance curve appears to shift left. This implied that the geosynthetic-soil interface suffered more damage at low plastic strain levels, namely, that the interface approached failure despite low plastic strain as the submerging period increased. On the contrary, A decreased with the increase of the normal stress under the same chemical and submergence conditions, which demonstrated the shape of the disturbance curve shifted to the right. Hence, the geosynthetic-soil interface became more resistant against the damage at the small plastic strain level. It followed the same pattern as the previously examined shear stress degradation characteristics. The parameter Z increased with the increase of the submerging period and the normal stress, under the same chemical conditions. It implied that the slope of the curve tended to rise, indicating a rapid increase in material damage, i.e., the damage considerably increased with a small increase in plastic strain. The variation from the average value of A and Z was relatively small in case of neutral condition compared to the acid and basic conditions. All parameters are demonstrated in the previous study by Kwak (2014).

24.7 SUGGESTION OF NEW DISTURBANCE FUNCTION CURVES

Based on a series of cyclic simple shear tests and new disturbance function parameters suggested, new disturbance function curves corresponding to the chemical and normal stress conditions were evaluated. In the short-term behavior (30 days of submergence), acid and basic conditions were more vulnerable. The disturbance function curves of acid and basic conditions showed almost identical shapes under the short-term condition. Long-term behavior revealed more pronounced changes in the disturbance function value at a specific deviatoric plastic strain trajectory, ξ_D, than short-term values. The differences in disturbance curves at a given strain level can be detected easily by comparing curves on the same domain. The specific shape of the new function curves can be known by referring to Kwak (2014).

In all cases, it was discovered that the basic condition was the most vulnerable. It could be deduced that the damage of the filament of nonwoven fabric due to the vulnerability for the basic condition can cause the most significant damage of the geosynthetic-soil interface.

REFERENCES

Akpinar, M. V. and Benson, C. H., "Effect of temperature on shear strength of two geomembrane-geotextile interfaces," Geotext. Geomembr., 23, 5, 2005, 443–453.

Araei, A. A., Razeghi, H. R., Tabatabaei, S. H. and Ghalandarzadeh, A. "Evaluation of frequency content on properties of gravelly soils," Research Project, No. 1-1775-2008, BHRC, Iran, 2009.

Armaleh, S. H. and Desai, C. S. "Modeling include testing of cohesionless soils under disturbed state concept," Report to the NSF, Department of Civil Engineering and Engineering Mechanics, University of Arizona, Tuscon, Arizona, 1990.

Cassidy, P. E., Mores, M., Kerwick, D. J. and Koeck, D. J. "Chemical resistance of geosynthetic materials," *Geotextiles and Geomembranes*, 11, 1992, 61–98.

Castelli, F., Cavallaro, A. and Maugeri, M., "Laboratory tests for estimation of static and dynamic interface characteristics of geosynthetic," In Proceedings Sardinia 2001, 8th International Waste Management and Landfill Symposium, 3, 2001, 157–166.

Dach, J. and Jager, J., "Prediction of Gas and Temperature with the Disposal of Pretreated Residential Waste," Proceedings of the 5th International Waste Management and Landfill Symposium, Vol. 1, S. argherita di Pula, Cagliari, Italy, Oct 2–6, CISA, Faenza, Italy, 1995, 665–677.

De, A. and Zimmie, T. F., "Landfill Stability: Static and Dynamic Geosynthetic Interface Friction Value," Proceedings of Geosynthetic Asia '97, Bangalore, India, Nov 26–29, CRC Press, Boca Raton, FL, 1997, 271–278.

Desai, C. S. "A consistent finite element technique for work-softening behavior," Proceedings of International Conference on Computer Mechanics In Nonlinear Mechanics, University of Texas Press, Austin, Texas, 1974.

Desai, C. S. "Chapter 8: Constitutive Modeling Using the Disturbed State as Microstructure Self-Adjustment Concept," *Continuum Models for Material with Microstructure*, H.B. Muhlhaus, ed., John Wiley & Sons, Chichester, 1995.

Desai, C. S. and Ma, Y. "Modeling of joints and interfaces using the disturbed state concept," *J. Numer. Anal. Methods. Geomech.*, 16, 9, 1992, 623–653.

Desai, C. S. *Mechanics of Materials and Interfaces: The Disturbed State Concept*. Boca Raton, FL: CRC. 2001.

Furukawa, G. T., McCoskey, R. E., and King, G. J., "Calorimetric Properties of Polytetrafluoroethylene (Teflon) from 0o to 365o K1," *J. Res. Natl. Bureau Stand.*, 49, 4, 1952, 273–278.

Grag, K. G., Saran, S., "Evaluation of soil–reinforcement interface friction." Proceedings of Indian Geotechnical Conference Bombay, India, 1, 1990, 27–31.

Hanson, J. L., Chrysovergis, T. S., Yesiller, N., and Manheim, D. C., "Temperature and moisture effects on GCL and 32 geotechnical testing journal textured geomembrane interface shear strength," *Geosynth. Int.*, 22, 1, 2015, 110–124.

Jafari, N. H., Stark, T. D., and Rowe, R. K., "Service life of HDPE geomembranes subjected to elevated temperatures," *J. Hazard., Toxic, Radioact. Waste*, 18, 1, 2014, 16–26.

Jones, D.R.V., Dixon, N., "Shear strength properties of geomembrane/geotextile interfaces," *Geotextiles and Geomembranes*, 16, 1998, 45–71.

Kaliakin, V.N., Li, J., "Insight into deficiencies associated with commonly used zero-thickness interface elements," *Computers and Geo-technics*, 17, 2, 1995, 225–252.

Klein, R., Baumann, T., Kahapka, E., and Niessner, R., "Temperature development in a modern municipal solid waste incineration (MSWI) bottom ash landfill with regard to sustainable waste management," *J. Hazard. Mater.*, 83, 3, 2001, 265–280.

Koerner, R. M., *Designing with Geosynthetics*, 3rd ed. Prentice Hall Inc., Englewood Cliffs, NJ, 1994.

Koerner, R. M., *Designing With Geosynthetics*, 6th ed., Vol. 2, Xlibris, Corp, Bloomington, IL, 2012

Koerner, G. R. and Koerner, R. M., "Long-term temperature monitoring of geomembranes at dry and wet landfills," *Geotext. Geomembr.*, 24, 1, 2006, 72–77.

Kwak, C. W., "Cyclic shear behaviors of geosynthetic–soil interface considering chemical effects," Ph.D Dissertation, Seoul National University, 2014.

Kwak, C. W., Park, I. J., and Park, J. B. "Modified cyclic shear test for evaluating disturbance function and numerical formulation of geosynthetic–soil interface considering chemical effect." *Geotec. Tesing J.*, 36, 4, 2013, 553–567.

Kwak, C. W., Park, I. J., and Park, J. B. "Dynamic shear behavior of concrete-soil interface based on cyclic simple shear test." *KSCE J. Civil Eng.*, 18, 3, 2014, 787–793.

Kwak, C. W., Park, I. J., and Park, J. B. "Development of modified interface apparatus and prototype cyclic simple shear test considering chemical and thermal effects." *Geotec. Tesing J.*, 39, 1, 2016, 20–34.

Lee, K. M. and Manjunath, V. R. "Experimental and numerical studies of geosynthetic-reinforced sand slopes loaded with a footing," *Canadian Geotec J.*, 37, 4, 2000, 828–842.

Lefebvre, X., Lanini, S., and Houi, D., "The role of aerobic activity on refuse temperature rise, 1. landfill experimental study," *Waste Manage. Res.*, 18, 5, 2000, 444–452.

Ma, Y., "Disturbed state concept for rock joints," Ph.D Dissertation, Department of Civil Engineering and Engineering Mechanics, University of Arizona, Tuscon, Arizona, 2000.

Navayogarajah, N. and Desai, C. S., "Hierarchical single-surface model for static and cyclic behavior of interfaces," *J. Eng. Mech.*, 118, 5, 1992. 990–1011.

Park, I. J., "Disturbed state modeling for dynamic and liquefaction analysis," Ph.D Dissertation, Department of Civil Engineering and Engineering Mechanics, University of Arizona, Tuscon, Arizona, 1997.

Park, I. J. and Desai, C. S., "Cyclic behavior and liquefaction of sand using disturbed state concept," *J. Geotec and Geoenviron. Eng.*, ASCE, 126, 9, 2000, 834–846.

Park, I. J., Kwak, C. W. and Kim, J. K., "The characteristics of dynamic behaviors for geosynthetic-soil interface considering chemical influence factors," *J. Korean Geo-Environ. Soc.*, 11, 11, 2010, 47–54.

Park, I. J., Yoo, J. H., and Kim, S. I., "Disturbed state modeling for dynamic analysis of soil-structure interface," *Journal of the KGS*, 16, 3, 2000, 5–13.

Rigby, D. B. and Desai, C.S., "Testing, modeling, and application of saturated interfaces in dynamic soil-structure interaction," Report to the NSF, Department of Civil Engineering and Engineering Mechanics, University of Arizona, Tuscon, Arizona, 1995.

Rowe, R. K., "Geosynthetics and the Minimization of Contaminant Migration Through Barrier Systems Beneath Solid Waste," Proceedings of the 6th International Conference on Geosynthetics, Atlanta, GA, March 25–29, IFAI, Roseville, MN, 1998, 27–102.

Rowe, R. K. and Islam, M. Z., "Impact of landfill liner time-temperature history on the service life of HDPE geomembranes," *Waste Manage.*, 29, 10, 2009, 2689–2699.

Rowe, R. K. and Sangam, H. P., "Durability of HDPE Geomembranes," Geotextiles and Geomembranes, No. 20, Elsevier Science Publishers, Ltd., England, 2002, 77–95.

Rowe, R. K., Islam, M. Z., and Hsuan, Y. G., "Leachate chemical composition effects on OIT depletion in an HDPE geomembrane," *Geosynth. Int.*, 15, 2, 2008, 136–151.

Rowe, R. K., Rimal, S., and Sangam, H., "Ageing of HDPE geomembrane exposed to air, water and leachate at different temperatures," *Geotext. Geomembranes*, 27, 2, 2009, 137–151.

Schnabel, W., Polymer Degradation: Principles and Practical Applications, Hanser International, Toronto, 227, 1981.

Seo, M. W., Park, I. J., and Park J. B., "Development of displacement-softening model for interface shear behavior between geosynthetics," *Soils and Foundations*, 44, 6, 2004, 27–38.

Seo, M. W., Park, J. B., and Park I. J., "Evaluation of interface shear strength between geosynthetics under wet condition," *Soils and Foundations*, 47, 5, 2007, 845–856.

Serway, R. A. and Jewett, J. W., Physics for Scientists and Engineers, 6th ed., 1, Cengage, Independence, KY, 132, 2004,

Shallenberger, W. C. and Filz, G. M., "Interface strength determination using a large displacement shear box". In Proceedings of the Second International Congresson Environmental Geotechnics, Osaka, Japan, 1, 1996, 147–152.

Shibuya, S., Mitachi, T., Fukuda, F., and Degoshi, T., "Strain-rate effects on shear modulus and damping of normally consolidated clay," *Geotec Testing J.*, 18, 3, 1995, 365–375.

Silva, W. J., Turcotte, T., and Moriwaki, Y., "Soil response to earthquake ground motion," EERI Report NP-5747, Electric Power Research Institute, Palo Alto, CA, 1988.

Stark, T. D., Williamson, T. A., and Eid, H. T., "HDPE geomembrane/geotextile interface shear strength," Journal of Geotechnical Engineering, 122, 3, 1996, 197–203.

Stark, T. D., Akhtar, K., and Hussain, M., "Stability Analysis for a Landfill Experiencing Elevated Temperatures," Proceedings of GeoFlorida 2010: Advances in Analysis, Modeling & Design, West Palm Beach, Florida, February 20–24, ASCE, Reston, VA, 2010, 3110–3119.

Triplett, E. J. and Fox, P. J., "Shear strength of HDPE geomembrane/geosynthetic clay liner interfaces," *J. Geotech. Geoenviron. Eng.*, 127, 10, 2001, 939–944.

Vucetic, M. and Lacasse, S., "Specimen Size Effect in Simple Shear Test," *J. Geotech. Eng. Div.*, 108, 12, 1982, 1564–1585.

Weibull, W., "A Statistical Distribution Function of Wide Applicability," *J. Appl. Mech.*, 18, 1951, 293–297.

Yegian, M. K., Yee, Z. Y., and Harb, J. N., "Seismic response of geosynthetic/soil systems." Geoenvironment 2000, vol. 45. Geotechnical Special Publication ASCE, 1995, 1113–1125.

Yesiller, N., Hanson, J. L., and Liu, W., "Heat Generation in Municipal Solid Waste Landfills," *ASCE J. Geotech. Geoenviron. Eng.*, 131, 11, 2005, 1330–1344.

Zhi-ling, S. and Chao, X., "Experimental research on shearing resistance property of geogrid–sand interface under normal cyclic load," In International Symposium on Deformation Characteristics of Geomaterials, Seoul, Korea, 2011.

Zhou, Y. and Rowe, R. K., "Development of a technique for modelling clay liner desiccation," *Int. J. Numer. Anal Methods. Geomech.*, 27, 6, 2003, 473–493.

25 Unsaturated Soil Interfaces

Gerald A. Miller
School of Civil Engineering and Environmental Science,
University of Oklahoma

Tariq B. Hamid
Dulles Geotechnical and Materials Testing Services, Inc.

Charbel N. Khoury
MC Squared LLC

Arash Hassanikhah
California Department of Transportation (Caltrans)

Kianoosh Hatami
School of Civil Engineering and Environmental Science,
University of Oklahoma

25.1 INTRODUCTION AND BACKGROUND – UNSATURATED INTERFACES

There are many soil–structure interaction problems that involve interfaces including those associated with deep foundations, shallow foundations, retaining walls and associated reinforcement, buried pipelines, and many others. The interface between soil and another material can be considered as a thin layer of soil through which stresses are transferred between the soil and the structure (e.g. Desai 2001, Desai et al. 1986, 1984, Frantziskonis et al. 1986, Ma and Desai 1990). In the case of a rough structural surface or counterface, an interface can be relatively strong and stiff, with shear strength and stiffness comparable to that of the neighboring soil, whereas a smooth counterface may lead to a relatively weak interface that permits significant relative displacement between the neighboring materials. Often, the interface exists in soils that are in an unsaturated condition. Understanding the behavior of the interface is important for understanding the global behavior in soil–structure interaction problems. It follows that developing realistic numerical models of boundary value problems involving soil–structure interfaces requires interface constitutive models that capture the essential behavior of the interface. Thus, it is important in the case of unsaturated soils to consider the moisture state of soil and matric suction, as well as the net normal stress acting on the interface.

In this chapter, an interface model developed from the hierarchical single-surface model for static behavior of interfaces (Desai 2001, Desai et al. 1986, 1984, Frantziskonis et al. 1986, Ma and Desai 1990, Navayogarajah et al. 1992) and modified for application in unsaturated soil conditions is presented. The adequacy of the model is demonstrated using the results of interface direct shear tests involving low-plasticity silt and rough and smooth steel plates, cohesionless sandy silt and woven geotextile, and a clay soil and geomembrane.

DOI: 10.1201/9781003362081-32

25.2   DSC/HISS MODELING OF AN UNSATURATED INTERFACE

25.2.1   Observations from Interface Testing with Unsaturated Soils

Interface tests were conducted in a modified direct shear apparatus (Miller and Hamid 2007) involving rough and smooth steel interfaces and low-plasticity fine-grained soil. Tests were conducted under suction control where pore air and pore water pressure were independently controlled. Shearing of the interface was conducted under constant normal stress and constant suction. Figures 25.1–25.4 show results that illustrate the influence of matric suction ($s = u_a - u_w$; $u_a =$ pore air pressure, $u_w =$ pore water pressure) and net normal stress ($\sigma_n = \sigma - u_a$; $\sigma =$ total normal stress)

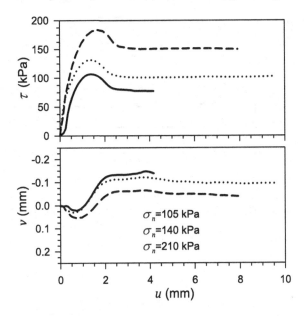

FIGURE 25.1   Horizontal displacement (u) versus shear stress (τ) and vertical displacement (v) for rough interface at three different net normal stresses, σ_n, and constant matric suction, $s = 100\,\mathrm{kPa}$.

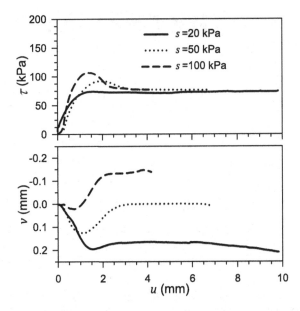

FIGURE 25.2   Horizontal displacement (u) versus shear stress (τ) and vertical displacement (v) for rough interface at three different matric suctions, s, and constant net normal stress, $\sigma_n = 105\,\mathrm{kPa}$.

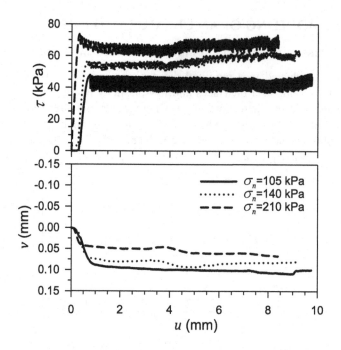

FIGURE 25.3 Horizontal displacement (u) versus shear stress (τ) and vertical displacement (v) for smooth interface at three different net normal stresses, σ_n, and constant matric suction, $s = 50\,\text{kPa}$.

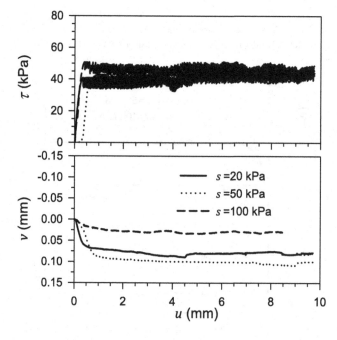

FIGURE 25.4 Horizontal displacement (u) versus shear stress (τ) and vertical displacement (v) for smooth interface at three different matric suctions, s, and constant net normal stress, $\sigma_n = 105\,\text{kPa}$.

on shearing behavior of rough and smooth interfaces. Observations regarding rough and smooth interface behavior follow.

Rough:

1. The maximum shearing stress increased as both net normal stress and matric suction increased, with the influence of net normal stress being more pronounced.
2. The residual shearing stress increased as net normal stress increased but was changed very little with increase in matric suction.
3. Dilation and strain-softening behavior was strongly influenced by suction. At low suction, compression behavior was predominant with little tendency for dilation and strain soften-ing. For the same net normal stress, at the highest suction, significant dilation and strain softening occurred.
4. There was a decrease in the tendency for dilation with increasing net normal stress.
5. Volume change behavior and shear stresses appeared to reach steady state upon reaching the residual shear stress.

Smooth:

1. The maximum shearing stress increased as the net normal stress increased, while a much smaller increase was associated with increasing matric suction.
2. After yielding, stick-slip behavior occurred as indicated by the cyclic variation in peak shear stress.
3. Residual shearing behavior was practically nonexistent and steady-state behavior generally appeared to coincide with yielding.
4. Volumetric compression was small compared to the rough interface, less predictable, occurred during initial loading, and became relatively constant upon yielding.
5. Interestingly, volumetric compression was slightly lower at higher net normal stress.
6. Volumetric compression was slightly less at the highest suction.

Noting the similarities in the behavior of the rough and smooth interfaces in unsaturated soil with behavior predicted by the interface models proposed by Desai and colleagues (Desai 2001, Desai et al. 1986, 1984, Frantziskonis et al. 1986, Ma and Desai 1990, Navayogarajah 1990, Navayogarajah et al. 1992), Hamid and Miller (Hamid 2005, Hamid and Miller 2008) extended the model to capture the essential behavior of the unsaturated interface as a function of matric suction. The model was used to develop the incremental stress-displacement relationships for the unsaturated interface tests based on model parameters determined from the unsaturated test results. Later, Khoury (Khoury 2010, Khoury et al. 2011) applied the model to the results of interface direct shear tests under suction control con-ducted on an unsaturated sandy silt–woven geotextile interface. In a similar fashion, Hassanikhah (2016) applied the model to unsaturated clay–geomembrane interface test results. Although there are some deviations between observed and modeled behavior, generally, the model appears to do an excel-lent job of capturing the essential features of the stress-displacement and volume change behavior of these unsaturated interfaces. In the following section, the essential aspects of the model for determin-ing the incremental stress-displacement behavior of the unsaturated interface are presented.

25.2.2 MODEL FORMULATION

25.2.2.1 Development of Incremental Stress-Displacement Relationship

A model for unsaturated soil interfaces (Hamid and Miller 2008) was developed by modifying the hierarchical elastoplastic model for interfaces presented by Desai and colleagues (Desai 2001, Desai et al. 1986, 1984, Frantziskonis et al. 1986, Ma and Desai 1990, Navayogarajah 1990, Navayogarajah et al. 1992). The interface is considered to be a thin zone between two bodies, the counterface and the soil, which is subjected to a two-dimensional displacement field described by deformations, u and v, and stresses, σ_n and τ, normal and tangential to the interface, respectively. A general constitutive model that relates incremental net normal ($d\sigma_n$) and shear ($d\tau$) stresses to incremental normal (du) and shear displacements (dv) is given by Eq. (25.1),

$$\left\{ \begin{array}{c} d\sigma_n \\ d\tau \end{array} \right\} = \left[\begin{array}{cc} C_{11} & C_{12} \\ C_{21} & C_{22} \end{array} \right] \left\{ \begin{array}{c} du \\ dv \end{array} \right\} \tag{25.1a}$$

or

$$\{d\sigma\} = [C]\{du\} \tag{25.1b}$$

where $[C]$ is the constitutive matrix.

The incremental displacements are decomposed into elastic and plastic components as in Eq. (25.2),

$$\left\{ \begin{array}{c} du \\ dv \end{array} \right\} = \left\{ \begin{array}{c} du^e \\ dv^e \end{array} \right\} + \left\{ \begin{array}{c} du^p \\ dv^p \end{array} \right\} \tag{25.2a}$$

or

$$\{du\} = \{du^e\} + \{du^p\} \tag{25.2b}$$

where subscripts "e" and "p" denote elastic and plastic, respectively. Both the incremental elastic and plastic deformations depend on the current stress state. Incremental elastic displacement and incremental stress can be expressed in terms of the elastic constitutive matrix, $[C^e]$, as given in Eq. (25.3).

$$\left\{ \begin{array}{c} d\sigma_n \\ d\tau \end{array} \right\} = [C^e] \left\{ \begin{array}{c} du^e \\ dv^e \end{array} \right\} = \left[\begin{array}{cc} C_{11}^e & C_{12}^e \\ C_{21}^e & C_{22}^e \end{array} \right] \left\{ \begin{array}{c} du^e \\ dv^e \end{array} \right\}$$

$$= \left[\begin{array}{cc} K_n & 0 \\ 0 & K_s \end{array} \right] \left\{ \begin{array}{c} du^e \\ dv^e \end{array} \right\} \tag{25.3a}$$

or

$$\{d\sigma\} = [C^e]\{du^e\} \tag{25.3b}$$

Note, off-diagonal terms in $[C^e]$ are equal to zero because elastic normal and shear behavior are assumed to be uncoupled. Constants K_n and K_s represent the normal and shear elastic stiffness for the interface. Noting that $\{du^e\} = \{du\} - \{du^p\}$, and substituting in Eq. (25.3b) gives,

$$\{d\sigma\} = [C^e](\{du\} - \{du^p\}) \tag{25.4}$$

To model plastic behavior, a nonassociative flow rule is adopted from which the plastic strain increment is defined by the potential function, Q, and is given as,

$$\{du^p\} = d\lambda \frac{\partial Q}{\{\partial\sigma\}} \tag{25.5}$$

where $\dfrac{\partial Q}{\{\partial\sigma\}}$ is a vector normal to the potential function in stress space and $d\lambda$ is a scalar quantity of proportionality that defines the length of the normal vector. If the stress point under consideration does not coincide with the yield surface, i.e., the yield function, F is less than or greater (not permissible) than zero, then $d\lambda = 0$, and if $F = 0$ then $d\lambda > 0$. Combining Eqs. (25.4) and (25.5) gives Eq. (25.6) for the incremental stress-displacement equation.

$$\{d\sigma\} = \left[C^e\right]\{du\} - d\lambda\left[C^e\right]\frac{\partial Q}{\{\partial\sigma\}} \tag{25.6}$$

The consistency condition for the incremental stress condition during loading must be satisfied. That is, during plastic loading, the stress point remains on the yield surface, either at a single point or moving along it, and the yield function and its derivative must be zero. Thus, the yield function is used to represent the conditions of plastic loading in the general form,

$$F(\{\sigma\} + \{d\sigma\}) = 0 = F(\{\sigma\}) + dF \tag{25.7}$$

where $dF = \nabla F^T \{d\sigma\} = 0$. Since $F(\{\sigma\}) = 0$, $dF = 0$, which implies that during plastic loading a change in stress occurs only tangential to the yield surface. Substituting Eq. (25.6) into the consistency condition equation leads to:

$$dF = \nabla F^T\{d\sigma\} = \left(\frac{\partial F}{\{\partial\sigma\}}\right)^T\{d\sigma\} = \left(\frac{\partial F}{\{\partial\sigma\}}\right)^T\left(\left[C^e\right]\{du\} - d\lambda\left[C^e\right]\frac{\partial Q}{\{\partial\sigma\}}\right) = 0 \tag{25.8}$$

If isotropic hardening is considered, the consistency condition becomes,

$$dF = \left(\frac{\partial F}{\{\partial\sigma\}}\right)^T\{d\sigma\} + H d\lambda = 0 \tag{25.9a}$$

or

$$dF = \left(\frac{\partial F}{\{\partial\sigma\}}\right)^T\left(\left[C^e\right]\{du\} - d\lambda\left[C^e\right]\frac{\partial Q}{\{\partial\sigma\}}\right) + H d\lambda = 0 \tag{25.9b}$$

where H is the plastic modulus that describes the growth of the yield surface with an increment of stress. Solving Eq. (25.9a) for $d\lambda$ gives Eq. (25.10).

$$d\lambda = \frac{\left(\dfrac{\partial F}{\{\partial\sigma\}}\right)^T\left[C^e\right]\{du\}}{\left(\dfrac{\partial F}{\{\partial\sigma\}}\right)^T\left[C^e\right]\dfrac{\partial Q}{\{\partial\sigma\}} - H} \tag{25.10}$$

Substituting Eq. (25.10) back into Eq. (25.6) provides the incremental stress-displacement relationship, Eq. (25.11).

$$\{d\sigma\} = \left[C^e\right]\left(\{du\} - \frac{\left(\frac{\partial F}{\{\partial\sigma\}}\right)^T \left[C^e\right]\{du\}\left(\frac{\partial Q}{\{\partial\sigma\}}\right)}{\left(\frac{\partial F}{\{\partial\sigma\}}\right)^T \left[C^e\right]\left(\frac{\partial Q}{\{\partial\sigma\}}\right) - H}\right) \tag{25.11a}$$

or

$$\{d\sigma\} = \left(\left[C^e\right] - \frac{\left[C^e\right]\left(\frac{\partial Q}{\{\partial\sigma\}}\right)\left(\frac{\partial F}{\{\partial\sigma\}}\right)^T \left[C^e\right]}{\left(\frac{\partial F}{\{\partial\sigma\}}\right)^T \left[C^e\right]\left(\frac{\partial Q}{\{\partial\sigma\}}\right) - H}\right)\{du\} \tag{25.11b}$$

or

$$\{d\sigma\} = \left[C^{ep}\right]\{du\} \tag{25.11c}$$

In Eq. (25.11c), $[C^{ep}]$ is the elastoplastic constitutive matrix for loading. The plastic modulus, H, is given by Eq. (25.12).

$$H = \left(\frac{\partial F}{\partial \xi_v}\right)\left|\frac{\partial Q}{\partial \sigma_n}\right| + \left(\frac{\partial F}{\partial \xi_D}\right)\left|\frac{\partial Q}{\partial \tau}\right| \tag{25.12}$$

In Eq. (25.12), $\xi_v = \int|dv^p|$ and $\xi_D = \int|du^p|$, are the volumetric and deviatoric plastic displacement trajectories, respectively. dv^p and du^p are the incremental volumetric and deviatoric plastic displacements normal and tangential, respectively, to the shearing surface.

25.2.2.2 Yield and Potential Functions *F* and *Q*

During interface shearing, plastic deformations will occur when the normal and shear stress state cause yielding. A yield function based on the associative flow rule presented by Desai and Fishman (1991) for a planar interface representing rock joints was proposed to model interfaces in contact with soil (Desai 2001, Desai et al. 1986, 1984, Frantziskonis et al. 1986, Ma and Desai 1990, Navayogarajah et al. 1992). The yield function, F, given by Eq. (25.13) includes interface parameters γ and n and hardening function, α. Hardening function α describes the growth of the yield function during deformation, γ is related to the peak shear strength, and n defines the transition from contractive to dilative volumetric strain.

$$F = \tau^2 + \alpha\sigma^n - \gamma\sigma^2 = 0 \tag{25.13}$$

Incorporating the concept of bonding stress $R(s)$ as presented by Geiser et al. (2000) to represent the effect of matric suction in a yield function for unsaturated soil, Hamid and Miller (2008) proposed Eq. (25.14) as a modification to Eq. (25.13) for interfaces in unsaturated soil.

$$F = \tau^2 + \alpha(s)\left[\sigma_n + R(s)\right]^n - \gamma(s)\left[\sigma_n + R(s)\right]^2 = 0 \tag{25.14}$$

In Eq. (25.14), (s) indicates the dependence of the parameters on matric suction, s, and net normal stress, σ_n, in the interface. In essence, $R(s)$ is related to the change in shear strength due to matric suction, which also depends on the normalized interface roughness, R_n.

   Hardening function, $\alpha(s)$, proposed for unsaturated soil interfaces is similar to that from previous models (Desai 2001, Desai et al. 1986, 1984, Frantziskonis et al. 1986, Ma and Desai 1990, Navayogarajah et al. 1992) except for the dependence of model parameters on suction noted in Eq. (25.15).

$$\alpha(s) = \gamma(s)\exp\left(-a(s)\xi_v\right)\left(\frac{\xi_D^* - \xi_D}{\xi_D^*}\right)^{b(s)} \tag{25.15}$$

The hardening function depends on failure parameter $\gamma(s)$, material parameters $a(s)$ and $b(s)$, and volumetric and deviatoric plastic displacement trajectories, $\xi_v = \int\left|dv^p\right|$ and $\xi_D = \int\left|du^p\right|$, respectively. Parameter ξ_D^* is the value of ξ_D when the shear stress is maximum.

   As noted by Desai and colleagues (Desai 2001, Desai et al. 1986, 1984, Frantziskonis et al. 1986, Ma and Desai 1990, Navayogarajah et al. 1992), an associative flow rule utilizing the yield function to define incremental plastic displacements typically leads to over prediction of dilation behavior. They proposed a nonassociative flow rule by introducing a modification to the hardening parameter, to develop an expression for the plastic potential function, Q, which Hamid and Miller (2008) adopted for unsaturated interfaces as in Eq. (25.16).

$$Q = \tau^2 + \alpha_Q(s)\left[\sigma_n + R(s)\right]^n - \gamma(s)\left[\sigma_n + R(s)\right]^2 = 0 \tag{25.16}$$

Unsaturated interface test results indicate that the volumetric strain approaches a steady state and shear stress reaches a peak followed by softening to a residual value at large displacements. These observations suggest $\alpha_Q(s)$ may incorporate a damage function (D) to capture the strain-softening behavior. Following on work by Desai and colleagues (Desai et al. 1984, 1986, Frantziskonis et al. 1986, Ma and Desai 1990), Navayogarajah et al. (1992) proposed Eq. (25.17):

$$\alpha_Q(s) = \alpha(s) + \alpha_{ph}\left(1 - \frac{\alpha(s)}{\alpha_i}\right)\left[1 - \kappa\left(1 - \frac{D}{D_u}\right)\right] \tag{25.17}$$

Parameter α_{ph} is the value of $\alpha(s)$ in the yield function, F, at the phase change point marking the transition from compression to dilation ($v = 0$). At this point, yield and potential functions are both equal to zero. By differentiating the yield function F with respect to $\sigma_n + R(s)$, the value of α_{ph} is expressed as Eq. (25.18).

$$\alpha_{ph} = \frac{2\gamma(s)}{n}\left[\sigma_n + R(s)\right]^{2-n} \tag{25.18}$$

Parameter α_i is the value of α at the beginning of nonassociative behavior, which is the point at which shear stress is applied for an isotropic interface under normal stress. For this condition, α_i is given by Eq. (25.19) (Navayogarajah et al. 1992).

$$\alpha_i = \gamma(s)\left[\sigma_n + R(s)\right]^{2-n} \tag{25.19}$$

The term κ in Eq. (25.17) is a material parameter, also referred to as the "nonassociative" parameter, related to the normalized roughness and state of stress in the interface. To capture the strain-softening behavior, disturbance function, D, is introduced in Eq. (25.17). This approach is consistent with the disturbed state concept (DSC) model developed by Dr. Desai and his colleagues (e.g. Desai 2001). Disturbance function D as given in Eq. (25.20) decomposes the average stress state at the interface into normal and shear stresses associated with intact and damaged parts. The damaged parts are considered the fully adjusted parts of the soil. Intact parts can sustain increases in net normal and shear stress as indicated by the first term on the right, whereas fully adjusted parts have zero shearing resistance as indicated by the second term on the right. The value of D at any point during shearing gives the proportion of fully adjusted parts of the interface; when the entire interface approaches a fully adjusted state, D approaches 1.

$$\left\{ \begin{array}{c} \sigma_n \\ \tau \end{array} \right\} = (1-D)\left\{ \begin{array}{c} \sigma_n \\ \tau \end{array} \right\} + D\left\{ \begin{array}{c} \sigma_n \\ 0 \end{array} \right\} \tag{25.20}$$

Damage during shearing is related to the accumulated plastic deformation represented by the deviatoric plastic displacement trajectory as in Eq. (25.21) (e.g. Desai 2001, Frantziskonis and Desai, 1987).

$$D = 0 \quad \text{for} \quad \xi_D < \xi_D^* \tag{25.21a}$$

$$D = D_u - D_u \exp\left[-A\left(\xi_D - \xi_D^*\right)^2\right] \quad \text{for} \quad \xi_D \geq \xi_D^* \tag{25.21b}$$

where $D_u = (\tau_p - \tau_r)/\tau_p$ and τ_p and τ_r are the peak and residual shear stresses, respectively. Peak shear stress, $\tau_p = \gamma(s)^{1/2}[\sigma_n + R(s)]$, and residual shear stress, $\tau_r = \mu_0(\sigma_n)$, with μ_0 being a function of the interface normalized roughness, R_n. Parameter A is a nonassociative parameter representing strain-softening behavior.

25.2.2.3 Determination of Model Parameters

Determination of the model parameters is demonstrated based on the work of Hamid and Miller (Hamid 2005, Miller and Hamid 2007, Hamid and Miller 2008). The suction-controlled interface tests described in Section 25.2.1 were conducted using low-plasticity clayey silt in contact with smooth and rough stainless steel counterfaces. Following the approach of Uesugi and Kishida (1986), the surface roughness was characterized by normalized roughness, R_n, which is the ratio of maximum distance between peak and valley along the counterface (R_{max}) to the diameter of soil grains corresponding to 50% finer (D_{50}). The D_{50} of the test soil was 0.05 mm and the rough stainless steel plate with manufactured saw-tooth roughness had an R_{max} of 0.38 mm giving $R_n = 7.6$. For the smooth polished stainless steel interface, R_{max} of 0.0025 mm was estimated based on the information published by the American National Standards Institute (ANSI 1985) giving $R_n = 0.05$. Model parameters were determined as a function of normalized roughness, net normal stress, and matric suction using results of interface tests sheared under constant suction and net normal stress.

Determination of Bonding Stress, $R(s)$: The bonding stress term in the yield function is related to the increase in shear strength due to suction. It is a function of the normalized roughness and suction as shown in Figure 25.5 and is calculated using Eq. (25.22).

$$R(s) = \lambda(s)(u_a - u_w) + \lambda_1 R_n + \lambda_2 \tag{25.22}$$

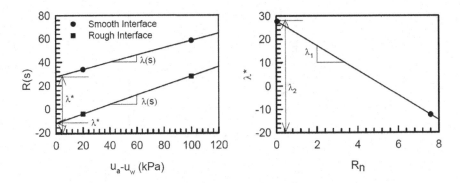

FIGURE 25.5 Determination of bonding stress $R(s)$ based on interface direct shear tests on rough and smooth interfaces for $s = 20$ kPa and $s = 100$ kPa; (a) determination of $\lambda(s)$ and (b) determination of λ_1 and λ_2 (Hamid, T. B., & Miller, G. A. (2008). A constitutive model for unsaturated soil interfaces. *International Journal for Numerical and Analytical Methods in Geomechanics*, 32(13), 1693–1714.).

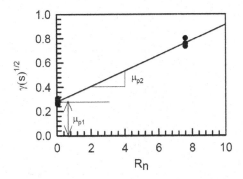

FIGURE 25.6 Determination of the relationship between $\gamma(s)$ and R_n based on interface direct shear tests on rough and smooth interfaces for $\sigma_n = 105$ kPa and 210 kPa and $s = 20$ kPa and 100 kPa. (Hamid, T. B., & Miller, G. A. (2008). A constitutive model for unsaturated soil interfaces. *International Journal for Numerical and Analytical Methods in Geomechanics*, 32(13), 1693–1714.)

Phase Change Parameter, n: The phase change parameter, n, corresponds to the state of stress where the volumetric displacement passes through a state of zero volume change. The shape of the yield surface is primarily governed by n.

Ultimate Parameter, $\gamma(s)$: When the peak shear stress, τ_p, is reached during interface shearing, $\alpha(s) = 0$ in the yield function, Eq. 25.14, which leads to $\gamma(s)^{1/2} = \tau_p/(\sigma_n - R(s))$. The $R(s)$ represents the increase in the strength of the unsaturated interface with the increase in suction. Values of $\gamma(s)^{1/2}$ calculated from the results of interface direct shear tests for smooth and rough interfaces encompassing different combinations of net normal stress and matric suction are plotted against normalized roughness in Figure 25.6 (from Hamid and Miller 2008). From this, an expression relating $\gamma(s)^{1/2}$ to R_n is given as:

$$\gamma(s)^{1/2} = \mu_{p1} + \mu_{p2}R_n \tag{25.23}$$

where the parameters μ_{p1} and μ_{p2} represent the intercept and slope, respectively.

Hardening Parameters ξ_D^*, a and b: Accumulated plastic deformation, ξ_D^*, in Eq. (25.15), corresponding to the peak shear stress is determined from experimental results by subtracting the calculated elastic displacement from the total horizontal displacement at the peak shear stress. It is

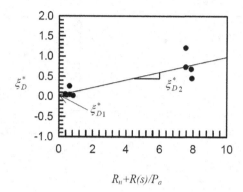

FIGURE 25.7   Determination of relationship between ξ_D^* and $R_n+R(s)/P_a$ based on interface direct shear tests on rough and smooth interfaces for $\sigma_n = 105\,\text{kPa}$ and $210\,\text{kPa}$ and $s = 20\,\text{kPa}$ and $100\,\text{kPa}$. (Hamid, T. B., & Miller, G. A. (2008). A constitutive model for unsaturated soil interfaces. *International Journal for Numerical and Analytical Methods in Geomechanics*, *32*(13), 1693–1714.)

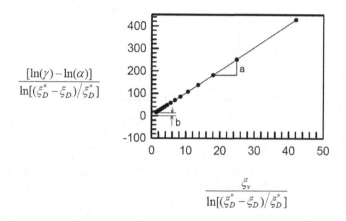

FIGURE 25.8   Typical plot for rough interface to determine parameters a and b obtained from experimental results for values of ξ_D between 0 and ξ_D^*. (Hamid, T. B., & Miller, G. A. (2008). A constitutive model for unsaturated soil interfaces. *International Journal for Numerical and Analytical Methods in Geomechanics*, *32*(13), 1693–1714.)

plotted against $R_n+R(s)/P_a$ in Figure 25.7, where P_a is atmospheric pressure. The best fit line is used to define the slope and intercept giving the relationship for ξ_D^* in Eq. (25.24).

$$\xi_D^* = \xi_{D1}^* + \xi_{D2}^* \left(R_n + \frac{R(s)}{P_a} \right) \tag{25.24}$$

Parameters a and b in Eq. (25.15) are determined by generating plots as shown in Figure 25.8 for each interface roughness, and each test with different net normal stress and matric suction. Data for each plot correspond to values of ξ_D between 0 and ξ_D^*. Average values from all net normal stress and matric suction results for a given roughness were found to provide satisfactory predictions.

Nonassociative Parameter, κ: The parameter κ is computed using the slope of the curve relating volumetric displacement to shear displacement corresponding to the peak shear stress, or where $\xi_D = \xi_D^*$, and is given by Eq. (25.25).

$$\kappa = -\gamma^{\frac{1}{2}} \frac{dv^p}{du}\bigg|_{\xi_D=\xi_D^*} = \kappa_1 + \kappa_2 \left(R_n + \frac{R(s)}{P_a} \right) \qquad (25.25)$$

By plotting values of κ computed using experimental data against $R_n+R(s)/P_a$, the terms κ_1 and κ_2 can be determined from the graph shown for example in Figure 25.9.

Residual Shear Stress Parameter, μ_{01}: The residual shear stress parameter, μ_0, is calculated using experimental data, where $\mu_0=\tau_r/\sigma_n$, and plotted against normalized roughness as shown in Figure 25.10. Parameters μ_{01} and μ_{02} are determined and used to calculate μ_0 in the model. In two of the studies discussed subsequently, the suction at a given normal stress had minimal influence on the residual shear stress. However, in the study involving geomembranes, the suction had some influence on the residual shear stress and so μ_0 was defined as $\mu_0=\tau_r/(\sigma_n+R(s))$, and calibrated accordingly.

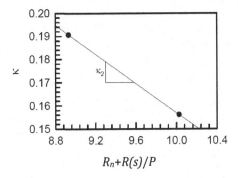

FIGURE 25.9 Typical plot for the determination of parameters κ_1 (intercept) and κ_2 (slope) for experimental results for determining nonassociative parameter κ (Hamid, T. B. (2005). *Testing and modeling of unsaturated interfaces*. Ph.D. Dissertation. The University of Oklahoma.).

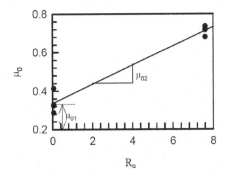

FIGURE 25.10 Plot for determination of parameters μ_{01} (intercept) and μ_{02} (slope) for experimental results for determining nonassociative parameter μ_0. (Hamid, T. B. (2005). *Testing and modeling of unsaturated interfaces*. Ph.D. Dissertation. The University of Oklahoma.)

Nonassociative Strain-Softening Parameter A: Parameter A controls the slope of the shear stress-shear displacement curve between the peak and residual shear stress. In this work, values were simply selected by trial and error to achieve the best match between the model and experimental data.

Elastic Stiffness, K_n and K_s: The elastic constants K_n and K_s in Eq. 25.3a were determined by unloading–reloading behavior observed during application of net normal stress and shear stress cycles during interface testing. The same elastic constants were used for both smooth and rough interfaces for all values of suction and net normal stress tested. While it is expected that there would be some dependency of elastic stiffness on the suction and net normal stress state, sensitivity analysis (Hamid 2005) indicated that predictions were not overly sensitive to K_s and relatively insensitive to K_n for shearing under constant net normal stress. However, as discussed in the next section, some improvements in model predictions can be achieved by using K_s values that depend on the stress state.

25.3   APPLICATIONS OF DSC/HISS UNSATURATED INTERFACE MODELS

The DSC/HISS model was applied to three different interfaces during the course of three different studies. In each case, drained suction-controlled interface direct shear tests were conducted under constant net normal stress for different levels of net normal stress and matric suction. The model parameters are provided and simulations of shear stress and volume change behavior as a function of shear displacement are presented. Characteristics of the soils and counterfaces used in each study are presented in Tables 25.1 and 25.2, respectively. The calibrated model parameters for interfaces used in each study are summarized in Table 25.3.

25.3.1   MODELING ROUGH AND SMOOTH SOIL–STEEL INTERFACES

Suction-controlled, drained interface direct shear tests were conducted using unsaturated Minco Silt and rough and smooth stainless steel counterfaces (Hamid 2005, Miller and Hamid 2007).

TABLE 25.1
Characteristics of Test Soils Used in Each Study

Study No. Soil Name	1 Minco Silt	2 Silica and Glass Beads	3 Chickasha Clay
USCS	CL	ML	CL
LL	28	NP	38
PI	8	NP	18
% fines	73	52	89
% sand	27	48	11
D_{50} (mm)	0.05	0.7	0.005
γ_{dmax} (kN/m³)[a]	17.7	16.3	17.3
$W_{opt}(\%)$[a]	12.8	16.5	18
G_s	2.68	2.65	2.75
$\gamma_d\left(kN/m^3\right)$[b]	15.7	15.4	16.4
$w\ (\%)^2$	20±1	17.2±1	21
Counterface Types	Rough and Smooth Stainless Steel	Woven Geotextile	Textured and Smooth Geomembrane

Notes:

[a] Based on standard Proctor.

[b] Used for sample preparation

TABLE 25.2
Characteristics of Counterfaces used in Each Study

Study No.	1	2	3
Counterface 1	Saw-tooth Stainless Steel	Woven PP Geotextile	Textured HDPE Geomembrane
Rmax (mm)	0.38	0.30	0.45
$R_n = R_{max}/D_{50}$	7.6	0.43	90
Counterface 2	Polished Stainless Steel	NA	Smooth HDPE Geomembrane
Rmax (mm)	0.0025	NA	0.0045
$R_n = R_{max}/D_{50}$	0.05	NA	0.9

TABLE 25.3
Model Parameters Determined from Experimental Data for Each Study

Model Parameter	Calibrated Parameter	Rough Steel	Smooth Steel	Geotextile	Textured Geomembrane	Smooth Geomembrane
R_n		7.6	0.05	4.2	90	0.9
$R(s)$	$\lambda(s)$	0.3990	0.3990	0.1454	0.0500	0.0200
	λ_1	−5.2285	−5.2285	−13.2680	−0.1000	−0.100
	λ_2	29.4865	29.4865	77.0	10.0	10.0
n		8, 4[a]	7	2.3	2.1	2.1
$\gamma(s)$	μ_{p1}	0.2796	0.2796	0.3995	0.3500, 0.4100[c]	0.2800, 0.3800[c]
τ_p	μ_{p2}	0.0635	0.0635	0.0639	0.0010	0.0010
ξ_D^* (mm)	ξ_{D1}^*	0.0318	0.0318	0.1714	0.5100	0.5100
	ξ_{D2}^*	0.0951	0.0951	0.0715	0.0070	0.0070
a		17.4	56.0	34.0	60.0	60.0
b		2.85	1.00	4.100	4.0	4.0
κ	κ_1	−2.9927	−0.2308, 0.0229[b]	0.0962	2.600	0.0100
	κ_2	0.4004	0.000, −0.0318[b]	0.0323	−0.0280, −0.0250[c]	−0.1100, 0.0700[c]
μ_0	μ_{01}	0.3479	0.3479	0.4682	0.2600	0.2600
τ_r	μ_{01}	0.0490	0.0490	0.0755	0.0020	0.0020
A		1	1	3	1	1
K_n (kPa/mm)		1000	1000	1050	365	365
K_s (kPa/mm)		150	150	282	35	35

Notes:
[a] $n = 8$ for $s = 20$ kPa, $n = 4$ for $s = 50$ & 100 kPa.
[b] First value for $s = 20$ & 50 kPa, second value for $s = 100$ kPa.
[c] First value for $s = 200$ kPa, second value for $s = 400$ kPa.

Interface direct shear tests were conducted for net normal stresses of 105, 140, and 210 kPa with matric suctions of 20, 50, and 100 kPa. Pore air, pore water, and net normal stress were maintained constant during shearing. Soil and counter face characteristics are presented in Tables 25.1 and 25.2 and model parameters are presented in Table 25.3. In Figures 25.11–25.13, experimental data and model results are shown for net normal stresses of 105, 140, and 210 kPa for suctions of 20, 50, and 100 kPa for the rough stainless steel interface. In Figures 25.14–25.16, experimental data and model

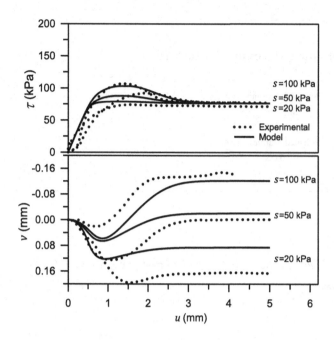

FIGURE 25.11   Experimental and model results for $\sigma_n = 105\,$kPa for rough interface.

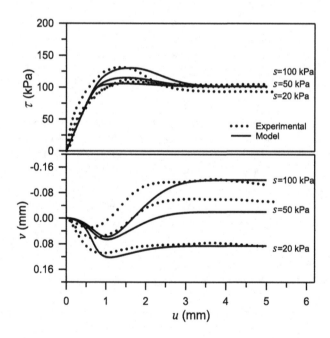

FIGURE 25.12   Experimental and model results for $\sigma_n = 140\,$kPa for rough interface.

results are shown for suctions of 20 and 100 kPa and net normal stresses of 105 and 210 kPa for the smooth interface.

As observed, the model is capable of capturing the overall features of the evolution of interface shear stress (τ) and volumetric displacement (ν) as a function of the shear displacement (u).

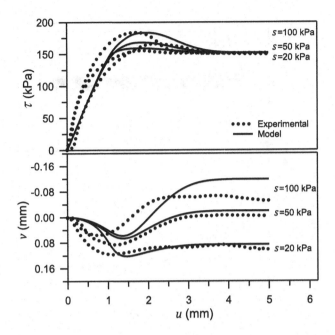

FIGURE 25.13 Experimental and model results for $\sigma_n = 210\,\text{kPa}$ for rough interface.

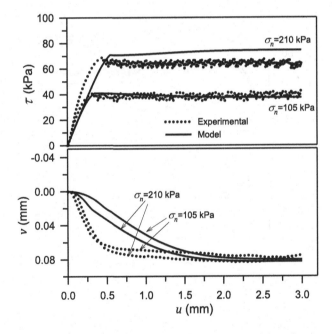

FIGURE 25.14 Experimental and model results for $s = 20\,\text{kPa}$ for smooth interface.

The volume change behavior is particularly sensitive to the exponent n, which marks the transition from contractive to dilative behavior and hardening parameter κ. The calibration procedure revealed that these parameters exhibit some dependency on the matric suction, which is reflected in the values determined from the calibration exercise in Table 25.3 for the rough and smooth interfaces.

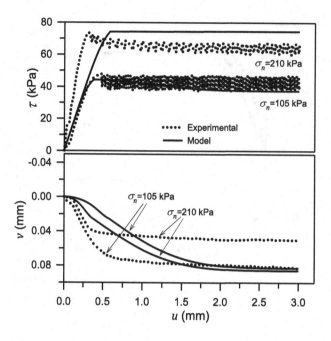

FIGURE 25.15   Experimental and model results for $s = 50\,\mathrm{kPa}$ for smooth interface.

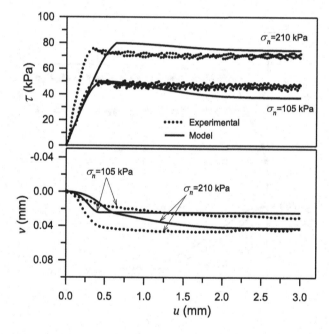

FIGURE 25.16   Experimental and model results for $s = 100\,\mathrm{kPa}$ for smooth interface.

Looking at Figures 25.11–25.16, there are some differences between the model and experimental data that deserve further discussion. At the highest net normal stress (210 kPa) and suction (100 kPa) for both the rough and smooth interfaces, the initial slope of the shear displacement versus shear stress curve is lower for the model compared to the experimental data. This is the result of using the same elastic shear stiffness, K_s, for all stress conditions. Using the same initial elastic shear stiffness

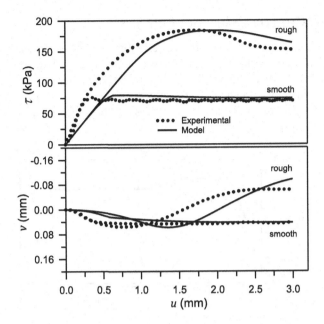

FIGURE 25.17 Experimental and model results for $sn = 210\,kPa$ and $s = 100\,kPa$ for smooth and rough interfaces with $K_s = 150$ kPa/mm.

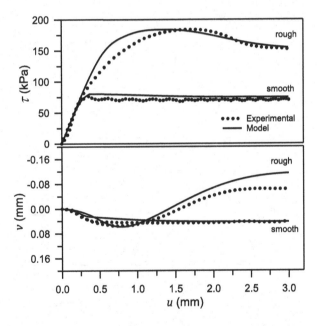

FIGURE 25.18 Experimental and model results for $sn = 210\,kPa$ and $s = 100\,kPa$ for smooth and rough interfaces with $K_s = 300$ kPa/mm.

also contributes to the lack of agreement between the model and experimental volumetric compression data at higher net normal stress, for example, as seen in Figures 25.13 and 25.16. Using higher elastic shear stiffness for higher stress conditions results in a better match as shown in Figures 25.17 and 25.18, where the results for smooth and rough interfaces are shown for $K_s = 150$ (original) and 300 kPa/mm, respectively.

25.3.2 MODELING GEOTEXTILE–SANDY SILT INTERFACES

Suction-controlled, drained interface direct shear tests were conducted using a manufactured sandy silt and woven geotextile counterface (Khoury 2010, Khoury et al. 2011). Soil and counterface characteristics are presented in Tables 25.1 and 25.2 and model parameters are presented in Table 25.3. In Figure 25.19, experimental data and model results are shown for net normal stress of 100 kPa for suctions of 25 and 100 kPa, and in Figure 25.20, results are shown for net normal stresses of 50 and 100 kPa for suction of 100 kPa.

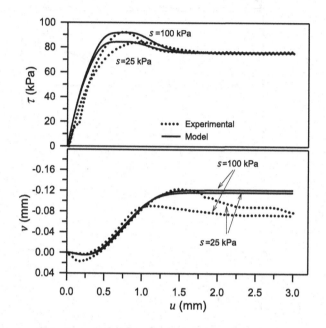

FIGURE 25.19   Experimental and model results for $\sigma_n = 100$ kPa for geotextile interface.

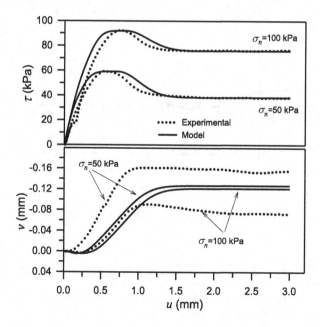

FIGURE 25.20   Experimental and model results for $s = 100$ kPa for geotextile interface.

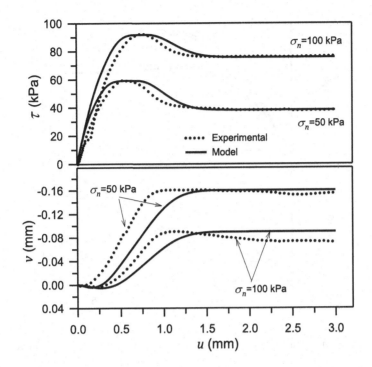

FIGURE 25.21 Experimental and model results for $s = 100\,\mathrm{kPa}$ for geotextile interface with $\kappa_2 = 0.0450$ and 0.0200 for $\sigma_n = 50$ and $100\,\mathrm{kPa}$, respectively.

As observed, the model is capable of capturing the overall features of the evolution of interface shear stress (τ) and volumetric displacement (ν) as a function of the shear displacement (u). The predicted volume change behavior was similar for all values of net normal stress and suction while the experimental values were somewhat inconsistent and appear more sensitive to net normal stress than predicted by the model as shown in Figure 25.20. On the other hand, the predicted shear stress versus displacement curves agreed very well with the experimental results. By varying the hardening parameter κ as a function of net normal stress, improvement in the predicted volume change behavior can be achieved as demonstrated in Figure 25.21, as compared to the volume change behavior in Figure 25.20 for suction of $100\,\mathrm{kPa}$. Specifically, κ_2 was changed from the original value of 0.0323–to values of 0.0450 and 0.0200,respectively, for net normal stress of 50 and $100\,\mathrm{kPa}$.

25.3.3 Modeling Geomembrane–Clay Interfaces

Suction-controlled, drained interface direct shear tests were conducted using a clay soil and textured and smooth High Density Polyethylene (HDPE) geomembranes (Hassanikhah 2016, Hassanikhah et al. 2020). Soil and counterface characteristics are presented in Tables 25.1 and 25.2 and model parameters are presented in Table 25.3. In Figures 25.22 and 25.23, experimental data and model results are shown for net normal stress of $50\,\mathrm{kPa}$ for suctions of 200 and $400\,\mathrm{kPa}$, for textured and smooth geomembranes. Similar to the previous two studies, the model does a good job simulating the shear stress and volume change behavior of the interfaces, particularly for the textured interface. While the predicted volume change of the smooth interface in Figure 25.23 is reasonably good, the shear stress versus displacement behavior can be improved with more refined calibration of the model parameters.

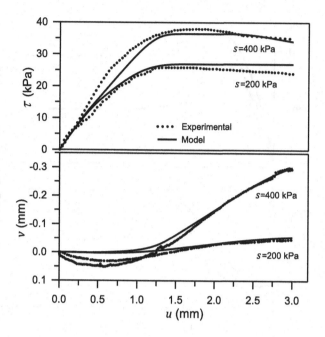

FIGURE 25.22 Experimental and model results for $\sigma_n = 50\,\text{kPa}$ for textured geomembrane interface.

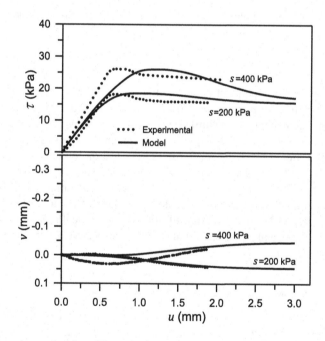

FIGURE 25.23 Experimental and model results for $\sigma_n = 50\,\text{kPa}$ for smooth geomembrane interface.

25.4 USEFULNESS OF DSC/HISS UNSATURATED INTERFACE MODELS

Development of interface constitutive models for unsaturated soils has gained attention in recent years including the DSC/HISS model described here, as well as elastoplastic models based on critical state theory (Lashkari and Kadivar 2016) and bounding surface plasticity (Zhou et al. 2020), and discrete element modeling (Liu et al. 2021). All of these modeling techniques have their advantages and can provide reasonable simulations of interface behavior as demonstrated by the model comparisons of Liu et al. (2021), which compared simulations of the rough interface test results for Minco Silt (from Figure 25.1). The model for unsaturated interfaces built on the DSC/HISS framework of Dr. Desai and his colleagues (Desai et al. 1984, 1986, Ma and Desai 1990, Frantziskonis et al. 1986, Navayogarajah et al. 1992) and presented in this chapter is shown to be capable of modeling different unsaturated soil interfaces with various soil types and materials. Proper interface constitutive models are important for use with the finite element method to realistically predict soil and interface behavior for complex boundary value problems. For example, Peng et al. (2022) incorporated a DSC/HISS model similar to the one presented here for unsaturated interfaces using interface elements in a Finite Element Model (FEM) simulation of a pull-out shear test on soil nails. The model simulations provided good agreement with the pull-out behavior of the nails with respect to the displacement and shear stress under different net normal stresses and matric suctions.

25.5 CONCLUSIONS

A DSC/HISS model developed for unsaturated soil interfaces (Hamid and Miller 2008) was developed by modifying the hierarchical elastoplastic model for interfaces developed by Dr. Desai and his colleagues (Desai et al. 1984, 1986, Ma and Desai 1990, Frantziskonis et al. 1986, Navayogarajah et al. 1992). In this chapter, the incremental stress-displacement formulation and calibration procedures were presented. The ability of the model was demonstrated using results of suction-controlled interface tests on low-plasticity clayey silt with rough and smooth steel counterfaces (Hamid 2005, Miller and Hamid 2007), sandy silt with woven geotextile counterface (Khoury et al. 2011), and clay with textured and smooth geomembrane counterfaces (Hassanikhah et al. 2020). Some general conclusions follow:

1. The model is capable of simulating essential features of the interface behavior including strain-hardening prior to failure, strain softening after the peak shear stress is reached, and volumetric compression and dilation.
2. The model provides excellent simulation of the shear stress-displacement behavior.
3. The model generally provides good simulation of volume change behavior but predictions were not as consistent with experimental results in comparison to the shear stress-displacement behavior. There are two primary reasons for this. First, the predicted volume change behavior is more sensitive to the model parameters governing compression and dilation obtained during calibration. Second, experimental results for volume change behavior are less consistent in some cases with respect to suctions and net normal stresses used during testing. This is likely due in part to inconsistencies in test sample heterogeneity and accuracy of measurements of vertical displacements, which are quite small in some cases.
4. As demonstrated by others using a similar DSC/HISS model for unsaturated interfaces, when implemented for interface elements in a finite element formulation, accurate simulations of soil–structure interaction are achievable considering variations in net normal stress and suction conditions.

REFERENCES

American National Standards Institute. Standards Committee B46, Classification, & Designation of Surface Qualities. (1985). *Surface Texture: Surface Roughness, Waviness and Lay*. American Society of Mechanical Engineers.

Desai, C. S. (2001). *Mechanics of Materials and Interfaces: The Disturbed State Concept*. CRC Press, Boca Raton, FL.

Desai, C. S., & Fishman, K. L. (1991, January). Plasticity-based constitutive model with associated testing for joints. In *International Journal of Rock Mechanics and Mining Sciences & Geomechanics Abstracts* (Vol. 28, No. 1, pp. 15–26). Pergamon.

Desai, C. S., Somasundaram, S., & Frantziskonis, G. (1986). A hierarchical approach for constitutive modelling of geologic materials. *International Journal for Numerical and Analytical Methods in Geomechanics*, *10*(3), 225–257.

Desai, C. S., Zaman, M. M., Lightner, J. G., & Siriwardane, H. J. (1984). Thin-layer element for interfaces and joints. *International Journal for Numerical and Analytical Methods in Geomechanics*, *8*(1), 19–43.

Frantziskonis, G., & Desai, C. S. (1987). Elastoplastic model with damage for strain softening geomaterials. *Acta Mechanica*, *68*(3), 151–170.

Frantziskonis, G., Desai, C. S., & Somasundaram, S. (1986). Constitutive model for nonassociative behavior. *Journal of Engineering Mechanics*, *112*(9), 932–946.

Geiser, F, Laloui, L. & Vulliet, L. (2000). Modelling the behaviour of unsaturated silt. In *Experimental Evidence and Theoretical Approaches in Unsaturated Soils* (pp. 163–184). CRC Press, Boca Raton, FL.

Hamid, T. B. (2005). *Testing and modeling of unsaturated interfaces*. Ph.D. Dissertation. The University of Oklahoma.

Hamid, T. B., & Miller, G. A. (2008). A constitutive model for unsaturated soil interfaces. *International Journal for Numerical and Analytical Methods in Geomechanics*, *32*(13), 1693–1714.

Hassanikhah, A. (2016). *Impact of Moisture Changes in Unsaturated Soil Engineering Applications*. Ph.D. Dissertation. The University of Oklahoma.

Hassanikhah, A., Miller, G. A., & Hatami, K. (2020). Laboratory investigation of unsaturated clayey soil-geomembrane interface behavior. *Geosynthetics International*, *27*(4), 379–393.

Khoury, C. N. (2010). *Influence of hydraulic hysteresis on the mechanical behavior of unsaturated soils and interfaces*. Ph.D. Dissertation. The University of Oklahoma.

Khoury, C. N., Miller, G. A., & Hatami, K. (2011). Unsaturated soil–geotextile interface behavior. *Geotextiles and Geomembranes*, *29*(1), 17–28.

Lashkari, A., & Kadivar, M. (2016). A constitutive model for unsaturated soil–structure interfaces. *International Journal for Numerical and Analytical Methods in Geomechanics*, *40*(2), 207–234.

Liu, X., Zhou, A., Shen, S. L., Li, J., & Arulrajah, A. (2021). Modelling unsaturated soil-structure interfacial behavior by using DEM. *Computers and Geotechnics*, *137*, 104305.

Ma, Y., and Desai, C.S. (1990). Constitutive modeling of joints and interfaces by using the Disturbed State Concept. *Report to NSF and AFOSR*, Dept. of Civil Eng., and Eng., Mechanics, Univ. of Arizona, Tucson, AZ, USA.

Miller, G. A., & Hamid, T. B. (2007). Interface direct shear testing of unsaturated soil. *Geotechnical Testing Journal*, *30*(3), 182–191.

Navayogarajah, N. (1990). *Constitutive modeling of static and cyclic behavior of interfaces and implementation in boundary value problems*. Ph.D. Dissertation. The University of Arizona.

Navayogarajah, N., Desai, C. S., & Kiousis, P. D. (1992). Hierarchical single-surface model for static and cyclic behavior of interfaces. *Journal of Engineering Mechanics*, *118*(5), 990–1011.

Peng, F., Li, X., Lv, M. F., Li, Y. H., & Guo, Y. C. (2022). A hierarchical single-surface model for an unsaturated soil-structure interface. *KSCE Journal of Civil Engineering*, *26*(6), 2675–2684.

Uesugi, M., & Kishida, H. (1986). Frictional resistance at yield between dry sand and mild steel. *Soils and Foundations*, *26*(4), 139–149.

Zhou, C., Tai, P., & Yin, J. H. (2020). A bounding surface model for saturated and unsaturated soil-structure interfaces. *International Journal for Numerical and Analytical Methods in Geomechanics*, *44*(18), 2412–2429.

Index

Note: **Bold** page numbers refer to tables; *italic* page numbers refer to figures.

Printed in the United States
Baker & Taylor Publisher Services